Heat, Energy, or Work Equivalents

(ft)(lb$_f$)	kWh	hp-hr	Btu	calorie*	Joule
0.7376	2.773×10^{-7}	3.725×10^{-7}	9.484×10^{-4}	0.2390	1
7.233	2.724×10^{-6}	3.653×10^{-6}	9.296×10^{-3}	2.3438	9.80665
1	3.766×10^{-7}	5.0505×10^{-7}	1.285×10^{-3}	0.3241	1.356
2.655×10^6	1	1.341	3.4128×10^3	8.6057×10^5	3.6×10^6
1.98×10^6	0.7455	1	2.545×10^3	6.4162×10^5	2.6845×10^6
74.73	2.815×10^{-5}	3.774×10^{-5}	9.604×10^{-2}	24.218	1.0133×10^2
3.086×10^3	1.162×10^{-3}	1.558×10^{-3}	3.9657	1×10^3	4.184×10^3
7.7816×10^2	2.930×10^{-4}	3.930×10^{-4}	1	2.52×10^2	1.055×10^3
3.086	1.162×10^{-6}	1.558×10^{-6}	3.97×10^{-3}	1	4.184

*The thermochemical calorie = 4.184 J.

Pressure Equivalents

mm Hg	in. Hg	bar	atm	kPa	psia
1	3.937×10^{-2}	1.333×10^{-3}	1.316×10^{-3}	0.1333	1.934×10^{-2}
25.40	1	3.386×10^1	3.342×10^{-2}	3.386	0.4912
750.06	29.53	1	0.9869	100.0	1.415×10^{-3}
760.0	29.92	1.013	1	101.3	14.696
75.02	0.2954	1.000×10^{-2}	9.872×10^{-3}	1	0.1451
51.71	2.036	6.893×10^{-2}	6.805×10^{-2}	6.893	1

Ideal Gas Constant R

1.987 cal/(g mol)(K)
1.987 Btu/(lb mol)(°R)
10.73 (psia)(ft^3)/(lb mol)(°R)
8.314 (kPa)(m^3)/(kg mol)(K) = 8.314 J/(g mol)(K)
82.06 (cm^3)(atm)/(g mol)(K)
0.08206 (L)(atm)/(g mol)(K)
21.9 (in Hg)(ft^3)/(lb mol)(°R)
0.7302 (ft^3)(atm)/(lb mol)(°R)

Miscellaneous Conversion Factors

To convert from	To	Multiply by
angstrom	meter	1.000×10^{-10}
barrel (petroleum)	gal	42
centipoise	(newton)(s)/m^2	1.000×10^{-3}
torr (mm Hg, 0°C)	newton/meter^2	1.333×10^2
fluid oz	cm^3	29.57

BASIC PRINCIPLES AND CALCULATIONS IN CHEMICAL ENGINEERING

SIXTH EDITION

BASIC PRINCIPLES AND CALCULATIONS IN CHEMICAL ENGINEERING

David M. Himmelblau
University of Texas

For book and bookstore information

http://www.prenhall.com

Prentice Hall PTR
Upper Saddle River, NJ 07458

Library of Congress Cataloging-in-Publication Data

Himmelblau, David Mautner
    Basic principles and calculations in chemical engineering / David
M. Himmelblau. —6th ed.
        p.   cm.
    Includes bibliographical references and index.
    ISBN 0–13–305798–4
    1. Chemical engineering—Tables.   I. Title.
TP151.H5   1996
660′.2—dc20   95–20656
                                                            CIP

Acquisitions editor: Bernard Goodwin
Cover designer: Scott Weiss
Cover design director: Jerry Votta
Manufacturing buyer: Alexis R. Heydt
Compositor/Production services: Pine Tree Composition, Inc.

The publisher offers discounts on this book when ordered in bulk quantities.

For more information contact:

  Corporate Sales Department
  Prentice Hall PTR
  One Lake Street
  Upper Saddle River, New Jersey 07458

  Phone: 800–382–3419
  Fax: 201–236–7141
  email: corpsales@prenhall.com

Printed in the United States of America
10   9   8

ISBN: 0-13-305798-4

Prentice-Hall International (UK) Limited, *London*
Prentice-Hall of Australia Pty. Limited, *Sydney*
Prentice-Hall of Canada, Inc., *Toronto*
Prentice-Hall Hispanoamericana, S. A., *Mexico*
Prentice-Hall of India Private Limited, *New Delhi*
Prentice-Hall of Japan, Inc., *Tokyo*
Prentice-Hall Asia Pte. Ltd., *Singapore*
Editora Prentice-Hall do Brasil, Ltda., *Rio de Janeiro*

To Betty (one more time)

CONTENTS

PREFACE

PURPOSE OF THE BOOK

This book is intended to serve as an introduction to the principles and techniques used in the field of chemical, petroleum, and environmental engineering. Although the range of subjects deemed to be in the province of "chemical engineering" has broadened over the last decade, the basic principles involved in chemical engineering remain the same. This book lays a foundation of certain information and skills that can be repeatedly employed in subsequent courses as well as in professional life.

A good introductory book to chemical engineering principles and calculations should (1) explain the fundamental concepts in not too stilted language together with generous use of appropriate equations and diagrams; (2) provide sufficient examples with detailed solutions to clearly illustrate (1); (3) present ideas in small packages that are easily identified as part of a larger framework; (4) include tests and answers that enable the reader to evaluate his or her accomplishments; and (5) provide the instructor with a wide selection of problems and questions to evaluate student competence. All of these features have been built into the sixth edition.

I kept in mind four major objectives for a reader in preparing this sixth edition:

1. to **develop systematic problem solving skills,** enhance confidence, and generate careful work habits;
2. to **learn what material balances are,** how to formulate and apply them, and how to solve them;
3. to **learn what energy balances are** and how to apply them;
4. to learn how to **deal with the complexity of big problems.**

In addition to accomplishing these goals, a reader is exposed to background information on units and measurements of **physical properties,** basic laws about the **behavior of gas, liquids, and solids,** and some **basic mathematical tools.** Other objectives that an instructor may want to include in a course, such as programming and communication skills, information about professional activities, developing a professional attitude, establishing personal goals, developing social awareness, and so on must be implemented by the instructor from other sources. Economic feasibility, a major factor in engineering decision making, costing, and optimization have been omitted because of lack of space. To provide an appreciation of what processing equipment really looks like and how it works, in the Supplement on the CD disk are numerous pictures of the equipment described in the worked out problems.

If this book is used as part of a scheduled course, the role of the teacher must be something more than just communicating the subject matter. The job of the teacher is to arouse emotional reactions of feeling good in connection with the content being conveyed. Creating positive feelings so that a student enjoys the subject makes a teacher effective.

SCOPE AND PARTS OF THE BOOK

The central themes of the book involve (1) learning how to formulate and solve (a) material balances, (b) energy balances, and (c) both simultaneously; (2) developing problem solving skills; and (3) becoming familiar with the use of units, physical properties, and the behavior of gases and liquids.

The chapters in sequence and their general contents are

Chapter 1: Background information

Chapter 2: Problem solving skills and tools

Chapter 3: Material balances

Chapter 4: Gases, liquids, and solids

Chapter 5: Energy balances

Chapter 6: Combined material and energy balances in large-scale problems

Chapter 7: Unsteady state material and energy balances

GENERAL FEATURES OF THE BOOK

I have selected, arranged, and presented the material in this book with care based on past teaching experience. All sections are divided into Objectives, Looking Ahead (a preview), Main Concepts, Additional Details (text containing nonessential information), Looking Back (a summary of the section), Key Ideas, Key Words, Self-Assessment Tests, Thought Problems, and Discussion Questions. Some other features common to all the chapters are:

- The book is **self-contained** except for some homework problems that deliberately require outside information.
- The presentation is **detailed** enough so that reference to other books can be omitted.
- The **examples are simple** and concrete to make the book teachable and useful for self-instruction.
- The chapters are largely **independent,** providing flexibility in teaching.
- The book has been reviewed for **readability.**
- The examples and homework problems support **good learning principles.**
- **Numerous illustrations** enhance learning.
- **Subheadings** clearly distinguish successive topics.
- **Thought problems and discussion problems** have been included for class discussion.
- A **table of contents** is listed **at the beginning of each chapter** to show the contents of the chapter.
- **Vital words and concepts** are in boldfaced type.
- At the end of each chapter **references** and numerous **supplementary references** are included.
- **Solutions** to about one-quarter of the **problems** in the problem sets are in the Appendix.
- **Data and computer codes** for solving problems have been provided in an accompanying CD.

At the beginning of each section is a **list of objectives** to be achieved by the reader, stated in such a way that attainment can be readily measured. We often present our objectives by such broad, fuzzy statements that neither the student nor the teacher can ascertain whether students have achieved them. (Unfortunately, this situation does not seem to inhibit the testing of students.) Each set of objectives is quite concrete and has a corresponding set of self-assessment questions and problems at the end of the respective section.

Piaget has argued that human intelligence proceeds in stages from the concrete to the abstract and that one of the big problems in teaching is that the teachers are formal reasoners (using abstraction) while many students are still concrete thinkers or at best in transition to formal operational thinking. I believe that this is true. Consequently, most topics are initiated with simple illustrations that illustrate the basic ideas. In this book the **topics are presented in order of easy assimilation** rather than in a strictly logical order. The organization is such that easy material is alternated with difficult material in order to give a "breather" after passing over each hump. For example, discussion of unsteady-state balances has been deferred until the final chapter because experience has shown that most students lack the mathematical and engineering maturity to absorb these problems simultaneously with the steady-state balances.

A principle of educational psychology is to reinforce the learning experience by providing detailed guided practice following each new principle. We all have found from experience that there is a vast difference between understanding a principle and in establishing our ability to apply it. For example, can you learn how to play piano from a series of lectures? By the use of numerous detailed **examples** following each brief section of text, it is hoped that straightforward, orderly methods of procedure can be instilled along with some insight into the principles involved. Furthermore, the wide variety of **problems** at the end of each chapter, about one-fourth of which are accompanied by **answers,** offer practice in the application of the principles explained in the chapter.

After all these years a perplexing problem still remains for an author in preparing a new edition, namely, the extent to which SI units should be used. I believe that SI is an important system of measurement that chemical engineers must be able to deal with, but also feel that chemical engineering students in the United States must be familiar with a variety of systems for some years to come. As a compromise, a little more than one-half of the text, examples, and problems and most of the tables employ SI units. For convenience, some of the crucial tables, such as the steam tables, are presented in both American Engineering and SI units.

Self-assessment tests have been included to provide readers with questions and answers that assist them in appraising and developing their knowledge about a particular topic. Self-assessment is intended to be an educational experience for a student. The availability of answers to the self-assessment questions together with supplementary reading citations for further study is an inherent characteristic of self assessment. To help the reader think about the concepts and decide whether to study further is one reason for having appraisal questions.

Let me now mention some of the new features of the sixth edition, features that were not present in earlier editions.

NEW FEATURES IN THE BOOK

For the sixth edition I have added a number of new features (and deleted some old ones) that make both teaching and self-study easier.

In this edition special attention has been devoted to presenting a consistent **sound strategy for solving material balance and energy balance** problems, one that can be used again and again as a framework for solving word problems. All the examples showing how to solve material and energy balances have been reformulated according to this strategy (see Table 3.1). In teaching I ask students to learn the strategy and apply it in all their homework problems and exams. I discourage the use of self-devised heuristic algorithms, or "cookbook" methods, pointing out that they may be successful for one class of problems but fail quite dismally for others. By this means a student is guided into forming generalized patterns of attack in problem solving that can be used successfully in connection with unfamiliar types of problems. The text is designed to acquaint the student with a sufficient number of fundamental concepts so that he or she can (1) continue with his or her training, and (2) start finding solutions to new types of problems on his or her own. It offers practice in finding out what the problem is, defining it, collecting data, analyzing and breaking down information, assembling the basic ideas into patterns, and, in effect, doing everything but testing the solution experimentally.

A major problem in any book is to what extent and in what manner should problems involving the use of computer codes be introduced into the text. If the use of the computer is to be integrated into the classroom successfully, it is wise to start early in the game. The selection of appropriate problems and the illustration of good computer habits, pointing out instances in which computer solutions are not appropriate or efficient as well as instances when they are, are important. What I have observed is a shift in paradigm from teaching programming skills to teaching the use of specialized software (such as Polymath or Matlab) to solve problems. Consequently, no reference to programming is made in the book, but Section 2.1 explains briefly how to use current software packages. Some Fortran programs that solve linear and nonlinear equations, retrieve the properties of water and steam, and of air-water mixtures, calculate the vapor pressures of pure substances, and calculate enthalpy changes from heat capacity equations, can be found on the **CD containing codes in a pocket in the back of the book.** As a result, the portions of the book formerly treating graphical integration, trial and error solutions, lever arm principles, and graphical solution methods have been drastically reduced or entirely eliminated.

Other new features are:

1. **Discussion questions added** at the end of each section. These are open-ended problems requiring collection of information not in the book, and can be used for class discussion, required or optional written reports, group assignments, and so forth.
2. **A section on problem solving has been added** (in Chapter 2). Several techniques of solving open-ended problems are discussed.

3. A section (Section 2.2) has been organized to treat **methods of solving problem with the help of computer software.**

4. **Chapter 5** on energy balances **has been completely revised, reduced in length, and simplified.**

5. **The problems** at the end of each section **have been augmented** to include features of safety, semiconductor processing, and biotechnology.

6. **The chapter on material balances, Chapter 3, has been completely revised** to provide a consistent algebraic approach to the formulation of problems that is carried out in all subsequent chapters.

7. **A summary of key concepts** has been added to the end of each section.

8. **Lists of key words and the page numbers where first introduced** have been added to the end of each section.

To provide high-quality software to aid readers in solving problems, in addition to the specialized codes that accompanied the fifth edition, in this sixth edition you will find on the CD in the back of the book two new significant additions:

1. Polymath, self-documented widely used software that runs on PCs, can solve linear and nonlinear equations and regression problems, and carry out matrix operations.

2. A database supplied by Professor Yaws of Lamar University, Beaumont, Texas that contains retrievable physical properties (such as vapor pressures, and heat capacities and enthalpies) for 700 compounds.

With these tools, execution of the solution phase of any problem becomes easily managed.

SUGGESTIONS TO THE READER AS TO HOW TO USE THIS BOOK

How should you study using this book? Read the objectives before and after studying each section. Read the text, and when you get to an example, cover up the solution and try to solve the stated problem. Some people, those who learn by reading concrete examples, should look at the examples first, and then read the text. Memorization is minimal in achieving the stated objectives, but practice in problem solving is essential, hence, after reading a section, solve some of the problems at the end of the chapter listed under that section number. R. P. Feynman, the Nobel laureate in physics, made the point: "You do not know anything until you have practiced." Whether you solve the problems using hand calculators or computer programs is up to you, but use the systematic and general steps listed in Table 3.1. Use the supplement on the CD in the back of the book (print it out if you need to) as a source of examples of additional solved problems to practice solving problems. Finally, when you are confident that you have achieved the stated objectives for a section, complete the self-assessment test at the end of the section (the answers are in the Appendix).

SUGGESTIONS FOR A COURSE

This book can be used in a variety of learning environments besides the traditional lectures, such as self-paced instruction, group study or discussion groups, and individual study. More topics have been included in the text than can be covered in one semester, so that an instructor has some choice as to pace of instruction and topics to include. In a lecture course for students with a background of having completed only freshman courses, Chapters 1, 3, 4, and 5 form the basis of an ample course. For more experienced students, if computer flow sheeting programs are available, perhaps Chapter 1 can be skimmed and Chapters 2 and 6 included. Chapter 7 will probably take at least one week (and probably two weeks) if you want to include it as well.

ACKNOWLEDGMENTS

I am indebted to many of my former teachers, colleagues, and students who directly or indirectly helped me in preparing this book, and in particular the present edition of it. Special thanks go to Kim Mathews and Carrie Anderson for preparing a number of new homework problems and putting the manuscript in its final form. Professor Donald Woods was most helpful in providing information about sound techniques of problem solving. I also want to thank Professor C. L. Yaws for his kindness in making available the physical properties database that is on the CD in the back of this book, and also thanks to Professors M. B. Cutlip and M. Shacham who graciously made the Polymath software available. Far too many instructors using the text have contributed their corrections and suggestions for me to list them all by name. However, I do wish to express my appreciation for their kind assistance. Any further comments and suggestions for improvement of the book would be gratefully received.

David M. Himmelblau
Austin, Texas

BASIC PRINCIPLES AND CALCULATIONS IN CHEMICAL ENGINEERING

INTRODUCTION TO CHEMICAL ENGINEERING CALCULATIONS

1

What do chemical engineers do? Although their backgrounds and professional skills are similar, chemical engineers work in a wide variety of industries, in addition to chemicals and petroleum, such as:

Biotechnology

Consulting

Drugs and pharmaceuticals

Fats and oils

Fertilizer and agricultural chemicals

Foods and beverages

Government

Lime and cement

Man-made fibers

Metallurgical and metal products

Paints, varnishes, and pigments

Pesticides and herbicides

Plastic materials and synthetic resins

Solid state materials

Chemical engineers focus on design, operation, control, troubleshooting, research, management, and even politics—the latter because of environmental and economic concerns. This book is not an introduction to chemical engineering as a profession. Instead, it is an introduction to the types of calculations made by chemical engineers in their everyday work. For you to learn how to appreciate and treat the problems that will arise in modern technology, and especially in the technology of the future, it is necessary to learn certain basic principles and practice their application. This text describes how to make material and energy balances, and illustrates their application in a wide variety of ways.

We begin the book by reviewing in Chapter 1 certain background information. You have already encountered most of these concepts in your basic chemistry and physics courses. Why, then, the need for a review? First, from experience we have found it necessary to restate these familiar basic concepts in a somewhat more precise and clearer fashion; second, you will need practice to develop your ability to analyze and work engineering problems. If, because of an incomplete background, when encountering new material, instead of focusing on it, you flounder over little gaps in your skills or knowledge, you will find the path rough going. To read and understand the principles discussed in this chapter is relatively easy; to apply them to different unfamiliar situations is not. An engineer becomes competent in his or her profession by mastering the techniques developed by ones predecessors—thereafter comes the time to pioneer new ones.

The chapter begins with a discussion of units, dimensions, and conversion factors, and then goes on to review some terms you should already by acquainted with, including:

(1) Mole and mole fraction
(2) Density and specific gravity
(3) Measures of concentration
(4) Temperature
(5) Pressure

Finally, we review the principles of stoichiometry.

1.1   UNITS AND DIMENSIONS

> *Your objectives in studying this section are to be able to:*
>
> 1. Add, subtract, multiply, and divide units associated with numbers.
> 2. Specify the basic and derived units in the SI and American Engineering systems for mass, length, volume, density, and time, and their equivalences.
> 3. Convert one set of units in a function or equation into another equivalent set for mass, length, area, volume, time, energy, and force.

4. Explain the difference between weight and mass.

5. Define and know how to use the gravitational conversion factor g_c.

6. Apply the concepts of dimensional consistency to determine the units of any term in a function.

"Take care of your units and they will take care of you."

LOOKING AHEAD

In this section we review the SI and American Engineering systems of units, show how conversions between units can be accomplished efficiently, and discuss the concept of dimensional consistency. We also provide some comments with respect to the number of significant figures to retain in your calculations.

MAIN CONCEPTS

At some time in every student's life comes the exasperating sensation of frustration in problem solving. Somehow, the answers or the calculations do not come out as expected. Often this outcome arises because of inexperience in the handling of units. The use of units or dimensions along with the numbers in your calculations requires more attention than you probably have been giving to your computations in the past. The proper use of dimensions in problem solving is not only sound from a logical viewpoint—it will also be helpful in guiding you along an appropriate path of analysis from what is at hand through what has to be done to the final solution.

1.1-1 Units and Dimensions

What are units and dimensions and how do they differ?

Dimensions are our basic concepts of measurement such as *length, time, mass, temperature,* and so on; **units** are the means of expressing the dimensions, such as *feet* or *centimeters* for length, or *hours* or *seconds* for time. By attaching units to all numbers that are not fundamentally dimensionless, you get the following very practical benefits:

(1) Diminished possibility of inadvertent inversion of any portion of the calculation.

(2) Reduced intermediate calculations and time in problem solving.

(3) A logical approach to the problem rather than remembering a formula and plugging numbers into it.

(4) Easy interpretation of the physical meaning of the numbers you use.

Every freshman knows that what you get from adding apples to oranges is fruit salad! The rules for handling units are essentially quite simple:

Addition, Subtraction, Equality

You can add, subtract, or equate numerical quantities only if the units of the quantities are the same. Thus the operation

$$5 \text{ kilograms} + 3 \text{ joules}$$

cannot be carried out because the dimensions as well as the units of the two terms are different. The numerical operation

$$10 \text{ pounds} + 5 \text{ grams}$$

can be performed (because the dimensions are the same, mass) *only* after the units are transformed to be the same, either pounds, or grams, or ounces, and so on.

Multiplication and Division

You can multiply or divide unlike units at will such as

$$50(\text{kg})(\text{m})/(\text{s})$$

but you cannot cancel or merge units unless they are identical. Thus, 3 m²/60 cm can be converted to 3 m²/0.6m and then to 5 m. The units contain a significant amount of information content that cannot be ignored. They also serve as guides in efficient problem solving, as you as you will see shortly.

EXAMPLE 1.1 Dimensions and Units

Add the following:

(a)   1 foot + 3 seconds

(b)   1 horsepower + 300 watts

Solution

The operation indicated by

$$1 \text{ ft} + 3 \text{ s}$$

has no meaning since the dimensions of the two terms are not the same. One foot has the dimension of length, whereas 3 seconds has the dimension of time. In the case of

$$1 \text{ hp} + 300 \text{ watts}$$

the dimensions are the same (energy per unit time) but the units are different. You must transform the two quantities into like units, such as horsepower or watts before the addition can be carried out. Since 1 hp = 746 watts,

$$746 \text{ watts} + 300 \text{ watts} = 1046 \text{ watts}$$

Table 1.1 list the SI units that are employed in this book. Table 1.2 lists similar units in the American Engineering system.

TABLE 1.1   SI Units Encountered in this Book

Physical Quantity	Name of Unit	Symbol for Unit*	Definition of Unit
	Basic SI Units		
Length	metre, meter	m	
Mass	kilogramme, kilogram	kg	
Time	second	s	
Temperature	kelvin	K	
Amount of substance	mole	mol	
	Derived SI Units		
Energy	joule	J	$kg \cdot m^2 \cdot s^{-2}$
Force	newton	N	$kg \cdot m \cdot s^{-2} \rightarrow J \cdot m^{-1}$
Power	watt	W	$kg \cdot m^2 \cdot s^{-3} \rightarrow J \cdot s^{-1}$
Density	kilogram per cubic meter		$kg \cdot m^{-3}$
Velocity	meter per second		$m \cdot s^{-1}$
Acceleration	meter per second squared		$m \cdot s^{-2}$
Pressure	newton per square meter, pascal		$N \cdot m^{-2}$, Pa
Heat Capacity	joule per (kilogram · kelvin)		$J \cdot kg^{-1} \cdot K^{-1}$
	Alternative Units		
Time	minute, hour, day, year	min, h, d, y	
Temperature	degree Celsius	°C	
Volume	litre, liter (dm^3)	L	
Mass	tonne, ton (Mg), gram	t, g	

*Symbols for units do not take a plural form, but plural forms are used for the unabbreviated names.

TABLE 1.2   American Engineering System Units Encountered in this Book

Physical Quantity	Name of Unit	Symbol
	Basic Units	
Length	feet	ft
Mass	pound (mass)	lb_m
Force	pound (force)	lb_f
Time	second, hour	s, hr
Temperature	degree Rankine	°R
	Derived Units	
Energy	British thermal unit, foot pound (force)	Btu, $(ft)(lb_f)$
Power	horsepower	hp
Density	pound (mass) per cubic foot	lb_m/ft^3
Velocity	feet per second	ft/s
Acceleration	feet per second squared	ft/s^2
Pressure	pound (force) per square inch	$lb_f/in.^2$
Heat capacity	Btu per pound (mass) per degree F	$Btu/(lb_m)(°F)$

The distinction between uppercase and lowercase letters should be followed even if the symbol appears in applications where the other lettering is in uppercase style. Unit abbreviations have the same form for both the singular and plural, and they are not followed by a period (except in the case of inches). One of the best features of the SI system is that (except for time) units and their multiples and submultiples are related by standard factors designated by the prefix indicated in Table 1.3. Prefixes are not preferred for use in denominators (except for kg).

When a compound unit is formed by multiplication of two or more other units, its symbol consists of the symbols for the separate units joined by a centered dot (e.g., N · m for newton meter). The dot may be omitted in the case of familiar units such as watt-hour (symbol Wh) if no confusion will result, or if the symbols are separated by exponents, as in N · m^2kg^{-2}. Hyphens should not be

TABLE 1.3   SI Prefixes

Factor	Prefix	Symbol	Factor	Prefix	Symbol
10^9	giga	G	10^{-1}	deci	d
10^6	mega	M	10^{-2}	centi	c
10^3	kilo	k	10^{-3}	milli	m
10^2	hecto	h	10^{-6}	micro	μ
10^1	deka	da	10^{-9}	nano	n

used in symbols for compound units. Positive and negative exponents may be used with the symbols for the separate units either separated by a solidus or multiplied by using negative powers (e.g., m/s or m · s⁻¹ for meters per second). **However, we do not use the center dot for multiplication in this text.** A dot can easily get confused with a period or missed entirely in handwritten calculations. Instead, we will use parentheses or vertical rules, whichever is more convenient, for multiplication and division. Also, the SI convention of leaving a space between groups of numbers such as 12 650 instead of inserting a comma, as in 12,650, will be ignored to avoid confusion in handwritten numbers.

1.1-2 Conversion of Units and Conversion Factors

In this book, to help you follow the calculations and emphasize the use of units, we frequently make use of a special format in the calculation, as shown in Example 1.2 below, that contains the units involved as well as the numbers. The concept is to multiply any number and its associated units with dimensionless ratios termed **conversion factors** to arrive at the desired answer and its associated units. Conversion factors are statements of equivalent values of different units in the same system or between systems of units. Inside the front cover you will find tables of conversion factors. Memorize a few of the common ones to save time looking them up. It takes less time to use conversion factors you know than to look up direct conversion factors in a handbook.

EXAMPLE 1.2 Conversion of Units

If a plane travels at twice the speed of sound (assume that the speed of sound is 1100 ft/s), how fast is it going in miles per hour?

Solution

$$\frac{2}{} \left| \frac{1100 \text{ ft}}{\text{s}} \right| \frac{1 \text{ mi}}{5280 \text{ ft}} \left| \frac{60 \text{ s}}{1 \text{ min}} \right| \frac{60 \text{ min}}{1 \text{ hr}} = 1500 \ \frac{\text{mi}}{\text{hr}}$$

Note the format of the calculations in Example 1.2. We have set up the calculations with vertical lines separating each ratio. These lines retain the same meaning as an · or multiplication sign × placed between each ratio. We will use this form throughout most of this text to enable you to keep clearly in mind the significance of units in problem solving. We recommend that you always write down the units next to the associated numerical value (unless the calculation is very simple) until you become quite familiar with the use of units and dimensions and can carry them in your head.

At any point in the dimensional equation you can determine the consolidated net units and see what conversions are still required. If you want, you can do this formally as shown below by drawing slanted lines below the dimensional equation and writing the consolidated units on these lines, or it may be done by eye, mentally canceling and accumulating the units, or, you can strike out pairs of units as you proceed:

$$\frac{2 \times 1100 \text{ ft}}{\text{s}} \left| \frac{1 \text{ mi}}{5280 \text{ ft}} \right| \frac{60 \text{ s}}{1 \text{ min}} \left| \frac{60 \text{ min}}{1 \text{ hr}} \right.$$

$$\frac{\text{ft}}{\text{s}} \qquad \frac{\text{mi}}{\text{s}} \qquad \frac{\text{mi}}{\text{min}}$$

Consistent use of dimensional equations throughout your professional career will assist you in avoiding silly mistakes such as converting 10 centimeters to inches by multiplying by 2.54:

$$\frac{10 \text{ cm}}{} \left| \frac{2.54 \text{ cm}}{\text{in.}} \right. \ne 2.54 \text{ in.} \qquad \text{but instead} = 25.4 \; \frac{\text{cm}^2}{\text{in.}}$$

Note how easily you discover that a blunder has occurred by including the units in the calculations.

Here is another example of the conversion of units.

EXAMPLE 1.3 Use of Units

Change 400 in.^{3}/day to cm^3/min.

Solution

$$\frac{400 \text{ in.}^3}{\text{day}} \left| \left(\frac{2.54 \text{ cm}}{1 \text{ in.}} \right)^3 \right| \frac{1 \text{ day}}{24 \text{ hr}} \left| \frac{1 \text{ hr}}{60 \text{ min}} \right. = 4.56 \frac{\text{cm}^3}{\text{min}}$$

In this example note that not only are the numbers raised to a power, but the units also are raised to the same power.

Conversion of SI units is simpler than conversions in the American Engineering system. We can use Newton's law to compare the respective units:

$$F = Cma \qquad (1.1)$$

where F = force
   C = a constant whose numerical value and units depend on those selected for F, m, and a
   m = mass
   a = acceleration

In the SI system in which the unit of force is defined to be the newton (N), if $C = 1$ N/(kg)(m)/s², then when 1 kg is accelerated at 1 m/s²

$$F = \frac{1 \text{ N}}{\frac{(\text{kg})(\text{m})}{\text{s}^2}} \left| \frac{1 \text{ kg}}{} \right| \frac{1 \text{ m}}{\text{s}^2} = 1 \text{ N}$$

A conversion factor is required to achieve the end result of newtons, but the value associated with the conversion factor is 1 so that the conversion factor seems simple, even nonexistent.

In the American Engineering system a conversion factor is also required, but a constraint exists. We require that the numerical value of the force and the mass be essentially the same at the earth's surface. Hence, if a mass of 1 lb_m is accelerated at g ft/s², where g is the acceleration of gravity (about 32.2 ft/s² depending on the location of the mass), we can make the force be 1 lb_f by choosing the proper numerical value and units for C:

$$F = (C) \frac{1 \text{ lb}_m}{} \left| \frac{g \text{ ft}}{\text{s}^2} = 1 \text{ lb}_f \right. \tag{1.2}$$

Observe that for Eq. (1.2) to hold, the units of C have to be

$$C \rightarrow \frac{\text{lb}_f}{\text{lb}_m \left(\dfrac{\text{ft}}{\text{s}^2} \right)}$$

A numerical value of 1/32.174 has been chosen for the numerical value of the constant because 32.174 is the numerical value of the average acceleration of gravity (g) at sea level at 45° latitude when g is expressed in ft/s². The acceleration of gravity, you may recall, varies by a few tenths of 1% from place to place on the surface of the earth. The inverse of the conversion factor with the numerical value 32.174 included is given the special symbol g_c

$$g_c = 32.174 \frac{(\text{ft})(\text{lb}_m)}{(\text{s}^2)(\text{lb}_f)}$$

Division by g_c achieves exactly the same result as multiplication by C in Newton's law. You can see, therefore, that in the American Engineering system we have the convenience that the numerical value of a pound mass is also that of a pound force if the numerical value of the ratio g/g_c is equal to 1, as it is approximately in most cases

$$F = \left(\frac{1(\text{lb}_f)(\text{s}^2)}{32.174(\text{lb}_m)(\text{ft})} \right) \left(\frac{1 \text{ lb}}{} \left| \frac{g \text{ ft}}{\text{s}^2} \right. \right) = 1 \text{ lb}_f$$

Furthermore, a one pound mass is said to weigh one pound if the mass is in static equilibrium on the surface of earth. We can define **weight** as the opposite of the force required to support a mass. For the concept of weight for masses that are not stationary at the earth's surface, or are affected by the earth's rotation (a factor of only 0.3%), or are located away from the earth's surface as in a rocket or satellite, consult your physics text.

To sum up, always keep in mind that the two quantities g and g_c, are **not** the same. Also, **never forget that the pound (mass) and pound (force) are not the same units in the American Engineering system** even though we speak of *pounds* to express force, weight, or mass. Nearly all teachers and writers in physics, engineering, and related fields in technical communications are careful to use the terms "mass," "force," and "weight" properly. On the other hand, in ordinary language most people, including scientists and engineers, omit the designation of "force" or "mass" associated with the pound but pick up the meaning from the context of the statement. No one gets confused by the fact that a man is 6 feet tall but only has two feet. **In this book, we will not subscript the symbol lb with m (for mass) or f (for force)** unless it becomes essential to do so to avoid confusion. **We will always mean by the unit lb without a subscript the quantity pound mass.**

EXAMPLE 1.4 Use of g_c

One hundred pounds of water is flowing through a pipe at the rate of 10.0 ft/s. What is the kinetic energy of this water in $(ft)(lb_f)$?

Solution

$$\text{Kinetic energy} = K = \tfrac{1}{2} mv^2$$

Assume that the 100 lb of water means the mass of the water.

$$K = \frac{1}{2} \left| \frac{100 \ lb_m}{} \right| \left| \frac{\left(\dfrac{10 \ ft}{s}\right)^2}{} \right| \frac{1}{32.174 \dfrac{(ft)(lb_m)}{(s^2)(lb_f)}} = 155 \ (ft)(lb_f)$$

EXAMPLE 1.5 Use of g_c

What is the potential energy in $(ft)(lb_f)$ of a 100 lb drum hanging 10 ft above the surface of the earth with reference to the surface of the earth?

Solution

$$\text{Potential energy} = P = mgh$$

Assume that the 100 lb means 100 lb mass; g = acceleration of gravity = 32.2 ft/s^2.

$$P = \frac{100\ lb_m}{} \frac{32.2\ ft}{s^2} \frac{10\ ft}{} \frac{1}{32.174 \frac{(ft)(lb_m)}{(s^2)(lb_f)}} = 1001\ (ft)(lb_f)$$

Notice that in the ratio of g/g_c, or 32.2 ft/s^2 divided by 32.174 (ft/s^2)(lb$_m$/lb$_f$), the numerical values are almost equal. A good many people would solve the problem by saying that 100 lb × 10 ft = 1000 (ft)(lb) without realizing that in effect they are canceling out the numbers in the g/g_c ratio.

1.1-3 Dimensional Consistency

Now that we have reviewed some background material concerning units and dimensions, we can immediately make use of this information in a very practical and important application. **A basic principle exists that equations must be dimensionally consistent.** What the principle requires is that each term in an equation must have the same net dimensions and units as every other term to which it is added or subtracted or equated. Consequently, dimensional considerations can be used to help identify the dimensions and units of terms or quantities in an equation.

The concept of dimensional consistency can be illustrated by an equation that represents gas behavior and is known as van der Waals equation, an equation to be discussed in more detail in Chapter 4:

$$\left(p + \frac{a}{V^2}\right)(V - b) = RT$$

Inspection of the equation shows that the constant a must have the units of [(pressure)(volume)^2] in order for the expression in the first set of parentheses to be consistent throughout. If the units of pressure are atm and those of volume are cm^3, a will have the units specifically of [(atm)(cm)^6]. Similarly, b must have the same units as V, or in this particular case the units of cm^3. If T is in K, what must be the units of R? Check your answer by looking up R inside the front cover of the book. All equations must exhibit dimensional consistency.

EXAMPLE 1.6 Dimensional Consistency

Your handbook shows that microchip etching roughly follows the relation

$$d = 16.2 - 16.2e^{-0.021t} \qquad t < 200$$

where d is the depth of the etch in microns (micrometers; μm) and t is the time of the etch in seconds. What are the units associated with the numbers 16.2 and 0.021? Convert the relation so that d becomes expressed in inches and t can be used in minutes.

Solution

Both values of 16.2 must have the units of microns. The exponential must be dimensionless so that 0.021 must have the units of 1/seconds.

$$d_{\text{in.}} = \frac{16.2\ \mu\text{m}}{} \left|\frac{1\text{m}}{10^6\ \mu\text{m}}\right.\left|\frac{39.37\ \text{in.}}{1\ \text{m}}\right.\left[1 - \exp\left.\frac{-0.021}{\text{s}}\right|\frac{60\ \text{s}}{1\ \text{min}}\right|t_{\text{min}}\right]$$

$$= 6.38 \times 10^{-4}(1 - e^{-1.26t\text{min}})$$

Groups of symbols may be put together, either by theory or based on experiment, that have no net units. Such collections of variables or parameters are called **dimensionless** or **nondimensional groups**. One example is the Reynolds number (group) arising in fluid mechanics.

$$\text{Reynolds number} = \frac{D\upsilon\rho}{\mu} = N_{\text{Re}}$$

where D is the pipe diameter, say in cm; υ is the fluid velocity, say in cm/s; ρ is the fluid density, say in g/cm^3; and μ is the viscosity, say in centipoise, units that can be converted to g/(cm)(s). Introducing the consistent set of units for D, υ, ρ, and μ into $D\upsilon\rho/\mu$, we find that all the units cancel out.

$$\frac{\cancel{\text{cm}}}{}\left|\frac{\cancel{\text{cm}}}{\cancel{\text{s}}}\right.\left|\frac{\cancel{\text{g}}}{\text{cm}^3}\right.\left|\frac{(\cancel{\text{cm}})(\cancel{\text{s}})}{\cancel{\text{g}}}\right.$$

ADDITIONAL DETAILS

Before proceeding to the next section, we should mention briefly some aspects of significant figures, accuracy, and precision of numbers. Measurements collected by process instruments can be expected to exhibit some random error and may also be biased. The accuracy of the results of a calculation depends on the proposed application of the results. The question is: How close is close enough? For example, in income tax forms you do not need to include cents, whereas in a bank statement, cents (two decimals) must be included. In engineering calculations, if the costs of inaccuracy are great (failure, fire, downtime, etc.), knowledge of the uncertainty in the calculated variables is vital. On the other hand, in determining how much fertilizer to put on your lawn in the summer, being off by 10 to 20 pounds out of 100 pounds is not important.

Several options exist (besides common sense) in establishing the degree of certainty in a number evolving from calculations. Three common decision criteria are: (1) absolute error, (2) relative error, and (3) statistical analysis.

First, consider the **absolute error** in a number. We assume that the last significant figure in an inexact number represents the associated uncertainty. Thus, the number 100.3 carries along the implication of 100.3 ± 0.1 meaning 100.3 could be 100.4 or 100.2. The number 100.300 usually does not imply additional significant figures with the additional zeros because in scientific notation the number is 1.003×10^2. Similarly 100,300 implies only four significant figures because it is 1.003×10^5 in scientific notation. In the product (1.47)(3.0926) = 4.54612, since 1.47 has only three significant figures, the answer can be truncated to 4.55 to avoid exceeding the presumed precision.

Absolute errors are easy to track and compute, but they can lead to gross distortions in the specified uncertainty of a number. Let us look at **relative error**. Suppose we divide one number by another number close to it such as 1.01/1.09 = 0.9266, and select 0.927 as the answer. The uncertainty in the answer based on the previous analysis is presumed to be 0.001/0.927 or about 0.1%, whereas there was (0.01/1.09)100, or about a 1% uncertainty in the original numbers. Should the relative uncertainty of the answer be fixed at about 1%, that is, truncate the answer to 0.93 rather than 0.927? Such would be the case if the concept of relative error were applied.

A more rigorous but more complicated way to treat uncertainty in numbers is to apply statistics in the calculations. What is involved is the concept of confidence limits for the starting numbers in a calculation, and the propagation of errors step by step through each stage of the calculations to the final result. But even a statistical analysis is not exact because we deal with nonlinear ratios of numbers. You should refer to a book on statistics for further information about this approach.

In this book we base most answers on absolute error, but may use one or two extra figures in certain intermediate calculations. Keep in mind that some numbers are exact, such as the 1/2 in $K = 1/2\, mv^2$ and the 2 in the superscript. You will also encounter integers such as 1, 2, 3 and so on, which in some cases are exact (2 reactors, 3 input streams) but in other cases are shortcut substitutes for presumed measurements in problem solving (3 moles, 10 kg). You can assume that 10 kg infers a reasonable number of significant figures in relation to the other values of parameters stated in an example or problem, such as 10 kg $\rightarrow$ 10.00 kg. You will also occasionally encounter fractions such as $\frac{2}{3}$, which can also be treated as 0.6667 in relation to the accuracy of other values in a problem.

Feel free to round off parameters such as $\pi = 3.1416$, $\sqrt{2} = 1.414$, or Avogadro's number $N = 6.02 \times 10^{23}$. In summary, be sure to round off your answers to problems even though in the intermediate calculations numbers are carried out to 10 or more digits in your computer, because the final answers can be no more accurate than the accuracy of the numbers introduced into the problem during its solution. In this text for convenience we will use 273 K for the temperature equivalent to 0°C instead of 273.15 K, thus introducing a relative error of 0.15/273.15 = 0.00055 into a temperature calculation (or an absolute error of 0.15). This is such a small error relative to the other known or presumed errors in your calculations that it can be neglected in almost all instances. Keep in mind, however, that in addition, subtraction, multiplication, and division, the errors you introduce propagate into the final answer.

LOOKING BACK

In this section we have reviewed some of the essential background you need with respect to units and dimensions.

Key Ideas

1. You can add, subtract, or equate the same units, but not unlike units.
2. You can multiply or divide unlike units but not cancel them.
3. Chemical engineers should be able to carry out calculations using both SI and American Engineering units.
4. Always carry units along with numbers in your calculations (on paper, in the computer, or in your head).
5. g_c is required as a conversion factor in the American Engineering system of units.
6. All valid equations require dimensional consistency.

Key Terms

American engineering (p. 6)	Mass (p. 10)
Conversion factors (p. 7)	SI (p. 5)
Conversion of units (p. 7)	Significant figures (p. 13)
Dimensional consistency (p. 11)	Units (p. 3)
Dimensions (p. 3)	Weight (p. 10)
Force (p. 8)	

Self-Assessment Test[1]

1. Convert 10 gal/hr to m^3/s; 50 $lb_f/in.^2$ to N/m^2.
2. On Phobos, the inner moon of Mars, the acceleration of gravity is 3.78 ft/sec^2. Suppose that an astronaut was walking around on Phobos, and this person plus the space suit and equipment had an earth weight of 252 lb_f.
   (a) What is the mass in lb_m of the astronaut plus suit and equipment?
   (b) How much would the space-suited astronaut weigh in pounds on Phobos?

---

[1]Answers to the self-assessment test problems are given in Appendix A.

3. Prepare a table in which the rows are: length, area, volume, mass, and time. Make two columns, one for the SI and the other for the American Engineering systems of units. Fill in each row with the name of the unit, and in a third column, show the numerical equivalency (i.e., 1 ft = 0.3048 m).

4. An orifice meter is used to measure the flow rate in pipes. The flow rate is related to the pressure drop by an equation of the form

$$u = c\sqrt{\frac{\Delta p}{\rho}}$$

   where u = fluid velocity
   Δp = pressure drop
   ρ = density of the flowing fluid
   c = constant of proportionality

   What are the units of c in the SI system of units?

5. What are the value and units of g_c?

6. In the SI system of units, the weight of a 180 lb man standing on the surface of the earth is approximately:
   (a) 801N   **(c)** Neither of these
   (b) 81.7 kg   **(d)** Both of these

7. The thermal conductivity k of a liquid metal is predicted via the empirical equation $k = A \exp (B/T)$, where k is in J/(s)(m)(K) and A and B are constants. What are the units of A? Of B?

Thought Problems

1. Comment about what is wrong with the following two statements from a textbook:
   (a) Weight is the product of mass times the force of gravity.
   (b) A 67 kg person on Earth will weigh only 11 kg on the moon.

2. A state representative is proposing a bill in the legislature that declares: "No person shall discharge into the atmosphere any gaseous effluent containing radioactive materials." Would you be in favor of this bill? Is it feasible?

3. An article in the newspaper said, "Your July 12 page-one article concerning the world standard of weight, 'Le Gran K,' failed to distinguish between mass and weight. It was not clear whether Le Gran K had lost 75-billionths of a kilogram of mass or 75-billionths of a kilogram of weight in the past 100 years." Le Gran K is the world's reference mass for the kilogram. What are three possible reasons for the reported loss?

Discussion Questions

1. In a letter to the editor, the letter writer says:

   I believe SI notation might be improved so as to make it mathematically more useful by setting SI-sanctioned prefixes in, for example, **bold face** type. Then one would write, **1 c = 10 m** without any ambiguity [**c** = 10^{-2}, **m** = 10^{-3}] and the meaning of "mm" would be at once clear to any mathematically literate, if scientifically illiterate citizen: namely, either 10^{-3} m [**m**m], 10^{-6} [**mm**], or (after Gauss and early algebraists) m^2 [mm].

With respect to the "mm" problem and remarks regarding the difference between "one square millimeter" [(**m**m)^2] and "one milli squaremeter" [**m**(m^2)], these difficulties are analogous to the confusion between a "camel's-hair brush" and a "camel's hairbrush."

What do you think of the author's proposal?

2. In spite of the official adoption of the SI system of units in Europe, people still use kg as a unit of weight and force as well as mass. For example, an individual buys 10 kg of potatoes and not 98 newtons of potatoes, and weighs the amount purchased so that the weight is 10 kg. Automobile tire pressure is expressed in kg/cm^2 (or just kg) and not newtons per square centimeter, or pascals. Why does this usage occur?

1.2   THE MOLE UNIT

Your objectives in studying this section are to be able to:

1. Define a kilogram mole, pound mole, and gram mole.
2. Convert from moles to mass and the reverse for any chemical compound given the molecular weight.
3. Calculate molecular weights from the molecular formula.

LOOKING AHEAD

What is a mole? The best answer is that a **mole** is a certain number of molecules, atoms, electrons, or other specified types of particles. In this section we review the concept of the mole and molecular weight.

MAIN CONCEPTS

The word mole appears to have been introduced by William Ostwald in 1896 who took it from the Latin word *moles* meaning "heap" or "pile." If you think of the mole as a large heap of particles you will have the general idea. A more precise definition was set out by the 1969 International Committee on Weights and Measures, which approved the mole (symbol mol in the SI system) as being "the amount of a substance that contains as many elementary entities as there are atoms in 0.012 kg of carbon 12." The entities may be atoms, molecules, ions, or other particles.

In the SI a mole is composed of 6.02×10^{23} molecules. However, for convenience in calculations, we can make use of other nonstandard specifications for moles and such as *pound mole* (lb mol, comprised of $6.02 \times 10^{23} \times 453.6$ molecules), the *kg mol* (kilomole, kmol, comprised of 1,000 moles), and so on. Such abbreviations in the definition of a mole help avoid excess details in many calculations. To keep the units straight, we will use the designation of g mol for the SI mole.

To convert the number of moles to mass, we make use of the **molecular weight**—the mass per mole:

$$\text{the g mol} = \frac{\text{mass in g}}{\text{molecular weight}}$$

$$\text{the lb mol} = \frac{\text{mass in lb}}{\text{molecular weight}}$$

or

$$\text{mass in g} = (\text{mol. wt.})(\text{g mol})$$

$$\text{mass in lb} = (\text{mol. wt.})(\text{lb mol})$$

Furthermore, there is no reason why you cannot carry out computations in terms of ton moles, kilogram moles, or any corresponding units if they are defined analogously, even if they are not standard units.

Values of the molecular weights (relative molar masses) are built up from the tables of atomic weights based on an arbitrary scale of the relative masses of the elements. **Atomic weight** of an element is the mass of an atom based on the scale that assigns a mass of exactly 12 to the carbon isotope ^{12}C, whose nucleus contains 6 protons and 6 neutrons. The terms atomic "weight" and molecular "weight" are universally used by chemists and engineers instead of the more accurate terms atomic "mass" or molecular "mass." Since weighing was the original method for determining the comparative atomic masses, as long as they were calculated in a common gravitational field, the relative values obtained for the atomic "weights" were identical with those of the atomic "masses."

Appendix B lists the atomic weights of the elements. On this scale of atomic weights, hydrogen is 1.008, carbon is 12.01, and so on. (In most of our calculations we shall round these off to 1 and 12, respectively, for convenience.)

A compound is composed of more than one atom, and the molecular weight of the compound is nothing more than the sum of the weights of atoms of which it is composed. Thus H_2O consists of 2 hydrogen atoms and 1 oxygen atom, and the molecular weight of water is $(2)(1.008) + 16.000 = 18.02$. These weights are all relative to the ^{12}C atom 12.0000, and you can attach any unit of mass you desire to these weights; for example, H_2 can be 2.016 g/g mol, 2.016 lb/lb mol, 2.016 ton/ton mol, and so on.

You can compute average molecular weights for mixtures of constant composition even though they are not chemically bonded if their compositions are known accurately. Later (Example 1.10) we show how to calculate the fictitious quantity called the average molecular weight of air. Of course, for a material such as fuel oil or coal whose composition may not be exactly known, you cannot determine an exact molecular weight, although you might estimate an approximate average molecular weight good enough for engineering calculations. Keep in mind that the symbol lb refers to lb_m unless otherwise stated.

EXAMPLE 1.7 Calculation of Molecular Weight

Since the discovery of superconductivity almost 100 years ago scientists and engineers have speculated about how it can be used to improve the use of energy. Until recently most applications were not economically viable because the niobium alloys used had to be cooled below 23 K by liquid He. However, in 1987 superconductivity in Y-Ba-Cu-O material was achieved at 90 K, a situation that permits the use of inexpensive liquid N_2 cooling.

What is the molecular weight of the following cell of a superconductor material? (The figure represents one cell of a larger structure.)

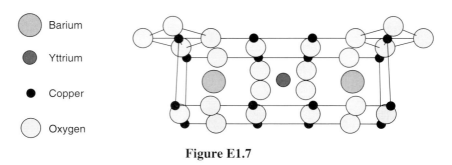

Barium

Yttrium

Copper

Oxygen

Figure E1.7

Solution

Assume that one cell is a molecule. By counting the atoms you can find

Element	Number of atoms	Atomic Weight	Mass
Ba	2	137.34	2(137.34)
Cu	16	63.546	16(63.546)
O	24	16.00	24(16.00)
Y	1	88.905	1(88.905)
		Total	1763.3

The molecular weight of the cell is 1763.3 g/g mol.

EXAMPLE 1.8 Use of Molecular Weights

If a bucket holds 2.00 lb of NaOH (mol. wt. = 40.0), how many

(a) Pound moles of NaOH does it contain?
(b) Gram moles of NaOH does it contain?

Solution

(a) $\dfrac{2.00 \text{ lb NaOH} \mid 1 \text{ lb mol NaOH}}{40.0 \text{ lb NaOH}} = 0.050 \text{ lb mol NaOH}$

(b₁) $\dfrac{2.00 \text{ lb NaOH} \mid 1 \text{ lb mol NaOH} \mid 454 \text{ g mol}}{40.0 \text{ lb NaOH} \mid 1 \text{ lb mol}} = 22.7 \text{ g mol}$

or

(b₂) $\dfrac{2.00 \text{ lb NaOH} \mid 454 \text{ g} \mid 1 \text{ g mol NaOH}}{1 \text{ lb} \mid 40.0 \text{ g NaOH}} = 22.7 \text{ g mol}$

EXAMPLE 1.9 Use of Molecular Weights

How many pounds of NaOH are in 7.50 g mol of NaOH?

Solution

Basis: 7.50 g mol of NaOH

$\dfrac{7.50 \text{ g mol NaOH} \mid 1 \text{ lb mol} \mid 40.0 \text{ lb NaOH}}{454 \text{ g mol} \mid 1 \text{ lb mol NaOH}} = 0.661 \text{ lb NaOH}$

LOOKING BACK

We have explained that a mole is an enumerated number of entities, and that although the official designation of *mole* in SI units means 6.02×10^{23} molecules, for convenience we will also in this book use lb mol, kg mol, etc. The molecular weight is the same value in any system of units, namely the mass per mole.

Key Ideas

1. The mole in SI units (g mol in this book) is a quantity of material equal to about 6.02×10^{23} molecules.
2. In this book for convenience in avoiding conversions we also use lb mol ($453.6 \times 6.02 \times 10^{23}$ molecules), kg mol, i.e., kilo mol or kmol ($1,000 \times 6.02 \times 10^{23}$ molecules).
3. Molecular weight (really mass) is just the mass of a compound or element per mole.

Key Terms

Atomic weight (p.17)   Molecular weight (p. 17)
Mole (p. 16)

Self-Assessment Test

1. What is the molecular weight of acetic acid (CH_3COOH)?
2. What is the difference between a kilogram mole and a pound mole?
3. Convert 39.8 kg of NaCl per 100 kg of water to kilogram moles of NaCl per kilogram mole of water.
4. How many pound moles of $NaNO_3$ are there in 100 lb?
5. One pound mole of CH_4 per minute is fed to a heat exchanger. How many kilograms is this per second?

Thought Problem

1. There is twice as much copper in 480 g of copper as there is in 240 g of copper, but is there twice as much copper in 480 g of copper as there is silver in 240 g of silver?

Discussion Questions

1. In the journal *Physics Education* (July, 1977, p. 276) McGlashan suggested that the physical quantity we call the mole is not necessary. Instead, it would be quite feasible to use molecular quantities, that is, number of molecules or atoms, directly. Instead of pV = nRT where n denotes the number of moles of a substance, we should write pV = NkT where N denotes the number of molecules and k is the Boltzmann constant

[1.380×10^{-23}J/(molecule)(K)]. Thus, for example 3.0×10^{-29} m^3/molecule would be the molecular volume for water, a value used instead of 18 cm^3/g mol. Is the proposal a reasonable idea?

2. In asking the question "What is meant by a mole?", the following answers were obtained. Explain which are correct and which are not, and why.
   (a) A mole is the molecular weight expressed in grams.
   (b) A mole is the quantity of material in one gram.
   (c) A mole is a certain number of cm^3 of one substance or another.
   (d) A mole is the weight of a molecule expressed in grams.
   (e) A mole is the number of molecules in one gram of a substance.

1.3   CONVENTIONS IN METHODS OF ANALYSIS AND MEASUREMENT

> ***Your objectives in studying this section are to be able to:***
>
> 1. Define density and specific gravity.
> 2. Calculate the density of a liquid or solid given its specific gravity and the reverse.
> 3. Look up and interpret the meaning of density and specific gravity of a liquid or solid in reference tables.
> 4. Specify the common reference material(s) used to determine the specific gravity of liquids and solids.
> 5. Convert the composition of a mixture from mole fraction (or percent) to mass (weight) fraction (or percent) and the reverse.
> 6. Transform a material from one measure of concentration to another, including mass/volume, moles/volume, ppm, and molarity.
> 7. Calculate the mass or number of moles of each component in a mixture given the percent (or fraction) composition, and the reverse, and compute the pseudo average molecular weight.
> 8. Convert a composition given in mass (weight) percent to mole percent, and the reverse.
> 9. Understand the measures of flammability and toxicity.

LOOKING AHEAD

In this section we review some of the common conventions used in reporting physical properties including density, specific gravity, measures of concentration, flammability, and toxicity.

MAIN CONCEPTS

Certain definitions and conventions exist that we review in this section since they will be used constantly throughout the book. If you memorize them now, you will immediately have a clearer perspective and save considerable trouble later.

1.3-1   Density

Density is the ratio of mass per unit volume, as, for example, kg/m^3 or lb/ft^3. It has both a numerical value and units. To determine the density of a substance, you must find both its volume and its mass. Densities for liquids and solids do not change significantly at ordinary conditions with pressure, but they do change with temperature as shown in Figure 1.1.

Figure 1.2 illustrates how density also varies with composition. In the winter you may put antifreeze in your car radiator. The service station attendant checks the concentration of antifreeze by measuring the specific gravity and, in effect, the density of the radiator solution after it is mixed thoroughly. A little thermometer exists in the hydrometer kit in order to be able to read the density corrected for temperature.

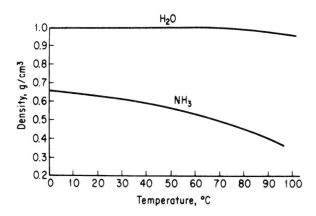

Figure 1.1   Densities of liquid H_2O and NH_3 as a function of temperature.

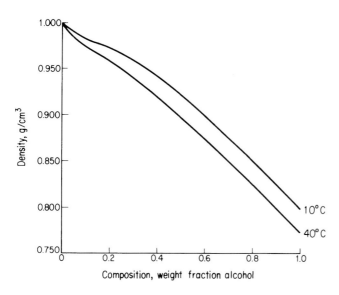

Figure 1.2   Density of a mixture of ethyl alcohol and water as a function of composition.

1.3-2 Specific Gravity

Specific gravity is commonly thought of as a dimensionless ratio. Actually, it should be considered as the ratio of two densities—that of the substance of interest, A, to that of a reference substance—each of which has associated units. In symbols:

$$\text{sp gr} = \textbf{specific gravity} = \frac{\left(\text{lb} / \text{ft}^3\right)_A}{\left(\text{lb} / \text{ft}^3\right)_{\text{ref}}} = \frac{\left(\text{g} / \text{cm}^3\right)_A}{\left(\text{g} / \text{cm}^3\right)_{\text{ref}}} = \frac{\left(\text{kg} / \text{m}^3\right)_A}{\left(\text{kg} / \text{m}^3\right)_{\text{ref}}}$$

The reference substance for liquids and solids is normally water. Thus, the specific gravity is the ratio of the density of the substance in question to the density of water. The specific gravity of gases frequently is referred to air, but may be referred to other gases. To be precise when referring to specific gravity, state the temperature at which each density is chosen. Thus

$$\text{sp gr} = 0.73\frac{20°}{4°}$$

can be interpreted as follows: the specific gravity when the solution is at 20°C and the reference substance (water) is at 4°C is 0.73. In case the temperatures for which the sp gr is stated are unknown, assume ambient temperature and 4°C, respectively. Since the density of water at 4°C is very close to 1.0000 g/cm³, in the SI system the numerical values of the specific gravity and density in this system are essentially equal. Since densities in the American Engineering system are expressed in lb/ft^3 and the density of water is about 62.4 lb/ft^3, it can be seen that the specific gravity and density values are not numerically equal in the American Engineering system. Yaws and colleagues[2] (1991) is a source for values of liquid densities, and the disk in the back of the book contains Yaw's database.

In the petroleum industry the specific gravity of petroleum products is usually reported in terms of a hydrometer scale called °API. The equation for the API scale is

$$°API = \frac{141.5}{sp\ gr\ \dfrac{60°}{60°}} - 131.5 \tag{1.3}$$

or

$$sp\ gr\ \frac{60°}{60°} = \frac{141.5}{°API\ +\ 131.5} \tag{1.4}$$

The volume and therefore the density of petroleum products vary with temperature, and the petroleum industry has established 60°F as the standard temperature for volume and API gravity. The °API is being phased out as SI units are accepted for densities.

EXAMPLE 1.10 Density and Specific Gravity

If dibromopentane (DBP) has a specific gravity of 1.57, what is the density in (a) g/cm^3? (b) lb$_m$/ft^3? and (c) kg/m^3?

Solution

No temperatures are cited for the dibromopentane or the reference compound (presumed to be water, hence we assume that the temperatures are the same and that water has a density of 1.00×10^3 kg/m^3 (1.00 g/cm^3).

(a)
$$\frac{1.57\ \dfrac{g\ DBP}{cm^3}}{1.00\ \dfrac{g\ H_2O}{cm^3}} \left| 1.00\ \frac{g\ H_2O}{cm^3} \right. = 1.57\ \frac{g\ DBP}{cm^3}$$

(b)
$$\frac{1.57\ \dfrac{lb_m\ DBP}{ft^3}}{1.00\ \dfrac{lb_m\ H_2O}{ft^3}} \left| 62.4\ \frac{lb_m\ H_2O}{ft^3} \right. = 97.97\ \frac{lb_m\ DBP}{ft^3}$$

---

[2]Yaws, C.L., H C. Yang, J. R. Hopper, and W. A. Cawley, "Equation for Liquid Density," *Hydrocarbon Processing.* (Jan. 1991):103–106.

(c) $\dfrac{1.57 \text{ g DBP}}{\text{cm}^3} \left| \left(\dfrac{100 \text{ cm}}{1 \text{ m}}\right)^3 \right| \dfrac{1 \text{ kg}}{1000 \text{ g}} = 1.57 \times 10^3 \dfrac{\text{kg DBP}}{\text{m}^3}$

or

$$\dfrac{1.57 \dfrac{\text{kg DBP}}{\text{m}^3} \left| \dfrac{1.00 \times 10^3 \text{ kg } H_2O}{\text{m}^3} \right.}{1.00 \dfrac{\text{kg } H_2O}{\text{m}^3}} = 1.57 \times 10^3 \dfrac{\text{kg DBP}}{\text{m}^3}$$

Note how the units of specific gravity as used here clarify the calculations.

Do **not** attempt to get an average specific gravity or average density for a mixture of solids or liquids by multiplying the individual component specific gravities or densities by the respective mass fractions of the components in the mixture and summing the products. The proper way to use specific gravity is demonstrated in the next example.

EXAMPLE 1.11 Application of Specific Gravity

In the production of a drug having a molecular weight of 192, the exit stream from the reactor flows at the rate of 10.3 L/min. The drug concentration is 41.2% (in water), and the specific gravity of the solution is 1.025. Calculate the concentration of the drug (in kg/L) in the exit stream, and the flow rate of the drug in kg mol/min.

Solution

For the first part of the problem, we want to transform the mass fraction of 0.412 into mass per liter of the drug. Take 1.000 kg of exit solution as a basis for convenience. See Figure E1.11.

Basis: 1.000 kg solution

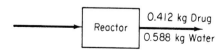

Figure E1.11

How do we get mass per volume (the density) from the given data, which is in terms of mass per mass? Use the specific gravity of the solution.

$$\text{density of solution} \;=\; \dfrac{1.025\,\dfrac{\text{g soln}}{\text{cm}^3}\;\bigg|\;1.000\,\dfrac{\text{g H}_2\text{O}}{\text{cm}^3}}{1.000\,\dfrac{\text{g H}_2\text{O}}{\text{cm}^3}} \;=\; 1.025\,\dfrac{\text{g soln}}{\text{cm}^3}$$

Next

$$\dfrac{0.412\text{ kg drug}}{1.000\text{ kg soln}}\;\bigg|\;\dfrac{1.025\text{ g soln}}{1\text{ cm}^3}\;\bigg|\;\dfrac{1\text{ kg}}{10^3\text{ g}}\;\bigg|\;\dfrac{10^3\text{ cm}^3}{1\text{ L}} \;=\; 0.422\text{ kg drug/L soln}$$

To get the flow rate, we take a different basis, 1 minute.

$$\text{Basis: 1 min} \equiv 10.3\text{ L solution}$$

$$\dfrac{10.3\text{ L soln}}{1\text{ min}}\;\bigg|\;\dfrac{0.422\text{ kg drug}}{1\text{ L soln}}\;\bigg|\;\dfrac{1\text{ kg mol drug}}{192\text{ kg drug}} \;=\; 0.0226\text{ kg mol/min}$$

1.3-3 Specific Volume

The specific volume of any compound is the inverse of the density, that is, the volume per unit mass or unit amount of material. Units of specific volume might be ft^3/lb$_m$, ft^3/lb mol, cm^3/g, bbl/lb$_m$, m^3/kg, or similar ratios.

1.3-4 Mole Fraction and Mass (Weight) Fraction

Mole fraction is simply the moles of a particular substance divided by the total number of moles present. This definition holds for gases, liquids, and solids. Similarly, the **mass (weight) fraction** is nothing more than the **mass (weight)** m of the substance divided by the total mass (weight) of all of the substances present. Although the mass fraction is what is intended to be expressed, ordinary engineering usage employs the term **weight fraction**. Mathematically, these ideas can be expressed as

$$\text{mole fraction of } A = \frac{\text{moles of } A}{\text{total moles}}$$

$$\text{mass (weight) fraction of } A = \frac{\text{mass (weight) of } A}{\text{total mass (weight)}}$$

Mole percent and weight are the respective fractions times 100.

EXAMPLE 1.12   **Mole Fraction and Mass (Weight) Fraction**

An industrial-strength drain cleaner contains 5.00 kg of water and 5.00 kg of NaOH. What are the mass (weight) fraction and mole fraction of each component in the drain cleaner container?

Solution

Basis: 10.0 kg of total solution

Component	kg	Weight fraction	Mol. wt.	kg mol	Mole fraction
H_2O	5.00	$\dfrac{5.00}{10.0} = 0.500$	18.0	0.278	$\dfrac{0.278}{0.403} = 0.69$
NaOH	5.00	$\dfrac{5.00}{10.0} = 0.500$	40.0	0.125	$\dfrac{0.125}{0.403} = 0.31$
Total	10.00	1.000		0.403	1.00

The kilogram moles are calculated as follows:

$$\frac{5.00 \text{ kg } H_2O}{} \left| \frac{1 \text{ kg mol } H_2O}{18.0 \text{ kg } H_2O} \right. = 0.278 \text{ kg mol } H_2O$$

$$\frac{5.00 \text{ kg NaOH}}{} \left| \frac{1 \text{ kg mol NaOH}}{40.0 \text{ kg NaOH}} \right. = 0.125 \text{ kg mol NaOH}$$

Adding these quantities together gives the total kilogram moles.

1.3-5 Analyses

Some confusion exists when you need an analysis of a mixture because you may be uncertain as to whether the numbers presented represent mass (weight) fraction or mole fraction.

In this book, the composition of gases will always be presumed to be given in mole percent or fraction unless specifically stated otherwise.

Analyses of liquids and solids are usually given by mass (weight) percent or fraction, but occasionally by mole percent.

In this text, the analyses of liquids and solids will always be assumed to be weight percent unless specifically stated otherwise.

For example, Table 1.4 lists the detailed composition of dry air. Note the heading for the analysis.

We will often use as the composition of air 21% O_2 and 79% N_2—the N_2 includes Ar and the other components and has a pseudo molecular weight of 28.2.

Table 1.5 lists the chemical analysis of some wastes. Are the percents mass or mole percent?

A useful quantity in many calculations is the pseudo average molecular weight of air which can be calculated on the basis of 100 mol of air:

Component	Moles = percent	Mol. wt.	lb or kg	Weight %
O_2	21.0	32	672	23.17
N_2	79.0	28.2	2228	76.83
Total	100		2900	100.00

The average molecular weight is 2900 lb/100 lb mol = 29.0, or 2900 kg/100 kg mol = 29.0.

1.3-6 Concentrations

Concentration means the quantity of some solute per specified amount of solvent, or solution, in a mixture of two or more components; for example:

TABLE 1.4   Composition of Clean, Dry Air near Sea Level

Component	Mole Percent	Component	Mole percent
Nitrogen	78.084	Xenon	0.0000087
Oxygen	20.9476	Ozone	
Argon	0.934	Summer	0-0.000007
Carbon dioxide	0.0350	Winter	0-0.000002
Neon	0.001818	Ammonia	0-trace
Helium	0.000524	Carbon monoxide	0-trace
Methane	0.0002	Iodine	0-0.000001
Krypton	0.000114	Nitrogen dioxide	0-0.000002
Nitrous oxide	0.00005	Sulfur dioxide	0-0.001
Hydrogen	0.00005		

TABLE 1.5   Chemical Analyses of Various Wastes by Percent

Material	Raw Paper	Charred Paper	Tire Rubber	Dry Sewage Sludge	Charred Sewage Sludge	Charred Animal Manure	Garbage Composite A	Garbage Composite B
Moisture	3.8	0.8	0.5	13.6	1.2	0.0	3.4	12.3
Hydrogen*	6.9	3.1	4.3	6.7	1.4	5.4	6.6	7.0
Carbon	45.8	84.9	86.5	28.7	48.6	41.2	57.3	44.4
Nitrogen	—	0.1	—	2.6	3.7	1.5	0.5	0.4
Oxygen*	46.8	8.5	4.6	26.5	—	26.0	22.1	42.1
Sulfur	0.1	0.1	1.2	0.6	—	0.4	0.4	0.2
Ash	0.4	2.5	3.4	34.9	45.7	25.5	10.2	5.9

*The hydrogen and oxygen values reflect that due to both the presence of water and that contained within the moisture-free material.

(1) Mass per unit volume (lb_m of solute/ft^3, g of solute/L, lb_m of solute/bbl, kg of solute/m^3).

(2) Moles per unit volume (lb mol of solute/ft^3, g mol of solute/L, g mol of solute/cm^3).

(3) Parts per million (ppm); parts per billion (ppb)—a method of expressing the concentration of extremely dilute solutions. Ppm is equivalent to a weight fraction for solids and liquids because the total amount of material is of a much higher order of magnitude than the amount of solute; it is a mole fraction for gases. Why?

(4) Other methods of expressing concentration with which you should be familiar are molarity (g mol/L) and normality (equivalents/L).

A typical example of the use of some of these concentration measures is the set of guidelines by which the Environmental Protection Agency defines the extreme levels at which the five most common air pollutants could harm people over stated periods of time.

(1) *Sulfur dioxide:* 365 μ g/m^3 averaged over a 24-hr period

(2) *Particulate matter:* 260 μ g/m^3 averaged over a 24-hr period

(3) *Carbon monoxide:* 10 mg/m^3 (9 ppm) when averaged over an 8-hr period; 40 mg/m^3 (35 ppm) when averaged over 1 hr

(4) *Nitrogen dioxide:* 100 μg/m^3 averaged over 1 year

EXAMPLE 1.13 Use of ppm

The current OSHA 8 hour limit for HCN in air is 10.0 ppm. A lethal dose of HCN in air (from the Merck Index) is 300 mg/kg of air at room temperature. How many mg HCN/kg air is the 10.0 ppm? What fraction of the lethal dose is 10.0 ppm?

Solution

$$\text{Basis: 1 kg mol of the air/HCN mixture}$$

The 10.0 ppm is $\dfrac{10.0 \text{ g mol}}{\left(10^6 \text{ air} + \text{HCN}\right) \text{ g mol}} = \dfrac{10.0 \text{ g mol HCN}}{10^6 \text{ g mol air}}$ because the amount of

HCN is so small. Then

$$\frac{10.0 \text{ g mol HCN}}{10^6 \text{ g mol air}} \left| \frac{27.03 \text{ g HCN}}{1 \text{ g mol HCN}} \right| \frac{1 \text{ g mol air}}{29 \text{ g air}} \left| \frac{1000 \text{ mg HCN}}{1 \text{ g HCN}} \right.$$

(a) $\times \ \dfrac{1000 \text{ g air}}{1 \text{ kg air}} = 9.32 \text{ mg HCN / kg air}$

(b) $\dfrac{9.32}{300} = 0.031$

It is important to remember that in an ideal solution, such as gases or a simple mixture of hydrocarbon liquids or compounds of like chemical nature, the volumes of the components may be added without great error to get the total volume of the mixture. **For the so-called nonideal mixtures this rule does not hold**, and the total volume of the mixture is bigger or smaller than the sum of the volumes of the pure components.

In Chapter 3 we will use **stream flows** and composition in making material balances. To calculate the **mass flow rate**, $\dot{m}$, from a known **volumetric flow rate**, q, you multiply the volumetric flow rate by the mass concentration thus

$$\frac{q \text{ m}^3}{\text{s}} \left| \frac{\rho \text{ kg}}{\text{m}^3} \right. = \dot{m} \frac{\text{kg}}{\text{s}}$$

How would you calculate the volumetric flow rate from a known mass flow rate?

From the volumetric flow rate you can calculate the **average velocity**, v, in a pipe if you know the area, A, of the pipe from the relation

$$q = Av \tag{1.5}$$

ADDITIONAL DETAILS

Flammability, Toxicity, and Autoignition

The **flash point** of a substance is the temperature in air at which the vapor above the substance will flash or explode in the presence of a flame. Thus, the flash point is one of the factors that you must always take into consideration in handling liquids. Other properties of materials are also important. City fire codes require chemicals to be classified and handled according to their (1) toxicity rating, (2) flammability rating, (3) reactivity rating, and (4) flash point. Table 1.6 lists a few of these parameters for some common compounds. Amounts stored over certain minimum amounts are reportable under the code. For example, for flammable materials, you must report

Rating	Flammable	Toxic or Reactive
4	20L	10 g or 10 cm^3
3	200L	1 kg or 1L
2	2000L	200 kg or 200L

The **toxicity** ratings (health hazard) are based on human exposure by any route, and are: 4 (death), 3 (major temporary or permanent injury that may threaten life), 2 (minor temporary or permanent injury), 1 (minor injury, readily reversible), and 0 (no toxic effect unless overwhelming dosage occurs). The other ratings have analogous classifications. **Flammability** is based on experiments, and includes: 4 (forms explosive mixtures in air with dusts, mists, liquid droplets), 3 (can be ignited under ambient temperatures, burn with rapidity via

Table 1.6   Common Chemicals and Ratings

Chemical	Toxicity Rating	Flammability Rating	Reactivity Rating	Flash point (°F)
Acetaldehyde (ℓ)	2	4	1	−38
Acetone (ℓ)	1	4	0	−4
Benzene (ℓ)	2	4	0	12
Chlorine (g)	3	0	0	—
Ethanol (ℓ, 36–112 proof)	0	3	0	73–100
Gasoline (ℓ, 100 octane)	1	4	0	−50

self-contained oxygen, contains materials that ignite spontaneously when exposed to air), 2 (materials that must be moderately heated or exposed to high temperatures to ignite, or may release hazardous vapor mixtures in air), and 0 (noncombustible—will not burn in air at 1500°F [815°C] for a period of 5 minutes). **Reactivity** ratings are somewhat similar.

Another factor you must consider in engineering is the **autoignition temperature** (AIT) of a combustible vapor or material. AIT is the lowest temperature at which a mixture will ignite in the absence of a spark or flame. Figure 1.3 illustrates the concept. At elevated temperatures, the oxygen in the air begins to interact with the combustible material, resulting in an exothermic reaction. In addition to hazard analysis, studies of autoignition are important in designing combustion engines where knock autoignition can occur. Egolf and Jurs (1992)[3] describe models that can be used to predict the autoignition temperature for mixtures of gases.

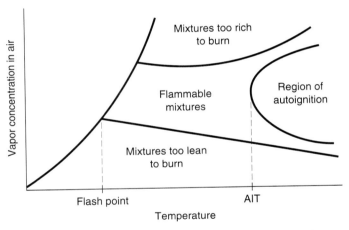

Figure 1.3 Concept of flammability and autoignition regions for a combustible substance in air.

EXAMPLE 1.14 Hazard Analysis

At a given temperature most gases have an upper and lower limit of flammability in oxygen. Within those limits a flame can self-propagate (explode) after ignition takes place. Figure E1.14 is a triangular diagram showing the mole fractions of O_2, N_2, and CH_4 in a gas mixture. From any point in the diagram you can read along the solid lines as follows to get the concentration of the individual components. For example, from

---

[3]Egolf, L. M. and P. C. Jurs, *Ind. Eng. Chem. Res.* **31** (1992):1798–1807.

the point marked A to get the mole fraction CH_4 read along the line parallel to the N_2 boundary (to get 0.40), to get the mole fraction O_2 read along the line parallel to the CH_4 boundary (to get 0.20), and get the mole fraction N_2 by reading along the line parallel to the O_2 boundary (to get 0.40). The region in which explosions can take place falls within the triangle bounded by the heavy lines.

To avoid the possibility of explosion in a vessel containing gas having the composition of 40% N_2, 45% O_2, and 15% CH_4, the recommendation is to dilute the gas mixture by adding an equal amount of pure N_2. Will this action do the job?

Solution

We need to compute the composition of the gas mixture after addition of the N_2. The basis is 100 moles of initial gas.

Comp.	original mixture mol = %		added N_2	final mixture mol	mol fr.
N_2	40	100	140	0.70	
O_2	45		45	0.23	
CH_4	15		15	0.07	
Total	100		200	1.00	

Based on the final mixture, the composition still falls within the triangle, so that the suggested dilution will still leave an explosive gas mixture.

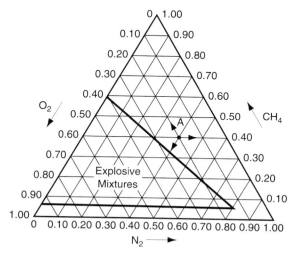

Figure E1.14

LOOKING BACK

We have reviewed the concepts and given examples of applications of density, specific gravity, mass (weight) and mole fraction, and various measures of concentration. Become accustomed to using all of these quantities so that their use becomes instinctive and you can focus on solving problems.

Key Ideas

1. Density is the mass of a substance per unit volume.
2. Specific gravity is the ratio of two densities. Given the value of the reference density, you can determine the density of the desired compound.
3. In this book the compositions of liquids and solids are denoted in weight (mass) fraction or percent whereas the compositions of gases are in mole fraction or percent.
4. A pseudo average molecular weight can be calculated for a mixture of pure components.
5. A wide variety of ways to express concentration exists.
6. In safety analysis, key measures of hazardous properties are toxicity, flammability, reactivity, flash point, and the autoignition temperature.

Key Terms

Average molecular weight (p. 28)   Mass fraction (p. 26)
Autoignition (p. 32)   Mole fraction (p. 26)
Concentration (p. 28)   ppm (p. 29)
Density (p. 22)   Reactivity (p. 32)
Flammability (p. 31)   Specific gravity (p. 23)
Flash point (p. 31)   Specific volume (p. 26)
Mass flow (p. 30)   Toxicity (p. 31)

Self-Assessment Test

1. Answer the following questions true (T) or false (F).
   (a) The density and specific gravity of mercury are the same.
   (b) The inverse of density is the specific volume.
   (c) Parts per million denotes a concentration that is a mole ratio.
   (d) Concentration of a component in a mixture does not depend on the amount of the mixture.
2. A cubic centimeter of mercury has a mass of 13.6 g at the earth's surface. What is the density of the mercury.
3. What is the approximate density of water at room temperature?

4. For liquid HCN, a handbook gives: sp gr 10°C/4°C = 1.2675. What does this mean?

5. For ethanol, a handbook gives: sp gr 60°F = 0.79389. What is the density of ethanol at 60°F?

6. Commercial sulfuric acid is 98% H_2SO_4 and 2% H_2O. What is the mole ratio of H_2SO_4 to H_2O?

7. A container holds 1.704 lb of HNO_3/lb of H_2O and has a specific gravity of 1.382 at 20°C. Compute the composition in the following ways:
   (a) Weight percent HNO_3
   (b) Pounds HNO_3 per cubic foot of solution at 20°C
   (c) Molarity at 20°C

8. The specific gravity of steel is 7.9. What is the volume in cubic feet of a steel ingot weighing 4000 lb?

9. A solution contains 25% salt by weight in water. The solution has a density of 1.2 g/cm^3. Express the composition as:
   (a) Kilograms salt per kilogram of H_2O
   (b) Pounds of salt per cubic foot of solution

10. A liquefied mixture of n-butane, n-pentane, and n-hexane has the following composition in percent

$$
\begin{array}{ll}
n\text{-}C_4H_{10} & 50 \\
n\text{-}C_5H_{12} & 30 \\
n\text{-}C_6H_{14} & 20 \\
\end{array}
$$

For this mixture, calculate:
   (a) The weight fraction of each component
   (b) The mole fraction of each component
   (c) The mole percent of each component
   (d) The average molecular weight

Thought Problems

1. **"Drop in the bucket.** Service Station Operator Balks at Paying for 'Shrinking' Gasoline" was a recent headline in the newspaper. The way the station operator sees it, when you pay for a gallon of gasoline, you should get a gallon of gasoline. He contends he's paying for thousands of gallons of gasoline each year he doesn't get. He's suing to try to settle the issue.

   The issue is "shrinking" gasoline. According to the American Petroleum Institute, gasoline shrinks at the rate of six gallons per 10,000 gallons for every one-degree-Fahrenheit drop in temperature and expands at the same rate when the temperature rises. On a typical 8,000-gallon shipment from a refinery at, say, 75 degrees, that means a loss of 72 gallons when the gasoline gets stored in the station operator's 60-degree underground tanks and thereafter is sold to customers.

   "It may seem like a small matter," he says, "but I'm a high-volume dealer, and I lose

$3,000 to $4,000 a year to shrinkage." Major oil companies adjust their billing for temperature changes in big shipments to each other, but "We don't bill that way because changes in temperature in winter and summer balance each other out, so the dealer comes out pretty much even," an oil company spokesman says. **"Dealers Don't Agree"** was the next heading. "Dealers say that's nonsense." Why do you think the dealers believe the lack of temperature billing is not fair?

2. The National Museum is considering buying a Maya plaque from Honduras that the seller claims to be jade. Jade is either jadite (sp gr 3.2 to 3.4) or nephrite (sp gr 3.0). What liquids would you recommend using to test whether the mask is jade?

3. A refinery tank that had contained gasoline was used for storing pentane. The tank overflowed when the level indicator said that it was only 85% full. The level indicator was a DP cell that measured weight of fluid. Can you explain what went wrong?

Discussion Questions

1. From *Chemical and Engineering News* (Oct. 12, 1992):10.

> Two Dutch scientists have won government and industry support to explore the possibility of raising the level of the ground in coastal areas of their low-lying country by converting subsurface limestone to gypsum with waste sulfuric acid.
> The scheme centers on the fact that gypsum, $CaSO_4 \cdot 2H_2O$, occupies twice the volume of a corresponding amount of calcium carbonate. The project envisions drilling holes as deep as 1 km at selected sites above limestone strata for injecting the acid. The resulting gypsum should raise the surface as much as several meters. Instances of ground swelling have already occurred from sulfuric acid spillage in the Netherlands at Pernis, an industrial region near Rotterdam.

What do you think of the feasibility of this idea?

2. It has been suggested that an alternative to using pesticides on plants is to increase the level of natural plant toxins by breeding or gene manipulation. How feasible is this approach from the viewpoint of a mutagenic and carcinogenic effects on human beings? For example, solamine and chaconine, some of the natural alkaloids in potatoes, are present at a level of 15,000 µg per 200 g potatoes. This amount is about 1/6 of the toxic level for human beings. Neither alkaloid has been tested for carcinogenicity. The man-made pesticides intake by humans is estimated to be about 150 µg/day. Only about one-half have been shown to be carcinogenic in test animals. The intake of known natural carcinogens is estimated to be 1 g per day from fruit and vegetables alone, omitting coffee (500 µg per cup), bread (185 µg per slice), and cola (2000 µg/bottle).
   Prepare a brief report ranking possible carcinogenic hazards from man-made and natural substances. List the possible exposure, source of exposure, carcinogenic dose per person, and the relative potency and risk of the carcinogen.

3. In 1990 Congress passed several major pieces of legislation governing the discharge

and treatment of industrial wastes. Some discharge limits for water set by the EPA are listed in table as follows:

Compound	Freshwater		Saltwater		Human Consumption	
	Max.	Contin.	Max.	Contin.	Max.	Contin.
Arsenic	360	190	69	3.6	0.08	0.14
Cadmium	3.9	1.1	43	9.3	16	170
Chromium (IV)	16	11	1,100	50	1,300	—
Lead	82	3.2	220	8.5	50	—
Mercury	2.4	0.012	2.1	0.025	0.014	0.15

Are the compounds in the discharge actually the elements shown in the table? What do you think the units associated with the numbers might be? Do the values in the column indicate relative toxicity of the waste components? Why do the specifications have both max. (maximum) and contin. (continues) values specified? Comment on how the specifications might be determined by the EPA.

4. The four layers of the atmosphere in ascending order (in a direction up from the earth) are the troposphere, stratosphere, mesosphere, and the thermosphere. A number of investigators have focused on the problems caused by depletion of the ozone layer, which occurs in the middle of the stratosphere at concentrations up to 2×10^{-4} g/m^3 in air having a density of 1 to 10^2 g/m^3 depending on the height. The ozone shields the surface of the earth from UV radiation in the range of 2300 to 3200 Ångstrom.

   Although skin cancer is the best documented effect of ozone depletion, it is not necessarily the most serious. What might be some of the other effects of decreased ozone concentration in the stratosphere?

1.4   BASIS

> ### Your objectives in studying this section are to be able to:
>
> 1. State the three questions useful in selecting a basis.
> 2. Apply the three questions to problems in order to select a suitable basis or sequence of bases.

LOOKING AHEAD

In this section we discuss how you choose a basis on which to solve problems.

MAIN CONCEPTS

Have you noted in previous examples that the word **basis** has appeared at the top of the computations? This concept of basis is vitally important both to your understanding of how to solve a problem and also to your solving it in the most expeditious manner. The basis is the reference chosen by you for the calculations you plan to make in any particular problem, and a proper choice of basis frequently makes the problem much easier to solve. The basis may be a period of time—for example, hours, or a given mass of material—such as 5 kg of CO_2, or some other convenient quantity. In selecting a sound basis (which in many problems is predetermined for you but in some problems is not so clear), you should ask yourself the following questions:

(1) What do I have to start with?
(2) What answer is called for?
(3) What is the most convenient basis to use?

These questions and their answers will suggest suitable bases. Sometimes when a number of bases seem appropriate, you may find it is best to use a unit basis of 1 or 100 of something, as, for example, kilograms, hours, moles, cubic feet. For liquids and solids in which a weight analysis is used, a convenient basis is often 1 or 100 lb or kg; similarly, 1 or 100 moles is often a good choice for a gas. The reason for these choices is that the fraction or percent automatically equals the number of pounds, kilograms, or moles, respectively, and one step in the calculations is saved.

EXAMPLE 1.15 Choosing a Basis

The dehydrogenation of the lower alkanes has been carried out using a ceric oxide catalyst. What is the mass fraction and mole fraction of Ce and O in the catalyst?

Solution

Since no specific amount of material is given, the question, what do I have to start with, does not apply. Neither does the question about the desired answer. A sensible and convenient basis would be to take 1 kg mol because we know the mol ratio of Ce to O in the compound. A basis of 1 kg is not convenient as the mass of the Ce to O is unknown. In fact, it must be calculated.

Basis: 1 kg mol CeO

Component	kg mol	Mole fraction	Mol. wt.	kg	Mass fraction
Ce	1	0.50	140.12	140.12	0.90
O	1	0.50	16.0	16.0	0.10
Total	2	1.00		156.1	1.00

EXAMPLE 1.16 Choosing a Basis

Most processes for producing high-energy-content gas or gasoline from coal include some type of gasification step to make hydrogen or synthesis gas. Pressure gasification is preferred because of its greater yield of methane and higher rate of gasification.

Given that a 50.0-kg test run of gas averages 10.0% H_2, 40.0% CH_4, 30.0% CO, and 20.0% CO_2, what is the average molecular weight of the gas?

Solution

The obvious basis is 50.0 kg of gas ("what I have to start with"), but a little reflection will show that such a basis is of no use. You cannot multiply *mole percent* of this gas times kg and expect the answer to mean anything. Thus, the next step is to choose a "convenient basis," which is 100 kg mol or lb mol of gas, and proceed as follows:

Basis: 100 kg mol or lb mol of gas

Component	percent = kg mol or lb mol	Mol. wt.	kg or lb
CO_2	20.0	44.0	880
CO	30.0	28.0	840
CH_4	40.0	16.04	642
H_2	10.0	2.02	20
Total	100.0		2382

$$\text{Average molecular weight} = \frac{2382 \text{ kg}}{100 \text{ kg mol}} = 23.8 \text{ kg / kg mol}$$

It is important that your basis be indicated near the beginning of the problem so that you will keep clearly in mind the real nature of your calculations and so that anyone checking your problem solution will be able to understand on what basis your calculations are performed. If you change bases in the middle of the problem, a new basis should be indicated distinctly at that time.

EXAMPLE 1.17 Changing Bases

A medium-grade bituminous coal analyzes as follows

Component	Percent
S	2
N	1
O	6
Ash	11
Water	3

The residuum is C and H in the mole ratio H/C = 9. Calculate the weight (mass) fraction composition of the coal with the ash and the moisture omitted.

Solution

Take as a basis 100 kg of coal because then percent = kilograms.

$$\text{Basis: 100 kg of coal}$$

The sum of the S + N + O +ash +water is

$$2 + 1 + 6 + 11 + 3 = 23 \text{ kg}$$

Hence, the C + H must be 100 − 23 = 77 kg.

To determine the kilograms of C and H, we have to select a new basis. Is 77 kg satisfactory? No. Why? Because the H/C ratio is in terms of moles, not weight (mass). Pick instead:

$$\text{Basis: 100 kg mol of C + H}$$

Component	Mole fraction	kg mol	Mol. Wt.	kg
H	$\dfrac{9}{1+9} = 0.90$	90	1.008	90.7
C	$\dfrac{1}{1+9} = 0.10$	10	12	120
Total		1.00	100	210.7

Finally, to return to the original basis, we have

$$\text{H:} \quad \frac{77 \text{ kg}}{} \cdot \frac{90.7 \text{ kg H}}{210.7 \text{ kg total}} = 33.15 \text{ kg H}$$

$$\text{C:} \quad \frac{77 \text{ kg}}{} \cdot \frac{120 \text{ kg H}}{210.7 \text{ kg total}} = 43.85 \text{ kg C}$$

and we can prepare a table summarizing the results on the basis of the coal ash free and water free.

Component	kg	Wt. fraction
C	43.85	0.51
H	33.15	0.39
S	2	0.02
N	1	0.01
O	6	0.07
Total	86.0	1.00

LOOKING BACK

We posed three questions to be used in helping you to select a basis, and gave example applications as well as showing how to change basis.

Key Ideas

The ability to choose the basis that requires the fewest steps in solving a problem can only come with practice. You can quickly accumulate the necessary experience if, as you look at each problem illustrated in this text, you determine first in your own mind what the basis should be and then compare your choice with the selected basis. By this procedure you will quickly obtain the knack of choosing a sound basis.

Key Terms

Basis (p. 38)   Selecting a basis (p. 38)
Changing basis (p. 40)

Self-Assessment Test

1. What are the three questions you should ask yourself in selecting a basis?
2. What would be good initial bases to select in solving Problems 1.6, 1.11, 1.30, and 1.39?

Thought Problem

1. Water-based dust-suppression systems are an effective and viable means of controlling dust and virtually eliminating the historic risk of fires and explosions in grain elevators. Water-based safety systems have resulted in cleaner elevators, improved respiratory atmospheres for employees, and reduced dust emissions into the environments surrounding storage facilities. However, some customers have complained that adding water to

the grain causes the buyer to pay too much for the grain. The grain elevators argue that all grain shipments unavoidably contain a weight component in the form of moisture. Moisture is introduced to grain and to grain products in a broad variety of practices.

What would you recommend to elevator operators and grain dealers to alleviate this problem?

1.5   TEMPERATURE

Your objectives in studying this section are to be able to:

1. Define temperature.
2. Explain the difference between absolute temperature and relative temperature.
3. Convert a temperature in any of the four scales (°C, K, °F, °R) to any of the others.
4. Convert an expression involving units of temperature and temperature difference to other units of temperature and temperature difference.
5. Know the reference points for the four temperature scales.

LOOKING AHEAD

In this section we discuss temperatures scales, both relative and absolute, and conversion of one temperature to another.

MAIN DETAILS

Our concept of temperature no doubt originated with our physical sense of hot and cold. **Temperature** can be rigorously defined once you have an acquaintance with thermodynamics, but here we simply paraphrase Maxwell's definition:

> *The temperature of a body is a measure of its thermal state considered in reference to its power to transfer heat to other bodies.*

Measurement of the thermal state can be accomplished through a wide variety of instruments as indicated in Figure 1.4. In this book we use four measures of temperature, those based on a relative scale, degrees **Fahrenheit** and **Celsius,** and those based on an absolute scale, degree **Rankine** and **kelvin**.

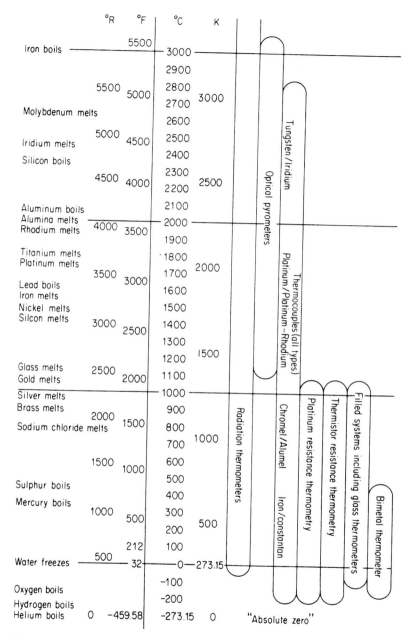

Figure 1.4 Temperature measuring instruments span the range from near absolute zero to beyond 3000 K. The chart indicates the preferred methods of thermal instrumentation for various temperature regions.

Absolute temperature scales have their zero point at the lowest possible temperature that we believe can exist. As you may know, this lowest temperature is related both to the ideal gas laws and to the laws of thermodynamics. The absolute scale that is based on degree units the size of those in the Celsius (centigrade) scale is called the *kelvin* scale; the absolute scale that corresponds to the Fahrenheit degree units is called the *Rankine* scale in honor of W. J. M. Rankine (1820–1872), a Scottish engineer. The relations between relative temperature and absolute temperature are illustrated in Figure 1.5. We shall round off absolute zero on the Rankine scale of $-459.67°$ to $-460°F$; similarly, $-273.15°C$ will be rounded off to $-273°C$. In Figure 1.5 all of the values of the temperatures have been rounded off, but more significant figures can be used. $0°C$ and its equivalents are known as *standard conditions of temperature.*

You should recognize that the unit degree (i.e., the unit temperature difference) on the kelvin-Celsius scale is not the same size as that on the Rankine-Fahrenheit scale. If we let $\Delta°F$ represent the unit temperature difference in the Fahrenheit scale, $\Delta°R$ be the unit temperature difference in the Rankine scale, and

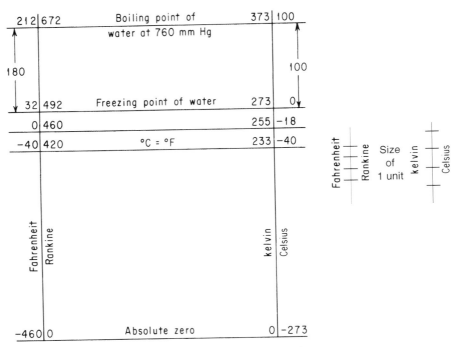

Figure 1.5   Temperature scales.

$\Delta°C$ and ΔK be the analogous units in the other two scales, you should be aware that

$$\Delta°F = \Delta°R$$
$$\Delta°C = \Delta K$$

Also, if you keep in mind that the $\Delta°C$ is larger than the $\Delta°F$,

$$\frac{\Delta°C}{\Delta°F} = 1.8 \quad \text{or} \quad \Delta°C = 1.8\Delta°F$$

$$\frac{\Delta K}{\Delta°R} = 1.8 \quad \text{or} \quad \Delta K = 1.8\Delta°R$$

Thus, when we cite the temperature of a substance we are stating the number of units of the temperature scale that occur (an enumeration of ΔTs) measured from the reference point.

Unfortunately, the symbols $\Delta°C$, $\Delta°F$, ΔK, and $\Delta°R$ are not in standard usage; the Δ symbol is suppressed. A few books try to maintain the difference between degrees of temperature (°C, °F, etc.) and the unit degree by assigning the unit degree the symbol C°, F°, and so on. But most journals and texts use the same symbol for the two different quantities. Consequently, **the proper meaning of the symbols °C, °F, K, and °R, as either the temperature or the unit temperature difference, must be interpreted from the context of the equation or sentence being examined.**

If you are not well acquainted with temperature conversion, be sure to practice conversions until they become routine. Many calculators and computers make the conversions automatically, but you should know that

$$T_{°R} = T_{°F}\left(\frac{1\,\Delta°R}{1\,\Delta°F}\right) + 460 \tag{1.6}$$

$$T_K = T_{°C}\left(\frac{1\,\Delta K}{1\,\Delta°C}\right) + 273 \tag{1.7}$$

Because the relative temperature scales do not have a common zero, as can be seen from Figure 1.5, the relation between °F and °C is

$$T_{°F} - 32 = T_{°C}\left(\frac{1.8\,\Delta°F}{1\,\Delta°C}\right) \tag{1.8}$$

EXAMPLE 1.18 Temperature Conversion

Convert 100°C to (a) K, (b) °F, and (c) °R.

Solution

(a)
$$(100 + 273)^\circ C \frac{1 \, \Delta K}{1 \, \Delta ^\circ C} = 373 \, K$$

or with suppression of the Δ symbol,

$$(100 + 273)^\circ C \frac{1 \, K}{1 \, ^\circ C} = 373 \, K$$

(b)
$$(100^\circ C) \frac{1.8^\circ F}{1^\circ C} + 32^\circ F = 212^\circ F$$

(c)
$$(212 + 460)^\circ F \frac{1^\circ R}{1^\circ F} = 672^\circ R$$

or

$$(373K) \frac{1.8^\circ R}{1 \, K} = 672^\circ R$$

Note the suppression of the Δ symbol.

EXAMPLE 1.19 Temperature Conversion

The thermal conductivity of aluminum at 32°F is 117 Btu/(hr)(ft²)(°F/ft). Find the equivalent value at 0°C in terms of Btu/(hr)(ft²)(K/ft).

Solution

Since 32°F is identical to 0°C, the value is already at the proper temperature. The "°F" in the denominator of the thermal conductivity actually stands for Δ°F, so that the equivalent value is

$$\frac{117 \, (Btu)(ft)}{(hr)(ft^2)(\Delta^\circ F)} \left| \frac{1.8 \, \Delta^\circ F}{1 \, \Delta^\circ C} \right| \frac{1 \, \Delta^\circ C}{1 \, \Delta \, K} = 211 \, Btu/(hr)(ft^2)(K/ft)$$

or with the Δ symbol suppressed,

$$\frac{117 \, (Btu)(ft)}{(hr)(ft^2)(^\circ F)} \left| \frac{1.8 \, ^\circ F}{1 \, ^\circ C} \right| \frac{1 \, ^\circ C}{1 \, K} = 211 \, Btu/(hr)(ft^2)(K/ft)$$

EXAMPLE 1.20 Temperature Conversion

The heat capacity of sulfuric acid in a handbook has the units J/(g mol)(°C) and is given by the relation

$$\text{heat capacity} = 139.1 + 1.56 \times 10^{-1}T$$

where T is expressed in °C. Modify the formula so that the resulting expression has the associated units of Btu/(lb mol)(°R) and T is in °R.

Solution

The units of °C in the denominator of the heat capacity are Δ °C, whereas the units of T are °C. First, substitute the proper relation in the formula to convert T in °C to T in °R, and then convert the units in the resulting expression to those requested.

$$\text{heat capacity} = \left\{139.1 + 1.56 \times 10^{-1}\left[(T_{°R} - 460 - 32)\frac{1}{1.8}\right]\right\}\frac{1}{(\text{g mol})(°C)}$$

$$\times \left|\frac{1 \text{ Btu}}{1055 \text{ J}}\right|\frac{454 \text{ g mol}}{1 \text{ lb mol}}\left|\frac{1°C}{1.8°R}\right| = 23.06 + 2.07 \times 10^{-2}T_{°R}$$

LOOKING BACK

In this section we have distinguished between relative temperature and absolute temperature scales. Examples of the conversion from one scale to another as well as applications indicate the need to be careful in applying the same symbol (°F, °C, °R, K) for both the temperature unit and the enumeration of the units relative to a reference point.

Key Ideas

1. Temperature is a measurement of the thermal state of a substance.
2. Both relative (°C, °F) and absolute (°R, K) temperature scales are used by engineers.
3. Conversion from one scale to another is still needed because no scale is accepted universally in the U.S.
4. The unit degree is represented by exactly the same symbol as the respective temperature (Δ°C is °C, Δ°F is °F, Δ°R is °R, and ΔK is K) so that you have to be very careful in converting temperatures and units expressed as "per degree".

Key Terms

Degree Celsius, °C (p. 42)   Standard conditions (p. 44)

Degree Fahrenheit, °F (p. 42)   Temperature (p. 42)

Degree kelvin, K (p. 42)   Temperature conversion (p. 45)

Degree Rankine, °R (p. 42)   Unit degree (p. 45)

Self-Assessment Test

1. What are the reference points of **(a)** the Celsius and **(b)** Fahrenheit scales?
2. How do you convert a *temperature difference,* Δ, from Fahrenheit to Celsius?
3. Is the unit temperature difference Δ°C a larger interval than Δ°F? Is 10°C higher than 10°F?
4. In Appendix E, the heat capacity of sulfur is C_p = 15.2 + 2.68T, where C_p is in J/g mol)(K) and T is in K. Convert so that C_p is in cal/(g mol)(°F) with T in °F.
5. Complete the following table with the proper equivalent temperatures:

°C	°F	K	°R
-40.0			
	77.0		
		698	
			69.8

6. Suppose that you are given a tube partly filled with an unknown liquid and are asked to calibrate a scale on the tube in °C. How would you proceed?
7. Answer the following questions:
   (a) In relation to absolute zero, which is higher, 1°C, or 1°F?
   (b) In relation to 0°C, which is higher, 1°C, or 1°F?
   (c) Which is larger, 1Δ°C or 1Δ°F?

Thought Problems

1. In reading a report on the space shuttle you find the statement that "the maximum temperature on reentry is 1482.2°C." How many significant figures do you think are represented by this temperature?
2. What temperature measuring devices would you recommend to make the following measurements?
   (a) Temperature of the thermal decomposition of oil shale (300° to 500°C)
   (b) Air temperature outside your home ($-20°$ to 30°C)
   (c) Temperature inside a freeze-drying apparatus ($-100°$ to 0°C)
   (d) Flame temperature of a Bunsen burner (2000° to 2500°C)
3. A vacuum tower used to process residual oil experienced severe coking (carbon formation) on the tower internals when it rained. Coking occurs because the temperature of the fluid gets to be too high. The temperature of the entering residual was controlled by a temperature recorder-controller (TRC) connected to a thermocouple inserted onto a Thermowell in the pipeline bringing the residual into the column. The TRC was oper-

ating at 700°F whereas the interior of the column was at 740°F (too hot). What might be the problem?

Discussion Questions

1. In the book by Eric Rogers, (*Physics for the Inquiring Mind*, Princeton, NJ: Princeton University Press, 1960), temperature is defined as: "temperature is the hotness measured in some definite scale." Is this correct? How would you define temperature?
2. In the kelvin or Rankine (absolute temperature) scales, the relation used for temperature is $T = n\Delta T$, where ΔT is the value of the unit temperature and n is the number of units enumerated. When $n = 0$, $T = 0$. Suppose that temperature is defined by the relation for $\ln(T) = n\Delta T$. Does $T = 0$ occur? What does $n = 0$ mean? Does the equivalent of absolute zero kelvin exist?

1.6   PRESSURE

> ### Your objectives in studying this section are to be able to:
>
> 1. Define pressure, atmospheric pressure, barometric pressure, standard pressure, and vacuum.
> 2. Explain the difference between absolute pressure and relative pressure (gauge pressure).
> 3. List four ways to measure pressure.
> 4. Convert from gauge pressure to absolute pressure and the reverse.
> 5. Convert a pressure measured in one set of units to another set, including kPa, mm Hg, in. H_2O, atm, in. Hg, and psi using the standard atmosphere or density ratios of liquids.
> 6. Calculate the pressure from the density and height of a column of fluid.

LOOKING AHEAD

In this section we review various measures of pressure, both relative and absolute, and illustrate converting from one set of pressure units to another.

MAIN CONCEPTS

Pressure is defined as "normal force per unit area." Examine Figure 1.6. Pressure is exerted on the top of the cylinder of water by the atmosphere, and on the bottom of the cylinder itself by the water.

The pressure at the bottom of the **static** (nonmoving) column of water exerted on the sealing plate is

$$p = \frac{F}{A} = \rho g h + p_0 \tag{1.9}$$

where p = pressure at the bottom of the column of the fluid
F = force
A = area
ρ = density of fluid
g = acceleration of gravity
h = height of the fluid column
p_0 = pressure at the top of the column of fluid

Suppose that the cylinder of fluid in Figure 1.6 is a column of mercury that has an area of 1 cm² and is 50 cm high. From Table D.1 we can find that the sp gr at 20°C, and hence the density of the Hg, is 13.55 g/cm³. Thus the force exerted by the mercury alone on the 1-cm² section of the bottom plate by the column of mercury is

$$F = \frac{13.55 \text{ g}}{\text{cm}^3} \left| \frac{980 \text{ cm}}{\text{s}^2} \right| 50 \text{ cm} \left| 1 \text{ cm}^2 \right| \frac{1 \text{ kg}}{1000 \text{ g}} \left| \frac{1 \text{ m}}{100 \text{ cm}} \right| \frac{1 \text{ N}}{\frac{1 \text{ (kg)(m)}}{\text{s}^2}}$$

$$= 6.64 \text{ N}$$

The pressure on the section of the plate covered by the mercury is the force per area of the mercury *plus* the pressure of the atmosphere

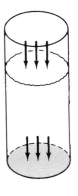

Figure 1.6. Pressure is the normal force per unit area. Arrows show the force exerted on the respective areas.

$$p = \frac{6.64 \text{ N}}{1 \text{ cm}^2} \left| \left(\frac{100 \text{ cm}}{1 \text{ m}} \right)^2 \right| \frac{(1 \text{ m}^2)(1 \text{ Pa})}{(1 \text{ N})} \left| \frac{1 \text{ kPa}}{1000 \text{ Pa}} \right. = 66.4 \text{ kPa} + p_0$$

If we had started with units in the American Engineering system, the pressure would be computed as (the density of mercury is $(13.55)(62.4)\text{lb}_m/\text{ft}^3$))

$$p = \frac{845.5 \text{ lb}_m}{1 \text{ ft}^3} \left| \frac{32.2 \text{ ft}}{s^2} \right| 50 \text{ cm} \left| \frac{1 \text{ in.}}{2.54 \text{ cm}} \right| \frac{1 \text{ ft}}{12 \text{ in.}} \left| \frac{1}{\frac{32.174(\text{ft})(\text{lb}_m)}{(s)^2(\text{lb}_f)}} \right. + p_0$$

$$= 1388 \frac{\text{lb}_f}{\text{ft}^2} + p_0$$

Sometimes in engineering practice, a liquid column is referred to as *head of liquid*, the head being the height of the column of liquid. Thus, the pressure of the column of mercury could be expressed simply as 50 cm Hg, and the pressure on the sealing plate at the bottom of the column would be 50 cm Hg + p_0 (in cm of Hg).

Pressures, like temperatures, can be expressed by either absolute or relative scales. **Whether relative or absolute pressure is measured in a pressure-measuring device depends on the nature of the instrument used to make the measurements.** For example, an open-end manometer (Figure 1.7a) would measure a **relative pressure (gauge pressure)**, since the reference for the open end is the pressure of the atmosphere at the open end of the manometer. On the other

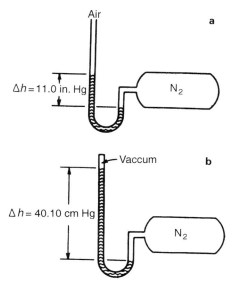

Figure 1.7 (a) Open-end manometer showing a pressure above atmospheric pressure. (b) Absolute pressure manometer.

hand, closing off the end of the manometer (Figure 1.7b) and creating a vacuum in the end results in a measurement against a complete vacuum, or against "no pressure"; p_0 in Eq. (1.9) is zero. This measurement is called **absolute pressure.** Since absolute pressure is based on a complete vacuum, a fixed reference point that is unchanged regardless of location, temperature, weather, or other factors, absolute pressure then establishes a precise, invariable value that can be readily identified. Thus, the zero point for an absolute pressure scale corresponds to a perfect vacuum, whereas the zero point for a relative pressure scale usually corresponds to the pressure of the air that surrounds us at all times, and as you know, varies slightly.

If a reading on a column of mercury occurs as illustrated in Figure 1.8, with the dish open to the atmosphere, the device is called a *barometer* and the reading of atmospheric pressure is termed **barometric pressure.**

In any of the pressure-measuring devices depicted in Figures 1.7 or 1.8, the fluid is at equilibrium, that is, a state of hydrostatic balance is reached in which the manometer fluid is stabilized, and the pressure exerted at the bottom of the U-tube in the part of the tube open to the atmosphere or vacuum exactly balances the pressure exerted at the bottom of the U-tube in the part of the tube connected to the tank of N_2. Water and mercury are commonly used indicating fluids for manometers; the readings thus can be expressed in "inches or cm of water," "inches or cm of mercury," and so on. (In ordinary engineering calculations we ignore the vapor pressure of mercury and minor changes in the density of mercury due to temperature changes in making pressure measurements.)

Another type of common measuring device is the visual *Bourdon gauge* (Figure 1.9), which normally (but not always) reads zero pressure when open to the atmosphere. The pressure-sensing device in the Bourdon gauge is a thin metal tube with an elliptical cross section closed at one end that has been bent into an arc. As the pressure increases at the open end of the tube, it tries to straighten out, and the movement of the tube is converted into a dial movement by gears and levers. Figure 1.10 indicates the pressure ranges for the various pressure-measuring devices.

Another term applied in measuring pressure, a term illustrated in

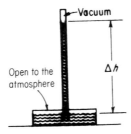

Figure 1.8   A barometer.

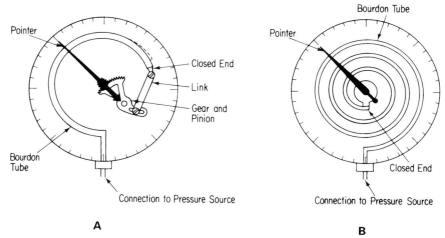

Figure 1.9 Bourdon gauge pressure-measuring devices. (**a**) "C" Bourdon; (**b**) Spiral Bourdon.

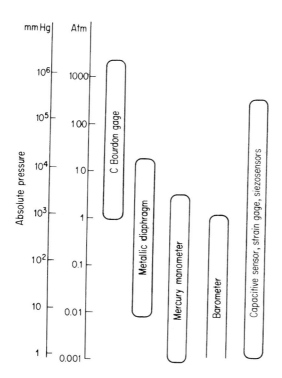

Figure 1.10 Ranges of application for pressure-measuring devices.

Figure 1.11, is **vacuum**. In effect, when you measure pressure as "inches of mercury vacuum," you reverse the usual direction of measurement and measure from the barometric pressure down to zero absolute pressure, in which case a perfect vacuum would be highest vacuum measure that you could achieve. The vacuum system of measurement of pressure is commonly used in an apparatus that operates at pressures less than atmospheric—such as a vacuum evaporator or vacuum filter. A pressure that is only slightly below barometric pressure may sometimes be expressed as a "draft" in inches of water, as, for example, in the air supply to a furnace or a water cooling tower.

Always keep in mind that **the reference point or zero point for the relative pressure scales is not constant.** The relationship between relative and absolute pressure is given by the following expression:

$$\textbf{gauge pressure} + \textbf{barometer pressure} = \textbf{absolute pressure} \qquad (1.10)$$

Examine Figure 1.12.

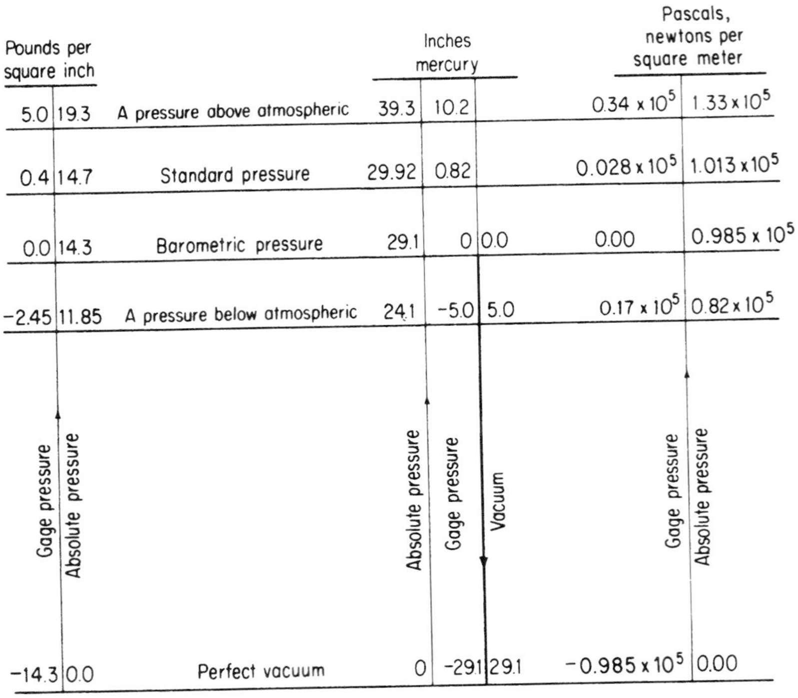

Figure 1.11   Pressure comparisons when barometer reading is 29.1 in.

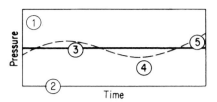

Figure 1.12   Pressure terminology. The standard atmosphere is shown by the heavy horizontal line. The dashed line illustrates the atmospheric (barometric) pressure, which changes from time to time. Point ① in the figure is a pressure of 19.3 psi referred to a complete vacuum or 5 psi referred to the barometric pressure: ② is the complete vacuum, ③ represents the standard atmosphere, and ④ illustrates a negative relative pressure or a pressure less than atmospheric. This type of measurement is described in the text as a vacuum type of measurement. Point ⑤ also indicates a vacuum measurement, but one that is equivalent to an absolute pressure above the standard atmosphere.

As to the units of pressure, Figure 1.11 shows three common systems: pounds (force) per square inch (psi), inches of mercury (in Hg), and pascals. Pounds per square inch absolute is normally abbreviated "psia," and "psig" stands for "pounds per square inch gauge." For other units, be certain to carefully specify whether they are gauge or absolute; for example, state "300 kPa absolute" or "12 cm Hg gauge." Other systems of expressing pressure exist; in fact, you will discover that there are as many different units of pressure as there are means of measuring pressure. Some of the other most frequently used systems are

(1) Millimeters of mercury (mm Hg)
(2) Feet of water (ft H_2O)
(3) Atmospheres (atm)
(4) Bars (bar): 100 kPa = 1 bar
(5) Kilograms (force) per square centimeter (kg_f/cm^2)—a common but theoretically prohibited measure in SI.[4]

You definitely must not confuse the standard atmosphere with atmospheric pressure. The *standard atmosphere* is defined as the pressure (in a standard gravitational field) equivalent to 1 atm or 760 mm Hg at 0°C or other equivalent value, whereas atmospheric pressure is a variable and must be obtained from

---

[4]Although the units of pressure in the SI system are pascal, you frequently find in common usage pressures denoted in the units of kg/cm^2 (i.e., kg_f/cm^2), units that have to be multiplied by

$$9.80 \times 10^4 \left| \frac{(kg_m)(m)}{(kg_f)(s^2)} \right| \left(\frac{cm^2}{m^2} \right) \quad \text{to get pascal}$$

a barometer each time you need it. The standard atmosphere may not equal the barometric pressure in any part of the world except perhaps at sea level on certain days, but it is extremely useful converting from one system of pressure measurement to another (as well as being useful in several other ways to be considered later). In a problem, if you are not given the barometric pressure, you usually *assume* the barometric pressure equals the standard atmosphere, but this assumption is only that, an assumption.

Expressed in various units, the *standard atmosphere* is equal to

1.000	atmospheres (atm)
33.91	feet of water (ft H_2O)
14.7	(14.696, more exactly) pounds per square inch absolute (psia)
29.92	(29.921, more exactly) inches of mercury (in. Hg)
760.0	millimeters of mercury (mm Hg)
1.013×10^5	pascal (Pa) or newtons per square meter (N/m^2); or 101.3 kPa

You can easily convert from one set of pressure units to another by using pairs of standard atmospheres as conversion factors as follows in which we convert 35 psia to inches of mercury and kPa by using ratios of the standard atmosphere to carry out the conversions:

$$\frac{35 \text{ psia}}{} \left| \frac{29.92 \text{ in. Hg}}{14.7 \text{ psia}} \right. = 71.24 \text{ in. Hg}$$

an identity

$$\frac{35 \text{ psia}}{} \left| \frac{101.3 \text{ kPa}}{14.7 \text{ psia}} \right. = 241 \text{ kPa}$$

EXAMPLE 1.21 Pressure Conversion

The pressure gauge on a tank of CO_2 used to fill soda-water bottles reads 51.0 psi. At the same time the barometer reads 28.0 in. Hg. What is the absolute pressure in the tank in psia? See Figure E1.21.

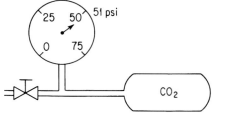

Figure E1.21

Solution

The pressure gauge is reading psig, not psia. From Eq. (1.10) the absolute pressure is the sum of the gauge pressure and the atmospheric (barometric) pressure expressed in the same units. We will change the atmospheric pressure to psia.

Basis: Barometric pressure = 28.0 in. Hg

$$\text{atmospheric pressure} = \frac{28.0 \text{ in. Hg}}{} \left| \frac{14.7 \text{ psia}}{29.92 \text{ in. Hg}} \right. = 13.76 \text{ psia}$$

The absolute pressure in the tank is

$$51.0 + 13.76 = 64.8 \text{ psia}$$

EXAMPLE 1.22 Pressure Conversion

Air is flowing through a duct under a draft of 4.0 cm H_2O. The barometer indicates that the atmospheric pressure is 730 mm Hg. What is the absolute pressure of the gas in inches of mercury? See Figure E1.22.

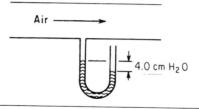

Air ⟶

4.0 cm H_2O

Figure E1.22

Solution

We can ignore the gas density above the manometer fluid. In the calculations we have to employ consistent units, and it appears in this case that the most convenient units are those of inches of mercury.

Basis: 730 mm Hg

$$\text{atmospheric pressure} = \frac{730 \text{ mm Hg}}{} \left| \frac{29.92 \text{ in. Hg}}{760 \text{ mm Hg}} \right. = 28.7 \text{ in. Hg}$$

Basis: 4.0 cm H_2O draft (under atmospheric)

$$\frac{4.0 \text{ cm } H_2O}{} \left| \frac{1 \text{ in.}}{2.54 \text{ cm}} \right| \frac{1 \text{ ft}}{12 \text{ in.}} \left| \frac{29.92 \text{ in. Hg}}{33.91 \text{ ft } H_2O} \right. = 0.12 \text{ in. Hg}$$

Since the reading is 4.0 cm H_2O draft (under atmospheric), the absolute reading in uniform units is

$$28.7 - 0.12 = 28.6 \text{ in. Hg absolute}$$

EXAMPLE 1.23 Vacuum Pressure Reading

Small animals such a mice can live at reduced air pressures down to 20 kPa (although not comfortably). In a test a mercury manometer attached to a tank as shown in Figure E1.23 reads 64.5 cm Hg and the barometer reads 100 kPa. Will the mice survive?

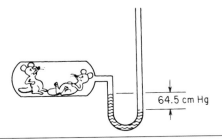

64.5 cm Hg

Figure E1.23

Solution

Basis: 64.5 cm Hg *below* atmospheric

We ignore any temperature corrections to the mercury density for temperature and also ignore the gas density above the manometer fluid. Then, since the vacuum reading on the tank is 64.5 cm below atmospheric, the absolute pressure in the tank is

$$100 \text{ kPa} \; - \; \frac{64.5 \text{ cm Hg}}{} \; \Big| \; \frac{101.3 \text{ kPa}}{76.0 \text{ cm Hg}} = 100 - 86 = 14 \text{ kPa absolute}$$

The mice probably will not survive.

In the examples so far we have ignored the gas in the manometer tube above the fluid. Is this OK? Examine Figure 1.13, which illustrates a manometer involving three fluids. When the columns of fluids are at equilibrium (it may take some time!), the relation between ρ_1, ρ_2 and the heights of the various columns of fluid is

$$p_1 + \rho_1 d_1 g = p_2 + \rho_2 g d_2 + \rho_3 g d_3 \qquad (1.11)$$

The level of fluid 2 at the bottom of d_1 is equal to the level of fluid 2 at the bottom of d_2, and serves as a reference level. We could introduce $\rho_2 \, g d_3$, where d_3 is the distance from d_1 or d_2 to the bottom of the manometer, on both sides of Eq. (1.11), and the equation would still be valid.

Can you show for the case in which $\rho_1 = \rho_3 = \rho$ that the manometer expression reduces to

$$p_1 - p_2 = (\rho_2 - \rho)g d_2 \qquad (1.12)$$

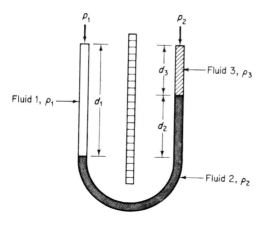

Figure 1.13 Manometer with three fluids.

Finally, suppose that fluids 1 and 3 are gases. Can you ignore the gas density relative to the manometer fluid density? For what types of fluids?

EXAMPLE 1.24 Calculation of Pressure Differences

In measuring the flow of fluids in a pipeline, a differential manometer, as shown in Fig. E1.24, can be used to determine the pressure difference across the orifice plate. The flow rate can be calibrated with the observed pressure drop. Calculate the pressure drop $p_1 - p_2$ in pascal for the manometer in Figure E1.24.

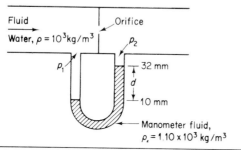

Figure E1.24

Solution

In this problem we cannot ignore the water density above the manometer fluid. Apply Eq. (1.12), as the densities of the fluids above the manometer fluid are the same.

$$p_1 - p_2 = (\rho_f - \rho)gd$$

$$= \frac{(1.10 - 1.00)10^3 \text{ kg}}{\text{m}^3} \left| \frac{9.807 \text{ m}}{\text{s}^2} \right| (22)(10^{-3})\text{m} \left| \frac{1 \text{ (N)(s}^2)}{\text{(kg)(m)}} \right| \frac{1 \text{ (Pa)(m}^2)}{1 \text{ (N)}}$$

$$= 21.6 \text{ Pa}$$

LOOKING BACK

In this section we defined pressure, discussed some of the ways to measure pressure, and showed how to convert from one set of pressure units to another via the standard atmosphere. We also emphasized the difference between the standard atmosphere and atmospheric pressure.

Key Ideas

1. Pressure is force per unit area.
2. Atmospheric pressure (barometric pressure) is the pressure of the air and the atmosphere surrounding us which changes from day to day.
3. Standard atmosphere is a constant reference atmosphere equal to 1.000 atmosphere or equivalent pressures in other units.
4. Absolute pressure is measured relative to a vacuum.
5. Gauge (relative) pressure is measured upward relative to atmospheric pressure.
6. Vacuum and draft pressures are measured downward from atmospheric pressure.
7. You can convert from one set of pressure measurements to another using the standard atmosphere.
8. A manometer measures a pressure difference in terms of heights of the fluids in the manometer tube.

Key Terms

Absolute pressure (p. 51)	Pressure (p. 49)
Barometric pressure (p. 52)	Pressure difference (p. 59)
Bourdon gauge (p. 53)	Relative pressure (p. 51)
Gauge pressure (p. 51)	Standard atmosphere (p. 55)
Manometer (p. 51)	Vacuum (p. 54)

Key Equations

(1.9)	(1.11)
(1.10)	

Self-Assessment Test

1. Write down the equation to convert gauge pressure to absolute pressure.
2. List the values and units of the standard atmosphere for six different methods of expressing pressure.
3. List the equation to convert vacuum pressure to absolute pressure.

4. Convert a pressure or 800 mm Hg to the following units:
   (a) psia   **(c)** atm
   (b) kPa   **(d)** ft H_2O

5. Your textbook lists five types of pressures: atmospheric pressure, barometric pressure, gauge pressure, absolute pressure, and vacuum pressure.
   (a) What kind of pressure is measured by Figure A?
   (b) What kind of pressure is measured by Figure B?
   (c) What would be the reading in Figure C assuming that the pressure and temperature inside and outside the helium tank are the same as in parts (a) and (b)?

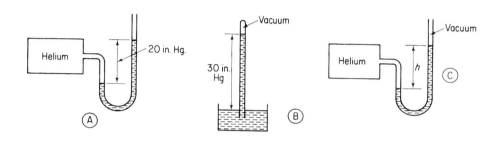

6. An evaporator shows a reading of 40 kPa vacuum. What is the absolute pressure in the evaporator in kilopascal?

7. Answer the following questions true (T) or false (F).
   (a) Air flows in a pipeline, and the manometer containing Hg that is set up as illustrated in Figure E1.24 shows a differential pressure of 14.2 mm Hg. You can ignore the effect of the density of the air on the height of the columns of mercury.
   (b) Lowering the He pressure 10% in Figure A will not cause the length of the column of Hg to decrease by 10%.

Thought Problems

1. A magic trick is to fill a glass with water, place a piece of paper over the top of the glass to cover the glass completely, and hold the paper in place as the glass is inverted 180°. On the release of your support of the paper, no water runs out! Many books state that the glass should be completely filled with water with no air bubbles present. Then the outside air pressure is said to oppose the weight of the water in the inverted glass. However, the experiment works just as well with a half-filled glass. The trick does not work if a glass plate is substituted for the piece of paper. Can you explain why?

2. A large storage tank was half full of a flammable liquid quite soluble in water. The tank roof needed maintenance. Since welding was involved, the foreman attached to the vent pipe on the top of the tank (in which there was a flame arrestor) a flexible hose and inserted the end of the hose into the bottom of a drum of water sitting on the ground to pick up any exhaust vapors. When the tank was emptied, the water rose up in the hose, and the tank walls collapsed inward. What went wrong in this incident?

3. Can a pressure lower than a complete vacuum exist?

4. If you fill a Styrofoam cup with water, and put a hole in the side, water starts to run out. However, if you hold your finger over the hole and then drop the cup from a height, water does not run out. Why?

Discussion Questions

1. Safety relief valves protect against overpressure in pipelines and process equipment. Overpressure can occur because of equipment failure in the process, fire, or human error. A valve must be selected of the proper size for anticipated scenarios, and the spring set for the proper relief pressure. Two different valves are proposed for use in a pipe line as illustrated figuratively in Figure DQ1.6–1. Which would you recommend for use?

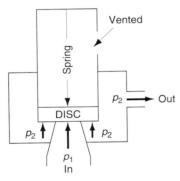

(A) This valve has the top of the closing disk open to the outlet (p_2)

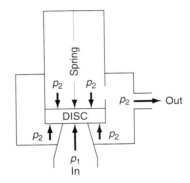

(B) This valve has the top and bottom of the closing disk open to the outlet (p_2)

Figure DQ1.6–1

2. Form a study group to investigate the possibility of raising the Titanic. The mystique of the sinking and attempts to raise the vessel have crept into the literature (*A Night to Remember*), the movie ("Raising the Titanic"), and magazines such as the *National Geographic*. Carry out a literature search to get the basic facts (4000 m deep, 4.86 × 108 N original weight, the density of sea water, and so on). Prepare a report giving

   (a) An executive summary including an estimate of feasibility.

   (b) The proposed method(s) of raising the ship.

   (c) A list of steps to execute to raise the ship.

   (d) A list of the equipment needed (include costs if possible).

   (e) A time schedule for the entire project (including obtaining the equipment and personnel).

(f) A list of (1) all assumptions made and (2) problems that might be encountered for which answers are not known.

(g) As an appendix, show all calculations made and references used.

1.7   THE CHEMICAL EQUATION AND STOICHIOMETRY

> ### Your objectives in studying this section are to be able to:
>
> 1. Write and balance chemical reaction equations.
> 2. Know the products of common reactions given the reactants.
> 3. Calculate the stoichiometric quantities of reactants and products given the chemical reaction.
> 4. Define excess reactant, limiting reactant, conversion, degree of completion, selectivity, and yield in a reaction.
> 5. Identify the limiting and excess reactants and calculate the percent excess reactant(s), the percent conversion, the percent completion, and yield for a chemical reaction with the reactants being in nonstoichiometric proportions.

LOOKING AHEAD

In this section we review some of the concepts related to chemical reactions, and define and apply a number by terms associated with complete and incomplete reactions.

MAIN CONCEPTS

As you already know, the chemical equation provides both qualitative and quantitative information essential for the calculation of the combining moles of materials involved in a chemical process. Take, for example, the combustion of heptane as shown below. What can we learn from this equation?

$$C_7H_{16} + 11\,O_2 \longrightarrow 7\,CO_2 + 8\,H_2O$$

It tells us about **stoichiometric ratios. First, make sure that the equation is balanced!** Then you can see that 1 mole (*not* lb_m or kg) of heptane will react with 11 moles of oxygen to give 7 moles of carbon dioxide plus 8 moles of water. These may be lb mol, g mol, kg mol, or any other type of mole. One mole of CO_2

is formed from each $\frac{1}{7}$ mole of C_7H_{16}. Also, 1 mole of H_2O is formed with each $\frac{7}{8}$ mole of CO_2. Thus the equation tells us in terms of moles (*not mass*) the ratios among reactants and products. The numbers that proceed the compounds are known as **stoichiometric coefficients:** 1 for C_7H_{16}, 11 for O_2, and so on.

Stoichiometry (stoy-kee-om-i-tree) deals with the combining of elements and compounds. The ratios obtained from the numerical coefficients in the chemical equation are the **stoichiometric ratios** that permit you to calculate the moles of one substance as related to the moles of another substance in the chemical equation. If the basis selected is to be mass (lb_m, kg) rather than moles, you should use the following method in solving problems involving the use of chemical equations: (1) Use the molecular weight to calculate the number of moles of the substance equivalent to the basis; (2) change this number of moles into the corresponding number of moles of the desired product or reactant by multiplying by the proper stoichiometric ratio, as determined from the chemical equation; and (3) then change the moles of product or reactant to a mass. All this can be done in one sequence.

For example, if 10 kg of C_7H_{16} react completely with the stoichiometric quantity of O_2, how many kg of CO_2 will be found as products? On the basis of 10 kg

$$\frac{10 \text{ kg } C_7H_{16}}{} \left| \frac{1 \text{ kg mol } C_7H_{16}}{100.1 \text{ kg } C_7H_{16}} \right| \frac{7 \text{ kg mol } CO_2}{1 \text{ kg mol } C_7H_{16}} \left| \frac{44.0 \text{ kg } CO_2}{1 \text{ kg mol } CO_2} \right| = 30.8 \text{ kg } CO_2$$

But you should keep in mind that the equation does not indicate how fast a reaction occurs or how much reacts, or even that a reaction occurs at all! For example, a lump of coal in air will sit unaffected at room temperature, but at higher temperatures it will readily burn. All the chemical equation indicates is the stoichiometric amounts required for the reaction and obtained from the reaction if it proceeds to completion in the manner written.

EXAMPLE 1.25 Use of the Chemical Equation

In the combustion of heptane, CO_2 is produced. Assume that you want to produce 500 kg of dry ice per hour and that 50% of the CO_2 can be converted into dry ice, as shown in Figure E1.25. How many kilograms of heptane must be burned per hour?

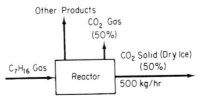

Figure E1.25

Solution

Basis: 500 kg of dry ice (equivalent to 1 hr)

Mol. wt. heptane = 100.1. The chemical equation is

$$C_7H_{16} + 11\,O_2 \longrightarrow 7\,CO_2 + 8\,H_2O$$

$$\frac{500 \text{ kg dry ice}}{} \left| \frac{1 \text{ kg } CO_2}{0.5 \text{ kg dry ice}} \right| \frac{1 \text{ kg mol } CO_2}{44.0 \text{ kg } CO_2} \left| \frac{1 \text{ kg mol } C_7H_{16}}{7 \text{ kg mol } CO_2} \right.$$

$$\left| \frac{100.1 \text{ kg } C_7H_{16}}{1 \text{ kg mol } C_7H_{16}} \right. = 325 \text{ kg } C_7H_{16}$$

Since the basis of 500 kg of dry ice is identical to 1 hr, 325 kg of C_7H_{16} must be burned per hour. Note that kilograms are first converted to moles, then the chemical equation is applied, and finally moles are converted to kilograms again for the final answer.

EXAMPLE 1.26 Stoichiometry

A limestone analysis

$CaCO_3$	92.89%
$MgCO_3$	5.41%
Isoluble	1.70%

(a) How many pounds of calcium oxide can be made from 5 tons of this limestone?
(b) How many pounds of CO_2 can be recovered per pound of limestone?
(c) How many pounds of limestone are needed to make 1 ton of lime?

Solution

Read the problem carefully to fix in mind exactly what is required. Lime will include all the impurities present in the limestone that remain after the CO_2 has been driven off. Next, draw a picture of what is going on in this process. See Figure E1.26.

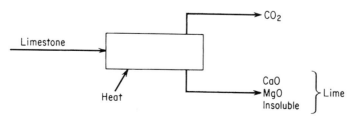

Figure E1.26

To complete the preliminary analysis you need the following chemical equations:

$$CaCO_3 \longrightarrow CaO + CO_2$$

$$MgCO_3 \longrightarrow MgO + CO_2$$

Additional data:

	$CaCO_3$	$MgCO_3$	CaO	MgO	CO_2
Mol. wt.:	100.1	84.32	56.08	40.32	44.0

Basis: 100 lb of limestone

This basis was selected because pounds = percent.

	Limestone		Products		
Component	lb = percent	lb mol	Solid Component	lb	CO_2(lb)
$CaCO_3$	92.89	0.9280	CaO	52.04	40.83
$MgCO_3$	5.41	0.0642	MgO	2.59	2.82
Insoluble	1.70		Insoluble	1.70	
Total	100.00	0.9920	Total	56.33	43.65

The quantities listed under Products are calculated from the chemical equations. For example

$$\frac{92.89 \text{ lb } CaCO_3}{} \left| \frac{1 \text{ lb mol } CaCO_3}{100.1 \text{ lb } CaCO_3} \right| \frac{1 \text{ lb mol CaO}}{1 \text{ lb mol } CaCO_3} \left| \frac{56.08 \text{ lb CaO}}{1 \text{ lb mol CaO}} \right| = 52.04 \text{ lb}$$

$$\frac{5.41 \text{ lb } MgCO_3}{} \left| \frac{1 \text{ lb mol } MgCO_3}{84.32 \text{ lb } MgCO_3} \right| \frac{1 \text{ lb mol MgO}}{1 \text{ lb mol } MgCO_3} \left| \frac{40.32 \text{ lb MgO}}{1 \text{ lb mol MgO}} \right| = 2.59 \text{ lb}$$

Note that the total pounds of products equal the 100 lb of entering limestone. Now to calculate the quantities originally asked for:

(a)   CaO produced $= \dfrac{52.04 \text{ lb CaO}}{100 \text{ lb stone}} \left| \dfrac{2000 \text{ lb}}{1 \text{ ton}} \right| \dfrac{5 \text{ ton}}{} = 5200 \text{ lb CaO}$

(b)   CO2 recovered $= \dfrac{43.65 \text{ lb } CO_2}{100 \text{ lb stone}} = 0.437 \text{ lb}$

(c)   Limestone required $= \dfrac{100 \text{ lb stone}}{56.33 \text{ lime}} \left| \dfrac{2000 \text{ lb}}{1 \text{ ton}} \right. = 3550 \text{ lb stone}$

An assumption implicit in the calculations above is that the reaction takes place exactly as written in the equation and proceeds to 100% completion. When reactants, products, or degree of completion of the actual reaction differ from the assumptions of the equation, additional data must be made available to predict the outcome of reactions.

In industrial reactors you will rarely find exact stoichiometric amounts of materials used. To make a desired reaction take place or to use up a costly reactant, excess reactants are nearly always used. This excess material comes out together with, or perhaps separately from, the product—and sometimes can be used again. Even if stoichiometric quantities of reactants are used, the reaction may not be complete or side reactions may occur so that the products will be accompanied by unused reactants as well as side products. In these circumstances some new definitions, as used in this book, must be understood. (Some of the following terms are sometimes used differently in other books, in chemistry, and in plant operations.)

1. Limiting reactant is the reactant that is present in the smallest stoichiometric amount. In other words, if two or more reactants are mixed and if the reaction were *to proceed according to the chemical equation to completion,* **whether it does or not,** the reactant that would first disappear is termed the limiting reactant. For example, using the equation in Example 1.25, if 1 g mol of C_7H_{16} and 12 g mol of O_2 are mixed, C_7H_{16} would be the limiting reactant even if the reaction does not take place.

As a shortcut to determining the limiting reactant, all you have to do is to calculate the mole ratio(s) of the reactants and compare each ratio with the corresponding ratio of the coefficients of the reactants in the chemical equation thus:

$$\frac{O_2}{C_7H_{16}} : \quad \overset{\textit{Ratio in feed}}{\frac{12}{1} = 12} \quad > \quad \overset{\textit{Ratio in chemical equation}}{\frac{11}{1} = 11}$$

If more than two reactants are present, you have to use one reactant as the reference substance, calculate the mole ratios of the other reactants in the feed relative to the reference, make pairwise comparisons versus the analogous ratios in the chemical equation, and rank each compound. For example, given the reaction

$$A + 3B + 2C \rightarrow \text{products}$$

and that 1.1 moles of *A*, 3.2 moles of *B*, and 2.4 moles of *C* are fed as reactants in the reactor, we choose *A* as the reference substance and calculate

Ratio in feed   *Ratio in chemical equation*

$$\frac{B}{A}: \quad \frac{3.2}{1.1} = 2.91 \quad < \quad \frac{3}{1} = 3$$

$$\frac{C}{A}: \quad \frac{2.4}{1.1} = 2.18 \quad > \quad \frac{2}{1} = 2$$

We conclude that B is the limiting reactant relative to A, and that A is the limiting reactant relative to C, hence B is the limiting reactant among the set of three reactants. In symbols we have $B < A$, $C > A$ (i.e., $A < C$), so that $B < A < C$.

2. Excess reactant is a reactant present in excess of the limiting reactant. The **percent excess** of a reactant is based on the amount of any excess reactant above the amount required to react with the limiting reactant according to the chemical equation or

$$\% \text{ excess} = \frac{\text{moles in excess}}{\text{moles required to react with limiting reactant}} (100)$$

where the moles in excess frequently can be calculated as the total available moles of a reactant less the moles required to react with the limiting reactant. In the first illustration in (1) above, the percent excess O_2 is

$$\% \text{ excess } O_2 = \frac{12 - 11}{11} 100 = 9.1\%$$

A common term, **excess air**, is used in combustion reactions; it means the amount of air available to react that is in excess of the air theoretically required to *completely* burn the combustible material. The *required* amount of a reactant is established by the limiting reactant and, can be calculated from the chemical equation for all other reactants. **Even if only part of the limiting reactant actually reacts, the required and excess quantities are based on the entire amount of the limiting reactant as if it had reacted completely.**

Three other terms that are used in connection with chemical reactions have more fuzzy definitions: conversion, selectivity, and yield. No universally agreed upon definitions exist for these terms—in fact, quite the contrary. Rather than cite all the possible usages of these terms, many of which conflict, we shall define them as follows:

3. Conversion is the fraction of the feed or some *key* material in the feed that is converted into products. Thus, percent conversion is

$$\% \text{ conversion} = 100 \frac{\text{moles (or mass) of feed (or a compound in the feed) that react}}{\text{moles (or mass) of feed (or a component in the feed) introduced}}$$

For example, in the first illustration in (1) above, if 14.4 kg of CO_2 are formed in the reaction of C_7H_{16}, we can calculate that 46.8% of the C_7H_{16} reacts thus

$$\begin{array}{c} C_7H_{16}\text{ equivalent} \\ \text{in the product} \end{array} : \frac{14.4\text{ kg }CO_2}{} \left|\begin{array}{c} 1\text{ kg mol }CO_2 \\ \hline 44.0\text{ kg }CO_2 \end{array}\right| \begin{array}{c} 1\text{ kg mol }C_7H_{16} \\ \hline 7\text{ kg mol }CO_2 \end{array} = 0.0468\text{ kg mol }C_7H_{16}$$

$$\begin{array}{c} C_7H_{16}\text{ in the} \\ \text{reactant} \end{array} : \frac{10\text{ kg }C_7H_{16}}{} \left|\begin{array}{c} 1\text{ kg mol }C_7H_{16} \\ \hline 100.1\text{ kg }C_7H_{16} \end{array}\right. = 0.0999\text{ kg mol }C_7H_{16}$$

$$\% \text{ conversion} = \frac{0.0468}{0.0999}100 = 46.8\%$$

What the basis in the feed is for the calculations and into what products the basis is being converted must be clearly specified or endless confusion results. Conversion is related to the **degree of completion** of a reaction, which is usually the percentage or fraction of the limiting reactant converted into products.

4. Selectivity is the ratio of the moles of a particular (usually the desired) product produced to the moles of another (usually undesired or by-product) product produced in a set of reactions. For example, methanol (CH_3OH) can be converted into ethylene (C_2H_4) or propylene (C_3H_6) by the reactions

$$2\,CH_3OH \rightarrow C_2H_4 + 2\,H_2O$$

$$3\,CH_3OH \rightarrow C_3H_6 + 3\,H_2O$$

Of course, for the process to be economical, the prices of methanol, ethylene, and propylene have to be appropriate. Examine the data for the concentrations of the products of the reactions in Figure 1.14 (note that by-products occur as well as C_2H_4 and C_3H_6). What is the selectivity of C_2H_4 relative to the C_3H_6 at 80% conversion of the CH_3OH? Proceed upwards at 80% conversion to get $y_{C_2H_4} \cong 0.19$ and $y_{C_3H_6} \cong 0.08$ so that the selectivity is 0.19/0.08 = 2.4 mol C_2H_4/mol C_3H_6.

5. Yield, for a single reactant and product, is the weight (mass) or moles of final product divided by the weight (mass) or moles of initial or key reactant (P lb of product A per R lb of reactant B) either fed or consumed. If more than one product and more than one reactant are involved, the reactant upon which the yield is to be based must be clearly stated. Suppose that we have a reaction sequence as follows:

$$A \longrightarrow B \longrightarrow C$$
$$\hspace{4.5cm} C$$

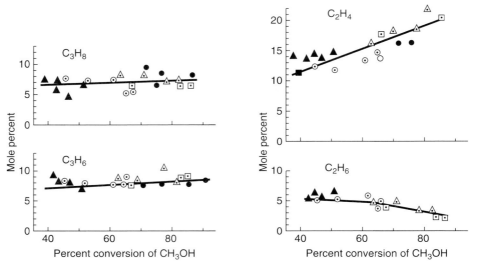

Figure 1.14   Products from the conversion of ethanol

with B the desired product and C the undesired one. The yield of B is the moles (or mass) of B produced divided by the moles (or mass) of A fed or consumed. The selectivity of B is the moles of B divided by the moles of C produced.

The terms "yield" and "selectivity" are terms that measure the degree to which a desired reaction proceeds relative to competing alternative (undesirable) reactions. As a designer of equipment you want to maximize production of the desired product and minimize production of the unwanted products. Do you want high or low selectivity? Yield?

The employment of these concepts is now illustrated by an example.

EXAMPLE 1.27 Incomplete Reaction

Antimony is obtained by heating pulverized stibnite (Sb_2S_3) with scrap iron and drawing off the molten antimony from the bottom of the reaction vessel.

$$Sb_2S_3 + 3Fe \longrightarrow 2Sb + 3FeS$$

Suppose that 0.600 kg of stibnite and 0.250 kg of iron turnings are heated together to give 0.200 kg of Sb metal. Determine:

(a) The limiting reactant

(b) The percentage of excess reactant

(c) The degree of completion (fraction)

(d) The percent conversion

(e) The yield

Solution

The molecular weights needed to solve the problem and the gram moles forming the basis are:

Component	kg	Mol. wt.	g mol
Sb_2S_3	0.600	339.7	1.77
Fe	0.250	55.85	4.48
Sb	0.200	121.8	1.64
FeS		87.91	

The process is illustrated in Figure E1.27

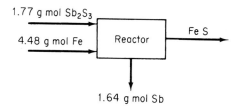

1.64 g mol Sb   **Figure E1.27**

(a) To find the limiting reactant, we examine the chemical reaction equation. The ratio of Sb_2S_3 to Fe in the equation is $1/3 = 0.33$. In the actual reaction the corresponding ratio is $1.77/4.48 = 0.40$, hence Sb_2S_3 is the excess reactant and Fe is the limiting reactant. The Sb_2S_3 required to react with the limiting reactant is $4.48/3 = 1.49$ g mol.

(b) The percentage of excess reactant is

$$\% \text{ excess} = \frac{1.77 - 1.49}{1.49} 100 = 18.8\% \text{ excess } Sb_2S_3$$

(c) Although Fe is the limiting reactant, not all the limiting reactant reacts. We can compute from the 1.64 g mol of Sb how much Fe actually does react.:

$$\frac{1.64 \text{ g mol Sb}}{} \left| \frac{3 \text{ g mol Fe}}{2 \text{ g mol Sb}} \right. = 2.46 \text{ g mol Fe}$$

If by the fractional degree of completion is meant the fraction conversion of Fe to products, then

$$\text{fractional degree of completion} = \frac{2.46}{4.48} = 0.55$$

(d) Let us assume that the percent conversion refers to the Sb_2S_3 since the reference compound is not specified in the question posed.

$$\frac{1.64 \text{ g mol Sb}}{} \left| \frac{1 \text{ g mol Sb}_2S_3}{2 \text{ g mol Sb}} \right. = 0.82 \text{ g mol Sb}_2S_3$$

$$\% \text{ conversion of Sb}_2S_3 \text{ to Sb} = \frac{0.82}{1.77}(100) = 46.3\%$$

(e) The yield will be stated as kilograms of Sb formed per kilogram of Sb_2S_3 that was fed to the reaction

$$\text{yield} = \frac{0.200 \text{ kg Sb}}{0.600 \text{ kg Sb}_2S_3} = \frac{1}{3}\frac{\text{kg Sb}}{\text{kg Sb}_2S_3} = \frac{0.33 \text{ kg Sb}}{1 \text{ kg Sb}_2S_3}$$

ADDITIONAL DETAILS

Note that in a reaction that is described by a number of reaction equations, the number of **independent reaction equations** is equal to the number of compounds reacting less the number of independent conservation equations that can be written for each element. For example, consider a system used in the manufacture of electronic materials (all gases except Si)

$$SiH_4, Si_2, H_6, SiH_2, H_2, Si$$

If we form an array (a matrix) in which the columns represent the species and the rows represent the elements, and the numbers entered into the array correspond to the number of the respective elements in each species, the difference between the number of species and the number of independent rows represents the number of independent reaction equations needed to represent the system. For example, the rows below represent the elements and the columns the compounds

	SiH_4	Si_2H_6	SiH_2	H_2	Si
Si:	1	2	1	0	1
H:	4	6	2	2	0

We have five species and two independent rows (the rank of the matrix is 2) in the array so that we only need $5 - 2 = 3$ equations to represent the system. From these equations, the other species can be calculated. Refer to the Appendix L under linear equations for more details. Which of the possible stoichiometric equations to select is beyond our scope here, but is discussed in books on thermodynamics.

LOOKING BACK

In this section we explained how the chemical equation is used to calculate quantitative relations between reactants and products. To use chemical equations, the quantities must be expressed in moles. We also defined a number of terms used by engineers in making calculations involving reactions.

Key Ideas

1. The chemical equation relates the moles of reactants to the moles of products.
2. The chemical equation does not indicate the true mechanism of the reaction or how fast or to what extent the reaction will take place.
3. The equation must be balanced to use it.

Key Terms

Conversion (p. 68)   Selectivity (p. 69)
Degree of completion (p. 69)   Stoichiometric ratio (p. 64)
Excess reactant (p. 68)   Stoichiometry (p. 64)
Independent reaction equations (p. 72)   Yield (p. 69)
Limiting reactant (p. 67)

Self-Assessment Test

1. Write balanced reaction equations for the following reactions:
   (a) C_9H_{18} and oxygen to form carbon dioxide and water
   (b) FeS_2 and oxygen to form Fe_2O_3 and sulfur dioxide
2. If 1 kg of benzene (C_6H_6) is oxidized with oxygen, how many kilograms of O_2 are needed to convert all the benzene to CO_2 and H_2O?
3. The electrolytic manufacture of chlorine gas from a sodium chloride solution is carried out by the following reaction:

$$2NaCl + 2H_2O \rightarrow 2NaOH + H_2 + Cl_2$$

How many kilograms of Cl_2 can one produce from 10 m³ of a brine solution containing 5% by weight of sodium chloride? The specific gravity of the solution relative to water at 4°C is 1.07.

4. Calcium oxide (CaO) is formed by decomposing limestone (pure $CaCO_3$). In one kiln the reaction goes to 70% completion.
   (a) What is the composition of the solid product withdrawn from the kiln?
   (b) What is the yield in terms of pounds of CO_2 produced per pound of limestone charged?

5. In problem 3, suppose that 50.0 kg of NaCl reacts with 10.0 kg of H_2O.

(a) What is the limiting reactant?
(b) What is the excess reactant?
(c) What components will the product solution contain if the reaction is 60% complete?

Thought Problems

1. An accident occurred in which one worker lost his life. A large steel evaporator in magnesium chloride service, containing internal heating tubes, was to be cleaned. It was shut down, drained, and washed. The next day two employees who were involved in the maintenance of the evaporator entered the vessel to repair the tubes. They were overcome, apparently from lack of oxygen. Subsequently, one employee recovered and escaped, but the other never regained consciousness, and died several days later. What in your opinion might have caused the accident (the lack of oxygen)?

2. A leaky pen will make itself known in bold color. Leaked information appears in the national news. But a leaking underground storage tank is extremely difficult to detect. Suggest some ways that can be used to detect leaky tanks. How effective are inventory control methods?

3. OSHA requires the use of breathing apparatus when working in or around tanks containing traces of solvents. While demolishing an old tank, a contractor purchased several cylinders of compressed air, painted gray. After two days he found that he needed more cylinders, and sent a truck for another cylinder. The driver returned with a black cylinder. None of the workers, including the person in charge of the breathing apparatus, noticed the change or, if they did, attached any importance to it. When the new cylinder was brought into use, a welder's facepiece caught fire. Fortunately, he pulled it off at once and was not injured. What would be the most likely cause of this accident?

4. A magazine reported that to degrade crude oil in a spill in seawater the oil-degrading bacteria require the dissolved oxygen in over 300,000 gal of air-saturated seawater to break down just 1 gal of crude. Can this estimate be correct?

Discussion Questions

1. Diesel pollutants pose a threat to the respiratory tract and are a potential cause of cause of cancer, according to the Environmental Protection Agency. The Clean Air Act of 1990 required that the sulfur content of diesel used on freeways must be lowered from 0.30 percent by weight to 0.05 percent—a substantial reduction. How might this be accomplished economically? In the oil well, at the refinery, at the service station, in the car, or what?

2. Among the important reactive nitrogen species in the troposphere 10 to 12 km above the earth surface are:

$$\left. \begin{array}{l} \text{Nitric oxide (NO)} \\ \text{Nitrogen dioxide } (NO_2) \end{array} \right\} \text{ Jointly referred to a } NO_x$$

Peroxyacetyl nitrate ($CH_3C(O)$-O-O-NO_2) known as PAN
Nitrous acid (HNO_2)
Dinitrogen pentoxide (N_2O_5)

Suggest some of the original sources of these compounds and their relative emissions per year. Hint: Some sources are not man made.

SUPPLEMENTARY REFERENCES

General

BHATT, B. I., and S. M. VORA. *Stoichiometry,* 2nd ed. New Delhi: Tata McGraw-Hill, 1976.

CLAUSEN, C. A., and G. MATTSON. *Principles of Industrial Chemistry.* New York: Wiley, 1979.

FELDER, R. M., and R. W. ROUSSEAU. *Elementary Principles of Chemical Processes,* 2nd ed. New York: Wiley, 1986.

LUYBEN, W. L., and L. A. WENZEL. *Chemical Process Analysis: Mass and Energy Balances.* Englewood Cliffs, NJ: Prentice-Hall, 1988.

MASSEY, B. S. *Measures in Science and Engineering.* New York: Halsted Press, 1986.

SHAHEEN, E. I. *Basic Practice of Chemical Engineering,* 2nd ed. Boston: Houghton Mifflin, 1984.

STOKER, H. S. *Introduction to Chemical Principles,* 2nd ed. New York: Macmillian, 1986.

Units and Dimensions

AMERICAN NATIONAL METRIC COUNCIL. *Metric Guide for Educational Materials.* Washington, DC: ANMC, 1977.

FRENCH, A. P. "Is g Really the Acceleration Due to Gravity?" *Phys. Teacher.* (November 1983): 528.

GLAVIC, P. "A Proposal for a Systematic Approach to Names and Symbols of Quantities Used in Chemical Engineering." *Chem. Biochem. Eng.,* **5** (1–2), (1991): 81–95.

HORVATH, A. L. *Conversion Tables of Units in Science and Engineering.* New York: Elsevier, 1986.

MOLYNEUX, P. "The Dimensions of Logarithmic Quantities." *J. Chem. Educ.,* **68** (6)(1991): 467–469.

NATIONAL INSTITUTE OF STANDARDS. *The International System of Units (SI).* NIST Special Publ. No. 330, U.S. Dept. Commerce, Gaithersburg, MD 20899 (1991).

REILLY, P. M. "A Statistical Look at Significant Figures." *Chem. Eng. Educ.* (Summer, 1992): 152–155.

SCHWARTZ, L. M. "Propagation of Significant Figures." *J. Chem. Educ.*, 62 (8) (1985): 693.

VATAVUK, W. M. "How Significant Are Your Figures." *Chem. Eng.*, (August 18, 1986): 97.

WHITNEY, H. "The Mathematics of Physical Quantities." *Am. Math. Mon.*, 75 (1968): 115, 227.

The Mole Unit

ALLSOP, R. T. "The Place and Importance of the Mole in Chemistry Courses." *Phys. Educ.*, (July, 1977): 285.

KOLB, D. "Chemical Principles Revisited." *J. Chem. Educ.*, 11, (1978): 728.

Temperature

BENZINGER, T. H., ed., *Temperature, Part 1, Arts and Concepts.* Stroudsburg, PA: Dowden, Hutchinson & Ross, 1977.

ROMER, R. H., "Temperature Scales." *The Physics Teacher* (October, 1982): 450.

THOMPSON, H. B., "Is 8°C Equal to 50°F?" *J. Chem. Educ.*, 68, (1991): 400.

Pressure

BENEDICT, R. P. *Fundamentals of Temperature, Pressure, and Flow Measurement,* 3rd ed. New York: Wiley, 1984.

DEMORSET, W. J., "Pressure Measurement." *Chem.. Eng.* (September 30, 1985): 560.

GILLUM, D. R. *Industrial Pressure Measurement.* Omega Engineering, Stamford, CT, 1992.

MULLIN, J., "(Bar)baric Standard Broached," *Chem. Eng.*, (February 1986): 41.

Conventions in Analysis and Measurement

SNIDER, E. H. *Methods of Analysis and Measurement*, Module MEB2, Vol. 1, Series F, AIChEMI. Modular Instruction Series, American Inst. Chem. Engr., New York, 1980.

Flammability and Hazards

AMERICAN INSTITUTE OF CHEMICAL ENGINEERS, *Guidelines for Chemical Reactivity Evaluation and Application to Process Design*, New York: AIChE, 1994.

IBID, *Guidelines for Hazard Evaluation Procedures with Worked Examples*, 2nd. ed., New York: AIChE, 1992.

IBID, *Safety, Health, and Loss Prevention in Chemical Processes: Problems for Undergraduate Engineering Curriculum.* New York: AIChE, 1990.

CROWL, D. A., and J. F. LOUVAR. *Chemical Process Safety: Fundamentals with Application,* Englewood Cliffs, NJ: Prentice-Hall, 1990.

NATIONAL FIRE PROTECTION ASSOCIATION. *National Fire Codes*, 1984.

PROBLEMS

Section 1.1

1.1. Carry out the following conversions:
   (a) How many m^3 are there in $1.00(\text{mile})^3$?
   (b) How many gal/min correspond to $1.00 \text{ ft}^3/\text{s}$?

1.2. In the American Engineering system of units, the viscosity can have the units of $(\text{lb}_f)(\text{hr})/\text{ft}^2$, while in the SI system the units are $(g)/(cm)(s)$. Convert a viscosity of $20.0\ (g)/(m)(s)$ to the given American Engineering units.

1.3. The permeability K of flow through a porous packed bed of particles is defined as

$$K = \frac{g_c D_p^2}{32}\left(\frac{Re}{f}\right)$$

   where D_p is the particle diameter, Re is dimensionless (the Reynolds Number), f is dimensionless (the friction factor). What are the units of K in the
   (a) American Engineering system?
   (b) SI?

1.4. A relation for a dimensionless variable called the compressibility (z) is $z = 1 + \rho B + \rho^2 C + \rho^3 D$ where ρ is the density in g mol/cm^3 What are the units of B, C, and D? Convert the coefficients in the equation for z so that the density can be introduced into the equation in the units of lb_m/ft^3 thus: $z = 1 + \rho^* B^* + (\rho^*)^2 C^* + (\rho^*)^3 D^*$ where ρ^* is in lb_m/ft^3. Give the units for B*, C*, and D*, and give the equations that relate B* to B, C* to C, and D* to D.

1.5. In a article describing an oil-shale retorting process, the authors say the retort: "could be operated at a solids mass flux well over 1,000 lb/h· ft^2 (48kPa/h)..." In several places they speak of the grade of their shale in the mixed unit "34 gal (129L)/ton." Does their report make sense?

1.6. Convert
   (a) $0.04\ g/(\text{min})(\text{in}^3)$ to $\text{lbm}/(\text{hr})(\text{ft}^3)$.
   (b) 2 L/s to ft^3/day.

   (c) $\dfrac{6(\text{in})(\text{cm}^2)}{(\text{yr})(s)(\text{lb}_m)(\text{ft}^2)}$ to all SI units.

1.7. Thermal conductivity in the American Engineering system of units is:

$$k = \frac{\text{Btu}}{(\text{hr})\left(\text{ft}^2\right)(^\circ F/\text{ft})}$$

Change this to:

$$\frac{\text{kJ}}{(\text{day})(m^2)(^\circ C/\text{cm})}$$

1.8. Water is flowing through a 2-inch diameter pipe with a velocity of 3ft/s.
    (a) What is the kinetic energy of the water in (ft)(lbf)/lb?
    (b) What is the flowrate in gal/min?

1.9. The contents of packages are often labeled in a fashion such as "net weight 250 grams." Is it correct to so label a package?

1.10. The following test will measure your SIQ. List the correct answer.
    (a) Which is the correct symbol?
        (1) nm   **(2)** °K   **(3)** sec   **(4)** N/mm
    (b) Which is the wrong symbol?
        (1) MN/m^2 **(2)** GHz/s
        (3) $kJ/(s)(m^3)$ **(4)** °C/M/s
    (c) Atmospheric pressure is about:
        (1) 100 Pa   **(2)** 100 kPa **(3)** 10 Mpa **(4)** 1 GPa
    (d) The temperature 0°C is defined as:
        (1) 273.15°K   **(2)** Absolute zero
        (3) 273.15 K   **(4)** The freezing point of water
    (e) Which height and mass are those of a petite woman?
        (1) 1.50 m, 45 kg   **(2)** 2.00 m, 95 kg
        (3) 1.50 m, 75 kg   **(4)** 1.80 m, 60 kg
    (f) Which is a recommended room temperature in winter?
        (1) 15°C   **(2)** 20°C   **(3)** 28°C   **(4)** 45°C
    (g) The watt is:
        (1) One joule per second   **(2)** Equal to 1 kg·m2/s3
        (3) The unit for all types of power **(4)** All of the above
    (h) What force may be needed to lift a heavy suitcase?
        (1) 24 N   **(2)** 250 N   **(3)** 25 kN   **(4)** 250 kN

1.11. A technical publication describes a new model 20-hp Stirling (air cycle) engine that drives a 68-kW generator. Is this possible?

1.12. A freeze-dried coffee is advertised as "97% caffeine-free." Is it possible for this coffee to contain only 0.14% caffeine? Explain.

1.13. Your boss announced that the speed of the company Boeing 727 is to be cut from 525 mi/hr to 475 mi/hr to "conserve fuel," thus cutting consumption from 2200 gal/hr to 2000 gal/hr. How many gallons are saved in a 1000-mi trip?

1.14. The heat capacity of SO_2 is given in a handbook as $C_p = 6.945 + 10.01 \times 10^{-3} T - 3.794 \times 10^{-6} T^2$ where C_p is in (cal)/(g mol) (K) and T is in K. Modify the equation so that the units of C_p are Btu/(lb$_m$) (°F) and T can be inserted in the equation in °F.

1.15. What is meant by a scale that shows a weight of "21.3 kg"?

1.16. A tractor pulls a load with a force equal to 800 lb (4.0 kN) with a velocity of 300 ft/min (1.5 m/s). What is the power required using the given American Engineering system data? The SI data?

1.17. What is the kinetic energy of a vehicle with a mass of 2300 kg moving at the rate of 10.0 ft/sec in Btu? 1 Btu = 778.2 (ft) (1b$_f$).

1.18. A pallet of boxes weighing 10 tons is dropped from a lift truck from a height of 10 feet. The maximum velocity the pallet attains before hitting the ground is 6 ft/s. How much kinetic energy does the pallet have in (ft) ($1b_f$) at this velocity?

1.19. The velocity in a pipe in turbulent flow is expressed by the following equation

$$u = k\left[\frac{\tau}{\rho}\right]^{1/2}$$

where τ is the shear stress in N/m^2 at the pipe wall, ρ is the density of the fluid in kg/m^3, u is the velocity, and k is a coefficient. You are asked to modify the equation so that the shear stress can be introduced in the units of τ^1 which are $1b_f$/ft^2, and the density be ρ^1 for which the units are $1b_m$/ft^3 so that the velocity u^1 comes out in the units of ft/s. Show all calculations, and give the final equation in terms of u^1, τ^1, and ρ^1 so a reader will know that American Engineering units are involved in the equation.

1.20. The American Petroleum Institute has published a correlation for determining the hydrocarbon emissions from fixed-roof storage tanks (Evaporation from Fixed Roof Tanks, API Bulletin 2518, June 1962).

$$L_y = \frac{24}{1,000}\left(\frac{p}{14.7 - p}\right)^{0.68} D^{1.73} H^{0.51} T^{0.50} F_p C$$

where: L_y is breathing emissions, bbl/yr; p is true vapor pressure at the bulk temperature, psia; D is the tank diameter, ft; H is the height in ft; T is the average tank outage corrected for roof volume, ft; F_p is a dimensionless paint factor (a function of the age and the color of the paint on the exterior of the tank); and C is a dimensionless adjustment factor. For tanks larger than 30 ft dia., $C = 1$. For smaller tanks:

$$C = -0.00132D^2 + 0.07714D - 0.13344$$

Is the emissions equation dimensionally consistent?

1.21. Without integrating, select the proper answer for

$$\int \frac{dx}{x^2 + a^2} = \begin{cases} a & \arctan & (ax) \\ a & \arctan & (x/a) \\ (1/a) & \arctan & (x/a) \\ (1/a) & \arctan & (ax) \end{cases} + \text{constant}$$

where x = length and a is a constant.

1.22. In many plants the analytical instruments are located some distance from the equipment being monitored. Thus, some delay exists before detecting a process change and the activation of an alarm.

In a chemical plant, air samples from a process area are continuously drawn through a 1/4 in. diameter tube to an analytical instrument located 125 ft from the process area. The 1/4 in. tubing has an outside diameter of 0.25 in. (6.35 mm) and a wall thickness of 0.030 in.(0.762 mm). The sampling rate is 10 cm³/sec under ambient conditions of 22°C and 1.0 atm. The pressure drop in the transfer line can be considered negligible. Chlorine gas is used in the process, and if it leaks from the process, it can poison workers who might be in the area of the leak. Determine the time required to detect a leak of chlorine in the process area with the equipment currently installed. You may assume the analytical equipment takes 5 sec to respond once the gas reaches the instrument. You may also assume that samples travel through the instrument sample tubing without dilution by mixing with the air ahead of the sample. Is the time excessive? How might the delay be reduced? (Adapted from Problem 13, in *Safety Health and Loss Prevention in Chemical Processes* published by the American Institute of Chemical Engineers, New York (1990).

1.23. In 1916 Nusselt derived a theoretical relation for predicting the coefficient of heat transfer between a pure saturated vapor and a colder surface:

$$h = 0.943 \left(\frac{k^3 \rho^2 g \lambda}{L \mu \Delta T} \right)^{1/4}$$

where h = mean heat transfer coefficient, Btu/(hr) (ft²) (Δ°F)
         k = thermal conductivity, Btu/(hr) (ft) (Δ°F)
         ρ = density, lb/ft³
         g = acceleration of gravity, 4.17×10^8 ft/(hr)²
         λ = enthalpy change, Btu/lb
         L = length of tube, ft
         μ = viscosity, lb_m/(hr) (ft)
         ΔT = temp difference, Δ°F

What are the units of the constant: 0.943?

1.24. The equation for the velocity of a fluid stream measured with a Pitot tube is

$$v = \sqrt{\frac{2 \Delta p}{\rho}}$$

where v = velocity
         Δp = pressure drop
         ρ = density of fluid

Is the equation dimensionally consistent? Answer yes or no, and explain your answer. If the pressure drop is 15mm Hg, and the density of the fluid is 1.20 g/cm³, calculate the velocity in ft/s.

1.25. Explain in detail whether the following equation for flow through a rectangular weir is dimensionally consistent. (This is the modified Francis formula.)

$$q = 0.415 \, (L - 0.2 \, h_o) h_o^{1.5} \sqrt{2g}$$

where q = volumetric flow rate, ft3/s
L = crest height, ft
h_o = weir head, ft
g = acceleration of gravity, 32.2 ft/(s)^2

1.26. An experimental investigation of the rate of mass transfer of SO_2 from an air stream into water indicated that the mass transfer coefficient could be correlated by an equation of the form: $k_x = Ku^{0.487}$, where k_x is the mass transfer coefficient in mol/(cm^2)(s) and u is the velocity in cm/s. Does the constant K have dimensions? What are they? If the velocity is to be expressed in ft/s, and we want to retain the same form of the relationship, what would the units of K' have to be if k_x is still mol/(cm^2)(s), where K' is the new coefficient in the formula.

1.27. A useful dimensionless number called the *Reynolds number* is $\dfrac{DU\rho}{\mu}$ where
where D = diameter or length
U = some characteristic velocity
ρ = fluid density
μ = fluid viscosity

Calculate the Reynolds number for the following cases:

	1	2	3	4
D	2 in.	20 ft	1 ft	2 mm
U	10 ft/s	10 mi/hr	1 m/s	3 cm/s
ρ	62.4 lb/ft^3	1 lb/ft^3	12.5 kg/m^3	25 lb/ft^3
μ	0.3	0.14 × 10^{-4}	2 ×10^{-6}	1 × 10^6
	lb$_m$/(hr)(ft)	lb$_m$/(s)(ft)	centipoise (cP)	centipoise

1.28. Computers are used extensively in automatic plant process control systems. The computers must convert signals from devices monitoring the process, evaluate the data using the programmed engineering equations, and then feed back the appropriate control adjustments. The equations must be dimensionally consistent. Therefore, a conversion factor must be part of the equation to change the measured field variable into the proper units. Crude oil pumped from a storage unit to a tanker is to be expressed in tons/hr, but the field variables of density and the volumetric flow rate are measured in lb/ft^3 and gal/min, respectively. Determine the units and the numerical values of the factors necessary to convert the field variables to the desired output.

Section 1.2

1.29. The following was a letter to *The Chemical Engineer* (a British publication).

In reply to Dr J. B. Morris in the February issue, of course the symbols g mole and kg mole used to exist, and they still can if you want them to, but not in the SI. Anyway, what is wrong with the mole (mol) and kilomole (kmol)? They are easier both to say and to write. We are all aware of the apparent inconsistency in the choice of the mole rather than the kilomole as the basic SI unit for amount of substance, but the controversy is now and it is sterile to pursue the matter.

Dr Morris is treading on dangerous ground when he attempts "to remind us that the mole has the dimensions of mass". The mole is certainly related to mass, but this does not confer dimensions of mass on it. The amount of substance is proportional to mass divided by the relative molecular mass, a dimensionless ratio formerly known as the molecular weight. If SI units are involved, a dimensional constant of proportionality numerically equal to 10^3 is normally chosen, but there is no fundamentally compelling reason why we should do so.

Explain what is correct and what is not correct about this letter.

1.30. (a) What is the molecular weight of $CaCO_3$?
   (b) How many g mol are in l0 g of $CaCO_3$?
   (c) How many lb mol are in 20 lb of $CaCO_3$?
   (d) How many g are in 2 lb mol of $CaCO_3$?

1.31. Explain the differences between mole, molecule, and molecular weight.

1.32. What is wrong, or correct, about each of the following answers to the question: What is a mole?
   (a) A mole is found in a certain number of cm_3 of one substance or another.
   (b) A mole is the weight of a molecule expressed in grams.
   (c) A mole is the number of molecules in one gram of a substance.
   (d) A mole is the sum of atomic weights.
   (e) A mole is the molecular weight of an element.

1.33. Convert the following:
   (a) 4 g mol of $MgCl_2$ to g
   (b) 2 lb mol of C_3H_8 to g
   (c) 16 g of N_2 to lb mol
   (d) 3 lb of C_2H_6O to g mol

1.34. A textbook states: "A mole is a quantity of material whose weight is numerically equal to the molecular weight." State whether this statement is correct, incorrect, or partially correct, and explain in *no more* than three sentences the reasoning behind your answer.

1.35. What does the unit mol^{-1} mean? Can a unit be $mol^{1/3}$?

1.36. A solid compound was found to contain 42.11%C, 51.46% O, and 6.43% H. Its molecular weight was about 341. What is the formula for the compound?

1.37. The structural formulas in Figure P1.37 are for vitamins:
   (a) How many pounds of compound are contained in each of the following (do for each vitamin):
      (1) 2.00 g mol   **(2)** 16 g
   (b) How many grams of compound are contained in each of the following (do for each vitamin):
      (1) 1.00 lb mol   **(2)** 12 lb

Vitamin	Structural formula	Dietary sources	Deficiency symptoms
Vitamin A	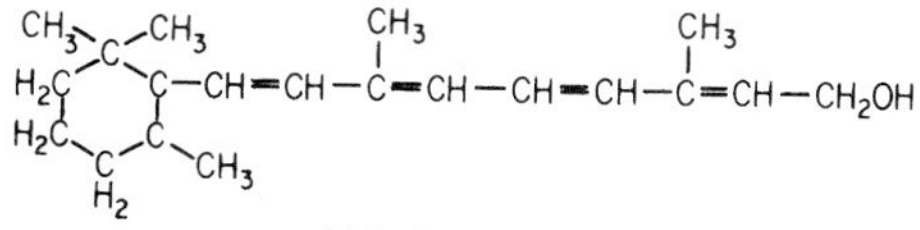 Retinol	Fish liver oils, liver, eggs, fish, butter, cheese, milk; a precursor, β-carotene, is present in green vegetables, carrots, tomatoes, squash	Night blindness, eye inflammation
Ascorbic acid (vitamin C)	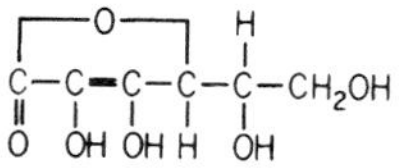	Citrus fruit, tomatoes, green peppers, strawberries, potatoes	Scurvy
Vitamin D	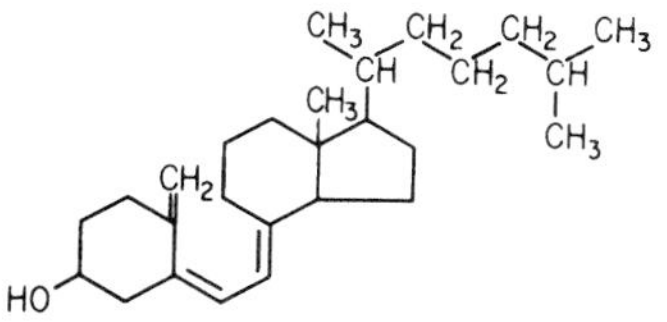 Vitamin D_3	Fish liver oils, butter, vitamin-fortified milk, sardines, salmon; the body also obtains this compound when ultraviolet light converts 7-dehydrocholesterol in the skin to vitamin D	Rickets, osteomalacia, hypoparathyroidism

Figure P1.37

Section 1.3

1.38. The specific gravity of acetic acid is 1.049. What is the density in lb_m/ft^3?

1.39. The specific gravity of a fuel oil is 0.82. What is the density of the oil in lb/ft^3? Show all units.

1.40. You are asked to decide what size containers to use to ship 1,000 lbs of cottonseed oil of specific gravity equal to 0.926. What would be the minimum size drum expressed in gallons?

1.41. You are asked to make up a laboratory solution of 0.10 molar H_2SO_4 (0.10 mol H_2SO_4/L) from concentrated (96.0%) H_2SO_4 You look up the specific gravity of 96.0% H_2SO_4 and find it is 1.858. Calculate

(a) the weight of 96.0% acid needed per L of solution.

(b) the volume of 96.0% acid used per L of solution.

(c) the density of the 0.1 molar solution.

(d) the ppm of H_2SO_4 in the 0.1 molar solution

1.42. A bartender claims that his special brand of rum is so strong that ice cubes sink in it. Is this possible?

1.43. The specific gravity of a solution of KOH at 15°C is 1.0824 and contains 0.813 lb KOH per gal of solution. What are the mass fractions of KOH and H_2O in the solution?

1.44. The density of benzene at 60°F is 0.879 g/cm^3. What is the specific gravity of benzene at 60°F/60°F?

1.45. A liquid has a specific gravity of 0.90 at 25°C. What is its
  (a) Density at 25°C in kg/m^3?
  (b) Specific volume at 25°C in ft^3/lbm?
  (c) If the liquid is placed in a 1.5-L bottle that has a mass of 232 g, how much will the full bottle weigh?

1.46. Given a water solution that contains 1.704 kg of HNO$_3$/kg H$_2$O and has a specific gravity of 1.382 at 20°C, express the composition in the following ways:
  (a) Weight percent HNO$_3$
  (b) Pounds HNO$_3$ per cubic foot of solution at 20°C
  (c) Molarity (gram moles of HNO$_3$ per liter of solution at 20°C)

1.47. The Federal Water Pollution Control Act, P.L. 92-500, specifies legally acceptable methods for wastewater analysis. Analysis for cyanide is done according to the method outlined in "Standard Methods for the Examination of Water Wastewater." Mercuric chloride is used in this method to decompose complex cyanides, and 200 mg are used per analysis.

  The Illinois Pollution Control Board has established Water Quality Standards that limit mercury (as Hg) to 0.0005 ppm in any effluent. Permit holders are required to submit daily reports on their effluent. Will a permit holder discharging 100,000 gal/day be in violation of the cited standard if one analysis is made?

1.48. OSHA (Occupational Safety and Health Administration) has established limits for the storage of various toxic or hazardous chemicals (OSHA 29 CFR 1910.119, Appendix A). The maximum limit for acetaldehyde is 113.0 kg and for nitrogen dioxide 110 kg. What is the minimum size spherical vessel that can be used to store each of these respective liquids at room temperature?

1.49. The table lists values of Fe, Cu, and Pb in Christmas wrapping paper for two different brands. Convert the ppm to mass fractions on a paper free basis.

	Concentration, ppm		
	Fe	Cu	Pb
Brand A	1310	2000	2750
Brand B	350	50	5

1.50. For the purpose of permit compliance, all hazardous materials are categorized into three hazard categories: toxicity, flammability, and reactivity, and assigned numbers in each category from 0 to 4 (most severe). Methyl alcohol (methanol, CH$_3$OH) has the code 1, 4, 0 in the liquid state. For the toxic category, any amount of stored toxic material of category 1 over 0.35 oz. must be reported by city ordinance. Must a one-half liter bottle of methanol (sp.gr. = 0.792) be reported?

1.51. The density of a certain solution is 8.80 lb/gal at 80°F. How many cubic feet will be occupied by 10,010 lb of this solution at 80°F?

1.52. How many ppb are there in 1 ppm? Does the system of units affect your answer? Does it make any difference if the material for which the ppb are measured is a gas, liquid, or solid?

1.53. Five thousand barrels of 28°API gas oil are blended with 20,000 bbl of 15°API fuel oil. What is the density of the mixture in lb/gal and lb/ft^3? Assume that the volumes are additive. 1 bbl = 42 gal. The density of water at 60°F = 0.999 g/cm^3.

$$\text{Specific gravity} \ \frac{60°\text{F}}{60°\text{F}} = \frac{141.5}{°\text{API} + 131.5}$$

1.54. In a handbook you find that the conversion between °API and density is 0.800 density = 45.28°API. Is this a misprint?

1.55. Harbor sediments in the New Bedford, Massachusetts, area contain PCBs at levels up to 190,000 ppm according to a report prepared by Grant Weaver of the Massachusetts Coastal Zone Management Office *(Environ. Sci. Technol.,* **16**(9) (1982):491A. What is the concentration in percent?

1.56. NIOSH sets standards for CCl_4 in air at 12.6 mg/m^3 of air (a time weighted average over 40 hr). The CCl_4 found in a sample is 4800 ppb (parts per billion; billion = 10^9). Does the sample exceed the NIOSH standard? Be careful!

1.57. The following table shows the annual inputs of phosphorus to Lake Erie:

	Short tons/yr
Source	
Lake Huron	2,240
Land drainage	6,740
Municipal waste	19,090
Industrial waste	2,030
	30,100
Outflow	4,500
Retained	25,600

(a) Convert the retained phosphorus to concentration in micrograms per liter assuming that Lake Erie contains 1.2×1014 gal of water and that the average phosphorus retention time is 2.60 yr.

(b) What percentage of the input comes from municipal water?

(c) What percentage of the input comes from detergents, assuming they represent 70% of the municipal waste?

(d) If 10 ppb of phosphorus triggers nuisance algal blooms, as has been reported in some documents, would removing 30% of the phosphorus in the municipal waste and all the phosphorus in the industrial waste be effective in reducing the eutrophication (i.e., the unwanted algal blooms) in Lake Erie?

(e) Would removing all the phosphate in detergents help?

1.58. Solubility of formaldehyde in a solvent is measured by a spectrophotometer operating at 570 nm. The data collected are optical density versus concentration expressed in grams per liter.

Concentration, c	Optical Density, d
100	0.086
300	0.269
500	0.445
600	0.538
700	0.626

Based on Beer's law, obtain the best estimates of the coefficients for a linear relation

$$c = b_0 + b_1 d$$

and plot the equation so obtained together with the data. Refer to Appendix M.

Section 1.4

1.59. You have 130 kg of gas of the following composition: 40% N_2, 30% CO_2, and 30% CH_4 in a tank. What is the average molecular weight of the gas?

1.60. You have 25 lb of a gas of the following composition:

$$
\begin{array}{ll}
CH_4 & 80\% \\
C_2H_4 & 10\% \\
C_2H_6 & 10\%
\end{array}
$$

What is the average molecular weight of the mixture? What is the weight (mass) fraction of each of the components in the mixture?

1.61. You analyze the gas in 100 kg of gas in a tank at atmospheric pressure, and find the following:

CO_2: 19 .3% N_2: 72.1% O_2: 6.5% H_2O: 2.1%

What is the average molecular weight of the gas?

1.62. Two hundred kg of liquid contains 40% butane, 40% pentane, and 20% hexane. Determine the mole fraction composition of the liquid, and the mass fraction composition of the liquid on a hexane free basis.

1.63. The proximate and ultimate analysis of a coal is given in the table. What is the composition of the "Volatile Combustible Matter (VCM)"? Present your answer in the form of mass percent of each element in the VCM.

Proximate Analysis (%)		Ultimate Analysis (%)	
Moisture	3.2	Carbon	79.90
Volatile Combustible		Hydrogen	4.85
matter (VCM)	21.0	Sulfur	0.69
Fixed Carbon	69.3	Nitrogen	1.30
Ash	6.5	Ash	6.50
		Oxygen	6.76
Total	100.0	Total	100.00

1.64. A fuel gas is reported to analyze, on a mole basis, 20% methane, 5% ethane, and the remainder CO_2. Calculate the analysis of the fuel gas on a mass percentage basis.

1.65. A gas mixture consists of three components: argon, *B,* and *C.* The following analysis of this mixture is given:

$$40.0 \text{ mol } \% \text{ argon}$$
$$18.75 \text{ mass } \% \text{ } B$$
$$20.0 \text{ mol } \% \text{ } C$$

The molecular weight of argon is 40 and the molecular weight of *C* is 50. Find:
(a) The molecular weight of B
(b) The average molecular weight of the mixture.

1.66. Two engineers are calculating the average molecular weight of a gas mixture containing oxygen and other gases. One of them uses the correct molecular weight of 32 for oxygen and determines the average molecular weight as 39.2. The other uses an incorrect value of 16 and determines the average molecular weight as 32.8. This is the only error in his calculations. What is the percentage of oxygen in the mixture expressed as mol %?

Section 1.5

1.67. **"Japan, U.S. Aim for Better Methanol-Powered Cars,"** read the headline in the *Wall Street Journal.* Japan and the U.S. plan to join in developing technology to improve cars that run on methanol, a fuel that causes less air pollution than gasoline. An unspecified number of researchers from Japanese companies will work with the EPA to develop a methanol car that will start in temperatures as low as minus 10 degrees Celsius. What is this temperature in degrees Rankine, kelvin, and Fahrenheit?

1.68. Can negative temperature measurements exist?

1.69. The heat capacity C_p of acetic acid in J/(g mol)(K) can be calculated from the equation

$$C_p = 8.41 + 2.4346 \times 10^{-5}T$$

where *T* is in K. Convert the equation so that *T* can be introduced into the equation in °R instead of K.

1.70. Convert the following temperatures to the requested units:
(a) 10°C to °F
(b) 10°C to °R
(c) −25°F to K
(d) 150K to °R

1.71. Heat capacities are usually given in terms of polynomial functions of temperature. The equation for carbon dioxide is

$$C_p = 8.4448 + 0.5757 \times 10^{-2}T - 0.2159 \times 10^{-5}T^2 + 0.3059 \times 10^{-9}T^3$$

where T is in °F and C_p is in Btu/(lb mol)(°F). Convert the equation so that T can be in °C and C_p will be in J/(g mol)(K).

1.72. In a report on the record low temperatures in Antarctica, *Chemical and Engineering News* said at one point that "the mercury dropped to –76°C." In what sense is that possible? Mercury freezes at –39°C.

1.73. "Further, the degree Celsius is exactly the same as a kelvin. The only difference is that zero degree Celsius is 273.15 kelvin. Use of Celsius temperature gives us one less digit in most cases" from [*Eng. Educ.*, (April 1977):678]. Comment on the quotation. Is it correct? If not, in what way or sense is it wrong?

1.74. Calculate all temperatures from the one value given:

	(a)	(b)	(c)	(d)	(e)	(f)	(g)	(h)
°F	140				1000			
°R			500			1000		
K		298					1000	
°C				-40				1000

1.75. The emissive power of a blackbody depends on the fourth power of the temperature and is given by

$$W = A T^4$$

where W = emissive power, Btu/(ft²)(hr)
        A = Stefan-Boltzmann constant, 0.171×10^{-8} Btu/(ft²)(hr)(°R)⁴
        T = temperature, °R

What is the value of A in the units J/(m²)(s)(K⁴) ?

Section 1.6

1.76. **"Holiday drivers stranded, Icicles blossom in Florida; 50 cities set record lows"** was the headline in the December 26 newspaper. After citing all of the cities with new low temperatures, the paper went on to say: "As the full force of the cold wave descended on the East it was 5 degrees Sunday morning in New York City. As one example of the strength of the cold air mass flowing from northwest Canada, the highest pressure ever recorded in the United States—31.42 inches of mercury—was measured on Christmas Eve in Miles City, Mont." Can this last statement be true?

1.77. Suppose that a submarine inadvertently sinks to the bottom of the ocean at a depth of 1000 m. It is proposed to lower a diving bell to the submarine and attempt to enter the conning tower. What must the minimum air pressure be in the diving bell at the level of the submarine to prevent water from entering into the bell when the opening valve at the bottom is cracked open slightly? Give your answer in absolute kilopascal. Assume that seawater has a constant density of 1.024 g/cm³.

1.78. Air in a scuba diver's tank shows a pressure of 300 kPa absolute. What is the pressure in:

(a) psi   (c) atm   (e) in. Hg   (g) kg_f/cm^2
(b) kPa   (d) bar   (f) ft H_2O

1.79. Flat-roof buildings are a popular architectural style in dry climates because of the economy of materials of construction. However, during the rainy season water may pool up on the roof decks so that structural considerations for the added weight must be taken into account. If 25 cm of water accumulates on a 10-m by 10-m area during a heavy rain storm, determine:

(a) The total added weight the building must support
(b) The force of the water on the roof in psi.

1.80. A problem with concrete wastewater treatment tanks set below ground was realized when the water table rose and an empty tank floated out of the ground. This buoyancy problem was overcome by installing a check valve in the wall of the tank so that if the water table rose high enough to float the tank, it would fill with water. If the density of concrete is 2080 kg/m^3, determine the maximum height at which the valve should be installed to prevent a buoyant force from raising a rectangular tank with inside dimensions of 30 m by 27 m and 5 m deep. The walls and floor have a uniform thickness of 200 mm.

1.81. A centrifugal pump is to be used to pump water from a lake to a storage tank that is 148 ft above the surface of the lake. The pumping rate is to be 25.0 gal/min, and the water temperature is 60°F. The pump on hand can develop a pressure of 50.0 psig when it is pumping at a rate of 25.0 gal/min. (Neglect pipe friction kinetic energy effects, or factors involving pump efficiency.)

(a) How high (in feet) can the pump raise the water at this flow rate and temperature?
(b) Is this pump suitable for the intended service?

1.82. A manometer uses kerosene, sp gr 0.82, as the fluid. A reading of 5 in. on the manometer is equivalent to how many millimeters of mercury?

1.83. The pressure gauge on the steam condenser for a turbine indicates 26.2 in. Hg of vacuum. The barometer reading is 30.4 in. Hg. What is the absolute pressure in the condenser in psia?

1.84. A pressure gauge on a process tower indicates a vacuum of 3.53 in. Hg. The barometer reads 29.31 in. Hg. What is the absolute pressure in the tower in millimeters of mercury?

1.85. John Long says he calculated from a formula that the pressure at the top of Pikes Peak is 9.75 psia. John Green says that it is 504 mm Hg because he looked it up in a table. Which John is right?

1.86. The floor of a cylindrical water tank was distorted into 7-in. bulges due to the settling of improperly stabilized soil under the tank floor. However, several consulting engineers restored the damaged tank to use by placing plastic skirts around the bottom of the tank wall and devising an air flotation system to move it to an adjacent location. The tank was 30.5 m in diameter and 13.1 m deep.

The top, bottom, and sides of the tank were made of 9.35-mm-thick welded steel sheets. The density of the steel is 7.86 g/cm³.
   (a) What was the gauge pressure in kPa of the water at the bottom of the tank when it was completely full of water?
   (b) What would the air pressure have to be in kPa beneath the empty tank in order to just raise it up for movement?

1.87. A pressure instrument has failed on a process line that requires constant monitoring. A bell-type gauge as shown in Figure P1.87 is available that has oil (density of 0.800 g/cm³) as a sealant liquid. Construction of the gauge limits the sealant liquid's travel to 12.7 cm before blowout of the oil occurs. What maximum pressures can this gauge measure in kPa?

1.88. Examine Figure P1.88. Oil (density = 0.91 g/cm³) flows in a pipe, and the flow rate is measured via a mercury (density = 13.546 g/cm³) manometer. If the difference in height of the two legs of the manometer is 0.78 in., what is the corresponding pressure difference between points A and B in mm Hg? At which point, A or B, is the pressure higher? The temperature is 60°F.

1.89. A student connected a Bourdon gauge designed to read gauge pressure to a vacuum line and, on finding out that it would not give a reading, decided to adjust the instrument so that it read 0 psi for a high vacuum, and read 14.7 psi at atmospheric pressure. However, on the day when this adjustment was made, the barometer read 735 mm Hg, so that his reading of 14.7 psi was slightly in error,

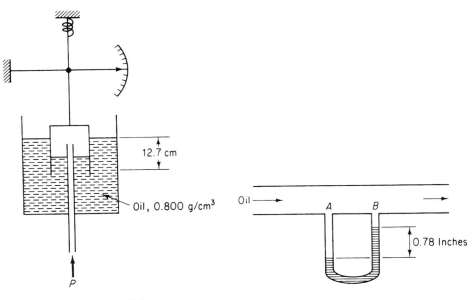

Figure P1.87   **Figure P1.88**

12.7 cm

Oil, 0.800 g/cm³

Oil

A   B

0.78 Inches

P

This gauge was later used to measure a pressure in a tank of air, and read 51 psig. What should the proper gauge reading be? The barometer at that time was 750 mm Hg. State any assumptions needed to solve this problem.

1.90. Suppose you prepare a barometer as follows: You fill a long (> 40 cm) glass tube sealed at the bottom end completely full of mercury, and carefully invert it and place the open end (hold your finger over it) into a beaker of Hg so that no air gets into the tube. Explain how this barometer works.

1.91. A manufacturer of large tanks calculates the mass of fluid in the tank by taking the pressure measurement at the bottom of the tank in psig, and then multiplies that value by the area of the tank in square inches. Can this procedure be correct?

1.92. Examine Figure P1.92. The barometer reads 740 mm Hg. Calculate the pressure in the tank in psia.

1.93. In Figure P1.93, Hg is used to measure the pressure between the points M and N. Water occupies the space above the Hg. What is Δp between M and N? Which pressure is higher, p_M or p_N?

1.94. Express p_B as a function of h based on the data given in Figure P1.94. The non-shaded tube contains oil.

1.95. A Bourdon pressure gauge is connected to a large tank and reads 440 kPa when the barometer reads 750 mm Hg. What will the gauge reading be if the atmospheric pressure increases to 765 mm Hg?

1.96. A vacuum gauge connected to a tank reads 31.5 kPa. What is the corresponding absolute pressure if the barometer reads 98.2 kPa?

1.97. A U-tube (Figure P1.97.) with one leg 20 inches high is filled with mercury to a depth of 12 inches. The end of the 20 inch leg is then closed, and mercury is added to the other leg until the level of mercury is 14 inches deep in the closed leg. How deep is the mercury in the open leg?

1.98. Examine Figure P1.98. Given that a = 7.5 in. and b = 12 in., what is the height of the water in the right hand vessel?

1.99. Hydrostatic tank gaging (HTG) uses on-line pressure measurements to calculate the density of a fluid. If the two pressure instruments in the tank are

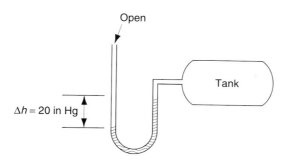

Figure P1.92

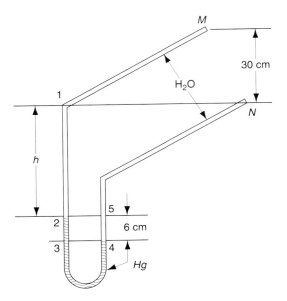

Figure P1.93

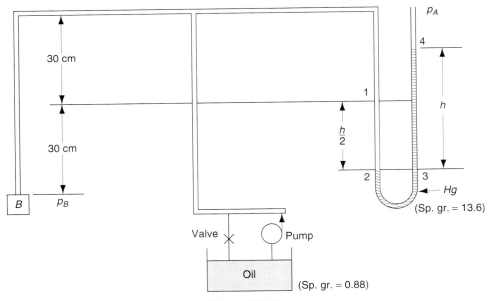

Figure P1.94

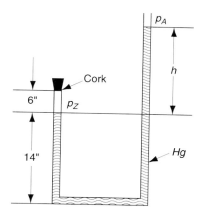

Figure P1.97

separated by 72.8 cm, and the larger pressure gauge reading is 267 mm Hg while the smaller reading is 211 mm Hg, what is the density of the fluid in the tank?

1.100. A differential pressure transmitter (DPT) can be used to sense the liquid level in a tank. See Figure P1.100.

How much error is made in not taking into account the height of the vapor above the liquid?

1.101. Deflagration is quick but progressive combustion (with or without explosion) as distinguished from a detonation which is the instantaneous decomposition of combustible material. (The National Fire Protection Association in Standard NFPA 68 has some more precise technical definitions.) For dust, the rate of deflagration depends on the particle size of the dust. NFPA gives a formula to cal-

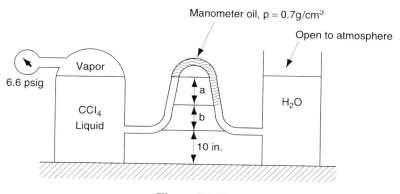

Figure P1.98

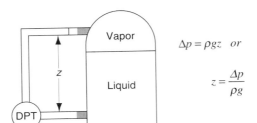

$$\Delta p = \rho g z \quad or$$

$$z = \frac{\Delta p}{\rho g}$$

Figure P1.100

culate the area for venting a building to prevent structural damage on deflagration

$$A_v = CA_s/p^{0.5}$$

where A_v is the necessary vent area (ft^2 or m^2), C is a constant that must be determined by experiment (NFPA68 lists some values), A_s is the surface area of the building, and p is the maximum internal (gauge) pressure that can be sustained by the building without damage occurring.

In a proposed grain storage elevator 8m in diameter and 10m tall, the vented area (between the roof and top walls) is 10.4m^2. Is this area sufficient if the elevator is designed to have a maximum overpressure of 7.5 kPa (7.5 kPa difference between the inside and outside)? Assume for grain dust that $C = 0.41(kPa)^{0.5}$. Note the vents should be distributed in practice uniformly throughout the structure.

1.102. Pressure in a gas cell is measured with an inverted manometer, as shown in Figure Pl.102. The scale on the far right-hand side of the figure shows the distances in mm (not to scale) of the interfaces of the liquids in the manometer. What is the pressure in the cell?

1.103. The indicating liquid in the manometer shown in Figure P1.103 is water, and the other liquid is benzene. These two liquids are essentially insoluble in each other. If the manometer reading is $\Delta Z = 36.3$ cm water, what is the pressure difference in kPa? The temperature is 25°C.

1.104. Examine Figure P1.104. The barometric pressure is 720 mm Hg. The density of the oil is 0.80 g/cm^3. The Bourdon gauge reads 33.1 psig. What is the pressure in kPa of the gas?

Section 1.7

1.105. $BaCl_2 + Na_2SO_4 \rightarrow BaSO4 + 2NaCl$
  (a) How many grams of barium chloride will be required to react with 5.00 g of sodium sulfate?

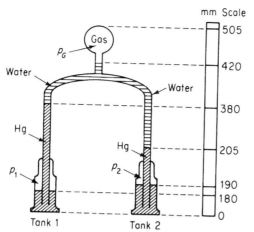

Figure P1.102

(b) How many grams of barium chloride are required for the precipitation of 5.00 g of barium sulfate?

(c) How many grams of barium chloride are needed to produce 5.00 g of sodium chloride?

(d) How many grams of sodium sulfate are necessary for the precipitation of the barium of 5.00 g of barium chloride?

(e) How many grams of sodium sulfate have been added to barium chloride if 5.00 g of barium sulfate is precipitated?

(f) How many pounds of sodium sulfate are equivalent to 5.00 lb of sodium chloride?

(g) How many pounds of barium sulfate are precipitated by 5.00 lb of barium chloride?

(h) How many pounds of barium sulfate are precipitated by 5.00 lb of sodium sulfate?

(i) How many pounds of barium sulfate are equivalent to 5.00 lb of sodium chloride?

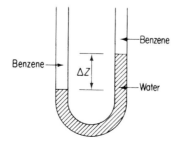

Figure P1.103

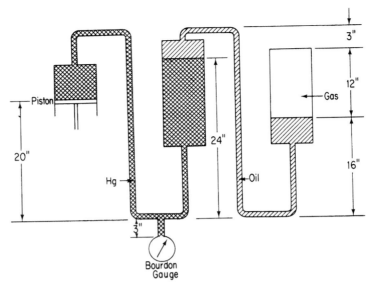

Figure P1.104

1.106. $AgNO_3 + NaCl \rightarrow AgCl + NaNO_3$

    (a) How many grams of silver nitrate will be required to react with 5.00 g of sodium chloride?

    (b) How many grams of silver nitrate are required for the precipitation of 5.00 g of silver chloride?

    (c) How many grams of silver nitrate are equivalent to 5.00 g of sodium nitrate?

    (d) How many grams of sodium chloride are necessary for the precipitation of the silver of 5.00 g of silver nitrate?

    (e) How many grams of sodium chloride have been added to silver nitrate if 5.00 g of silver chloride is precipitated?

    (f) How many pounds of sodium chloride are equivalent to 5.00 lb of sodium nitrate?

    (g) How many pounds of silver chloride are precipitated by 5.00 lb of silver nitrate?

    (h) How many pounds of silver chloride are precipitated by 5.00 lb of sodium chloride?

    (i) How many pounds of silver chloride are equivalent to 5.00 lb of silver nitrate?

1.107. A plant makes liquid CO_2 by treating dolomitic limestone with commercial sulfuric acid. The dolomite analyzes 68.0% $CaCO_3$, 30.0% $MgCO_3$, and 2.0% SiO_2; the acid is 94% H_2SO_4 and 6% H_2O. Calculate:

    (a) Pounds of CO2 produced per ton dolomite treated.

(b) Pounds of commercial acid required per ton of dolomite treated. Assume the reactions are complete.

1.108. The formula for vitamin C is as follows:

How many pounds of this compound are contained in 2 g mol?

1.109. Removal of CO_2 from a manned spacecraft has been accomplished by absorption with lithium hydroxide according to the following reaction:

$$2LiOH(s) + CO_2(g) \rightarrow Li_2CO_3(s) + H_2O(l)$$

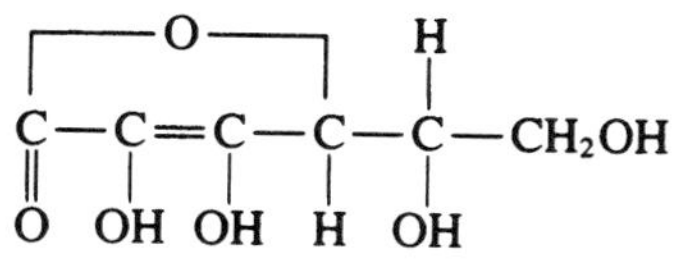

Figure P1.108

(a) If 1.00 kg of CO2 is released per day per person, how many kilograms of LiOH are required per day per person?

(b) What is the penalty (i.e., the percentage increase) in weight if the cheaper NaOH is substituted for LiOH?

1.110. Sulfuric acid can be manufactured by the contact process according to the following reactions:

(1) $S + O_2 \rightarrow SO_2$
(2) $2SO_2 + O_2 \rightarrow 2SO_3$
(3) $SO_3 + H_2O \rightarrow H_2SO_4$

You are asked as part of the preliminary design of a sulfuric acid plant with a design capacity of 2000 tons/day of 66° Be (Baumé) (93.2% H2SO4 by weight) to calculate the following:

(a) How many tons of pure sulfur are required per day to run this plant?

(b) How many tons of oxygen are required per day?

(c) How many tons of water are required per day for reaction (3)?

1.111. Seawater contains 65 ppm of bromine in the form of bromides. In the Ethyl-Dow recovery process, 0.27 lb of 98% sulfuric acid is added per ton of water, together with the theoretical Cl_2 for oxidation; finally, ethylene (C_2H_4) is united with the bromine to form $C_2H_4Br_2$. Assuming complete recovery and using a basis of 1 lb of bromine, find the weights of acid, chlorine, seawater, and dibromide involved.

$$2Br^- + Cl_2 \rightarrow 2Cl^- + Br_2$$
$$Br_2 + C_2H_4 \rightarrow C_2H_4Br_2$$

1.112. Acidic residue in paper from the manufacturing process causes paper based on wood pulp to age and deteriorate. To neutralize the paper, a vapor-phase treatment must employ a compound that would be volatile enough to permeate the fibrous structure of paper within a mass of books but that would have a chemistry that could be manipulated to yield a mildly basic and essentially non-volatile compound. George Kelly and John Williams successfully attained this objective in 1976 by designing a mass-deacidification process employing gaseous diethyl zinc (DEZ).

At room temperature, DEZ is colorless liquid. It boils at 117°C. When it is combined with oxygen, a highly exothermic reaction takes place:

$$(C_2H_5)_2Zn + 7O_2 \rightarrow ZnO + 4CO_2 + 5H_2O$$

Because liquid DEZ ignites spontaneously when exposed to air, a primary consideration in its use is the exclusion of air. In one case a fire caused by DEZ ruined the neutralization center.

Is the equation shown balanced? If not, balance it. How many kg of DEZ must react to form 1.5 kg of ZnO? If 20 cm³ of water are formed on reaction, and the reaction was complete, how many grams of DEZ reacted?

1.113.

BID EVALUATION

TO: _J. Coadwell_ DEPT: _Water Waste Water_ DATE: _9-29_
BID INVITATION: _0374-AV_

REQUISITION: _135949_ COMMODITY: _Ferrous Sulfate_

DEPARTMENT EVALUATION COMMENTS

It is recommended that the bid from VWR of $83,766.25 for 475 tons of Ferrous Sulfate Heptahydrate be accepted as they were the low bidder for this product. It is further recommended that we maintain the option of having this product delivered either by rail in a standard carload of 50 tons or by the alternate method by rail in piggy-back truck trailers.

What would another company have to bid to match the VWR bid if the bid they submitted was for ferrous sulfate ($FeSO_4 \cdot H_2O$)? For ($FeSO_4 \cdot 4H_2O$)?

1.114. Three criteria must be met if a fire is to occur: (1) There must be fuel present; (2) there must be an oxidizer present; and (3) there must be an ignition source. For most fuels, combustion takes place only in the gas phase. For example, gasoline does not burn as a liquid. However, when gasoline is vaporized, it burns readily.

A minimum concentration of fuel in air exists that can be ignited. If the fuel concentration is less than this lower flammable limit (LFL) concentration, ignition will not occur. The LFL can be expressed as a volume percent, which is equal to the mole percent under conditions at which the LFL is measured (atmospheric pressure and 25°C). There is also a minimum oxygen concentration required for ignition of any fuel. It is closely related to the LFL and can be calculated from the LFL. The minimum oxygen concentration required for ignition can be estimated by multiplying the LFL concentration by the ratio of the number of moles of oxygen required for complete combustion to the number of moles of fuel being burned.

Above the LFL, the amount of energy required for ignition is quite small. For example, a spark can easily ignite most flammable mixtures. There is also a fuel concentration called the upper flammable limit (UFL) above which the fuel-air mixture cannot be ignited. Fuel-air mixtures in the flammable concentration region between the LFL and the UFL can be ignited. Both the LFL and the UFL have been measured for most of the common flammable gases and volatile liquids. The LFL is usually the more important of the flammability concentrations because if a fuel is present in the atmosphere in concentrations above the UFL, it will certainly be present within the flammable concentration region at some location. LFL concentrations for many materials can be found in the NFPA Standard 325M, "Properties of Flammable Liquids," published by the National Fire Protection Association.

Estimate the maximum permissible oxygen concentration for *n*-butane. The LFL concentration for *n*-butane is 1.9 mole percent. This problem was originally based on a problem in the text *Chemical Process Safety: Fundamentals with Applications*, by D.A. Crowl and J.F. Louvar, published by Prentice Hall, Englewood Cliffs, NJ, and has been adapted from Problem 10 of the AIChE publication, *Safety, Health, and Loss Prevention in Chemical Processes* by J.R. Welker and C. Springer, New York (1990).

1.115. Odors in wastewater are caused chiefly by the products of the anaerobic reduction of organic nitrogen and sulfur-containing compounds. Hydrogen sulfide is a major component of wastewater odors; however, this chemical is by no means the only odor producer since serious odors can also result in its absence. Air oxidation can be used to remove odors, but chlorine is the preferred treatment because it not only destroys H_2S and other odorous compounds but it also retards the bacteria that cause the compounds in the first place. As a specific example, HOCl reacts with H_2S as follows in low pH solutions

$$HOCl + H_2S \rightarrow S + HCl + H_2O$$

If the actual plant practice calls for 100% excess HOCl (to make sure of the destruction of the H_2S because of the reaction of HOCl with other substances), how much HOCl (5% solution) must be added to 1L of a solution containing 50 ppm H_2S?

1.116. In a paper mill, soda ash (Na_2CO_3) can be added directly in the causticizing process to form, on reaction with calcium hydroxide, caustic soda (NaOH) for pulping. The overall reaction is $Na_2CO_3 + Ca(OH)_2 \rightarrow 2NaOH + CaCO_3$. Soda ash also may have potential in the on-site production of precipitated calcium carbonate, which is used as a paper filler. The chloride in soda ash (which causes corrosion of equipment) is 40 times less than in regular grade caustic soda (NaOH) which can also be used, hence the quality of soda ash is better for pulp mills. However, a major impediment to switching to soda ash is the need for excess causticization capacity, generally not available at older mills.

Severe competition exists between soda ash and caustic soda produced by electrolysis. Average caustic soda prices are about $265 per metric ton FOB (free on board, i.e., without fees for delivery to or loading on carrier) while soda ash prices are about $130/metric ton FOB.

To what value would caustic soda prices have to drop in order to meet the price of $130/metric ton based on an equivalent amount of NaOH?

1.117. The burning of limestone, $CaCO_3 \rightarrow CaO + CO_2$, goes only 70% to completion in a certain kiln.
  (a) What is the composition (mass %) of the solid withdrawn from the kiln?
  (b) How many kilograms of CO2 are produced per kilogram of limestone fed?
    Assume that the limestone is pure CaCO3.

1.118. In the semiconductor industry, integrated circuit (IC) production begins with the mechanical slicing of silicon rod into wafers. Once the wafers are sliced, the surfaces are lapped and polished to uniform flat surfaces. Contaminants and microscopic defects (work damage) are then removed chemically by etching. A traditional etching solution consists of a 4: 1: 3 volumetric ratio of 49% hydrofluoric, 70% nitric, and 100% acetic acids, respectively. Although work damage is usually only 10 μm deep, overetching to 20 μm per side is common. The reaction for dissolving the silicon surface is

$$3Si + 4HNO_3 + 18HF \rightarrow 3H_2SiF_6 + 4NO + 8H_2O$$

Calculate the flow rate of the etching solution in kilograms per hour if 20 μm per side is to be etched for 6000 wafers per hour of 150 mm diameter. What is the limiting reagent? Assume the limiting reactant reacts completely.

Data	Mol. wt.
Si—2.33 g/cm^3	28.09
Sp gr 49% HF is 1.198	20.01
Sp gr 70% HNO$_3$ is 1.4134	63.01
Sp gr 100% CH$_3$CO$_2$H is 1.0492	60.05

1.119. One method of synthesizing the aspirin substitute, acetaminophen, involves a three-step procedure as outlined in Figure Pl.119. First, *p*-nitrophenol is catalytically hydrogenated in the presence of aqueous hydrochloric acid to the acid chloride salt of *p*-aminophenol with a 86.9% degree of completion. Next the salt is neutralized to obtain *p*-aminophenol with a 0.95 fractional conversion.

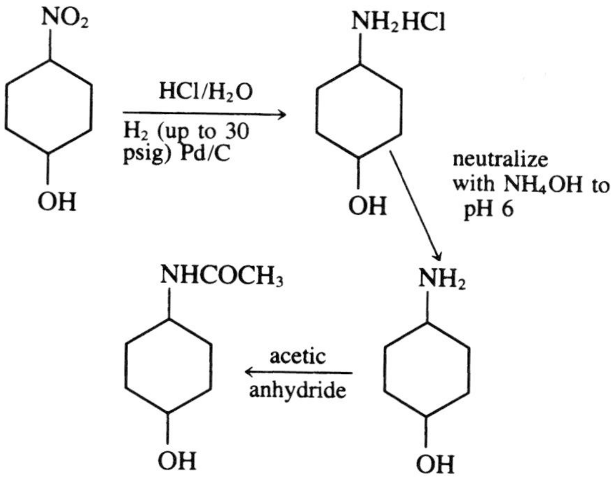

Figure P1.119

Finally, the *p*-aminophenol is acetalated by reacting with acetic anhydryde, resulting in a yield of 3 kg mol of acetaminophen per 4 kg mol. What is the overall conversion fraction of *p*-nitrophenol to acetaminophen?

1.120. The most economic method of sewage wastewater treatment is bacterial digestion. As an intermediate step in the conversion of organic nitrogen to nitrates, it is reported that the *Nitrosomonas* bacteria cells metabolize ammonium compounds into cell tissue and expel nitrite as a by-product by the following overall reaction:

$$5CO_2 + 55NH_4^+ + 76O_2 \rightarrow C_5H_7O_2N(tissue) + 54NO_2^- + 52H_2O + 109H^+$$

If 20,000 kg of wastewater containing 5% ammonium ions by weight flows through a septic tank inoculated with the bacteria, how many kilograms of cell tissue are produced, provided that 95% of the NH_4^+ is consumed?

1.121. One can view the blast furnace from a simple viewpoint as a process in which the principal reaction is

$$Fe_2O_3 + 3C \rightarrow 2Fe + 3CO$$

but some other undesired side reactions occur, mainly

$$Fe_2O_3 + C \rightarrow 2FeO + CO$$

After mixing 600.0 lb of carbon (coke) with 1.00 ton of pure iron oxide, Fe_2O_3, the process produces 1200.0 lb of pure iron, 183 lb of FeO, and 85.0 lb of Fe_2O_3. Calculate the following items:

(a) The percentage of excess carbon furnished, based on the principal reaction
(b) The percentage conversion of Fe_2O_3 to Fe
(c) The pounds of carbon used up and the pounds of CO produced per ton of Fe_2O_3 charged
(d) What is the selectivity in this process (of Fe with respect to FeO)?

1.122. A common method used in manufacturing sodium hypochlorite bleach is by the reaction

$$Cl_2 + 2NaOH \rightarrow NaCl + NaOCl + H_2O$$

Chlorine gas is bubbled through an aqueous solution of sodium hydroxide, after which the desired product is separated from the sodium chloride (a by-product of the reaction). A water-NaOH solution that contains 1145 lb of pure NaOH is reacted with 851 lb of gaseous chlorine. The NaOCl formed weighs 618 lb.

(a) What was the limiting reactant?
(b) What was the percentage excess of the excess reactant used?
(c) What is the degree of completion of the reaction, expressed as the moles of NaOCl formed to the moles of NaOCl that would have formed if the reaction had gone to completion?
(d) What is the yield of NaOCl per amount of chlorine used (on a weight basis)?

1.123. In the sulfate pulp process the alkaline cooking liquor is produced by the reactions:

$$Na_2SO_4 + 4C \rightarrow Na_2S + 4CO$$

$$Na_2S + CaO + H_2O \rightarrow NaOH + CaS$$

After the first reaction occurs, the sodium sulfide is leached from the mass and the solution treated with lime (CaO) in the form of a milk of lime suspension so that $Ca(OH)_2$ is actually the reactant. The precipitate of CaS and any unreacted $Ca(OH)_2$ is washed during filtration and removed. It is found that 12.7 lb of precipitate are formed for every 100 lb of cooking liquor prepared. The precipitate analyze CaS: 70.7%, $Ca(OH)_2$: 23.4%, and $CaCO_3$: 5.9%, the latter being present as an impurity in the lime used. The cooking liquor analyzes NaOH: 10.0%, Na_2S: 2.0%, and H_2O: 88.0%. Calculate

(a) the percent of excess lime used.
(b) the percent completion of the reaction.
(c) the composition of the lime used.

1.124. Ammonium sulfate is used as fertilizer. One method of manufacturing the compound is by the following sequence of reactions in water

$$2NH_3 + H_2O + CO_2 \rightarrow (NH_4)_2CO_3$$

$$(NH_4)_2CO_3 + CaSO_4 \rightarrow (NH_4)_2SO_4 + CaCO_3$$

After the reactions take place, the water slurry is filtered. The analysis shows the respective weight percents as *Filtrate*: 0.5% $(NH_4)_2CO_3$, 5.5% $(NH_4)_2SO_4$, 94.0% H_2O and *Filter Cake*: 8.2% $CaSO_4$, 91.8% $CaCO_3$. For the overall process, calculate

(a) the limiting reactant.

(b) the degree of completion.

(c) the percent excess reactant.

1.125. In a process for the manufacture of chlorine by direct oxidation of HCl with air over a catalyst to form Cl_2 and H_2O (only), the exit product is composed of HCl (4.4%), Cl_2 (19.8%), H_2O (19.8%), O_2 (4.0%), and N_2 (52.0%). What was

(a) the limiting reactant?

(b) the percent excess reactant?

(c) the degree of completion of the reaction?

1.126. A well known reaction to generate hydrogen from steam is the so called water gas shift reaction: $CO + H_2O \rightleftarrows CO_2 + H_2$. If the gaseous feed to a reactor consists of 30 moles of CO, 12 moles of CO_2, and 35 moles of steam per hour at 800°C, and 18 moles of H_2 are produced per hour, calculate

(a) the limiting reactant.

(b) the excess reactant.

(c) the fraction conversion of steam to H_2.

(d) the degree of completion of the reaction.

(e) the kg of H_2 yielded per kg of steam fed.

(f) the moles of CO_2 produced by the reaction per mole of CO fed.

PROBLEM SOLVING 2

Most of the literature on problem solving views a "problem" as a gap between some initial information (the initial state) and the desired information (the desired state). Problem solving is the activity of closing the gap between these two states. Of course, not all problems have a solution. Problems are often classified as (1) **open ended** versus (2) **closed ended**. The former means that the problem is not well posed and/or could have multiple solutions. The latter means that the problem is well posed and has a unique solution. At the end of each section in this book you will find open-ended discussion questions, and a few open-ended problems are included in the problems at the end of each chapter. However, all of the examples and the bulk of the problems in this book are closed ended—short, simple, and packaged with precisely the information needed, unlike those that occur in real life.

If you are going to become a professional, you will have to acquire a number of skills in problem solving such as:

- Formulating specific questions from vaguely specified problems
- Selecting effective problem-solving strategies
- Deciding when an estimate will suffice vs. an exact answer
- Using tables, graphs, spreadsheets, calculators, and computers to organize, solve, and interpret the results from solving problems
- Judging the validity of the work of others
- Estimating orders of magnitude to evaluate answers

To assist you in developing these and other skills in problem solving, in this chapter we briefly discuss (1) some ways to solve both open- and closed-ended problems, (2) computer tools to execute efficient solutions, and (3) sources of information that assist in problem solving.

2.1   TECHNIQUES OF PROBLEM SOLVING

*Your objectives in studying this
section are to be able to:*

1. Learn the components of effective problem solving.
2. Apply the components to different types of problems.

Weiler's Law: Nothing is impossible for the person who doesn't have to do it.

Howe's Law: Every person has a scheme which will not work.

The 90/90 Law: The first 10% of the task takes 90% of the time. The remaining 90% takes the remaining 10%.

Gordon's Law: If a project is not worth doing, it's not worth doing well.

Slack's Law: The least you will settle for is the most you can expect to get.

O'Toole's Commentary: Murphy was an optimist.[1]

One of the main objectives of this book is to enhance your **problem-solving** skills. If you can form good habits of problem solving early in your career, you will save considerable time and avoid many frustrations in all aspects of your work, in and out of school. Being able to solve material and energy balances means that in addition to learning basic principles, formulas, laws, and so on, *you must be able to apply them effectively*. Routine substitution of data into an appropriate equation will by no means be adequate to solve any material and energy balances other than the most trivial ones.

In working through this book, develop confidence in your problem-solving capabilities, become aware of your thought processes, get organized, manage

[1]Arthur Bloch. *Murphy's Law and Other Reasons Why Things Go Wrong!* New York: Price/Stern/Sloan, 1977.

your time effectively, and be flexible in seeking alternative solution strategies. Engineers believe that clear-headed, logical thinking is the way to solve real-life problems. Experience demonstrates that such thought processes are not natural, and, in fact, considerable practice must occur before an individual attains the necessary skills. Even for simple problems, the sequence of ideas is usually so tangled up that the interconnections cannot be easily discerned. None of the strategies described here is perfect for you or necessarily effective for all problems. You have to devise or imitate strategies of solving problems that you feel comfortable with and that have demonstrated validity. At the end of this chapter you will find a list of references offering numerous choices of problem-solving strategies. We will mention just a few here.

Polya[2] recommends the use of four steps for solving problems and puzzles: define, plan, carry out the plan, and look back. The key features of this strategy are the interaction among the steps and the interplay between critical and creative thinking. Fogler and LeBlanc[3] discuss a five step program: (1) define the problem (problem identification and exploration), (2) generate alternatives, (3) decide on a course of action, (4) carry through, and (5) evaluate the outcome(s). The McMaster five-step strategy developed by Woods[4] entails a similar set of steps: (1) define, (2) explore, (3) plan, (4) act, and (5) reflect.

The Kepner-Tregve (KT)[5] approach to problem solving is also an organized method, the detailed discussion of which can be found in several of the references at the end of this chapter. The KT strategy involves three phases: (1) analysis of the problem, (2) decision procedure, and (3) identification of potential future pitfalls associated with the problem solution. The first two are of interest here.

In the analysis phase, ask questions about the problem such as who, what, where, when, why, and how. For each question also ask what is the situation, what is not the situation, what is the difference between the two situations, and what are possible causes of the difference. For example, for the question "what," ask what is the problem, what is not the problem, what is the distinction between the "is" and "is not," and what cause(s) might arise as these questions are considered. For the question "where," ask what part of the process is affected, what is not affected, what is the distinction, and how do these questions lead to possible causes.

The goal of the decision phase is to select the best solution from the pro-

[2]G. Polya. *How to Solve It,* 2nd ed. New York: Doubleday, 1957.

[3]H. S. Fogler and S. E. LeBlanc. *Strategies for Creative Problem Solving.* Englewood Cliffs, NJ: Prentice-Hall, 1994.

[4]Woods, D. R. *Problem-Based Learning: How to Gain the Most from PBL.* D. R. Woods, McMaster University, Hamilton, Ont., 1994.

[5]Kepner, C. H. and B. B. Tregoe. *The New Rational Manager.* Princeton, NJ: Princeton Research Press, 1981.

posed alternatives, and possibly list the priority of alternatives. You can form a matrix (list the goals at the top) in which the heads of the columns designate possible actions to ameliorate the problem or possible choices to meet the stated goal in solving the problem. The rows of the matrix represent, first, the hard constraints, those factors that **must** be satisfied. Next would be listed the group of soft constraints, those that would be nice (but not essential) to satisfy. In each column for each option list the relative rating score (say on a range of 1 to 10) attributed to the degree of satisfaction of a constraint. Addition of the scores in each column helps in reaching a decision. Since many of the weights will be subjective, the final decision is not necessarily based solely on the total of the scores.

As an illustration of the KT approach, examine the following question: What is the maximum number of different rectangles can you build from 12 square tiles using all of the tiles in each rectangle? The options would be various numbers of rectangles: 1, 2, 3, 4, and so on. The hard constraints would be to (1) use all 12 tiles, (2) make only rectangles, and (3) obtain the maximum number of rectangles. Figure 2.1 demonstrates a possible matrix to use in solving the problem, and Figure 2.2 illustrates one way to carry out the analysis (via graphs). You will find considerable merit in going back and forth between visual, verbal, and symbolic representations of a problem.

Completely novel problems, of course, usually appear to be complex. Figure 2.3 outlines how your strategy for solving problems is influenced by the information you develop. Table 3.1 in the next chapter contains a checklist for solving various types of material and energy balance problems. For novel, and particularly for open-ended problems, we can for convenience classify the steps in problem solving into five phases (that do not necessarily have to be carried out serially):

1. Understand the problem and the goals
2. Formulate the options for solution
3. Consider the constraints

Constraints	Options (number of rectangles)						
	1	2	3	4	5	6	7
Hard:							
1. All 12 tiles used	10	10	10	10	10	10	0
2. Only rectangles formed	10	10	10	10	10	10	0
3. Maximum number achieved	0	0	0	0	0	10	0
Soft:							
1. Pleasing shape	0	0	0	0	0	0	0
Total	20	20	20	20	20	30	0

Figure 2.1 The matrix for problem solving.

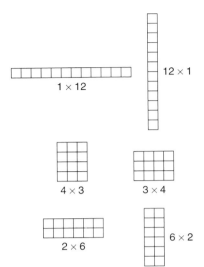

Figure 2.2   Problem analysis via graphics.

4. Execute the selected problem-solving strategy
5. Evaluate the procedure and results

In Figures 2.4, 2.5, 2.6, 2.7, and 2.8, we have listed for each of the five phases a set of suggested questions and/or activities to review while engaged in problem solving along with short list of verbs to help stimulate your thought processes.

Start with understanding the problem.

> It isn't that they can't see the solution. It is that they can't see the problem. G. K. Chesterton, *The Scandal of Father Brown.*

First, you must identify what results you are to achieve, that is, what the problem is. Then you must define the system, perhaps with the aid of a diagram. Various physical constraints will apply as well as the time available for you to work on the solution. In almost all cases you will have to look up data and make use of general laws. Finally, the results will have to be presented properly so that you can communicate them to someone else. You can work backward as well as forward in solving problems if the forward sequence of steps to take is not initially clear, and cycle back at will. Problems that are long and involved should be divided into parts and attacked systematically piece by piece.

From the viewpoint of understanding the problem, consider the open-ended problem of designing a foolproof way to prevent locking your keys in your car. If you think about this statement, you could, for example, examine options to prevent you from shutting the door with the keys still in the car. However, the problem

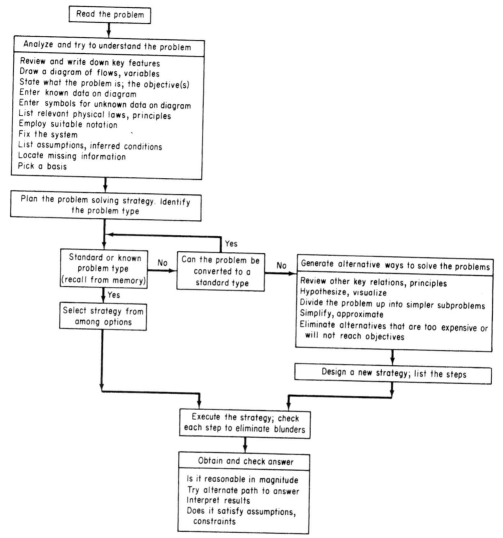

Figure 2.3   Selection of a problem-solving strategy.

could be revised to be: If I lock the car with the keys in it, how can I defeat the locking scheme and get into the car? This alternate viewpoint might lead to keeping a duplicate key in your wallet, and so on. By going down the concepts in Figure 2.4 and letting your imagination fly free, you can grasp what the problem is.

To have a good idea, you must have a lot of ideas.—Linus Pauling.

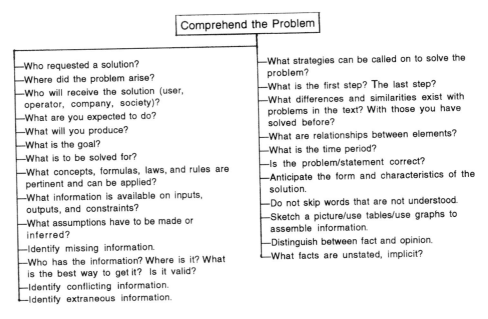

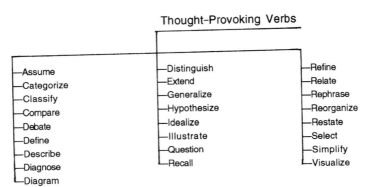

Figure 2.4   Phase 1: Understand the problem and the goals.

All of us would agree that looking at things from more than one perspective is valuable, outside as well as inside, the profession of engineering. Alex Osborn popularized *brainstorming* and spread creativity training from his advertising agency to business and engineering. Creativity is the process that produces new and valued responses to problems and at the same time facilitates learning, change, and innovation. Creativity means more that evolving bright ideas. It is more a way of thinking that sheds new light on old problems. It involves developing an ability to think in other than straight lines.

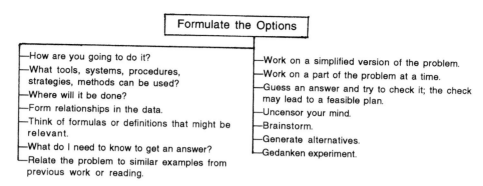

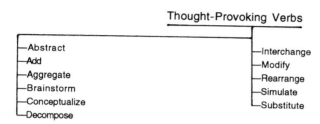

Figure 2.5   Phase 2: Formulate the options.

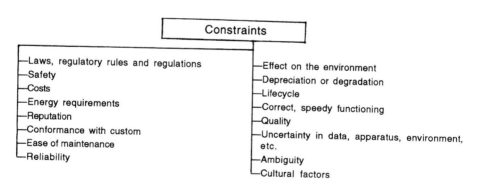

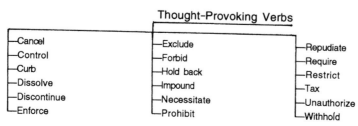

Figure 2.6   Phase 3: What are the constraints on the options?

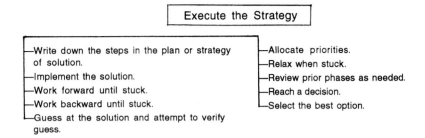

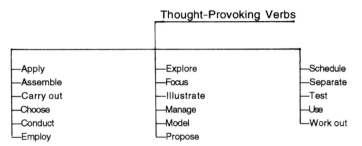

Figure 2.7   Phase 4: Execute the strategy.

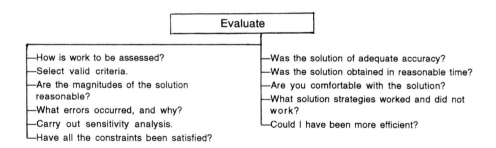

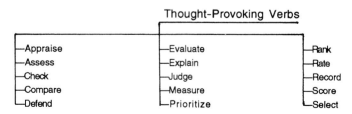

Figure 2.8   Phase 5: Evaluate the procedure and results.

By way of illustrating the development of options, let us return to the problem of avoiding locking your keys in the car. Some options might be

1. Have an alarm signal if any door opens with the key in the ignition lock,
2. Require that a car door be positively locked only from the outside with the ignition key,
3. Eliminate keys and use a key pad instead of a key,

and so on. Application of a little imagination will lead to many more options.

Once you have understood the problem, formulated some options, and ascertained the constraints, you need to select the problem-solving procedure and execute it. If the sequence of steps in the procedure is not obvious, take time to set up a plan. If one plan fails, try another.

After reaching a solution to a problem (or failing to reach a solution) you should evaluate what you did. Look at the questions in Figure 2.8, and determine if your problem-solving skills and judgment were satisfactory. Improvement is always possible if you carefully consider what worked and what did not.

What is the difference in the approach to problem solving between an expert and a novice? An expert proceeds in problem solving by abbreviated steps; many are done only mentally. A beginner should go through each step explicitly until he or she becomes experienced. For guidance, perhaps you should turn to Sherlock Holmes (as cited in Arthur Conan Doyle's "The Naval Treaty"):

"Do you see any clue?"

"You have furnished me with seven, but of course I must test them before I can pronounce upon their value."

"You must suspect someone?"

"I suspect myself."

"What?"

"Of coming to conclusions too rapidly."

Table 2.1 lists activities that an expert may use in problem solving. Table 2.2 contrasts the problem-solving habits of an expert with those of a novice. Table 2.3 is a checklist for self-assessment of your problem-solving traits. How many of the items in the table pertain to your problem-solving techniques? Practice visual thinking, stress management, and awareness of the process whereby you solve problems. Table 2.4 is a list of reasons why you may not have been successful in problem solving.

As formulated by Woods,[6] developing your awareness of your problem-solving skills is an important factor in improving them because

1. You can identify where you are when solving a problem.
2. You can develop a methodical approach.
3. You can, whenever you are stuck, identify the obstacle.
4. You can describe to others what you have done and any difficulties that you are encountering.
5. You become aware of what skills need improvement.
6. You increase your level of confidence.
7. You develop traits of carefulness.

If you can assimilate the procedures discussed above and make them a part of yourself—so that you do not have to think about the process of problem solving step by step—you will find that you will be able to materially improve your speed, performance, and accuracy in problem solving.

LOOKING BACK

In this section we briefly described several problem solving strategies for novel and open ended problems.

Key Ideas

1. Many problem-solving strategies have been proposed—you have to find one that meshes with your background and viewpoint.
2. All effective problem-solving strategies involve the phases of determining what the real problem is, generating alternative solutions, developing a precedence order for the solution, executing the solution, and evaluating the outcome.
3. Table 3.1 in the next chapter lists ten specific steps to be used in solving closed-ended material and energy balance problems.

Key Words

Brainstorming (p. 110)   Open ended (p. 104)
Closed ended (p. 104)   Polya (p. 106)
Fogler and LeBlanc (p. 106)   Problem solving (p. 105)
Kepner-Tregoe (p. 106)   Woods (p. 106)

[6] D. R. Woods. *Unit 1, Developing Awareness*, the McMaster Problem Solving Program, McMaster University, Hamilton, Ont., Canada, 1985.

TABLE 2.1   Techniques Used by Experts to Overcome Barriers to Problem Solving

Read the problem over several times but at different times. Be sure to understand all facets of it. Emphasize the different features each time.

Restate the problem in your own words. List assumptions.

Draw a comprehensive diagram of the process and enter *all* known information on the diagram. Enter symbols for unknown variables and parameters.

Formally write down what you are going to solve for: "I want to calculate . . ."

Choose a basis.

Relate the problem to similar problems you have encountered before, but note any differences.

Plan a strategy for solution; write it down if necessary. Consider different strategies.

Write down all the equations and rules that might apply to the problem.

Formally write down everything you know about the problem and what you believe is needed to execute a solution.

Talk to yourself as you proceed to solve the problem.

Ask yourself questions as you go along concerning the data, procedures, equations involved, etc.

Talk to other people about the problem.

Break off problem solving for a few minutes and carry out some other activity.

Break up the solution of the problem into more manageable parts, and start at a familiar stage. Write down the objective for each subproblem (i.e., convert mole fraction to mass fraction, find the pressure in tank 2, etc.).

Repeat the calculations but in a different order.

Work both forward and backward in the solution scheme.

Consider if the results you obtained are reasonable. Check both units and order of magnitude of the calculations. Are the boundary conditions satisfied?

Use alternative paths to verify your solution.

Maintain a positive attitude—you know the problem can be solved—just how is the question.

TABLE 2.2   A Comparison of the Problem-Solving Habits of a Novice and an Expert

A Novice:	An Expert:
Starts solving a problem before fully understanding what is wanted and/or what a good route for solution will be	Reviews the entire plan outlined in Figure 2.3, mentally explores alternative strategies, and clearly understands what result is to be obtained
Focuses only on a known problem set that he or she has seen before and tries to match the problem with one in the set	Concentrates on similarities to and differences from known problems; uses generic principles rather than problem matching
Chooses one procedure without exploring alternatives	Examines several procedures serially or in parallel
Emphasizes speed of solution, unaware of blunders	Emphasizes care and accuracy in the solution
Does not follow an organized plan of attack; jumps about, and mixes problem-solving strategies	Goes through the problem-solving process step by step, checking, reevaluating, and recycling from dead ends to other valid paths
Is unaware of missing data, concepts, laws	Knows what principles might be involved and where to get missing data
Exhibits bad judgment, makes unsound assumptions	Carefully evaluates the necessary assumptions
Gives up solving the problem because he or she does not know enough	Knows what the difficulty is and is willing to learn more that will provide the information needed
Gives up solving the problem because he or she does not have skills to branch away from a dead-end strategy	Aware that a dead end may exist for a strategy and has planned alternative strategies if a dead end is reached
Is unable to make approximations or makes bad ones	Makes appropriate approximations
Cannot conceive of disagreeing	Disagrees with other experts
Slavishly follows instructions; proceeds "by the book"	Breaks rules and makes exceptions
Does not know what to make of qualitative data	Is able to deal with qualitative data
Ignores possible limits	Recognizes limits
Fritters times way	Manages time well

TABLE 2.3   A Checklist of Personal Traits to Avoid in Problem Solving*

1. When I fail to solve a problem, I do not examine how I went wrong.

2. When confronted with a complex problem, I do not develop a strategy of finding out exactly what the problem is.

3. When my first efforts to solve a problem fail, I become uneasy about my ability to solve the problem (or I panic!).

4. I am unable to think of effective alternatives to solve a problem.

5. When I become confused about a problem, I do not try to formalize vague ideas or feelings into concrete terms.

6. When confronted with a problem, I tend to do the first thing I can think of to solve it.

7. Often I do not stop and take time to deal with a problem, but just muddle ahead.

8. I do not try to predict the overall result of carrying out a particular course of action.

9. When I try to think of possible techniques of solving a problem, I do not come up with very many alternatives.

10. When faced with a novel problem, I do not have the confidence that I can resolve it.

11. When I work on a problem, I feel that I am grasping or wandering, and not getting a good lead on what to do.

12. I make snap judgments (and regret them later).

13. I do not think of ways to combine different ideas or rules into a whole.

14. Sometimes I get so charged up emotionally that I am unable to deal with my problem.

15. I jump into a problem so fast, I solve the wrong problem.

16. I depend entirely on the worked-out sample problems to serve as models for other problems.

17. I do not plan my time.

18. I am afraid of losing face.

19. I fail to start on the easy (to me) problems first.

20. I ignore words I do not know.

21. I am easily distracted by the environment in which I work.

22. The stress of problem solving causes blocks and filters out good ideas.

23. Cultural blocks and lack of background information lead me down the wrong path.

*Based on the ideas in a questionnaire in P. P. Heppner, *P.S.I.*, Department of Psychology, University of Missouri-Columbia, 1982, plus The University of Texas Learning Aid.

TABLE 2.4   Diagnosis of Reasons for Failing to Solve Problems
(*"Experience is the name everyone gives to their mistakes."* **Oscar Wilde)**

Failure to work on a problem in a systematic rather than a scatter-brained way (start too soon; skip essential steps)

Failure to read/understand the problem thoroughly

Failure to draw a diagram and enter all data thereon and the symbols for the unknowns

Failure to ascertain the unknown

Fixing on the first, a poor, or an incorrect strategy of solution without considering alternative strategies

Selection of the wrong principle or equation to use (total moles instead of total mass, ideal gas instead of real gas) and solution of the wrong problem

Working with false information

Picking the wrong entry from a database, chart, or table (wrong sign, wrong units, decimal misplaced, etc.)

Entering incorrect inputs/parameters into calculations (transpose numbers, wrong units, etc.)

Failure to include units in each step of the calculations

Sloppy execution of calculations introduce errors (add instead of subtract, invert coefficients, etc.)

Difficulty in distinguishing new features in a problem that superficially looks familiar

Incorrect algebraic manipulations

Use of unsatisfactory computer code for the problem (too much error, premature termination)

Inability to locate needed data or coefficients by not reading the problem thoroughly or looking in the wrong database

Inability to estimate what the answer should be to use in comparison with the calculated answer

Inadequate (your database) (you have forgotten, or never learned, some essential laws, equations, values of coefficients, conversion factors, etc.)

Employment of only forward reasoning rather than both forward and backward reasoning

Emotional stress (fear of making a mistake, looking foolish or stupid)

Lack of motivation

Inability to relax

Self-Assessment Test

1. Prepare an information flow diagram showing the sequence (serial and parallel) of steps to be used in effective problem solving.

2. Take any example in Sec. 1.3, 1.5, or 1.6 and prepare an information flow diagram of your thought process in solving the example. Make a tree showing how the following classes of information are connected (put the solution as the last stage of the tree at the bottom):
   (a) Information stated in the problem
   (b) Information implied or inferred from the problem statement
   (c) Information from your memory (internal data bank!)
   (d) Information from an external data bank (reference source)
   (e) Information determined by reasoning or calculations

   Label each class with a different-type box (circle, square, diamond, etc.) and let arrows connect the boxes to show the sequence of information flow for your procedure.

3. What should you do if you experience the following difficulties in solving problems?
   (a) No interest in the material and no clear reason to remember
   (b) Cannot understand after reading the material
   (c) Read to learn "later"
   (d) Rapidly forget what you have read
   (e) Form of study is inappropriate

4. Apply the K-T method to solve the following problem.

   > A man has a raft and three cantaloupes. Each cantaloupe weighs a pound. However, the raft can hold only 202 pounds, and the man weighs 200 pounds himself. How does the man get to the other side of the river with the cantaloupes? (His weight includes his clothes and the cantaloupes can't be thrown across the river.)

2.2   COMPUTER-BASED TOOLS

> ### *Your objectives in studying this section are to be able to:*
>
> 1. Locate computer programs that will solve material and energy balances.
> 2. Use the programs for which you have access.

LOOKING AHEAD

In this section we point out some of the software that will assist in the solution of material and energy balances.

MAIN CONCEPTS

Software packages involving symbolic and numerical calculations along with graphics have over the last decade become essential tools for all engineers. The great potential of computers is their capacity to do anything that can be described mathematically as a series of operations and logical decisions—theoretically. From a practical viewpoint, you have to ask not merely whether it is feasible for a task to be performed on a computer, but whether it is sensible. Two criteria need to be applied to reach a decision: (1) Can the task be performed (or the problem solved) at all without the use of a computer? and (2) Is it better or cheaper or faster to use a computer to solve a problem than a hand-held calculator (or no machine at all)? To decide whether it is better or cheaper or faster to use a computer, you must consider, among other factors, the necessary investment of effort, time, and money. Is the problem to be solved just one time or many times? Is the accuracy of the solution enhanced through the use of the computer? Do you have a computer handy with the right program stored to solve the problem, or do you have to code an algorithm and debug it first? Answers to these and other questions will guide your choice of tools.

Will computer programs help you learn how to solve material and energy balance problems? Standard computing packages easily produce answers without the user understanding either the question or the solution process. You should be aware of

1. what assumptions and initial conditions are implicit in the program;
2. how accurate the information is that is called on by the code (in particular the physical properties);
3. errors you can make in introducing the problem into the software code.

Take care to maintain a positive but vigilant attitude toward the use of computers to solve problems. Be cautious, try different assumptions, compare your results on reference problems with known results, and evaluate the sensibility of your solution to small changes in the data used in the code.

As an example of what might go wrong, Shacham[7] points out how the solution of the equation that gives the fraction conversion in a chemical a reactor

---

[7]M. Shacham. "Recent Developments in Solution Techniques for Systems of Nonlinear Equations." In *Proceed. 2nd Intl. Conf. Foundations of Computer Aided Design*, (p. 891) A. W. Westerberg and H. H. Chien, eds., CACHE Corp., 1984.

$$f(x) = \frac{x}{(1-x)} - 5 \ln \frac{0.4(1-x)}{(0.4-0.5x)} + 4.45977$$

will converge to a solution only for initial guesses for x between 0.705 and 0.799! What happens if you pick an initial guess of $x = 1.00$?

In what follows we will mention the major classes of software that are or might be available to you. Most do not require any programming skills. Many of the classes of programs can be connected to each other so that calculations and graphics are not restricted to a single code (or its "add-ons").

In the back of this book you will find a disk that contains a number of computer codes that you can use to solve material and energy balances including the well known code Polymath.

Equation-Solving Programs

You can use inexpensive *equation-solving programs* (such as *TK Solver Plus*[8], *Gauss*[9]) available for microcomputers to solve linear and nonlinear material balances. Such codes may actually run faster than spreadsheet programs but not as fast as C or Fortran programs. A typical example of an equation-solving package is TK Solver. With TK Solver, you simply provide the defining list of equations or problem-solving rules, and provide values for your variables either from the keyboard or by importing a file. If you don't have exact values for your variables, TK Solver can work with your best guesses and derive either a range of solutions or an optimized solution.

TK Solver solves problems in terms of "models" composed of rules (equations and logic statements). The models you develop become "objects" that you can save, recall and edit, or combine to form even more complex models. You can list your numerical results, export them to a spreadsheet, or depict them as line, bar, or pie charts. The screen combines data, results, and graphics. Figure 2.9 illustrates two screens.

TK solver uses the Newton-Raphson method (see Appendix L) to solve nonlinear equations. No constraints can be specified to force the solution to be in a feasible region. You do not need to provide analytical derivatives—they are approximated by finite differences.

TK Solver Plus also provides a large set of library models (programs) that you can use as is or change for use in modified versions. Library models include: Roots of Equations, Differentiation and Integration, Differential Equations, Special Functions, Complex Variables, Optimization, Matrix Calculations, Arbitrary

[8]A trademark of Universal Technical Systems, Inc., 120 Rock St., Rockford, IL 61101.
[9]A trademark of Aptech Systems, Inc., 26250 196th Pl. S.E., Kent, WA 98042.

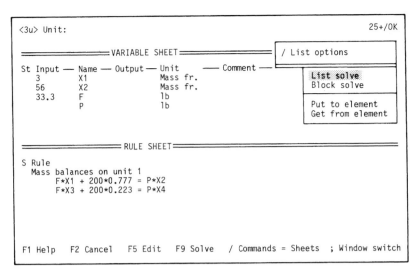

Figure 2.9   Two screens produced by TK Solver. Shown here are the Variable Sheet (top) and the Rule Sheet (bottom). After known values are entered into the Variable Sheet, the ! key is pressed to start the calculations. Once the solution is found, the values of the unknown variables appear in the output column spaces in the Variable Sheet.

Length Integers, Interactive Tables, Graphics, Statistics and Curve Fitting, Engineering and Science (mostly chemistry), and Finance.

Generic Computer Codes to Solve Sets of Equations

You can find software written in C or Fortran that can be used to solve linear and nonlinear equations in most computer libraries, even for PCs. For example, LA-PACK[10] is a package of codes written in Fortran 77 in the public domain that solves most common problems in linear algebra including linear equations. You can acquire individual routines from *netlib* via the Internet (as well as software that solves nonlinear equations). Another example is GINO,[11] which focuses on the solution of optimization problems, but also solves nonlinear equations. A longtime software package called IMSL is available for computers of all scales.

You will find the documentation and computer routines harder to understand, and take more time to absorb than most of the competing software tools, but the advantages of using such mathematical libraries are:

---

[10]E. Anderson, et al. *LAPACK*. Philadelphia: SIAM, 1992.

[11]J. Liebman, et al. *Modeling and Optimization with GINO*. Palo Alto, CA: Scientific Press, 1986.

1. You have available a broad range of reliable, robust, and efficient mathematical routines;
2. You get exposure to state-of-the-art software.
3. You get a better understanding of the limitations of software.
4. You avoid having to know how to program.

Spreadsheets

Spreadsheet software is probably the most widely used numerical tool in personal computing. For this reason, the capabilities of the programs has tended to be enhanced so that they take on features of the other computer tools discussed in this section rather than just perform calculations in a two-dimensional matrix of cells. An almost uncountable number of commercial spreadsheet packages exist so that it would be unrealistic to discuss any of them here. Instead, we mention some of the generic features of spreadsheet codes. Figure 2.10 shows the solution of a material balance using a spreadsheet. Figure 2.11 shows an Excel[12] template prepared for combustion analysis.

A discussion of the similarities among the commercial spreadsheets would perhaps fill a book. Each offers a scenario manager that lets you build several sets of assumptions into a single spreadsheet, and lets you switch from one set of assumptions to another quickly so that you can compare the solution of problems. All have a facility to create and manage libraries of point-and-click algorithms as well as 2- and 3-D graphics. In Excel, built-in functions solve linear and nonlinear equations, carry out optimization (including integer constraints), fit functions from data, and allow users to tailor their software via macros. Excel has a Visual Basic Application language to write subroutines and functions so that you can process and transfer data from other computer tools. Engineers find these features vastly simplify problem solving versus using or writing Fortran or C programs.

When you use spreadsheets for your calculations, try to avoid the following errors:

1. Erroneous formulas, particularly with respect to reference to the wrong cell.
2. Incorrect specification of the range of the formula.
3. Omission of a key variable or factor.
4. Entry of invalid or incorrect data.

---

[12]A trademark of the Microsoft Corp., P.O. Box 97017, Redmond, WA 98073.

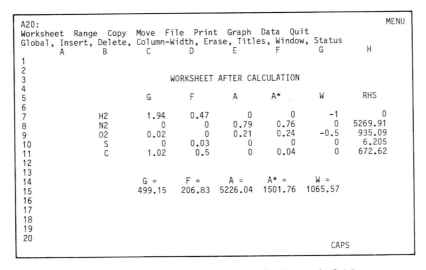

Figure 2.10 Spreadsheet solution for Example 3.16.

Flowsheeting Programs

In the 1960s, the chemical process industry initiated the use of large-scale programs for computer-aided process design. Such programs were frequently referred to as heat and material balancing programs, but now are generally called **flowsheeting programs**. Such programs accept information about a chemical process at the flowsheet level of detail, and make calculations that provide data

CP2Combustion2

	A	B	C	D E F G H I	J	K	L	M	N	O	P
1	Combustion Analyzer for Fuel Mixtures Containing Compounds of Carbon, Hydrogen, and Oxygen										
2											
3	Spreadsheet by Brice Carnahan: ChE 230 - Section 1 - Fall 1988										
4											
5											
6	Enter data into outlined cells onl All other cells are calculated										
7											
8											
9	Fuel mix: (enter up to 3 new names and formulas)				Percent Conversion to		CO CO2 H2O (for H2 fuel only)				
10											

Fuel 1 = propanol   C 3 H 8 O 1   50.0% 50.0%
Fuel 2 = acetone   C 3 H 6 O 1   50.0% 50.0%
Fuel 3 = iso-octane   C 8 H 18 O   50.0% 50.0%
Fuel 4 = carbon   C 1
Fuel 5 = hydrogen   H 2
Fuel 6 = carbon monoxide   C 1   O 1   (of initial CO)

Enter Type of Oxidant (air or oxygen) and Percent Excess Oxidant Used

Oxidant is: (enter 1 in appropriate box)   Air 1 O2 ☐
% excess air or oxygen =   50.0%

Combustion Reactions for Fuels 1,2,3 (optional - not needed for calculations)
                                                         Reactions
                                                            1
                                                            2
                                                            3
C + 0.5 O2 = CO;   C + O2 = CO2   4
H2 + 0.5 O2 = H2O   5
CO + 0.5 O2 = CO2   6

	Compound	Fuel Mixture moles	Oxidant moles	Delta Rxn moles	Product moles	Product mole %	Dry Basis mole %
				800			
				100			
propanol	100		−100.00	0.00	0.000	0.000
acetone	100		−100.00	0.00	0.000	0.000
iso-octane	100		−100.00	0.00	0.000	0.000
C	0		0.00	0.00	0.000	0.000
H2	0		0.00	0.00	0.000	0.000
CO	0		700.00	700.00	4.308	4.778
O2	0	3150.00	−1750.00	1400.00	8.615	9.556
N2	0	11850.00	0.00	11850.00	72.923	80.887
CO2	0		700.00	700.00	4.308	4.778
H2O	0		1600.00	1600.00	9.846	0.000
Total	300	15000	950	16250	100.000	100.000

Figure 2.11   A template prepared for combustion analysis to demonstrate the proper organization of a spreadsheet. Courtesy of Professor Brice Cannahan, Deptartment of Chemical Engineering, University of Michigan.

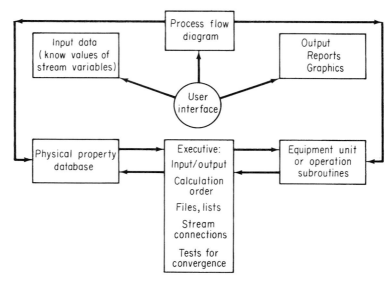

Figure 2.12   Structure of a generic flowsheeting code.

about not only material and energy flows but also about costs, pipe layout, time effects, and other useful information for design and operation. Figure 2.12 illustrates the structure of a flowsheeting code. Such codes can simulate the steady-state (and unsteady-state in some cases) performance of large integrated chemical plants consisting of interconnected process units with recycle streams, codes, and their sources.

Flowsheeting codes let you model an entire plant with as much complexity as you want. You can put the structure of the process into the computer via a graphical interface, and get the output either as flow diagrams or written reports. However, flowsheeting codes are substantially more complex to use than the other computational tools mentioned in this section. Read Section 6.2 for further details about flowsheeting codes.

Interactive Software

So-called interactive software has evolved from computer codes that just carried out matrix manipulations to rather comprehensive programs that combine matrix computations, data analysis, equation solving, graphics, and many other features. Two typical examples are Mathcad[13] and Matlab.[14] These codes are easier to use,

---

[13]A trademark of Mathsoft, Inc., One Kendall Square, Cambridge, MA 02139.
[14]A trademark of The Mathworks, 24 Prime Park Way, Natick, MA 01760.

and become skilled in using, than the generic codes, the flowsheeting codes, or the symbolic manipulators discussed below, but their scope is narrower. The primary reason these codes have been so successful is that the language used to formulate problems is simple and meshes well with the mathematics you already know. For example, in Matlab the command $\mathbf{x = A/b}$ gives the solution for the set of linear equations $\mathbf{Ax = b}$.

Mathcad screens behave like scratch pads. Equations and functions are treated in a free format mode so that mathematical symbols and operators appear on the screen as they would on hard copy. Any changes you make in the given data automatically are propagated so as to update all of your calculations. You do not have to have any experience in programming to use the programs. Mathcad contains a number of numerical algorithms and operations that can be called as well as both graphical and report writing methods of presenting results. Several applications packages are available including one for chemical engineering calculations. Figure 2.13 shows a screen set up to calculate the heat capacity.

In addition to solving numerical problems, Mathcad can carry out symbolic manipulations. It uses a menu of symbolic operators developed for the Maple code (see below). For example, it can take an equation and simplify it (if possible), solve for a variable, integrate symbolically, and so on. Mathcad can produce two- and three-dimensional graphs that can be viewed from various perspectives at adjustable scales. You can display arithmetic, semilog, and log-log plots with and without grid lines.

Matlab has established itself as a valuable tool in engineering because of its simplicity relative to C and Fortran codes (from the viewpoint of the users). For example, Figure 2.14 shows the simple commands you need to solve a set of linear equations. You can carry out various types of numerical analysis interactively with just a few keystrokes. As with Mathcad, problems to be solved and solutions are expressed almost exactly as they would be written mathematically. Matlab uses a matrix that does not require dimensioning as its basic data element. Matrix tools include eigenvalues, solution of linear equations, least squares, inverse, and many others. Also included are characteristic polynomial roots, residue calculation, and curve fitting. Graphics can be created as 3-D, linear, semilog, and polar data plots.

Matlab also has available a set of applications packages called "tool boxes" which are libraries of functions and programs to specific problems. For example, you can find tool boxes to solve problems in signal processing, image processing, symbolic math (based on Maple software), statistics, neural networks, spline approximations, control systems, and optimization. If these applications do not meet your needs, you can make changes to the Matlab algorithms or write your own code.

$$i := 1 .. n$$

Heat - capacity per mole calculation:

$$CP_i := a_i \cdot \frac{cal}{gmol \cdot K} + b_i \cdot 10^{-2} \cdot T \cdot \frac{cal}{gmol \cdot K^2} + c_i \cdot 10^{-5} \cdot T^2 \cdot \frac{cal}{gmol \cdot K^3} + d_i \cdot 10^{-9} \cdot T^3 \cdot \frac{cal}{gmol \cdot K^4}$$

$$da := \alpha_4 \cdot a_4 + \alpha_3 \cdot a_3 - \left[\alpha_2 \cdot a_2 + \alpha_1 \cdot a_1 \right] \qquad Temp_1 := T - 298 \cdot K$$

$$db := \left[\alpha_4 \cdot b_4 + \alpha_3 \cdot b_3 - \left[\alpha_2 \cdot b_2 + \alpha_1 \cdot b_1 \right] \right] \cdot 10^{-2} \qquad Temp_2 := T^2 - 298^2 \cdot K^2$$

$$dc := \left[\alpha_4 \cdot c_4 + \alpha_3 \cdot c_3 - \left[\alpha_2 \cdot c_2 + \alpha_1 \cdot c_1 \right] \right] \cdot 10^{-5} \qquad Temp_3 := T^3 - 298^3 \cdot K^3$$

$$dd := \left[\alpha_4 \cdot d_4 + \alpha_3 \cdot d_3 - \left[\alpha_2 \cdot d_2 + \alpha_1 \cdot d_1 \right] \right] \cdot 10^{-9} \qquad Temp_4 := T^4 - 298^4 \cdot K^4$$

Standard heat of reaction calculation at 298 K and 1 atm:

$$dH_{298} := \left[\alpha_3 \cdot dHf_Y + \alpha_4 \cdot dHf_Z - \left[\alpha_1 \cdot dHf_A + \alpha_2 \cdot dHf_B \right] \right]$$

$$dH_{298'} := dH_{298} \cdot 1000 \cdot \frac{cal}{kcal}$$

Standard heat of reaction calculation at T K and 1 atm:

$$DA := da \cdot \frac{cal}{gmol \cdot K} \qquad DB := \frac{db}{2} \cdot \frac{cal}{gmol \cdot K^2} \qquad DC := \frac{dc}{3} \cdot \frac{cal}{gmol \cdot K^3} \qquad DD := \frac{dd}{4} \cdot \frac{cal}{gmol \cdot K^4}$$

$$dH_{temp} := dH_{298'} + DA \cdot Temp_1 + DB \cdot Temp_2 + DC \cdot Temp_3 + DD \cdot Temp_4$$

Figure 2.13 Setup in Mathcad to calculate the heat capacity of a gas. T is the temperature, the reaction stoichiometric coefficients are α_i, and the Roman letters are the coefficients in the heat capacity equation.

Symbolic Manipulation

Several software packages are available that originated as symbolic manipulators as contrasted with numerical calculators. Tedious, error-prone mathematical derivations can be eliminated by such programs, which apply the rules of algebra, trigonometry, calculus, and matrix algebra to solve a wide range of problems. After a formula is entered (using standard operators and functions), it can be simplified, plotted, expanded, approximated, factored, placed over a common denominator, integrated, or differentiated, all symbolically. In addition, equations

A = [34 − 1;1 − 22; − 101]; b = [832]′;
xcomp = A\b

xcomp =
    1.0000
    2.0000   **Figure 2.14**   Solution of a set of
    3.0000   three linear equations by Matlab.

and inequalities can be solved analytically or approximately, and matrices can be added, multiplied, transposed, or inverted.

Three well known packages are Mathematica,[15] Maple,[16] and Derive.[17] All have the following features; they

- Execute symbolic math from algebra through calculus.
- Plot in both 2-D and 3-D.
- Solve equations exactly (symbolically).
- Manipulate vectors and matrices.
- Perform arithmetic to thousands of digits.
- Simplify, factor, and expand expressions.
- Handle exponential, logarithmic, trigonometric, hyperbolic, and probability functions.
- Carry out Taylor and Fourier series approximations.
- Permit recursive and iterative programming.
- Can generate Fortran, Pascal, and Basis code.
- Display accepted math notation.

By way of example, Figure 2.15 shows how simple the solution of a quadratic equation in Mathematica can be.

LOOKING BACK

In this section we examined briefly some features of equation solvers, generic codes, spreadsheets, symbolic manipulators, and flowsheeting codes that can be used to solve material (and energy) balances.

[15]A trademark of Wolfram Research, Inc., P.O. Box 6059, Champaign, IL 61821.

[16]A trademark of Waterloo Maple Software, 160 Columbia St. West, Waterloo, Ont., Canada N2L 3L3.

[17]A trademark of Sot Warehouse, Inc., 3660 Waialae Ave., Honolulu, HI 96816.

In[1]: = Solve [x ∧ 2 + 4x + 1 = = 0, x]

$$Out[1]: = \left\{ \left\{ x \rightarrow \frac{-4 + 2\ \text{Sqrt}\ [3]}{2} \right\}, \left\{ x \rightarrow \frac{-4 - 2\ \text{Sqrt}\ [3]}{2} \right\} \right\}$$

In[2]: = N[%]

Out[2]: = {{x → −0.26795}, {x → −3.73205}}

Figure 2.15 The symbolic and numerical solution of the quadratic equation $x^2 + 4x + 1 = 0$ by Mathematica.

Key Ideas

1. Numerous easy-to-use inexpensive software packages exist that will assist you in solving chemical engineering problems.
2. Be careful in using such software to avoid solving the wrong problem or believing an incorrect solution is valid.

Key Words

Equation-solving (p. 121)

Interactive software (p. 126)

Flowsheeting codes (p. 124)

Spreadsheets (p. 123)

Generic codes (p. 122)

Symbolic manipulators (p. 128)

Self-Assessment Test

1. Solve one or two of the examples in Chapter 3 using
   (a) A personal computer-based equation-solving code
   (b) A code from the disk in the pocket in the back of this book
   (c) A code taken from your computer center library
   (d) A spreadsheet program
   (e) A flowsheeting code

2.3   SOURCES OF DATA

> ### Your objectives in studying this section are to be able to:
>
> 1. Become familiar with sources of physical properties data.

LOOKING AHEAD

In this section we point out sources of physical property data available in books, journals, CD disks, and on the Internet.

MAIN CONCEPTS

Accurate values of physical properties are needed in almost all phases of chemical engineering design and analysis. Various ways to obtain data for the physical properties of components are:

Employer's databases   Professional society meetings
Design software (such as   Trade association meetings
    flowsheeting codes)   Continuing education courses
On-line databases   Other engineers in department
On-line bulletin boards/e-mail   Outside consultants
Personal files and books   Regulatory agencies
Departmental library   Raw material/equipment vendors
Employer's main library   Clients/customers
Outside library   Direct experimentation
Technical magazines/newsletters

Much of the data is available at little or no cost, particularly over the Internet. You will be interested in using physical property databases in one of three ways:

1. To retrieve an isolated value to be used in a calculation or in the calculation of other property values. Often a value is to be employed in hand calculations, or perhaps fed as input data to a computer program for further calculation.
2. To serve as a subroutine (such as a physical properties library) to another computer program to provide physical property data for process calculations.
3. To provide interactive capabilities for the rapid rendering of physical properties of substances of interest for parametric studies of process units.

Many of the materials we talk about and use every day are not pure compounds, but you can nevertheless obtain information about the properties of such materials. Data on materials such as coal, coke, petroleum products, and natural gas—which are the main sources of energy in this country—are available in reference books and handbooks. Examine Tables 2.5 and 2.6. In the back of this book

TABLE 2.5 Sources of Physical Property Data

American Chemical Society, *TAPD*, 17,000 compounds, on disk, ACS, Washington, DC, 1994.

American Chemical Society, *Chemical Abstracts Service*. Washington, DC: ACS. (Continuing printed, microform, and on-line electronic information service with over 12 million abstracts.)

American Gas Association, *Fuel Flue Gases*. New York: AGA, 1941.

American Petroleum Institute, *Technical Data Book—Petroleum Refining*. New York: Author, 1970.

Beilstein On-line; properties of 3.5 million chemical compounds. Also CD-disk; ongoing.

Chemsoft Inc., *Chemical Compounds Data Bank*. Houston: Gulf Publishing Company, 1987. (50 compounds on disk.)

Design Institute for Physical Property Data, *DIPPR®*, gives properties and prediction equations for over 1,400 compounds; available as printed book, tape, on-line; Amer. Inst. Chemical Engineers, New York, ongoing.

Dechema, *Chemistry Data Series*, Deutsche Gesellschaft für Chemisches Apparatewesen e.v., Berlin, Germany. (Continuing series on physical and thermodynamic properties.) 10,000 compounds.

Engineering Sciences Data Unit Ltd., *International Data Series*, London. (Continuing series of data and equations.)

Environmental Chemicals Data Information Network, *Ecdin*, data on 25,000 substances, registration of 103,000. Distributed by Technical Database Services (TDS).

Handbook of Physics and Chemistry. Boca Raton, FL: CRC Press, annual editions.

Lange's Handbook of Chemistry and Physics. New York: McGraw-Hill, issued periodically.

Lin, C. T., et al., "Data Bank for Synthetic Fuels"; *Hydrocarbon Process*, (May 1980):229.

National Engineering Laboratories, *PPDS2*, 1,600 compounds; ongoing, Glasgow, UK.

Natural Gas Processors Suppliers Association, *Engineering Data Book*. Tulsa, OK: Author. (Continuing editions.)

Perry, R. H. and D. Green, *Chemical Engineers' Handbook*, 6th ed. New York: McGraw-Hill, 1980.

PTB Laboratory, *Chemsafe*. 1,600 gases, liquids, dusts, that can explode; distributed on disk, tape, on-line; Braunschweig, Germany, 1995.

Reid, R. C., J. M. Prausnitz, and B. D. Poling, *The Properties of Gases and Liquids,* 4th ed. New York: McGraw-Hill, 1987.

STN International, *STN Express*, provides access to numerous databases on-line, Chemical Abstracts Service, Columbus Ohio, continuing.

Thermodynamic Research Center, Texas A&M University, *TRC Vapor Pressure Data Profile* for 5,500 chemicals; on disk, 1994.

Yaws, C. L. *Physical Properties, a Guide to the Physical, Thermodynamic and Transport Property Data of Industrially Important Chemical Compounds*. New York: McGraw-Hill, 1987.

TABLE 2.6   Professional Journals

Archival journals
  AIChE Journal
  Angewandte Chemie
  Canadian Journal of Chemical Engineering
  Chemical Engineering Communications
  Chemical Engineering Fundamentals
  Chemical Engineering Journal (Lausanne)
  Chemical Engineering Research and Design
  Chemical Engineering Science
  Chemical Reviews
  CODATA Bulletin
  Journal of Chemical and Engineering Data
  Journal of Chemical Engineering of Japan
  Journal of Chemical Technology and Biotechnology
  Journal of the Chinese Institute of Chemical Engineers

Other journals and magazines
  Chemical Engineering
  Chemical Engineering Progress
  Chemical Engineer (London)
  Chemical Processing
  Chemical Technology
  Chemical Technology Review
  Chemie-Ingenieur-Technik
  Chemistry and Industry (London)
  International Chemical Engineering

you will discover a number of appendices from which you can retrieve data necessary to solve most (but not all) of the problems at the end of the chapters. Scan through the appendices now. Also in the back of this book is a disk on which are stored tables and equations that yield physical properties.

LOOKING BACK

In this section we listed sources of physical property data you might need for your calculations.

Key Ideas

1. An enormous amount of information is available from books, handbooks, journals, on-line data bases, floppy disks, and CD-disks, much of it free.

2. In addition to data needed for the solution of problems in this book, you should be able

to search for and locate reliable information distributed by specialized professional or-
ganizations on a myriad of subjects, such as refrigerants, propellants, solvents, drugs,
insecticides, and so on.

Self-Assessment Test

1. What are five sources of data on physical properties from reference books? What are
two data banks that provide information on physical properties?

2. In what reference book might you find data on:
   (a) Boiling point of inorganic liquids?
   (b) Gas compositions for refinery gases?
   (c) Vapor pressures of organic liquids?
   (d) Chemical formula and properties of protocatechuic acid (3−, 4−)?
   List the page numbers on which the items described in question 2 can be found in
   the reference you select.

SUPPLEMENTARY REFERENCES

Problem Solving

BARAT, R. B. and N. ELLIOT. *The Complete Chemical Engineer: A Guide to Critical
Thinking.* Dubuque, IA: Kendall/Hunt, 1993.

BOYCE, A. J. "Teaching Engineering as the Science of Solving Word Problems," in *Proceed. 1991 ASEE Conf.,* (p. 1267), ASEE, 1991.

EIDE, A. R., et al., eds. *Engineering Fundamentals and Problem Solving,* 2nd ed. New
York: McGraw-Hill, 1986.

FRENSCH, P. A. and J. FUNKE. *Complex Problem Solving.* Hillsdale, NJ: Lawrence Erl-
baum, 1995.

LARSON, L. C. *Problem-Solving through Problems.* New York: Springer-Verlag, 1993.

LUMSDAINE, E. and M. LUMSDAINE. *Creative Problem Solving: An Introductory Course
for Engineering Students.* New York: McGraw-Hill, 1990.

RICKARDS, T. *Creativity and Problem Solving at Work.* Aldershot, U.K.: Gower, 1990.

RUBINSTEIN, M. F. and I. R. FIRSTENBERG. *Patterns of Problem Solving,* 2nd ed. Englewood Cliffs, NJ: Prentice-Hall, 1994.

SCARL, D. *How to Solve Problems,* 4th ed. Glen Cove, NY: Desoris, 1994.

WOODS, D. R. *Problem-Based Learning: How to Gain the Most from PBL.* Waterdown,
Ontario, Canada: Author, 1994.

Equation-Solving Codes

Borland International. *Eureka: The Solver.* Scotts Valley, CA: Author, 1987.

HUGHSON, R. V. *Chem. Eng.* (October 12, 1987): 123.

Generic Software

American Institute of Chemical Engineers. *Applications Software Survey for Personal Computers.* New York: Author, 1995.

ANDERSON, E., et al. *LAPACK: Users' Guide*, Philadelphia: SIAM, 1992.

CAE Consultants. *Chemical Engineering Software Guide*, Research Triangle Park, NC: LEDS Publishing Company.

GANAPATHY, V. *Basic Programs for Steam Plant Engineers.* New York: Marcel Dekker, 1986.

GRANDINE, T. A. *The Numerical Methods Programming Projects Book.* Oxford, England: Oxford University Press, 1990.

HOPKINS, T., and C. PHILLIPS. *Numerical Methods in Practice: Using the NAG Library.* Oxford, England: Addison-Wesley Publishers Ltd., 1988.

LIEBMAN, J., et al. *GINO.* Palo Alto, CA: Scientific Press, 1986.

WRIGHT, D. *Basic Programs for Chemical Engineers.* New York: Van Nostrand Reinhold, 1986.

Spreadsheets

JULIAN, F. M. "Flowsheets and Spreadsheets," *Chem. Eng. Progress*, **81** (1985): 35.

O'LEARY, T. J., et al. *Lotus 1-2-3,* New York: McGraw-Hill, 1995.

O'LEARY, T. J. *Quattro Pro 6.0*, New York: McGraw-Hill, 1995.

PITTER, K. *Introducing Microsoft Excel 5.0.* New York: McGraw-Hill, 1994.

SCHMIDT, W. P. and R. S. UPADHYE. "Material Balances on a Spreadsheet": *Chem. Engr.*, (Dec. 24, 1984): 67.

SCHUMAN, J. *Using Microsoft Excel 4.0 for Windows*, New York: McGraw-Hill, 1994.

SMITH, L. *First Look at Lotus 1-2-3 Release 4 for Windows*, New York: McGraw-Hill, 1994.

STIPES, J., and T. TRAINOR. *Introducing Quattro Pro 5.0 for Windows.* New York: McGraw-Hill, 1994.

WERSTLER, D. "Spreadsheet versus C Programming Solutions to Materials Problems." In *Proceed. 1991 ASEE Conf.* (p. 1126) ASEE, 1991.

Flowsheeting Codes

CLARK, S. M., and G. V. REKLAITIS. *Comput., Chem. Eng.*, **8** (1984): 205.

HUTCHINSON, H. P., D. J. JACKSON, and W. MORTON. "Equation Oriented Flowsheet Simulation, Design and Optimization." *Proc. Eur. Fed. Che. Eng. Conf. Comput., Appl. Chem. Eng.* Paris, April 1983; "The Development of an Equation-Oriented Flowsheet Simulation and Optimization Package." *Comput. Chem. Eng.*, **10** (1986): 19.

WESTERBERG, A. W., and H. H. CHIEN. *Comput. Chem. Eng.*, **9** (1986): 517.

WESTERBERG, A. W., H. P. HUTCHINSON, R. L. MOTARD, and P. WINTER. *Process Flowsheeting.* Cambridge: Cambridge University Press, 1979.

Also refer to Section 6.2.

Interactive Software

HIGHAM, N. J. *"MATLAB: A Tool for Teaching and Research." Mathematics and Statistics Newsletter of Computers in Teaching Initiative,* **1** (1990): 4.

Mathsoft, Inc. *Matcad 4.0: User's Guide,* 1993.

Symbolic Manipulators

CHAR, B. W. et al. MAPLE User's Guide, 4th ed., Waterloo, Ontario: WATCOM Publications Ltd., 1985.

Softwarehouse. *Derive User Manual, 4th ed.,* Honolulu: Soft Warehouse, Inc., 1993.

UHL, J. J. *"Mathematica and Me" Notices of the AMS,* **35** (1988): 1345.

WOLFRAM, S., *Mathematica—A System for Doing Mathematics by Computer,* Redwood City, CA: Addison-Wesley, 1988.

PROBLEMS

Section 2.1

2.1. I have always checked the condition of my battery by checking the specific gravity with a hydrometer. I recently purchased an Exide battery. The specific gravity of the battery stays about 1.225 whether the battery is fully charged or completely discharged. How is this possible?

2.2. Which is the greater amount, 1 dozen eggs, 6 watermelons, or 3 bars of gold?

2.3. What do Catherine the Great, Attila the Hun, and Eric the Red have in common?

2.4. Two problems that are posed alike can really be quite different, for example:
   1. It takes 1 man 5 days to dig a ditch. How long does it take 5 men to dig the ditch?
   2. It takes 1 ship 5 days to cross an ocean. How long does it take 5 ships to cross the ocean?

   These two problems are constructed exactly alike: Just substitute *ship* for *man* and *cross the ocean* for *dig a ditch.* Why then do the answers differ?

2.5. You have 64 meters of fence. What shape of dog pen should you construct for your dog?

2.6. Two phototimers are used to measure the time it takes for a ball to roll across a table (see Figure P2.6). If the timers are inaccurate and the meter stick used to measure the height of the table is inaccurate, how will the prediction of the location on the floor where ball hits be changed from that made with accurate instruments and ruler?

2.7. One day in Chicago when the air temperature was in the high 90s, a truck containing morpholine was in an accident and sprang a leak. The fire chief of the Chicago suburb involved appeared on the 10:00 o'clock news and explained that the air temperature that day was almost 100°F, and that 100°F is the flash point of mor-

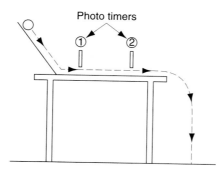

Figure P2.6

pholine. He then went on to explain that when the flash point of morphline is reached, the morpholoine will explode. He further said that it was very irresponsible for materials with such low flash points to be shipped on a hot day. Is the chief correct?

2.8.   **The Russian Vase**

Bursting through the double doors of the hotel kitchen, Kim Matthews leveled her gun at Philip Jacobs. Whipping away from the industrial stovetop to face Kim, Philip's apparent panic faded into a sinister smile.

"You'd like to arrest me, wouldn't you?" Philip looked around at his surroundings and then back at Kim. "But, whatever for?"

"For . . ." Kim began, but was interrupted by Detective Barry Stone, coming through doors behind Kim, "If he doesn't have the vase, you can't arrest him."

"That's right," Jacobs said, sauntering past Kim and out of the kitchen.

"I just don't get it, Barry," Kim said, "I saw him steal the crystal vase out of the Russian ambassador's exhibit in the lobby, and then I chased him in here. The vase wasn't that large, but it couldn't have been hidden that easily, that fast," Kim said, motioning to the kitchen, which was cluttered with the typical pots and utensils used in the hotel food industry. A butcher's block covered with fresh vegetables spanned the length of the right wall, blending into the stove top range with its large double doored oven, seared grill top, and a large bucket of cooking oil at the foot of the oven doors. Numerous deep sinks and counters used for washing dishes covered the left side of the kitchen, and in the middle stood the typical island cluttered with various knifes and other utensils.

"Who is this Jacobs, anyway?" Kim asked Detective Stone.

"Strangely enough, he's some optics professor from a local university who just cracked one day. Anyhow, I'll have my men search his place from top to bottom. In the mean time why don't you keep an eye on Jacobs."

"Actually, Barry, I think you'll have to do that. I'll go and arrest Jacobs, this time for real."

Why did Kim Matthews decide to arrest Jacobs?

2.9.   **The Laboratory Fire**

Kim Matthews waded through the sea of firefighters and policemen to reach Detective Barry Stone who was standing in cleared alcove next to the main counter in what remained of a chemistry laboratory.

"Kim, I'm glad you could come," Detective Stone said gesturing to the condition of the laboratory. "Obviously there has been very high temperature fire that melted steel and crumbled the concrete. From what the fireman can tell me, it seems to have been caused by a gas leak in the gas line to the hood. The gas line looks as though it got run into by a cart too many times, and due to the age of the line, it most likely just cracked easily. Anything could've caused the spark. During the fire, orange smoke billowed out of the lab, and it was impossible to put the fire out with the sprinkler system.

"Whose lab is this?" Kim asked.

"It used to belong to a Professor Bob Koker, and from the reaction of some passing students, he wasn't too popular. Here he comes now."

"All my work, gone," Koker began. "I can't believe my life's work has been destroyed by an insignificant gas leak," Koker exclaimed. "Now I will have to spend more of my time teaching those sniveling students!" Turning on his heel, Professor Koker took off towards the door of the lab.

"In his own world, I guess. Poor guy," Stone said while reaching for his notepad on the counter. "Ugh! What is this?" Stone exclaimed, noting the white powdery dust picked up on his notebook from the white dust covering the lab bench.

Turning, Kim scanned the floor by the bench, which she now observed was covered with the white powder.

With a sinister gleam in her eye, Kim said, "This explosion wasn't an accident, Stone, of that I'm sure. Let me have the dust analyzed." (The dust proved to be a mixture of aluminum oxide and ammonium nitrate.)

Later that week Matthews told Barry to start interrogating Koker's students. What made Matthews so sure that the explosion was not an accident?

2.10. Why does popcorn pop? Review the possibilities and carry out experimental observations to test hypotheses.

2.11. How can leaks from a gas pipeline be detected in practice?

2.12. One effect of potential global warming is the acceleration of the decomposition of organic material in the soil. How can you predict the rate of decay of organic matter in the soil?

Section 2.2

2.13. Solve Problem 6.28 using one or more of the computer-based tools discussed in Section 2.2. Change one or more of the input flows, and recalculate. Repeat for changes in the output flows. Repeat for one or more of the prespecified values in the diagram. Determine how sensitive one or more of the output variables are to changes of ±5% in one or more of the input variables.

2.14. Take any of the worked-out examples in the text, and resolve using several of the classes of computer codes mentioned in the chapter. Prepare a report comparing the relative difficulties of using the particular codes and the accuracy of the solution(s).

2.15. Find the solutions of the following equations by one of the computer codes on the disk in the back of this book.

a.
$$2x_1 + x_2 - 2x_3 = 0$$
$$3x_1 + 2x_2 + 2x_3 = 1$$
$$5x_1 + 4x_2 + 3x_3 = 4$$

b.
$$x_1 + x_2 - x_3 = 0$$
$$2x_1 + 4x_2 - x_3 = 0$$
$$3x_1 + 2x_2 + 2x_3 = 0$$

c.
$$x_1 + 2x_2 - 3x_3 = 4$$
$$x_1 + 3x_2 + x_3 = 11$$
$$2x_1 + 5x_2 - 4x_3 = 13$$
$$2x_1 + 6x_2 + 2x_3 = 22$$

d.
$$-2x_1 + 5x_2 + 7x_3 = 6$$
$$-x_1 + x_2 - 2x_3 = 1$$
$$x_1 + 2x_2 + x_3 = 3$$

Section 2.3

2.16. How would you estimate the rate at which a spherical water-filled capsule in outer space cools? List the types of information needed to solve such a problem, the assumptions that must be made, and what physical principles that you have studied in physics might be used to help solve the problem. Draw a picture of the process and indicate what the independent and dependent variables involved in the process might be. Where would you get the information from?

2.17. Visit your library and carry out data searches for the following information. Report the values or equations found and their respective sources in complete detail: names of authors, article name, book or journal name, volume (if applicable), page number, and date of publication.
(a) Density of lead thiocynate [$Pb(CNS)_2$] at 20°C
(b) Boiling point in °C of glyceryl tributyrate [(C2H5CH2CO3) C3H5] at atmospheric pressure
(c) Solubility of ammonium oxalate [$(NH) C O$] at 40°C in water
(d) Volume of methyl choride gas at 200°F and 6 psia

2.18. Does Perry's *Chemical Engineers' Handbook* contain information on densities of alcohol-water mixtures? osmotic pressure of sodium chloride solutions? corrosion properties of metals?

2.19. List four sources of physical property data and how they might be accessed on the Internet.

2.20. Prepare a short description (about five lines) for numbers _____ of the following terms. You may include some sketches if you want. List the name of the author, book, home page, CD disk, etc. that you use to prepare your report, and cite the reference in proper form.

(1) Furnace
(2) Boiler
(3) Heat exchanger
(4) Distillation column
(5) Absorption tower
(6) Adsorption
(7) Liquid-liquid extraction
(8) Leaching process
(9) Evaporator
(10) Chemical reactor

MATERIAL BALANCES

3

In this chapter we begin to study the concept of the **material balance.** Material balances are nothing more than the application of the conservation law for mass: "Matter is neither created nor destroyed." Just what this statement means in practice and how you can use the concept to solve problems of varying degrees of complexity requires some fairly extensive explanation.

Why study material balances as a separate topic? You will find that material balance calculations are almost invariably a prerequisite to all other calculations in the solution of both simple and complex chemical engineering problems. Furthermore, skills that you develop in analyzing material balances are easily transferred to other types of balances and other types of problems.

In approaching the solution of material balance problems, the first step is to consider how to analyze them in order to clarify the method and the procedure of solution. The aim will be to help you acquire a generalized approach to problem solving so that you may avoid looking upon each new problem, unit operation, or process as entirely new and unrelated to anything you have seen before. As you scrutinize the examples used to illustrate the principles involved in each section, explore the method of analysis, but avoid memorizing each example by rote, because, after all, they are only samples of the myriad of problems that exist or could be devised on the subject of material balances. Most of the principles we consider are of about the same degree of complexity as the law of compensation devised by some unknown, self-made philosopher who said: "Things are generally made even somewhere or some place. Rain always is followed by a dry spell, and dry weather follows rain. I have found it an invariable rule that when a man has one short leg, the other is always longer!"

In working these problems you will find it necessary to employ some engineering judgment. You think of mathematics as an exact science. For instance, suppose that it takes 1 man 10 days to build a brick wall; then 10 men can finish it in 1 day. Therefore, 240 men can finish the wall in 1 hr, 14,400 can do the job in 1 min, and with 864,000 men the wall will be up before a single brick is in place! Your password to success is to use some common sense in problem solving.

3.1 THE MATERIAL BALANCE

> ### Your objectives in studying this section are to be able to:
>
> 1. Define a system and draw the system boundaries for which the material balance is to be made.
> 2. Explain the difference between an open and a closed system.
> 3. Write the general material balance in words including all terms. Be able to apply the balance to simple problems.
> 4. Cite examples of processes in which no accumulation takes place; no generation or consumption takes place; no mass flow in and out takes place.
> 5. Explain the circumstances in which the mass of a compound entering the system equals the mass of the compound leaving the system. Repeat for moles.

LOOKING AHEAD

In this section we define a few terms and explain what the material balance is in words and some of its implications.

MAIN CONCEPTS

To make a material balance (or an energy balance as discussed in Chapter 5) for a process, you first need to specify what the system is for which you are making the balance, and outline its boundaries. According to the dictionary, a **process** is one or a series of actions or operations or treatments that result in an end [product]. Chemical engineering focuses on operations such as chemical reactions, fluid transport, size reduction and enlargement, heat generation and transport, distillation, gas absorption, bioreactors, and so on that cause physical and chemical change in materials.

The examples we use in this book often are based on abstractions of these processes to avoid confusing details. By **system** we mean any arbitrary portion or whole of a process set out specifically for analysis. Figure 3.1 shows a system in which flow and reaction take place; note particularly that the **system boundary** is formally circumscribed about the process itself to call attention to the importance of carefully delineating the system for each problem on which you work. An **open (or flow) system** is one in which material is transferred across the system boundary, that is, enters the system, leaves the system, or both. A **closed (or batch) system** is one in which there is no such transfer *during the time interval of interest.* Obviously, if you charge a reactor with reactants and take out the products, and the reactor is designated as the system, material is transferred across the system boundary. But you might ignore the transfer, and focus attention solely on the process of reaction that takes place only after charging is completed and before the products are withdrawn. Such a process would occur within a closed system.

A system boundary may be fixed with respect to the process equipment as in Figure 3.1, or the boundary may be an imaginary surface that grows or shrinks as the process goes on. Think of a tube of toothpaste that is squeezed. A fixed boundary might be the tube itself, in which case mass crosses the boundary as you squeeze the tube. Or, you can imagine a flexible boundary surrounding the toothpaste itself that follows the extruded toothpaste, in which case no mass crosses the boundary.

A **material balance** is nothing more than an accounting for material. Material balances are often compared to the balancing of checking accounts. Money is deposited and withdrawn, and the difference between the ending and beginning balances represents the accumulation (or depletion!) in the account.

Figure 3.2 illustrates a general system for which a material balance is to be made. Eq. (3.1) states the concept of the material balance in words.

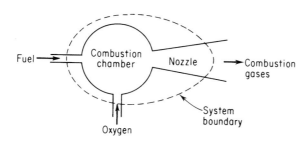

Figure 3.1   Flow (open) system defined by the dashed line.

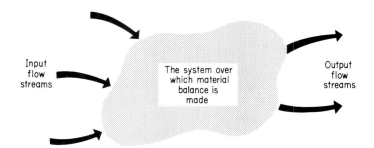

Figure 3.2   The figure illustrates a specified volume system, over which material balances are to be made. We are not concerned with the internal details, but only with the passage of material across the volume boundaries and the overall change of material inside the system.

$$\begin{Bmatrix} \textbf{accumulation} \\ \textbf{within} \\ \textbf{the} \\ \textbf{system} \end{Bmatrix} = \begin{Bmatrix} \textbf{input} \\ \textbf{through} \\ \textbf{system} \\ \textbf{boundaries} \end{Bmatrix} - \begin{Bmatrix} \textbf{output} \\ \textbf{through} \\ \textbf{system} \\ \textbf{boundaries} \end{Bmatrix}$$

$$+ \begin{Bmatrix} \textbf{generation} \\ \textbf{within} \\ \textbf{the} \\ \textbf{system} \end{Bmatrix} - \begin{Bmatrix} \textbf{consumption} \\ \textbf{within} \\ \textbf{the} \\ \textbf{system} \end{Bmatrix}$$

(3.1)

First, let us inquire as to what is balanced, that is, for what can the equation be applied? As a generic term, material balance can refer to a balance on a system for the

1. Total mass
2. Total moles
3. Mass of a chemical compound
4. Mass of an atomic species
5. Moles of a chemical compound
6. Moles of an atomic species
7. Volume (possibly)

Equation (3.1) applies to the first six categories. Why not number 7? Because the equation is based on the conservation of mass, and if the densities of the materials entering into each term are not the same, or mixing effects occur, then the volumes of the materials will not balance. Think of dissolving one liter of alcohol in one liter of water. Do you get two liters of the resulting solution?

Next, let us look at the meaning of each of the first three terms in Eq. (3.1). (We will defer consideration of the generation and consumption terms until Sec. 3.4.) In Eq. (3.1) the **accumulation** term refers to a change in mass or moles (plus or minus) within the system with respect to time, whereas the **transfer through the system boundaries** refers to inputs to and outputs of the system. Examine Figure 3.3.

Finally, we have to consider the time period for which the balance applies. When formulated for an instant of time, Eq. (3.1) is a differential equation. Consider for example the mass balance for water in Figure 3.4:

$$\frac{dm_{H_2O, \text{ within system}}}{dt} = \dot{m}_{H_2O, \text{ in}} - \dot{m}_{H_2O, \text{ out } 1} - \dot{m}_{H_2O, \text{ out } 2} \qquad (3.2)$$

where m_{H2O} denotes the mass of water according to the subscript, and $\dot{m}_{H_2O}$ denotes the mass flow rate of water (mass/time). Problems formulated as differential equations with respect to time represent **unsteady-state** (or transient) problems and are discussed in detail in Chapter 7. In contrast, in **steady-state** problems the values of the variables within the system do not change with time, hence the accumulation terms in Eqs. (3.1) and (3.2) are zero by definition.

In this book, except in Chapter 7, for convenience in treatment, we use an **integral balance** form of Eq. (3.1). What we do is take as a basis a time period such as one hour or minute, and integrate Eq. (3.1) with respect to time. For ex-

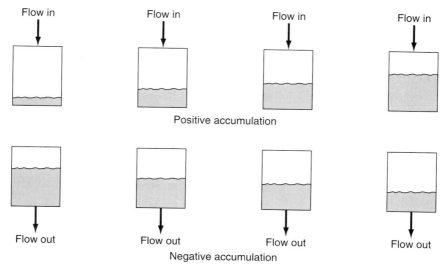

Figure 3.3 Pictorial representation of the terms for accumulation and flow through the system boundaries in the material balance.

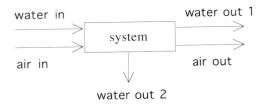

Figure 3.4   Process for a simple mass balance.

ample, for the specific application of Eq. (3.1) given as Eq. (3.2), the derivative (the left-hand side) in the differential equation when integrated becomes

$$\int_{t_1}^{t_2} \frac{dm_{H_2O, \text{ within system}}}{dt} \, dt = \int_{t_1}^{t_2} dm_{H_2O, \text{ within system}} = m_{H_2O}\Big|_{t_2} - m_{H_2O}\Big|_{t_1} = \Delta m_{H_2O}$$

where Δm is the difference between the water within the system at t_2 less that in the system at t_1. Integration of each of the terms in the right-hand side of Eq. (3.2) eliminates the rate such as for the first term

$$\int_{t_1}^{t_2} \dot{m}_{H_2O, \text{ in}} \, dt = m_{H_2O, \text{ in}}$$

where $m_{H_2O, \text{ in}}$ represents the entire quantity of water introduced into the system between t_1 and t_2. If the flow rate of H_2O into the system shown in Figure 3.4 is constant at the rate of 1200 kg/hr, by choosing a basis of one hour

$$\int_0^1 \frac{1200 \text{ kg}}{\text{hr}} \, \bigg| \, dt \text{ hr} = 1200 \text{ kg}(1 - 0) = 1200 \text{ kg} = m_{H_2O, \text{ in}}$$

Most, but not all, of the problems discussed in this chapter are steady-state problems. If no accumulation occurs in a problem, and the generation and consumption terms are omitted from consideration, the material balance reduces to a very simple relation for one compound or for the total material, a relation that can be stated in words briefly as: "What comes in must go out," or

$$\left\{ \begin{array}{c} \textbf{mass / mole input} \\ \textbf{through the} \\ \textbf{system boundaries} \end{array} \right\} = \left\{ \begin{array}{c} \textbf{mass / mole output} \\ \textbf{through the} \\ \textbf{system boundaries} \end{array} \right\} \qquad (3.3)$$

Material balances can be made for a wide variety of materials, for different sizes of systems, and in various degrees of complication. To obtain a perspective as to the scope of material balances, examine Figure 3.5, which shows a **flow-sheet** for a chemical plant that includes both mass and energy flows.

We now look at a simple example of the application of Eq. (3.1).

Process mass and energy balance

Basis: 2,000 lb. Phenol @ 60F

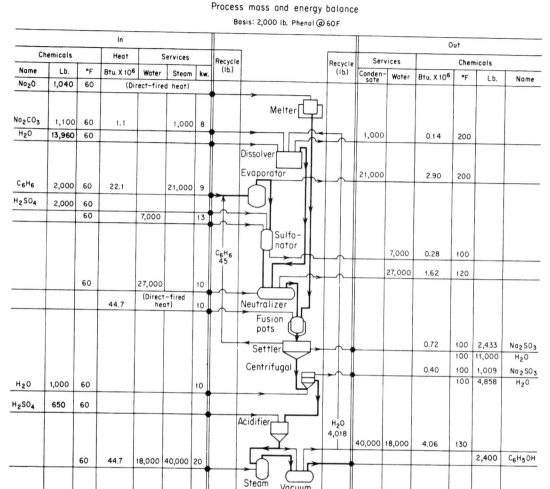

Name	Lb.	°F	Btu. X 10⁶	Water	Steam	kw.	Recycle (lb.)		Recycle (lb.)	Conden-sate	Water	Btu. X 10⁶	°F	Lb.	Name
Na₂O	1,040	60	(Direct–fired heat)												
Na₂CO₃	1,100	60	1.1		1,000	8									
H₂O	13,960	60								1,000		0.14	200		
										21,000		2.90	200		
C₆H₆	2,000	60	22.1		21,000	9									
H₂SO₄	2,000	60													
		60		7,000		13									
											7,000	0.28	100		
							C₆H₆ 45				27,000	1.62	120		
		60		27,000		10									
			44.7	(Direct–fired heat)		10									
												0.72	100	2,433	Na₂SO₃
													100	11,000	H₂O
												0.40	100	1,009	Na₂SO₃
H₂O	1,000	60				10							100	4,858	H₂O
H₂SO₄	650	60							H₂O 4,018						
										40,000	18,000	4.06	130		
		60	44.7	18,000	40,000	20								2,400	C₆H₅OH
Totals	21,700		114.6	52,000	62,000	80				62,000	52,000	10.12		21,700	Totals

Figure 3.5 Material (and energy) balances in the manufacture of phenol presented in the form of a ledger sheet. [Taken from *Chem. Eng.* (April 1961): 117 by permission.]

EXAMPLE 3.1 Total Mass Balance

A thickener in a waste disposal unit of a plant removes water from wet sewage sludge as shown in Figure E3.1. How many kilograms of water leave the thickener per 100 kg of wet sludge that enter the thickener? The process is in the steady state.

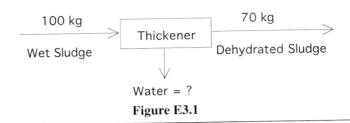

Figure E3.1

Solution

Basis: 100 kg wet sludge

The system is the thickener (an open system). No accumulation, generation, or consumption occur. The total mass balance is

$$\frac{In}{100 \text{ kg}} = \frac{Out}{70 \text{ kg} + \text{kg of water}}$$

Consequently, the water amounts to 30 kg.

EXAMPLE 3.2 Mass Balance

Silicon rods used in the manufacture of chips can be prepared by the Czochralski (LEC) process in which a cylinder of rotating silicon is slowly drawn from a heated bath. Examine Fig. E3.2. If the initial bath contains 62 kg of silicon, and a cylindrical ingot 17.5 cm in diameter is to be removed slowly from the melt at the rate of 3 mm per minute, how long will it take to remove one-half of the silicon? What is the accumulation of silicon in the melt?

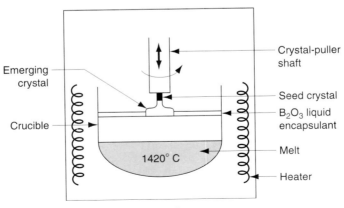

Figure E3.2

Solution

The density of the crystalline silicon in the cylinder is 2.4 g/cm³.

Basis: 62 kg silicon

The system is the melt, and there is no generation or consumption. Let Δm_t be the accumulation.

$$
\frac{Accumulation}{\Delta m_t} \quad = \quad \frac{Input}{0} \quad - \quad \frac{Output}{0.5(62 \text{ kg})}
$$

$$
\Delta m_t = -31 \text{ kg}
$$

Let t be the time in minutes to remove one-half of the silicon.

$$
\frac{2.4 \text{ g}}{\text{cm}^3} \left| \frac{\pi(17.5 \text{ cm})^2}{4} \right| \frac{0.3 \text{ cm}}{\text{min}} \left| \frac{t \text{ min}}{} \right. = \frac{1}{2}(62,000 \text{ g})
$$

$$
t = 179 \text{ min}
$$

LOOKING BACK

In this section we examined the meaning of the terms in the material balance, and explained what can be balanced. We also explained that the balances we used in this chapter are integral balances, that is, each term in the balance represents material flow or inventory change over a period of time.

Key Ideas

1. In a material balance, the words in Eq. (3.1) have to be translated into mathematical symbols and numerical values appropriate for each specific problem.
2. However, all material balances involve the same terms as in Eq. (3.1).
3. In most problems, one or more terms in Eq. (3.1) will be zero and are not involved in the problem solution.
4. A system and boundary must be specified for each problem.
5. Unsteady state processes involve accumulation; steady state processes do not.

Key Terms

Accumulation (p. 145)   Flowsheet (p. 146)
Closed system (p. 143)   Generation (p. 144)
Consumption (p. 144)   Input (p. 145)

Self-Assessment Test

1. Draw a sketch of the following processes and place a dashed line around the system:
   (a) Tea kettle
   (b) Fireplace
   (c) Swimming pool

2. Label the materials entering and leaving the systems in problem 1. Designate the time interval of reference and classify each system as open or closed.

3. Write down the general material balance in words. Simplify it for each process in problem 1 above, stating the assumptions made in each simplification.

4. Classify the following processes as (1) batch, (2) flow, (3) neither, or (4) both on a time scale of one day:
   (a) Oil storage tank at a refinery
   (b) Flush tank on a toilet
   (c) Catalytic converter on an automobile
   (d) Gas furnace in a home

5. What is a steady-state process?

6. Do the inputs and outputs of the chemicals in Figure 3.5 agree? Why not?

7. Define a material balance. A mass balance.

Thought Problems

1. Examine Figure TP3.1–1. A piece of paper is put into the bell in (1). In picture (2) we set fire to the paper. Ashes are left in (3). If everything has been weighed (the bell, the dish and the substances) in each case, we observe that:
   (a) Case 1 would have the larger weight.
   (b) Case 2 would have the larger weight.

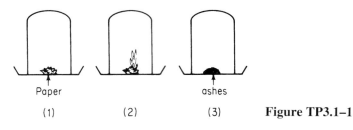

Paper   ashes

(1)   (2)   (3)   **Figure TP3.1–1**

(c) Case 3 would have the larger weight.

(d) None of the above.

Select your answer and explain it.

2. Certain critical processes require a minimum fluid flow for safe operation during the emergency shutdown of a plant. For example, during normal operation in one process, the chlorine is removed safely from the processing unit along with the flowing liquids. But during an emergency shutdown, the chlorine collects in the unit and its pipeline headers. Hence a minimum flow rate is needed to remove the chlorine. If the unit and pipelines are considered to one system, how can a minimum flow rate be obtained for safe operation if the electric power and controller fail?

Discussion Questions

1. Isotope markers in compounds are used to identify the source of environmental pollutants, investigate leaks in underground tanks and pipelines, and trace the theft of oil and other liquid products. Both radioactive and isotopic markers are used. Deuterium is typically used as a marker for organic compounds, by replacing three or more hydrogen atoms on the molecule in a reactor containing heavy water. However, isotopes of carbon and oxygen can also be used. The detection limit of tracers using a combination of gas chromatography and mass spectrometry is about 100 ppb in crude oil and about 20 ppb in refined products. Explain now such markers can be used in chemical processes.

2. Projects suggested to avoid climatic changes engendered by man's activities and in particular the increase in CO_2 in the atmosphere include dispersal of sulfate particles in the stratosphere to reflect sunlight and fertilizing the southern oceans with iron to stimulate phytoplankton growth. It is believed that low levels of iron limit the biological productivity of nutrient-rich southern oceans. Adding iron to these waters would increase the growth of phytoplankton, thus reducing CO_2 levels in the seawater and thereby altering the CO_2 balance between seawater and the atmosphere. Is this a realistic proposal from a scientific viewpoint? From a technological viewpoint? From a biological viewpoint?

3.2 PROGRAM OF ANALYSIS OF MATERIAL BALANCE PROBLEMS

> ### Your objectives in studying this section are to be able to:
>
> 1. Define what the term "solution of a material balance problem" means.
> 2. Ascertain that a unique solution exists for a problem using the given

data, and/or ascertain the number of degrees of freedom in a problem so that additional information can be obtained (and get it).

3. Decide which equations to use if you have redundant equations.

4. Solve a set of n independent equations containing n variables whose values are unknown.

5. Retain in memory and recall as needed the implicit constraints in a problem.

6. Prepare material flow diagrams from word problems.

7. Translate word problems and the associated diagrams into material balances with properly defined symbols for the unknown variables and consistent units for processes without chemical reaction.

8. Recite the ten steps used to analyze material balance problems so that you have an organized strategy for solving material balance problems.

LOOKING AHEAD

This section presents a logical methodology you can use to solve material balance problems. Descartes summed up the matter more than three centuries ago, when he wrote in his "Discours de la Methode": "Ce n'est pas assez d'avoir l'esprit bon, mais le principal est de l'appliquer bien."   In English: "It is not enough to have a good intelligence—the principal thing is to apply it well." We are going to discuss a strategy of analysis for material balance problems that will enable you to understand, first, how similar these problems are, and second, how to solve them in the most expeditious manner. For some types of problems the method of approach is relatively simple and for others it is more complicated. However, the important point is to regard problems in distillation, crystallization, evaporation, combustion, mixing, gas absorption, or drying not as being different from each other but as being related from the viewpoint of how to proceed to solve them.

MAIN CONCEPTS

An orderly method of analyzing problems and presenting their solutions represents training in logical thinking that is of considerably greater value than mere knowledge of how to solve a particular type of problem. Understanding how to approach these problems from a logical viewpoint will help you to develop those fundamentals of thinking that will assist you in your work as an engineer long after you have read this material. But keep in mind the old Chinese proverb:

None of the secrets of success will work unless you do.

First let us ask: "What does a solution to a material balance problem mean?" We are really interested in finding a **unique solution** because two or more solutions to a problem are almost as unsatisfactory as no solution at all. Appendix L summarizes the various conditions for no solution, a unique solution, and multiple solutions to exist for a set of linear equations. The essential principle in formulating a set of linear material balance equations is to make sure that the following necessary condition is satisfied:

> **The number of variables whose values are unknown equals the number of independent equations.**

The sufficient conditions are explained in Appendix L. There is usually no point in beginning to solve a set of material balances unless you can be certain that the equations have a unique solution.

You can easily count the number of variables in a material balance problem whose values are unknown. By experience you can formulate material balances so that the set of balances is comprised of independent equations. Appendix L explains how to calculate the rank of the coefficient matrix of a set of linear equations—**the rank equals the number of independent equations. Usually,** *but not always*, **in absence of chemical reactions the number of independent material balances equals the number of components involved in the process.**

It would be very convenient if we could make some statement about the existence and uniqueness of a solution for a problem that involves nonlinear equations. It is unlikely that there will ever be theorems for the case of nonlinear equations as exist for sets of linear equations. What we do know about the existence of a solution of a specific nonlinear problem rests on practical issues—the solution exists because of the physics of the process.

EXAMPLE 3.3 Determination of the Number of Independent Equations

One way to determine the number of independent linear equations in a problem is to determine the rank of the coefficient matrix for the set of equations. What is the rank of the following coefficient matrices? (The rank is equal to the order of the largest nonzero determinant in the matrix—see Appendix L for details.)

(a) $\begin{bmatrix} 4 & 0 & 0 \\ 0 & 5 & 0 \end{bmatrix}$

(2 equations, 3 variables)

(b) $\begin{bmatrix} 2 & 1 & 1 \\ 4 & 2 & 2 \\ 8 & 4 & 4 \\ -2 & -1 & -1 \end{bmatrix}$

(4 equations, 3 variables)

Solution

(a) The rank is 2 by inspection because the largest nonzero determinant of (a) is of order 2.

(b) The rank of (b) is only one! Try to form a nonzero determinant of order 2; or order 3. What do you get?

In this book we will ignore the uncertainty that exists in all process measurements. But you should be aware that repeated measurements of a variable will not necessarily yield the same value on each measurement. Consequently, the values of the variables and coefficients in the material balances made in practice are frequently only approximate, a material balance may not balance exactly, and the test for the rank of the coefficient matrix of a set of equations may not yield a clearcut answer.

In making steady-state mass (or mole) balances for a single component in a mixture, although the flows involved in the process are mass (or mole) flows of the component, two common ways exist to express these flows as variables. One is to just designate the mass flow by a symbol such as m_i, where the subscript i identifies the specific component. The other way is to use the product of a measure of concentration times the flow such as $\omega_i F$, where ω_i is the mass fraction of component i in F, and F is the total mass flow. Analogously, in representing moles, the symbol might be n_i which is equal to $x_i G$, where x_i would be the mole fraction and G the total number of moles. In any mixture of N components, N so called **stream variables** exist, either N values of m_i or n_i, as the case may be, or their equivalents, namely (N-1) values of ω_i (or x_i) plus the stream flow itself, F (or G).

Why (N-1) mass fractions rather than N? **Remember that implicit constraints (equations) exist in the problem formulation because of the definition of mass fraction,** namely that the sum of the mass fractions in each stream must be unity: $\sum \omega_i = 1$, as for example in a salt solution

$$\omega_{NaCl} + \omega_{H_2O} = 1$$

If the individual stream components are expressed as moles, a similar implicit equation exists for each stream—the mole fractions sum to unity: $\sum x_i = 1$, as for example

$$x_{NaCl} + x_{H_2O} = 1$$

(Additional stream variables such as pressure and temperature will be considered in subsequent chapters.)

To illustrate the concepts introduced above, examine Figure 3.6 which illus-

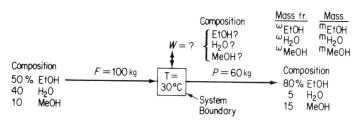

Figure 3.6   A steady state process involving three components without chemical reaction.

trates a process which has three components in each liquid stream. Assume the process is the steady state so that the accumulation term in Eq. (3.1) is zero. No reaction occurs so that the generation and consumption terms in Eq. (3.1) are also zero. All the known data has been placed on the streams in the figure. Keep in mind that the values recorded are mass percents, not mole percents. Do you know why? The objective is to calculate the values of the variables whose values are unknown.

First, let us count the number of variables whose values we do not know. They are W, ω_{EtOH}, ω_{H_2O}, and ω_{MeOH}. Consequently, we know that we need four independent equations to get a unique solution (the equations we solve will be linear). We can make three mass balances, one for each of the components, as follows, and use the summation of mole fractions to make the fourth independent equation

	In		*Out*
EtOH:	$(0.50)(100)$	$=$	$(0.80)(60) + \omega_{EtOH, \ w}(W)$
H_2O:	$(0.40)(100)$	$=$	$(0.05)(60) + \omega_{H_2O, \ w}(W)$
MeOH:	$(0.10)(100)$	$=$	$(0.15)(60) + \omega_{MeOH, \ w}(W)$

$$\omega_{EtOH,W} + \omega_{H_2O,W} + \omega_{MeOH,W} = 1$$

Note that the terms on the right-hand side of the component balances are nonlinear ($\omega \times W$). Consequently, to simplify the solution of the problem, it would be wise first to use the total mass balance

$$100 = 60 + W$$

and solve for W ($W = 40$), and then introduce W into the component balances so that the component balances become linear and **uncoupled**. Each component balance can then be solved independently. Note that the total mass balance is not an extra independent equation—it is just the sum of the three component balances with the summation of the mass fractions introduced—but it can be substituted for one of the independent balances so that the total number of independent balances remains at four.

To uncouple the equations during their formulation, we could have used as a set of variables m_{EtOH}, m_{H_2O} and m_{H_2O} plus W, where m stands for the mass flow of the component designated by the subscript. The variables whose values are unknown would be the three m's plus W, and three independent balances in kg to calculate the m_i would be

	In		Out	Solution for m_i
EtOH:	$(0.50)(100)$	=	$(0.80)(60) + m_{EtOH}$	2
H_2O:	$(0.40)(100)$	=	$(0.05)(60) + m_{H_2O}$	37
MeOH:	$(0.10)(100)$	=	$(0.15)(60) + m_{MeOH}$	1

Of course, to calculate the values of the unknown mass fractions we would have to calculate

$$W = m_{EtOH} + m_{H_2O} + m_{MeOH} = 2 + 37 + 1 = 40$$

and

$$\omega_{EtOH} = m_{EtOH} / W = 0.050$$
$$\omega_{H_2O} = m_{H_2O} / W = 0.925$$
$$\omega_{MeOH} = m_{MeOH} / W = 0.025$$
$$\Sigma \omega_i = 1.000$$

Note how the calculation of W is equivalent summing the mass fractions (What do you get when you divide each term of the equation to calculate W by W?).

Could we have made mole balances instead of mass balances for the problem illustrated in Figure 3.6? Certainly, but it would be inefficient to convert all of the mass fractions to mole fractions in order to write such balances. However, if the concentrations had been given in mole percent (which they were not because the flows were liquids) and F and P were stated in moles, mole balances would be more convenient to write than mass balances.

Are the material balance equations written for the process in Figure 3.6 independent equations? We can answer this question by determining the rank of the coefficient matrix for the set of linear equations involving m_i. Is it 3?

$$\begin{array}{c} \\ \text{Eq. 1} \\ \text{Eq. 2} \\ \text{Eq. 3} \end{array} \begin{array}{ccc} m_{EtOH} & m_{H_2O} & m_{MeOH} \\ \begin{bmatrix} 1 & 0 & 0 \\ 0 & 1 & 0 \\ 0 & 0 & 1 \end{bmatrix} \end{array}$$

So far we have focused attention on how to write down the material balances and the requirements that have to be met for the equations to have a solution. Now it is time to examine other important aspects of developing skills for successful problem solving. Table 3.1 lists the strategy we recommend you use in solving all the material and energy balance problems in this book.

If you use the steps in Table 3.1 as a mental checklist each time you start to work on a problem, you will have achieved the major objective of this chapter and substantially added to your professional skills. These steps do not have to be carried out in the order listed in Table 3.1, and you may repeat steps as the formulation of the solution of the problem become clearer. But the steps are all essential.

Here are some comments about each step.

1. *Read the problem* means **read the problem carefully** so that you understand it. Take this example.

A train is approaching the station at 105 cm/s. A man in one car is walking forward at 30 cm/s relative to the seats. He is eating a foot-long hot dog, which is entering

TABLE 3.1   Strategy for Analyzing Material Balance Problems
("*A problem recognized is a problem half-solved.*" Ann Landers)

1. Read the problem and clarify what is to be accomplished.
2. Draw a sketch of the process; define the system by a boundary.
3. Label with symbols the flow of each stream and the associated compositions and other information that is unknown.
4. Put all the known values of compositions and stream flows on the figure by each stream; calculate additional compositions and flows from the given data as necessary. Or, at least initially identify the known parameters in some fashion.
5. Select a basis.
6. Make a list by symbols for each of the unknown values of the stream flows and compositions, or at least mark them distinctly in some fashion, and count them.
7. Write down the names of an appropriate set of balances to solve; write the balances down with type of balance listed by each one. Do not forget the implicit balances for mass or mole fractions.
8. Count the number of independent balances that can be written; ascertain that a unique solution is possible. If not, look for more information or check your assumptions.
9. Solve the equations. Each calculation must be made on a consistent basis.
10. Check your answers by introducing them, or some of them, into any redundant material balances. Are the equations satisfied? Are the answers reasonable?

his mouth at the rate of 2 cm/s. An ant on the hot dog is running away from the man's mouth at 1 cm/s. How fast is the ant approaching the station?

A superficial analysis would take care to ignore the hot dog length but would calculate: $105 + 30 - 2 + 1 = 134$ cm/s for the answer. However, the problem states on more careful reading that the ant is moving away from the man's mouth at the rate of 1 cm/s. Because the man's mouth is moving toward the station at the rate of 135 cm/s, the ant is moving toward the station at the rate of 136 cm/s.

2. *Draw a sketch* means draw a simple box or circle to represent the process of interest. Draw arrows for any flows.

3. *Label* means to put a symbol for each variable, known or unknown, alongside the related arrow or box, such as F for feed or T for temperature.

4. *Put down known values* means to write next to or in place of each symbol its value if known. You can skip putting down symbols for known values, and just place the known values down as you gain experience. Think of a menu on a computer screen that you have to fill in. If you have insufficient data try

   (a) looking up data in a database;
   (b) calculating values of variables or parameters from known values of other data—insert zero where applicable;
   (c) making valid assumptions about possible values of selected variables or parameters.

5. *Select a basis.* Refer to Sec. 1.2.

6. *List symbols* means actually write down the symbols for the variables whose values are unknown, and count their number.

7. *Write down the independent equation* means begin by writing down the names of the independent equations you plan to use, and in conjunction with step 8 actually write the equations down.

8. *Count the number* means that literally you should ascertain if the number of variables whose values are unknown equals the number of independent equations. Make sure that the equations are indeed independent. What should you do if the count of independent equations and unknown variables does not match up? The best procedure is to review your analysis of the problem to make sure that you have not ignored some equation(s) or variable(s), double counted, forgotten to look up some missing data, or made error in your assumptions.

9. *Solve the equations.* For one to three linear equations you can solve by substitution. For more linear equations and for nonlinear equations you should use a computer program. Refer to the computer disk in the back of the book.

10. *Check the answer.* Everyone makes mistakes. When you obtain an answer, ask yourself if it makes sense. Use a redundant balance (such as the total mass balance) not used in the calculations to verify the values in the answers.

A surprising predicament exists: It is easier to find mistakes in someone else's calculations than your own! When you check your work, you tend to tread the same path whereas someone else looks at your work differently. Consequently, you have to think of different ways of problem solving and look at the problem from different angles if you want to catch a mistake.

You should also be concerned with efficient problem solving. You will discover many correct ways to solve a given problem; all will give the same correct answer if properly applied. Not all, however, will require the same amount of time and effort. For example, you may want to solve a problem initially stated in units of pounds by converting the given units to grams or gram moles, solving for the required quantities in SI units and converting the answer back to American Engineering units. Such a method is valid but can be quite inefficient—it consumes your time in unnecessary steps, and it introduces unnecessary opportunities for numerical errors to occur. You should, therefore, start developing the habit of looking for efficient ways to solve a problem, not just *a* way. A good example is to substitute the total mass balance for a component mass balance to reduce the number of unknown variables.

ADDITIONAL DETAILS

When you prepare the set of independent equations representing the material balances for a problem, and count the number of variables whose values are unknown, the counts may not agree. If you have not made any mistakes, what does this outcome mean? If more variables whose values are unknown exist than independent equations, an infinite number of solutions exists for a material balance problem—not a satisfactory outcome. Such problems are deemed **underspecified** (underdetermined). Either values of additional variables must be found to make up the deficit or the problem must be posed as an optimization problem, such as minimize some cost or revenue function subject to the constraints comprised of the material balances. On the other hand, if fewer values of the variables whose values are unknown exist than independent equations, the problem is **overspecified** (overdetermined), and no solution exists to the problem, as the equations are inconsistent. Again, the problem might be posed as an optimization problem, namely to minimize the sum of the squares of the deviations of the equations from zero (or their right-hand constants).

The difference between the number of variables whose values are unknown and the number of independent equations is known as the **number of degrees of**

freedom. If the degrees of freedom are positive, such as 2, you *must* seek out two additional independent equations or specifications of variables to get a unique solution to your material balance problem. If the degrees of freedom are negative, such as −1, you have too many equations or not enough variables in the problem. Perhaps you forgot to include one variable in setting up the information diagram for the problem. Perhaps some of the information you used was not correct. **Zero degrees of freedom means that the material balances problem is properly specified**, and you can then proceed to solve the equations for the variables whose values are unknown. (If the independent equations are nonlinear, possibly more than one solution exists as mentioned previously).

EXAMPLE 3.4 Determine the Number of Degrees of Freedom

Examine Figure E3.4a, which represents a simple flow sheet for a single unit. Only the value of D is known. What is the minimum number of other measurements that must be made to determine all the other stream and composition values?

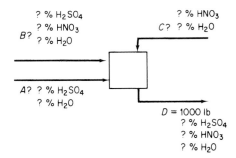

Figure E3.4a

Solution

What you are asked to find is the number of degrees of freedom for the problem in Figure E3.4a. You have to take into account that in any stream, one of the composition values can be determined by difference from 100%. Do you remember why? Here is the count of the minimum number of variables whose values are unknown:

Stream	Minimum number of unknown variables
A	2
B	3
C	2
D	2
Total	9

As an example of the count, in stream A specification of A plus one composition makes it possible to calculate the other composition and thus the mass flow of both the H_2SO_4 and the H_2O. Check the count yourself for the other streams. Are the values correct? In total only three independent material balance equations can be written (do you remember why?), leaving $9 - 3 = 6$ compositions and stream values that have to be specified.

Will specification of the values of any six variables in Figure E3.4a besides D do? No. Only those values should be specified that will leave a number of independent material balances equal to the number of unknown variables. As an example of a satisfactory set of measurements, choose one composition in stream A, two in B, one in C, and two in D, leaving the flows A, B, and C as unknowns.

What do you think of the specification of the following set of measurements: A, B, C, two compositions in D, and one composition in B? Draw a diagram of the information, as in Figure E3.4b(• = known quantity). Write down the three material balance equations. Are the three equations independent? You will find that they are not independent. Remember that the sum of the mass fractions is unity for each stream.

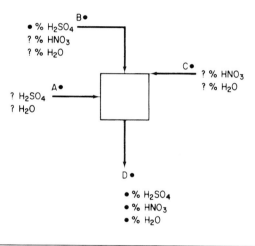

Figure E3.4b

LOOKING BACK

In this section we have presented a strategy of problem solving that applies to all types of material balance problems. We have also indicated that the main goal is to write down a number of independent equations just equal to the number of variables whose values are unknown.

Key Ideas

1. Use the same strategy to solve all material balance problems.
2. A tested strategy is outlined in Table 3.1 for you to follow.

3. In formulating the linear material balances to solve a problem, to obtain a unique solution you must have the same number of independent equations as you have variables whose values are unknown.

4. The number of linear equations that are independent is the same as the rank of the coefficient matrix of the equations.

Key Terms

Degrees of freedom (p. 159)   Uncoupled equations (p. 155)
Independent equations (p. 153)   Underspecified (p. 159)
Overspecified (p. 159)   Unique solution (p. 153)
Stream variables (P. 154)

Self-Assessment Test

1. What does the concept "solution of a material balance problem" mean?

2. (a) How many values of unknown variables can you compute from one independent material balance?
   (b) From three?
   (c) From four material balances, three of which are independent?

3. A water solution containing 10% acetic acid is added to a water solution containing 30% acetic acid flowing at the rate of 20 kg/min. The product P of the combination leaves at the rate of 100 kg/min. What is the composition of P? For this process,
   (a) Determine how many independent balances can be written.
   (b) List the names of the balances.
   (c) Determine how many unknown variables can be solved for.
   (d) List their names and symbols.
   (e) Determine the composition of P.

4. Can you solve these three material balances for F, D, and P?
$$0.1F + 0.3D = 0.2P$$
$$0.9F + 0.7D = 0.8P$$
$$F + \quad D = \quad P$$

5. Cite two ways to solve a set of linear equations.

6. If you want to solve a set of independent equations that contain fewer unknown variables than equations (the overspecified problem), how should you proceed with the solution?

7. What is the major category of implicit constraints (equations) you encounter in material balance problems?

8. If you want to solve a set of independent equations that contain more unknown variables than equations, what must you do to proceed with the solution?

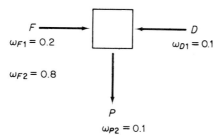

$$\omega_{F1} = 0.2$$
$$\omega_{D1} = 0.1$$
$$\omega_{F2} = 0.8$$
$$\omega_{P2} = 0.1$$

Figure AT3.2–9

9. How many values of the concentrations and flow rates in the process shown in the figure are unknown? List them. The streams contain two components, 1 and 2.

10. How many material balances are needed to solve problem 9? Is the number the same as the number of unknown variables? Explain.

Thought Problem

1. In the steady state flow process shown in the figure, a number of values of ω (mass fraction) are not given. Mary says that nevertheless the problem has a unique solution for the unknown values of ω. Kelly says that 4 values of ω are missing, that you can write 3 component material balances, and that you can use 3 relations for $\Sigma\omega_i = 1$, one for each stream, a total of 6 equations, so that a unique solution is not possible. Who is right? See Figure TP3.2–1.

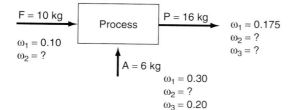

$F = 10$ kg
$\omega_1 = 0.10$
$\omega_2 = ?$

Process

$P = 16$ kg
$\omega_1 = 0.175$
$\omega_2 = ?$
$\omega_3 = ?$

$A = 6$ kg
$\omega_1 = 0.30$
$\omega_2 = ?$
$\omega_3 = 0.20$

Figure TP3.2–1

Discussion Problems

1. Consider the concept of zero discharge of liquid waste. It would seem to be a good idea both for the environment and the company. What are some of the arguments for and against the zero discharge of wastewater?

2. One proposed method of eliminating waste in solid, liquid, and gas streams is incineration. What are some of the pros and cons regarding disposal of waste by incineration?

3.3 SOLVING MATERIAL BALANCE PROBLEMS THAT DO NOT INVOLVE CHEMICAL REACTIONS

Your objectives in studying this section are to be able to:

1. Write a set of independent material balance equations for a process.
2. Solve a set of linear equations, or solve one or two simultaneous nonlinear equations.
3. Apply the 10-step strategy to solve problems without chemical reactions.

LOOKING AHEAD

In this section we will go through several examples in detail that involve the preparation and solution of material balances for a single system. Multiple connected systems will be treated in Sec. 3.5. If you hope to develop some skill and judgment in solving material balance problems, the way to proceed through this section is to first cover up the solution of the problem, then read the problem, then sketch out on a piece of paper your solution step by step, and only afterwards look at the solution that follows the problem. If you just read the problem and its solution, you will be deprived of the learning activity needed to become confident of your capabilities.

MAIN CONCEPTS

We illustrate a number of different processes and demonstrate how they all can be treated identically for analysis. You will find additional solved problems in the supplement to this book.

EXAMPLE 3.5 Membrane Separation

Membranes represent a relatively new technology for the separation of gases. One use that has attracted attention is the separation of nitrogen and oxygen from air. Figure E3.5 illustrates a nanoporous membrane that is made by coating a very thin layer of polymer on a porous graphite-supporting layer.

What is the composition of the waste stream if the stream amounts to 80% of the input?

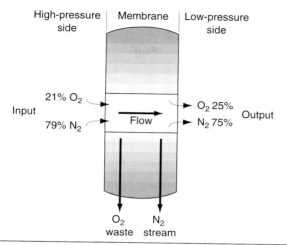

Figure E3.5

Solution

This is a steady state process without chemical reaction so that the accumulation term and the generation and consumption terms in Eq. (3.1) are zero. The system is the membrane. Let x_{O_2} be the mole fraction of oxygen, and x_{N_2} be the mole fraction of nitrogen, and let n_{O_2} and n_{N_2} be the respective moles.

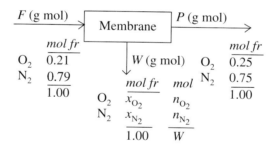

Figure E3.5b

Steps 1, 2, 3, and 4   All of the data and symbols have been placed in Figure E3.5b.

Step 5   Pick a convenient basis.

$$\text{Basis: } 100 \text{ g mol} = F$$

Then we know $W = 0.80(100) = 80$ mol.

Step 6   Three unknowns exist: P, x_{O_2}, and x_{N_2}, or P, n_{O_2}, and n_{N_2}.

Step 7   Two independent balances are the oxygen and nitrogen balances either as elements or as compounds. The third independent balance is $x_{O_2} + x_{N_2} = 1.00$, or $n_{O_2} + n_{N_2} = 80$.

Steps 8 and 9 The component balances are

	In		Out			In		Out

O_2: $\quad 0.21(100) = 0.25P + x_{O_2}(80) \qquad$ or $\qquad 0.21(100) = 0.25P + n_{O_2}$

N_2: $\quad 0.79(100) = 0.75P + x_{N_2}(80) \qquad$ or $\qquad 0.79(100) = 0.75P + n_{N_2}$

$$1.00 = x_{O_2} + x_{N_2} \qquad\qquad\qquad n_{O_2} + n_{N_2} = 80$$

The solution of these equations is $x_{O_2} = 0.20$, $x_{N_2} = 0.80$, and $P = 20$ g mol.
A simpler calculation involves the use of the total balance and one component balance because

$$F = P + W \quad \text{or} \quad 100 = P + 80$$

gives $P = 20$ straight off.

Step 10   Check. We can use the total balance as a check on the solution from the two component balances.

$$100 \overset{?}{=} 20 + 80 \qquad \text{ok}$$

EXAMPLE 3.6 Continuous Distillation

A novice manufacturer of alcohol for gasohol is having a bit of difficulty with a distillation column. The operation is shown in Figure E3.6. Technicians think too much alcohol is lost in the bottoms (waste). Calculate the composition of the bottoms and the mass of the alcohol lost in the bottoms.

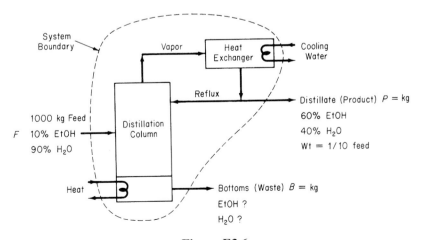

Figure E3.6

Solution

Although the distillation unit is comprised of two or three separate pieces of equipment, we have selected a system that includes inside the system boundary all of the equipment, and consequently can ignore all the internal streams. Let ω designate the mass fraction, and assume the process is in the steady state. No reaction occurs. Thus, the Eq. (3.1) reduces to In = Out in kg.

Steps 1, 2, 3, and 4   All of the symbols and known data have been placed on Figure E3.6.

Step 5   Select as the basis the given feed.

$$\text{Basis: } F = 1000 \text{ kg of feed}$$

Step 4   We are given that P is $\frac{1}{10}$ of F, so that $P = 0.1(1000) = 100$ kg.

Steps 6, 7, and 8   The remaining unknowns are $\omega_{EtOH,B}$, $\omega_{H_2O,B}$, and B. Two components exist yielding two independent component mass balances, and $\omega_{EtOH,B} + \omega_{H_2O,B} = 1$, so that the problem has a unique solution.

Step 9   Let us substitute the total mass balance $F = P + B$ for one of the component mass balances and calculate B by direct subtraction

$$B = 1000 - 100 = 900 \text{ kg}$$

The solution for the composition of the bottoms can be computed directly by subtraction.

	kg feed in		*kg distillate out*		*kg bottoms out*	*percent*
EtOH balance:	0.10(1000)	−	0.60(100)	=	40	4.4
H₂O balance:	0.90(1000)	−	0.40(100)	=	860	95.6
					900	100.0

Step 10   As a check we could use the total balance to calculate B, and the EtOH component balance to calculate $m_{EtOH, B}$ as 40 kg so that the

$$\text{mass H}_2\text{O in } B = 900 - 40 = 860 \text{ kg}$$

EXAMPLE 3.7 Mixing

Dilute sulfuric acid has to be added to dry charged batteries at service stations to activate a battery. You are asked to prepare a batch of new 18.63% acid as follows. A tank of old weak battery acid (H_2SO_4) solution contains 12.43% H_2SO_4 (the remainder is pure water). If 200 kg of 77.7% H_2SO_4 is added to the tank, and the final solution is to be 18.63% H_2SO_4, how many kilograms of battery acid have been made? See Figure E3.7.

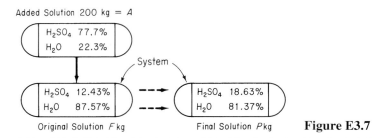

Original Solution F kg   Final Solution P kg   **Figure E3.7**

Solution

Steps 1, 2, 3, and 4   All of the values of the compositions are known and are on Figure E3.7. No reaction occurs. Should the process be treated as an unsteady-state process or a steady-state process? If the tank is the system, and the tank initially contains sulfuric acid solution, then a change occurs inside the system so that accumulation occurs in the system (the mass increases). This viewpoint calls for using the mass balance: *accumulation = in − out*.

From another viewpoint, you could regard the tank as initially being empty, the original solution is introduced into the system along with the 200 kg of 77.7% solution, the solutions are mixed, and finally the entire contents of the tank are removed leaving an empty tank. Then, the mass balance reduces to a steady state flow process: *in = out* because no accumulation occurs in the tank.

Let us first solve the problem with the mixing treated as an unsteady-state process, and then repeat the solution with the mixing treated as a steady-state process.

Step 5   Take 200 kg of A as the basis.

Step 6, 7 and 8   The two unknown quantities are F and P. Two components are present: H_2SO_4 and H_2O, hence you can write two independent mass balances, and a unique solution exists.

The balances will be in kilograms.

Type of Balance	Accumulation in Tank			A in		Out
	Final		Initial			
H_2SO_4	$P(0.1863)$	−	$F(0.1243)$	=	$200(0.777)$ −	0
H_2O	$P(0.8137)$	−	$F(0.8757)$	=	$200 (0.223)$ −	0
Total	P	−	F	=	200 −	0

Note that any pair of the three equations are independent, but coupled.

Step 9   Because the equations are linear and only two occur, you can take the total mass balance, solve it for F, and substitute for F in the H_2SO_4 balance to calculate P.

$$(P - 200)(0.1243) + 200(0.777) = P(0.1863)$$
$$P = 2110 \text{ kg acid}$$
$$F = 1910 \text{ kg acid}$$

Step 10 You can check the answer using the H_2O balance. Does the H_2O balance balance?

The problem could also be solved by considering the mixing to be a steady process with the initial solution F added to A in a vessel, and the resulting mixture removed from the vessel.

	A in		F in		P out
H_2SO_4	200(0.777)	+	F(0.1243)	=	P(0.1863)
H_2O	200(0.223)	+	F(0.8757)	=	P(0.8137)
Total	A	+	F	=	P

You can see by inspection that these equations are no different than the first set of mass balances except for the arrangement.

EXAMPLE 3.8 Drying

Fish caught by human beings can be turned into fish meal, and the fish meal can be used as feed to produce meat for human beings or used directly as food. The direct use of fish meal significantly increases the efficiency of the food chain. However, fish-protein concentrate, primarily for aesthetic reasons, is used mainly as a supplementary protein food. As such, it competes with soy and other oilseed proteins.

In the processing of the fish, after the oil is extracted, the fish cake is dried in rotary drum dryers, finely ground, and packed. The resulting product contains 65% protein. In a given batch of fish cake that contains 80% water (the remainder is dry cake), 100 kg of water is removed, and it is found that the fish cake is then 40% water. Calculate the weight of the fish cake originally put into the dryer.

Solution

Steps 1, 2, 3, and 4 Figure E3.8 is a diagram of the process. The process is a steady-state process without reaction. The system is the dryer.

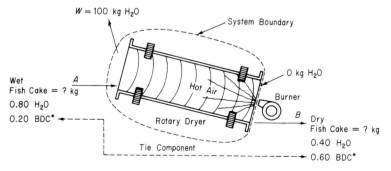

Figure E3.8

Step 5 Take a basis of what is given.

Basis: 100 kg of water evaporated $= W$

Step 6 The unknown stream flows are two: A and B. All the compositions are known.

Steps 7 and 8 Two independent balances can be written so that a unique solution exists.

We will use total mass balance plus the BDC (bone dry cake) balance (a tie component) rather than the water balance

$$0.80A = 0.40B + 100$$

because the BDC balance is slightly easier to use. The water balance can be used as a check on the calculations.

	In		Out	
Total balance:	A	$=$	$B + W = B + 100$	$\Big\}$ mass balance
BDC balance:	$0.20\,A$	$=$	$0.60\,B$	

The BDC balance gives the ratio of A to B: $B = 1/3A$. Introduce this relation into the total balance to get

$$A = 150 \text{ kg initial cake.}$$

Step 10 Check via water balance:

$$0.80(150) \overset{?}{=} 0.40(150)(1/3) + 100$$

$$120 = 120$$

EXAMPLE 3.9 Crystallization

A tank holds 10,000 kg of a saturated solution of Na_2CO_3 at 30°C. You want to crystallize from this solution 3000 kg of $Na_2CO_3 \cdot 10\,H_2O$ without any accompanying water. To what temperature must the solution be cooled?

Solution

This problem is a little more complicated to analyze than the previous problems because it not only requires a decision as to what the compounds are in the problem, but also it implies that the final solution is saturated at the final temperature without so stating. No reaction occurs. Although the problem could be set up as a steady-state problem with flows in and out of the system (the tank), it is equally justified to treat the process as an unsteady-state process. The major difficulty posed in this problem is to get all the necessary information about the compositions of the solutions and solid precipitate. If we can calculate the final concentration of the Na_2CO_3 in the tank, we can look up the corresponding temperature in a handbook containing solubility data. We will select Na_2CO_3 and H_2O as the components in the system rather than $Na_2CO_3 \cdot 10\ H_2O$ and H_2O because it takes fewer steps to make the required calculations.

Steps 1, 2, and 3   Figure E.3.9a is a diagram of the process.

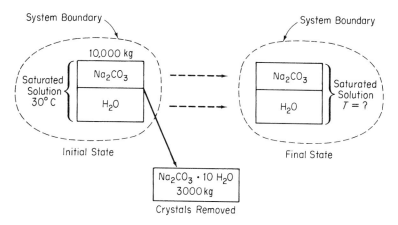

Figure E3.9a

Next, we need to get the compositions of the streams in so far as possible for each solution and the solid crystals of $Na_2CO_3 \cdot 10\ H_2O$

Steps 2, 3, and 4   We definitely need solubility data for Na_2CO_3 as a function of the temperature:

Temp. (°C)	Solubility (g Na_2CO_3/100 g H_2O)
0	7
10	12.5
20	21.5
30	38.8

Because the initial solution is saturated at 30°C, we can calculate the composition of the initial solution:

$$\frac{38.8\ \text{g Na}_2\text{CO}_3}{38.8\ \text{g Na}_2\text{CO}_3 + 100\ \text{g H}_2\text{O}} = 0.280 \text{ mass fraction } Na_2CO_3$$

Next, we need to calculate the composition of the crystals.

Basis: 1 g mol $Na_2CO_3 \cdot 10\ H_2O$

Comp.	mol	mol wt	mass	mass fr
Na_2CO_3	1	106	106	0.371
H_2O	10	18	180	0.629
Total			286	1.00

Step 5 Select a basis. The following is convenient; others can be used, such as 1000 g of $Na_2CO_3 \cdot 10\ H_2O$:

Basis: 10,000 kg of saturated solution at 30°C

Steps 2 and 3 (Repeated) We can now enter the known data concerning the compositions on the diagram as shown in Figure E3.9b

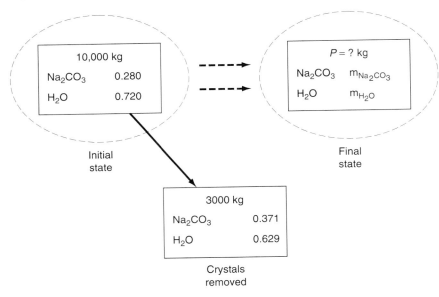

Figure E3.9b

Because we are treating this problem as an unsteady state problem Eq. (3.1) reduces to (the transport in = 0)

accumulation = − out

Steps 6, 7, 8, and 9 We have two components and hence two independent mass balances can be written. We have three unknowns: $m_{Na_2CO_3}$, m_{H_2O}, and P. The third independent balance is

$$m_{Na_2CO_3} + m_{H_2O} = P$$

so that the problem has a unique solution. The total and component balances (only two are independent) are (in kg)

Accumulation in Tank

	Final		_Initial_		_Transport out_
Na_2CO_3	$m_{Na_2CO_3}$	−	10,000 (0.280)	=	−3000 (0.371)
H_2O	m_{H_2O}	−	10,000 (0.720)	=	−3000 (0.629)
Total	P	−	10,000	=	−3000

The solution for the composition and amount of the final solution is

Component	kg
$m_{Na_2CO_3}$	1687
m_{H_2O}	5313
P (total)	7000

Step 10   Check on total

$$7,000 + 3,000 = 10,000$$

To find the temperature of the final solution, calculate the composition of the final solution in terms of grams of Na_2CO_3/100 grams of H_2O

$$\frac{1,687 \text{ g Na}_2\text{CO}_3}{5,313 \text{ g H}_2\text{O}} = \frac{31.8 \text{ g Na}_2\text{CO}_3}{100 \text{ g H}_2\text{O}}$$

Thus, the temperature to which the solution must be cooled is (using linear interpolation)

$$30°C - \frac{38.8 - 31.8}{38.8 - 21.5}(10.0°C) = 26°C$$

LOOKING BACK

In this section we have explained via examples how to analyze problems involving material balances in the absence of chemical reactions. The main difficulty in all of those problems is not the solution of the mass balances but their formulation, especially the preparation of the values of the compositions of the streams. If you followed the 10-step procedure outlined in Table 3.1, you should have learned a sound strategy of attacking each problem no matter what the process is.

Key Ideas

1. The general material balance, Eq. (3.1) can be applied to any process, and can usually be applied with some terms omitted.

2. The most difficult part of solving material balance problems is the collection and for-

mulation of the data specifying the compositions of the streams into and out of, or inside, the system.

Self-Assessment Test

1. A cellulose solution contains 5.2% cellulose by weight in water. How many kilograms of 1.2% solution are required to dilute 100 kg of 5.2% solution to 4.2%?

2. A cereal product containing 55% water is made at the rate of 500 kg/hr. You need to dry the product so that it contains only 30% water. How much water has to be evaporated per hour?

3. If 100 g of Na_2SO_4 is dissolved in 200 g of H_2O and the solution is cooled until 100 g of $Na_2SO_4 \cdot 10 H_2O$ crystallized out, find
   (a) The composition of the remaining solution (*mother liquor*)
   (b) The grams of crystals recovered per 100 g of initial solution.

4. Salt in crude oil must be removed before the oil undergoes processing in a refinery. The crude oil is fed to a washing unit where freshwater fed to the unit mixes with the oil and dissolves a portion of the salt contained in the oil. The oil (containing some salt but no water), being less dense than the water, can be removed at the top of the washer. If the "spent" wash water contains 15% salt and crude contains 5% salt, determine the concentration of salt in the "washed" oil product if the ratio of crude oil (with salt) to water used is 4:1.

Thought Problems

1. Although modern counterfeiters have mastered the duplication of the outside appearance of precious metals, some simple chemical/physical testing can determine their authenticity. Consult a reference book and determine the densities of gold, silver, copper, lead, iron, nickel, and zinc.
   (a) Could the density of pure gold be duplicated by using any of these metals?
   (b) Could the density of pure silver be duplicated by using any of these metals?
   (c) Assume that the volumes are conserved on mixing of the metals. What physical property makes any alloy an unlikely candidate for deception?

2. Incineration is one method of disposing of the sludge from sewage treatment plants. A lower limit for the combustion temperature exists to prevent ordorous materials from remaining in the flue gas, and an upper limit exists to avoid melting the ash. High-temperature operation of the incinerator will vaporize some of the heavy metals that cause air pollution. Thus, to prevent metals from vaporizing, it is preferable to operate at as low a temperature as possible.

In one run the following data were collected before and after combustion of the sludge.

	wt. % on a dry basis				mg/kg on a dry basis			
	Ash	Ig-loss	C	H	S	Cd	Pb	Cu	Zn
Sludge	67.0	33.0	15.1	2.2	1.1	169	412	384	1554
Residual ash	87.0	13.0	5.8	0	1.7	184	399	474	1943

Was the objective achieved of preventing vaporization of heavy metals?

Discussion Questions

1. Considerable concern has been expressed that the CO_2 generated from man's activities on earth has increased the CO_2 concentration in the atmosphere from 275 ppm in the last century to about 350 ppm currently. What are some of the important sources and sinks for CO_2 for the atmosphere on earth? Make a list. Estimate as best you can from news articles, books, and journals what the amount involved is for each source and sink. Estimate the accumulation of CO_2 per year in the atmosphere. Suggested references are *The Scientific American, Science, Nature, Chemical and Engineering News*, and various databases that can be accessed via CD disks and computer terminals.

2. Forests play an integral role in the dynamics of the global carbon cycle. Through photosynthesis, forests remove CO_2 from the atmosphere and accumulate some carbon over long periods of time. Decomposition of dead organic matter and fires release carbon back to the atmosphere. Tree harvesting also transfers carbon out of a forest. Prepare a report and figure(s) for the mass balance of carbon with a forest as the system. Include at a minimum growth, fire, mortality, harvesting, litter, decomposition, oxidation, and the soil and peat as components in the forest system. Designate pools of carbon that accumulate or are stable, and show the interface between the forest system and the surroundings (the atmosphere, sediments, usage by people, etc.).

3.4 SOLVING MATERIAL BALANCE PROBLEMS THAT INVOLVE CHEMICAL REACTIONS

Your objectives in studying this section are to be able to:

1. Define flue gas, stack gas, Orsat analysis, dry basis, wet basis, theoretical air (oxygen), required air (oxygen), and excess air (oxygen).

2. Given two of the three factors: entering gas (oxygen), excess air (oxygen), and required air (oxygen), compute the third factor.

3. Understand how to apply the material balance equation when chemical reactions occur.

4. Apply the 10-step strategy to solve problems involving reactions.

LOOKING AHEAD

In this section we analyze and solve material balances in problems in which chemical reactions occur. After defining a few specialized terms, we will analyze a number of examples in which material balances are used to solve the problems posed. The main new idea is that the generation and consumption terms can come into play in making component mole balances. A still valid concept is that balances on the elements themselves do not require use of the generation and consumption terms in Eq. (3.1).

MAIN CONCEPTS

Recall that Eq. (3.1) applies to total mass and component mass balances, and that the generation and consumption terms were zero in Sec. 3.3. In many instances we want to make balances on moles in lieu of element balances, but often in component and total balances, the moles will not necessarily balance unless the generation and consumption terms are taken into account. Take for example a simple steady state process of combustion shown in Figure 3.7.

The total moles in are equal to 2 and the total moles out are equal to 1. The moles of O_2 in are equal to 1 and the moles of O_2 out (as O_2) are equal to 0.

If we take into account the reaction equation and assume complete reaction of the O_2, then the generation term in Eq. (3.1) for O_2 has a value of 0, and the consumption term has a value of 1 mole. Thus, the material balance for O_2 in moles would be

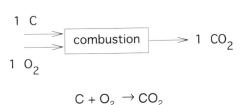

$$C + O_2 \rightarrow CO_2$$

Figure 3.7   Simple combustion.

	Accumulation		_In_		_Out_		_Generation_		_Consumption_
O_2	0	=	1	–	0	+	0	–	1

What would the mole balances on C and CO_2 as compounds be? Keep in mind that the balances for the elements C and O expressed in moles still hold with the generation and consumption terms having a value of zero

	Accumulation		_In_		_Out_		_Generation_		_Consumption_
C:	0	=	1	–	1	+	0	–	0
O:	0	=	2	–	2	+	0	–	0

Compare these two balances with the mass balances for the elements. How do the mass balances on the elements differ from the mole balances on the elements? Look at the next set of equations after you decide

	Accumulation		_In_		_Out_		_Generation_		_Consumption_
C:	0	=	12	–	12	+	0	–	0
O:	0	=	16	–	16	+	0	–	0

Observe that you can avoid using the generation and consumption terms in Eq. (3.1) if you make element balances.

Often for convenience the element balance in moles might be made on one C and two oxygens **denoted in this text as O2**, not meaning the compound O_2 but just two O's.

	Accumulation		_In_		_Out_		_Generation_		_Consumption_
C:	0	=	1	–	1	+	0	–	0
O2:	0	=	1	–	1	+	0	–	0

Before proceeding with some examples, we need to emphasize some terms commonly used in combustion problems. In dealing with problems involving combustion, you should become acquainted with those special terms:

1. **Flue or stack gas**   All the gases resulting from a combustion process including the water vapor, sometimes known as a _wet basis_.

2. **Orsat analysis or dry basis**   All the gases resulting from a combustion process _not including the water vapor_. Orsat analysis refers to a type of gas analysis apparatus in which the volumes of the respective gases are measured over and in equilibrium with water; hence each component is saturated with water vapor. The net result of the analysis is to eliminate water as a component being measured.

     Pictorially, we can express this classification for a given gas as in Figure 3.8. To convert from one analysis to another, you have to adjust the percentages for the components as explained in Sec. 1.4.

3. **Theoretical air** (or **theoretical oxygen**)   The amount of air (or oxygen) re-

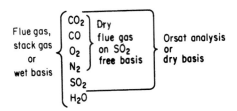

Figure 3.8   Comparison of gas analysis on different bases.

quired to be brought into the process **for complete combustion**. Sometimes this quantity is called the **required** air (or oxygen).

4. **Excess air** (or **excess oxygen**)   In line with the definition of excess reactant given in Chapter 1, excess air (or oxygen) would be the amount of air (or oxygen) **in excess of that required for complete combustion** as computed in (3).

   The calculated amount of *excess air does not depend on how much material is actually burned* **but what can be burned. Even if only partial combustion takes place,** as, for example, C burning to both CO and CO_2, **the excess air (or oxygen) is computed as if the process of combustion produced only CO_2.** The percent excess air is identical to the percent excess O_2 (often a more convenient calculation):

$$\% \text{ excess air} = 100\,\frac{\text{excess air}}{\text{required air}} = 100\,\frac{\text{excess } O_2 \,/\, 0.21}{\text{required } O_2 \,/\, 0.21} \qquad (3.4)$$

Note that the ratio $1/0.21$ of air to O_2 cancels out in Eq. (3.4). Percent excess air may also be computed as

$$\% \text{ excess air} = 100\,\frac{O_2 \text{ entering process} - O_2 \text{ required}}{O_2 \text{ required}} \qquad (3.5)$$

or

$$\% \text{ excess air} = 100\,\frac{\text{excess } O_2}{O_2 \text{ entering} - \text{excess } O_2} \qquad (3.6)$$

The precision achieved in these different relations to calculate the percent excess air may not be the same. If the percent excess air and the chemical equation are specified in a problem, you know how much air enters with the fuel, and hence the number of unknowns is reduced by one.

   Now, let us explore these concepts via some examples.

EXAMPLE 3.10 Excess Air

Fuels for motor vehicles other than gasoline are being eyed because they generate lower levels of pollutants than does gasoline. Compressed propane has been suggested as a source of economic power for vehicles. Suppose that in a test 20 kg of C_3H_8 is burned with 400 kg of air to produce 44 kg of CO_2 and 12 kg of CO. What was the percent excess air?

Solution

This is a problem involving the following reaction (is the reaction equation correctly balanced?)

$$C_3H_8 + 5O_2 \rightarrow 3CO_2 + 4H_2O$$

Basis: 20 kg of C_3H_8

Since the percentage of excess air is based on the *complete combustion* of C_3H_8 to CO_2 and H_2O, the fact that combustion is not complete has no influence on the definition of "excess air." The required O_2 is

$$\frac{20 \text{ kg } C_3H_8}{} \left| \frac{1 \text{ kg mol } C_3H_8}{44.09 \text{ kg } C_3H_8} \right| \frac{5 \text{ kg mol } O_2}{1 \text{ kg mol } C_3H_8} = 2.27 \text{ kg mol } O_2$$

The entering O_2 is

$$\frac{400 \text{ kg air}}{} \left| \frac{1 \text{ kg mol air}}{29 \text{ kg air}} \right| \frac{21 \text{ kg mol } O_2}{100 \text{ kg mol air}} = 2.90 \frac{\text{kg } O_2}{\text{mol}}$$

The percent excess air is

$$100 \times \frac{\text{excess } O_2}{\text{required } O_2} = 100 \times \frac{\text{entering } O_2 - \text{required } O_2}{\text{required } O_2}$$

$$\% \text{ excess air } = \frac{2.90 \text{ kg mol } O_2 - 2.27 \text{ kg mol } O_2}{2.27 \text{ kg mol } O_2} \left| \frac{100}{} \right. = 28\%$$

In calculating the amount of excess air, remember that the excess is the amount of air that enters the combustion process over and above that required for complete combustion. Suppose there is some oxygen in the material being burned. For example, suppose that a gas containing 80% C_2H_6 and 20% O_2 is burned in an engine with 200% excess air. Eighty percent of the ethane goes to CO_2, 10% goes to CO, and 10% remained unburned. What is the amount of the excess air per 100 moles of the gas? First, you can ignore the information about the CO and the unburned ethane because the basis of the calculation of excess air is *complete combustion.* Specifically C goes to CO_2, S to SO_2, H_2 to H_2O, CO goes to CO_2, and so on.

Second, the oxygen in the fuel cannot be ignored. Based on the reaction

$$C_2H_6 + \frac{7}{2}O_2 \rightarrow 2CO_2 + 3H_2O$$

80 moles of C_2H_6 require 3.5(80) = 280 moles of O_2 for complete combustion. However, the gas contains 20 moles of O_2, so that only $280 - 20 = 260$ moles of O_2 are needed in the entering air for complete combustion. Thus, 260 moles of O_2 are the required O_2, and the calculation of the 200% excess O_2 (air) is based on 260, not 280, moles of O_2:

Entering with air	Moles O_2
Required O_2:	260
Excess O_2 (2)(260):	520
Total O_2 (3)(260):	780

In the following problems each step cited in Table 3.1 will be identified so that you can follow the strategy of the solution.

EXAMPLE 3.11 Preventing Corrosion

Corrosion of pipes in boilers by oxygen can be alleviated through the use of sodium sulfite. Sodium sulfite removes oxygen from boiler feedwater by the following reaction:

$$2Na_2SO_3 + O_2 \longrightarrow 2\,Na_2SO_4$$

How many pounds of sodium sulfite are theoretically required (for complete reaction) to remove the oxygen from 8,330,000 lb of water (10^6 gal) containing 10.0 parts per million (ppm) of dissolved oxygen and at the same time maintain a 35% excess of sodium sulfite?

Solution

Steps 1, 2, 3, and 4   This is a steady-state process with reaction. The system is the pipe. The known data has been placed on Figure E3.11. The process is in the steady state with reaction

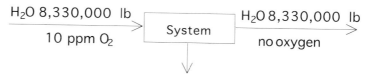

Figure E3.11

Step 5   A convenient basis is 10^6 gal, that is, 8,330,000 lb H_2O
Step 6   The unknown is F
Step 7, 8, and 9   The amount of O_2 entering is

$$\frac{8{,}330{,}000 \text{ lb } H_2O}{} \Bigg| \frac{10 \text{ lb } O_2}{(1{,}000{,}000 - 10 \text{ lb } O_2)\text{lb } H_2O} = 83.3 \text{ lb } O_2$$

effectively same as 1,000,000

The O_2 balance in lb is simple

_In_		_Out_		_Generation_		_Consumption_		_Accumulation_
83.3	–	0	+	0	–	m_{O_2}	=	0

$$m_{O_2} = 83.3 \text{ lb}$$

$$\frac{83.3 \text{ lb } O_2}{} \Bigg| \frac{1 \text{ lb mol } O_2}{32 \text{ lb } O_2} \Bigg| \frac{2 \text{ lb mol } Na_2SO_3}{1 \text{ lb mol } O_2} \Bigg| \frac{126 \text{ lb } Na_2SO_3}{1 \text{ lb mol } Na_2SO_3} \Bigg| 1.35$$

$$= 886 \text{ lb } Na_2SO_3$$

EXAMPLE 3.12 Combustion

Generation of methane-rich biogas is a way to avoid high waste-disposal costs, and burning it can meet up to 60% of the operating costs for such waste-to-energy plants. Further, in the European Community (EC) biogas plants are exempt from energy or carbon taxes, which will make them more attractive if the EC enforces its carbon tax proposal. Four full-scale demonstration projects are now up and running in Europe.

Let us consider only the combustion of methane. Figure E3.12 shows a simple combustion process, the mechanical details of which we can ignore.

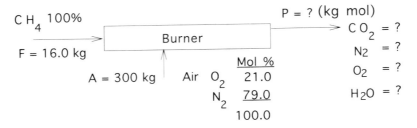

Figure E3.12

Solution

Steps 1, 2, 3, 4   This is a steady-state process with reaction; assume complete combustion. The system is the burner. All four steps have been completed with the data placed in the Figure E3.12. Because the process output is a gas, the composition will be mole fractions or moles, hence it is more convenient to use moles rather than mass in this problem even though the quantities of CH_4 and air are stated in kg.

$$\frac{300 \text{ kg A}}{29.0 \text{ kg A}} \left| \frac{1 \text{ kg mol A}}{} \right. = 10.35 \text{ kg mol A in}$$

$$\frac{16.0 \text{ kg C H}_4}{16.0 \text{ kg C H}_4} \left| \frac{1 \text{ kg mol CH}_4}{} \right. = 1.00 \text{ kg mol C H}_4 \text{ in}$$

$$\frac{10.35 \text{ kg mol A}}{1 \text{ kg mol A}} \left| \frac{0.21 \text{ kg mol O}_2}{} \right. = 2.17 \text{ kg mol O}_2 \text{ in}$$

$$\frac{10.35 \text{ kg mol A}}{1 \text{ kg mol A}} \left| \frac{0.79 \text{ kg mol N}_2}{} \right. = 8.18 \text{ kg mol N}_2 \text{ in}$$

Step 5   Since no particular basis is designated, we will pick a convenient basis

Basis: 16.0 kg CH_4 entering = 1 kg mol CH_4

Step 6   The unknowns are P and the four compositions in P: $x_{CO_2}^P$, $x_{N_2}^P$, $x_{O_2}^P$ and $x_{H_2O}^P$ (where x_i^P is the mole fraction of the component indicated by the subscript)

Steps 7 and 8   Let us use balances on the elements. An independent material balance can be made for each of the four elements C, O, N, and H. Also $\sum x_i^P = 1$ so that the problem has a unique solution. As an example of the balances, examine the C balance (remember Eq. (3.1) reduces to *In = Out* for the element balances here).

In		Out	
1 kg mol F	1.00 mol CH_4	$=$ P kg mol	$x^P_{CO_2}$ kg mol CO_2
	1 kg mol F		1 kg mol P

Other balances are similar. A summary of the balances is

	In			Out
	CH_4	Air		
C:	1(1.0)		$=$	$P\,(x^P_{CO_2})$
H2:	1 (2.0)		$=$	$P(x^P_{H_2O})$
O2:		2.17	$=$	$P\,(0.5\,x^P_{H_2O} + x^P_{O_2} + x^P_{CO_2})$
N2:		8.18	$=$	$P(x^P_{N_2})$
	$x^P_{CO_2} + x^P_{H_2O} + x^P_{O_2} + x^P_{N_2}$		$=$	1

Because this formulation leads to products of two variables whose values are unknown on the right-hand side of the equations, let us change variables and start with the unknowns as P and $n^P_{CO_2}$, $n^P_{N_2}$, $n^P_{O_2}$, and $n^P_{H_2O}$ (where n^P_i is the number of moles) so that we get

Balance	$CH_4\,in$	Air in		Out
C:	1		$=$	$n^P_{CO_2}$
H2:	2		$=$	$n^P_{H_2O}$
O2:		2.17	$=$	$0.5\,n^P_{H_2O} + n^P_{O_2} + n^P_{CO_2}$
N2:		8.18	$=$	$n^P_{N_2}$
	$n^P_{CO_2} + n^P_{H_2O} + n^P_{O_2} + n^P_{N_2}$		$=$	P

Observe that $\left(x^P_i\right)P = n^P_i$ and $x^P_i = \dfrac{n^P_i}{P}$

Step 9   Now we can easily solve the second set of equations sequentially (they become uncoupled)

			kg mol
$n^P_{CO_2}$	$=$	1	
$n^P_{H_2O}$	$=$	2	
$n^P_{N_2}$	$=$	8.18	
$n^P_{O_2}$	$=$	$2.17 - 0.5(2) - 1 = 0.17$	
P	$=$	$1 + 2 - 8.18 + 0.17 = 11.35$	

Then

$x^P_{CO_2}$	$=$	0.09	$x^P_{N_2}$	$=$	0.72
$x^P_{H_2O}$	$=$	0.18	$x^P_{O_2}$	$=$	0.01

Steps 7 and 8   revisited.

If we want to make mole balance on the compounds involved in the reaction, we need to draw information from the stoichiometric equation for the reaction

$$CH_4 + 2O_2 \rightarrow CO_2 + 2H_2O$$

Eq. (3.1) is

Compound	Accumulation		In		Out		Generation		Consumption
CH$_4$:	0	=	1.00	−	$n^P_{CH_4}$	+	0	−	1.00
O$_2$:	0	=	2.17	−	$n^P_{O_2}$	+	0	−	2(1.00)
N$_2$:	0	=	8.18	−	$n^P_{N_2}$	+	0	−	0
CO$_2$:	0	=	0	−	$n^P_{CO_2}$	+	1(1.00)	−	0
H$_2$O:	0	=	0	−	$n^P_{H_2O}$	+	2(1.00)	−	0

Solution of these equations gives the same results as shown above.

Which method of analysis is the best? Select which ever method what you feel most confident in using.

Step 10   The answer can be checked by adding the total mass in (316 kg) and determining if the total mass out is the same value, but we will omit this step here to save space.

EXAMPLE 3.13 Combustion with Nonprecise Data

The main advantage of catalytic incineration of odorous gases or other obnoxious substances over combustion is the lower cost. Catalytic incinerators operate at lower temperatures—500 to 900°C compared with 1100 to 1500°C for thermal incinerators—and use substantially less fuel. Because of the lower operating temperatures, materials of construction do not need to be as heat resistant, reducing installation and construction costs.

In a test run, a liquid that is proposed for use as a fuel in a flare and has the composition of 88% C and 12% H$_2$ is vaporized and burned with dry air to a flue gas (fg) of the following composition on a dry basis:

CO$_2$	13.4%
O$_2$	3.6%
N$_2$	83.0%
	100.0%

To help design the equipment of the continuous steady-state combustion device, determine how many kilogram moles of dry fg are produced per 100 kg of liquid feed. What was the percentage of excess air used?

Solution

This problem illustrates one common difficulty in solving combustion problems. The process is in the steady-state accompanied by chemical reaction. Let the system be the flare and associated equipment.

Steps 1, 2, 3, and 4   The necessary data are placed in Figure E3.13. Do not forget the water vapor and just use only the Orsat analysis for the exit gases!

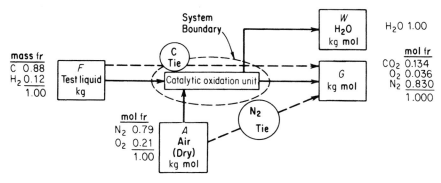

Figure E3.13

Step 5   A good basis is either F or G. Note the composition of F is mass fraction and that of G mole fraction. Let us make element balances. A good basis is then

Basis: 100 kg mol dry flue gas = G

Step 6   The variables whose values are unknown are F and A. All of the compositions are known.

Step 7   The number of element balances that can be written is 4: C, H, O, and N. Can one of these equations be redundant? We shall see.

Step 8   The atomic species balances (in moles) are (F is in kg; A, W, and G are in kg mol).

Balance	F In		A In		W out		G Out		Accumulation
C:	$\dfrac{(0.88)F}{12.0}$	+	0	−	0	−	$(0.134)100$	=	0
H2:	$\dfrac{(0.12)F}{2.016}$	+	0	−	W	−	0	=	0
O2:	0	+	$0.21A$	−	$\left(\tfrac{1}{2}\right)W$	−	$(0.134 + 0.036)100$	=	0
N2:	0	+	$0.79A$	−	0	−	$(0.830)100$	=	0

If you examine the coefficient matrix, you can note by inspection that the largest possible nonzero determinant is of order 3 so that at the most only 3 of the balances can be independent, and are, hence the problem has a unique solution.

Step 9 The equations are uncoupled and can be solved starting with C, then N_2, and finally H_2 (to get W, although W is not asked for).

$$\text{From C:} \qquad F = \frac{13.4(12)}{0.88} \qquad = \quad 182.73 \text{ kg}$$

$$\text{From } N_2: \qquad A = \frac{83.0}{0.79} \qquad = \quad 105.06 \text{ kg mol}$$

$$\text{From } H_2: \qquad W = \frac{182.73(0.12)}{2.016} \qquad = \quad 10.88 \text{ kg mol}$$

Step 10 Use the oxygen balance (a redundant balance) as a check.

$$105.06(0.21) \overset{?}{=} 10.88\left(\tfrac{1}{2}\right) + 17.00$$

$$22.06 \quad \overset{?}{=} \quad 22.44$$

An exact balance does not occur, but the answers agree reasonably well here. In many combustion problems, slight errors in the data will cause large differences in the calculated flows and percentage of excess air. Assuming that no mathematical mistakes have been made (it is wise to check), the better solution is the one involving the use of the most precise data. We will use both of these values for the oxygen, and determine what difference results in the calculation of the excess air.

We can now answer the requested questions.

$$\frac{G}{F} = \frac{100 \text{ kg mol}}{182.73 \text{ kg}} = \frac{54.73 \text{ kg mol } G}{100 \text{ kg } F}$$

The percent excess air can be calculated via two routes:

$$\% \text{ excess air} = 100 \times \frac{\text{excess O}_2}{\text{O}_2 \text{ entering} - \text{excess O}_2} \qquad \text{(a)}$$

or

$$\% \text{ excess air} = 100 \times \frac{\text{excess O}_2}{\text{required O}_2} \qquad \text{(b)}$$

The required O_2 is

$$C + O_2 \longrightarrow CO_2: \quad \frac{182.73(0.88)}{12} \quad = 13.40 \text{ kg mol}$$

$$H_2 + \tfrac{1}{2}O_2 \longrightarrow H_2O: \quad \frac{182.73(0.12)}{2.106}\left(\frac{1}{2}\right) = \frac{5.44 \text{ kg mol}}{18.84 \text{ kg mol}}$$

The excess O_2 is

$$22.44 - 18.44 = 3.60$$

$$\text{or} \quad 22.06 - 18.44 = 3.22$$

Basis: O_2 in = 22.44 mol	Basis: O_2 in = 22.06 mol
calculated from the	calculated from the
entering air	flue gas

By Eq. (a) % excess air: $100\,\dfrac{3.60}{22.44 - 3.60} = 19.1\%$ $100\,\dfrac{3.22}{22.06 - 3.22} = 16.7\%$

By Eq. (b) % excess air: $100\,\dfrac{3.60}{18.84} = 19.1\%$ $100\,\dfrac{3.22}{18.84} = 16.7\%$

Because we do not have to answer any questions about the water in the exit flue-gas steam, you can make use of the tie components shown in Figure E 3.13 as an alternative way to solve the problem. A tie component that related the test fluid to the dry flue gas and another one that related the air to the dry flue gas would be sufficient to solve the problem. In examining the data to determine whether a tie component exists, we see that the carbon goes directly from the test fluid to the dry flue gas, and nowhere else, so carbon will serve as one tie component. All the N_2 in the air exists in the dry flue gas, so N_2 can be used as another tie component. This approach is equivalent to using the C balance to solve for F and the N balance to solve for A.

EXAMPLE 3.14 Combustion of Coal

A local utility burns coal having the following composition on a dry basis. (Note that the coal analysis below is a convenient one for our calculations, but is not necessarily the only type of analysis that is reported for coal. Some analyses contain much less information about each element.)

Component	Percent
C	83.05
H	4.45
O	3.36
N	1.08
S	0.70
Ash	7.36
Total	100.0

The average Orsat analysis of the gas from the stack during a 24-hr test was

Component	Percent
$CO_2 + SO_2$	15.4
CO	0.0
O_2	4.0
N_2	80.6
Total	100.0

Moisture in the fuel was 3.90%, and the air on the average contained 0.0048 lb H_2O/lb dry air. The refuse showed 14.0% unburned coal, with the remainder being ash.

You are asked to check the consistency of the data before they are stored in a data base. Is the consistency satisfactory? What was the average percent excess air used?

Solution

Steps 1, 2, 3, and 4   This is a steady-state problem with reaction. The system is the furnace. The accumulation term in Eq. (3.1) is zero. All of the information given in the problem statement has been placed on Figure E3.14. Because the gas analysis is on a dry basis, we added a flow stream W for the exit water to the process diagram.

The composition of F and R are given in mass and those of P and A in moles. We will make the balances on the elements in moles so that the generation and consumption terms in Eq. (3.1) are zero.

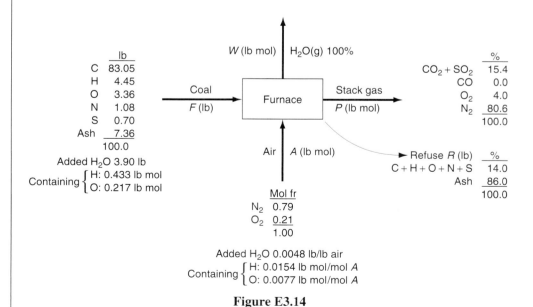

Figure E3.14

Step 5   Pick a basis of $F = 100$ lb as convenient.

Step 4 (repeated)   We first must add some extra information besides the compositions to the diagram.

In coal: H_2O

$$\frac{3.90 \text{ lb } H_2O}{} \left| \frac{1 \text{ lb mol } H_2O}{18 \text{ lb } H_2O} \right| \frac{2 \text{ lb mol } H}{1 \text{ lb mol } H_2O} = 0.433 \text{ lb mol } H$$

$$(0.217 \text{ lb mol } O)$$

In air:

$$\frac{0.0048 \text{ lb } H_2O}{\text{lb air}} \left| \frac{29 \text{ lb air}}{1 \text{ lb mol air}} \right| \frac{1 \text{ lb mol } H_2O}{18 \text{ lb } H_2O} = 0.0077 \frac{\text{lb mol } H_2O}{\text{lb mol air}}$$

$$(0.0154 \text{ lb mol } H/\text{lb mol } A)$$

$$(0.0077 \text{ lb mol } O/\text{lb mol } A)$$

Steps 6, 7, 8, and 9   We might neglect the C, H, O, N, and S in the refuse but will include the amount to show what calculations are necessary if the amounts of the elements were significant. The ash balance is (ash is a tie component)

$$7.36 = R(0.86)$$
$$R = 8.56 \text{ lb}$$

The unburned coal in the refuse is

$$8.56(0.14) = 1.20 \text{ lb}$$

If we assume that the combustibles in the refuse occur in the same proportions as they do in the coal (which may not be true), the quantities of the combustibles in R on an ash-free basis are:

Component	mass %	lb	lb mol
C	89.65	1.076	0.0897
H	4.80	0.058	0.0537
O	3.63	0.0436	0.0027
N	1.17	0.014	0.0010
S	0.76	0.009	0.0003
	100.00	1.20	0.1474

Step 6   The variables whose values are still unknown are: A, W, and P.

Step 7 and 8   We can write only four mole balances on the elements because S and C must be combined in as much as these two elements are combined in the stack gas analysis. Presumably one of the four equations will be redundant and can be used as a check on the calculations.

	In		Out		
	F	A	W	P	R

C + S: $\dfrac{83.05}{12.0} + \dfrac{0.70}{32.0} +$ 　　0　　$= 0 +$ 　$P(0.154)$　$+ 0.0897 + 0.0003$

H: 　$\dfrac{4.45}{1.008} + 0.433 +$ 　$0.0154\,A$ 　$= 2W +$ 　　0　　$+$ 　0.0537

O: 　$\dfrac{3.36}{16.0} + 0.217 + 0.21A(2) + 0.0077A = W + 2P(0.154 + 0.040) +$ 　0.0027

N: 　$\dfrac{1.08}{14.0} +$ 　　　$2(0.79A\,)$ 　　$= 0 +$ 　$2P(0.806)$ 　$+$ 　$2(0.001)$

Solve the C + S balance to get $P = 50$. Then solve the N balance to get $A = 45.35$. Next, solve the H balance to get $W = 2.746$. Finally, use the O balance to serve as a check: $19.8 \overset{?}{=} 20.0$. The difference is about 2%. Inasmuch as the data provided are actual measurements, in view of the random and possibly biased errors in the data, the round-off error introduced in the calculations, and possible leaks in the furnace, the data seem to be quite satisfactory. Try calculating W, a small number, from both the H and O balances. What size error do you find?

To calculate the excess air, because of the oxygen in the coal and the existence of unburned combustibles, we will calculate the total oxygen in and the required oxygen:

$$\% \text{ excess air} = 100 \times \frac{O_2 \text{ entering} - O_2 \text{ required}}{O_2 \text{ required}}$$

Component	Reaction	lb	lb mol	Required O_2 (lb mol)
C	$C + O_2 \rightarrow CO_2$	83.05	6.921	6.921
H	$H_2 + \frac{1}{2}O_2 \rightarrow H_2O$	4.45	4.415	1.104
O	—	3.36	0.210	(0.105)
N	—	—	—	—
S	$S + O_2 \rightarrow SO_2$	0.70	0.022	0.022
				7.942

and the oxygen in the air is $(45.35)(0.21) = 9.524$ lb mol.

$$\% \text{ excess air} = 100 \times \frac{9.524 - 7.942}{7.942} = 19.9\%$$

If you (incorrectly) calculated the % excess air from the wet stack gas analysis alone, you would get

$$100 \times \frac{4.0}{15.4 + 2.746\,/\,2} = 23.8\%$$

LOOKING BACK

In this section we applied Eq. (3.1) to processes involving reaction. If element balances are made, the generation and consumption terms in Eq. (3.1) are zero. If a balance is made on a compound, the terms are not zero, and you have to have some information about them, perhaps from the reaction equations and extent of reaction. In this book information about the generation and consumption terms for a chemical compound will be given a priori or can be inferred from the stoichiometric equations involved in the problem. Texts treating chemical reaction engineering describe how to calculate from basic principles gains and losses of chemical compounds.

Key Ideas

1. Eq. (3.1) can be applied to processes in which reaction occurs. The simple relation "the input equals the output" holds for *steady-state processes* (no accumulation) in the following circumstances.

	Equality Required for Input and Output of a Steady-State Process	
Type of Balance	Without Chemical Reaction	With Chemical Reaction
Total balances		
Total mass	Yes	Yes
Total moles	Yes	No
Component balances		
Mass of a pure compound	Yes	No
Moles of a pure compound	Yes	No
Mass of an atomic species	Yes	Yes
Moles of an atomic species	Yes	Yes

2. If Eq. (3.1) is to be applied to a compound, some information must be given about the stoichiometry involved and/or the extent of reaction.

3. Excess air for combustion is based on the assumption of complete reaction whether or not a reaction takes place.

Key Terms

Complete combustion (p. 178)
Dry basis (p. 177)
Excess air (oxygen) (p. 178)
Flue gas (p. 177)
Orsat analysis (p. 177)

Required air (oxygen) (p. 178)
Stack gas (p. 177)
Theoretical air (oxygen) (p. 177)
Wet basis (p. 178)

Self-Assessment Test

1. Explain the difference between flue gas analysis and Orsat analysis; wet basis and dry basis for a gas.

2. What does "SO_2 free basis" mean?

3. Write down the equation relating percent excess air to required air and entering air.

4. Will the percent excess air always be the same as the percent excess oxygen in combustion (by oxygen)?

5. In a combustion process in which a specified percentage of excess air is used, and in which CO is one of the products of combustion, will the analysis of the resulting exit gases contain more or less oxygen than if all the carbon had gone to CO_2?
   In solving the following problems, be sure to employ the ten steps listed in Table 3.1.

6. Pure carbon is burned in oxygen. The flue-gas analysis is:

   $$CO_2 \quad 75 \text{ mol } \%$$
   $$CO \quad \; 14 \text{ mol } \%$$
   $$O_2 \quad \;\; 11 \text{ mol } \%$$

   What was the percent excess oxygen used?

7. Toluene, C_7H_8 is burned with 30% excess air. A bad burner causes 15% of the carbon to form soot (pure C) deposited on the walls of the furnace. What is the Orsat analysis of the gases leaving the furnace?

8. Answer the following true (T) or false (F).
   (a) If a chemical reaction occurs, the total masses entering and leaving the system for a steady-state process are equal.
   (b) In combustion, all of the moles of carbon that enter a steady-state process exit from the process.
   (c) The number of moles of a chemical compound entering a steady-state process in which a reaction occurs with that compound can never equal the number of moles of the same compound leaving the process.

9. List the circumstances for a steady-state process in which the number of moles entering the system equals the number of moles leaving the system.

10. A synthesis gas analyzing CO_2: 6.4%, O_2: 0.2%, CO: 40.0%, and H_2: 50.8% (the balance is N_2) is burned with excess dry air. The problem is to determine composition of the flue gas. How many degrees of freedom exist in this problem, that is, how many additional variables have to have their values specified?

11. A coal analyzing 65.4% C, 5.3% H, 0.6% S, 1.1% N, 18.5% O, and 9.1% ash is burned so that all combustible is burnt out of the ash. The flue gas analyzes 13.00% CO_2, 0.76% CO, 6.17% O_2, 0.87% H_2, and 79.20% N_2. All of the sulfur burns to SO_2, which is included in the CO_2 figure in the gas analysis (i.e., $CO_2 + SO_2 = 13.00\%$). Calculate:
    (a) Pounds of coal fired per 100 lb mol of dry flue gas as analyzed;
    (b) Ratio of moles total combustion gases to moles of dry air supplied;

(c) Total moles of water vapor in the stack gas per 100 lb of coal if the air is dry;

(d) Percent excess air.

12. Hydrofluoric acid (HF) can be manufactured by treating calcium fluoride (CaF_2) with sulfuric acid (H_2SO_4). A sample of fluorospar (the raw material) contains 75% by weight CaF_2 and 25% inert (nonreacting) materials. The pure sulfuric acid used in the process is in 30% excess of that theoretically required. Most of the manufactured HF leaves the reaction chamber as a gas, but a solid cake is also removed from the reaction chamber that contains 5% of all the HF formed, plus $CaSO_4$, inerts, and unreacted sulfuric acid. How many kilograms of cake are produced per 100 kg of fluorospar charged to the process?

13. A hydrocarbon fuel is burnt with excess air. The Orsat analysis of the flue gas shows 10.2% CO_2, 1.0% CO, 8.4% O_2, and 80.4% N_2. What is the atomic ratio of H to C in the fuel?

Thought Problems

1. In a small pharmaceutical plant, it had not been possible for a period of two months to get more than 80% of rated output from a boiler rated at 120,000 lb of steam per hour. The boiler had complete flow metering and combustion control equipment, but the steam flow could not be brought to more than 100,000 lb/hr. What would you recommend be done to find the cause(s) of the problem and alleviate it?

2. An article advocating the planting of trees explains that a tree can assimilate 13 lb of carbon dioxide per year, or enough to offset the CO_2 produced by driving one car 26,000 miles per year. Can this statement be correct?

3. In connection with the concern about global warming, because of the increase in CO_2 concentration in the atmosphere, would you recommend the use of coal, ethanol, fuel oil, or natural gas as a fuel?

Discussion Questions

1. In situ biorestoration of subsurface materials contaminated with organic compounds is being evaluated by the EPA and by industry as one technique for managing hazardous wastes. The process usually involves stimulating the indigenous subsurface microflora to degrade the contaminants in place, although microorganisms with specialized metabolic capabilities have been added in some cases. The ultimate goal of biodegradation is to convert organic wastes into biomass and harmless byproducts of microbial metabolism such as CO_2, CH_4, and inorganic salts.

    Bioremediation of trichloroethylene, and *cis-* and *trans*-dichloroethylene was investigated in a test plot in a field trial. The aquifer was not pretreated with methane and oxygen to stimulate growth of the methanotrophs. The biotransformation of *trans*-dichloroethylene, *cis*-dichloroethylene, and trichloroethylene added at 50, 110, and 130 μg/L was 65, 45, and 25%, respectively, which suggests that the less-chlorinated compounds are more readily degraded than are the highly chlorinated compounds.

What other influences might have affected the results obtained?

2. On November 1, 1986, a fire at a Sandoz storehouse near Basel, Switzerland resulted in a substantial number of insecticides, pesticides, dyes, and other raw and intermediate materials being introduced into the Rhine River via runoff of about 15,000 m³ water used in fighting the fire. From Basel to the North Sea where the Rhine discharges is about 1200 km. The table lists some of the compounds discharged into the river along with the LC50 value (the concentration that will kill 50% of rainbow trout).

Compound	Estimated discharge (kg)	Estimated Concentration near discharge point (µg/L)	LC50 (µg/L)
Disulfostor (I)	3000 to 9000	600	6000
Thiometon (I)	1200 to 4000	500	8000
Ethoxyethyl mercury		12	
hydroxide (P)	18 to 200		3 to 1000
DNOC (P)	600 to 2000	100 to 430	66 to 1250
Endosulfan (P)	20 to 60	3 to 13	1.4

I = insecticide; P = pesticide

What were the probable consequences of the discharge along the river to the fish, biota, drinking water, and benthic organisms? Note that the Rhine has several dams to provide water for navigation. What would the concentration of these compounds be as a function of time at various towns downstream of Basel?

3. Suppose you are asked to serve as a consultant on the problem of how to produce oxygen on the moon as economically as possible. The raw material readily available is $FeTiO_3$, SiO_2, and/or FeO. The energy to carry out the reactions is presumably available from the sun or atomic energy that provides electricity or high pressure steam. Discuss some possible methods of O_2 generation and draw a simple flowsheet for the process.

   Some very useful references pertaining to this problem are (a) L. A. Taylor, "Rocks and Minerals in the Regolith of the Moon: Resources for a Lunar Base," in *Advanced Materials - Applications Mining and Metallurgical Processing Principles,* (pp. 29–47), ed. V. I. Lakshmanan, Soc. Mining, Mineral, & Exploration, Littleton, CO (1988); (b) L. A. Taylor, "Resources for a Lunar Base: Rocks, Minerals and Soils of the Moon," in *2nd Symp. on Lunar Base and Space Activities of the 21st Century,* ed. W. W. Mendell, Lunar & Planetary Inst., Houston, TX (1993); (c) L. A. Taylor and D. W. Carrier, "Oxygen Production Processes on the Moon: An Overview and Evaluation," *Resources in New-Earth Space.* University of Arizona Press, Tucson, AZ (1993).

3.5 SOLVING MATERIAL BALANCE PROBLEMS INVOLVING MULTIPLE SUBSYSTEMS

> ### *Your objectives in studying this section are to be able to:*
>
> 1. Write a set of independent material balances for a complex process involving more than one unit.
> 2. Solve problems involving several connected units by applying the 10-step strategy.

LOOKING FORWARD

In this section we are going to discuss how to treat and solve material balance problems for systems of coupled subsystems. You will be pleased to learn that principles employed in Secs. 3.3 and 3.4 still apply. All you have to do is apply the same 10-step strategy to individual subsystems and/or to the overall system.

MAIN CONCEPTS

As indicated by the flowsheet in Figure 3.5, plants in the process industries are comprised of many interconnected units. However, you can easily apply the same techniques discussed in the previous sections of this chapter to solve material balance problems in such plants. List and count up the number of variables whose values are unknown, making sure you do not count the same variable more than once, and then list and count up the number of *independent* balances you can make, making sure that balances for one unit do not render formerly independent balances for another unit into dependent balances. If the number of independent equations equals the number of variables whose values are unknown, at least for a set of linear equations, you can generally solve the equations for a unique answer.

If you ignore all the internal streams and variables within a set of connected subsystems, you can treat the overall system exactly as you treated a single system by drawing a boundary about the entire set of subsystems. From an overall viewpoint, the system of interest is indeed a single system.

What strategy should you use to solve the material balances for a sequence of connected units? Frequently, the best way to start is to make material balances for the **overall process** ignoring information about the internal **connections.** Then, you can make material balances for one or more subsystems. However,

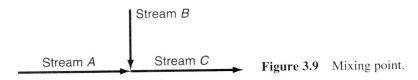

Stream A   Stream C

Figure 3.9   Mixing point.

since the overall balance is nothing more than the sum of the related balances about each subsystem, not all the balances you might write will be independent.

A subsystem does not necessarily have to be a piece of equipment. For example, a **mixing point** (junction of pipes) as illustrated in Figure 3.9 can be considered as a subsystem.

In Figure 3.10, multiple subsystems can be isolated such as a mixing point (junction of streams 1, 2, and 3), a **splitter** (junction of streams 5, 6, and 7), and a piece of equipment (represented by the box). The overall system designated by the dashed line involves streams 1, 2, 4, 6, and 7, but not 3 and 5.

Examine Figure 3.10. Which streams must have the same composition? Is the composition of stream 5 the same as the composition inside the unit? It will be the same if the contents of the unit are **well mixed,** the assumption in this text. Streams 5, 6, and 7 must have the same composition, and presumably if no reaction takes place, the output composition in stream 5 is the properly weighted average of the input compositions 3 and 4.

Now let us turn to the analysis of simple combinations of units. Suppose that a system is comprised of three subsystems as indicated in Figure 3.11. You can make material balances for the subsystems and overall system—just make sure that the balances selected for your solution are independent! How many values of the variables in Figure 3.11 are unknown? There will be seven in all: W, P, A, B, C, $\omega_{KCl,A}$ and $\omega_{H_2O,A}$.

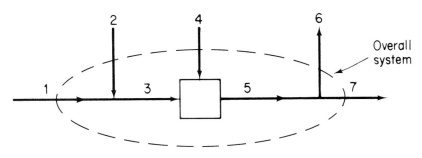

Figure 3.10   Multiple junction points connected serially.

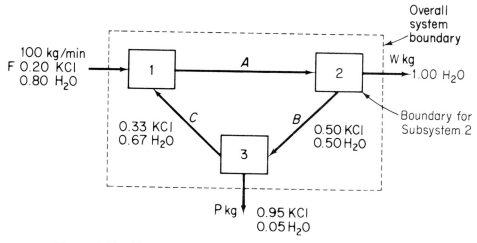

Figure 3.11   Flow diagram of a system comprised of three subsystems. No reaction takes place. The subsystem boundaries are the boxes.

How many independent equations must be written to obtain a unique solution? Seven. How many independent equations can you write? Two for each subsystem plus the sum of mass fractions for stream A. What are the names of such a set of equations? One set might be (on the basis of 1 min the units are in kg):

Unit 1, total	$100 + C = A$
Unit 1, KCl:	$(0.20)(100) + (0.33)(C) = (\omega_{KCl,A})$
Unit 2, total:	$A = W + B$
Unit 2, KCl:	$(\omega_{KCl,A})(A) = (0.50)B$
Overall, total:	$100 = W + P$
Overall, KCl	$(0.20)100 = (0.95)P$
$\Sigma\omega_i = 1$:	$\omega_{KCl,A} + \omega_{H_2O,A} = 1$

Other sets are possible. Write down a different set. Did you note that the set we have used has been chosen so as to include as few of the unknown variables as possible in a given equation; that is, we made component balances on KCl and not on H_2O. Keep in mind that the number of degrees of freedom summed for all the subsystems must be equal to zero to have a unique solution for the equations.

Keep in mind also that although you can make component or atomic species balances for each subsystem, the equations for each subsystem may not all be independent, and when you combine sets of equations from different subsystems, some of the equations may become dependent (redundant).

EXAMPLE 3.15 Independent Material Balances

Examine Figure E3.15. No reaction takes place. The composition of each stream is as follows:

    (1) Pure A

    (2) Pure B

    (3) A and B, concentrations known

    (4) Pure C

    (5) A, B, and C, concentrations known

    (6) Pure D

    (7) A and D, concentrations known

    (8) B and C, concentrations known

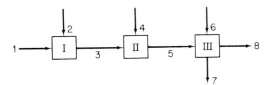

                                            Figure E3.15

What is the maximum number of independent mass balances that could be generated to solve this problem? How many can be used?

Solution

	Number of component balances
At unit I, two components are involved	2
At unit II, three components are involved	3
At unit III, four components are involved	4
Total	9

However, not all of the balances are independent. In the following list, all the concentrations are known values, and F represents the stream flow designated by the subscript.

Subsystem I

$$\text{Balances}\begin{cases} A\text{: } F_1(1.00) + F_2(0) = F_3\left(\omega_{F_3,A}\right) \\ B\text{: } F_1(0) + F_2(1.00) = F_3\left(\omega_{F_3,B}\right) \end{cases}$$

Subsystem II

$$\text{Balances} \begin{cases} A: F_3\left(\omega_{F_3,A}\right) + F_4(0) = F_5\left(\omega_{F_5,A}\right) \\ B: F_3\left(\omega_{F_3,B}\right) + F_4(0) = F_5\left(\omega_{F_5,B}\right) \\ C: F_3(0) + F_4(1.00) = F_5\left(\omega_{F_5,C}\right) \end{cases}$$

Subsystem III

$$\text{Balances} \begin{cases} A: F_5\left(\omega_{F_5,A}\right) + F_6(0) = F_7\left(\omega_{F_7,A}\right) + F_8(0) \\ B: F_5\left(\omega_{F_5,B}\right) + F_6(0) = F_7(0) + F_8\left(\omega_{F_8,B}\right) \\ C: F_5\left(\omega_{F_5,C}\right) + F_6(0) = F_7(0) + F_8\left(\omega_{F_8,C}\right) \\ D: F_5(0) + F_6(1.00) = F_7\left(\omega_{F_7,D}\right) + F_8(0) \end{cases}$$

If you take as a basis F_1, seven values of F_i are unknown, hence only seven independent equations need to be written. Can you show that the A and B balance can be reduced to just two independent equations so that among the entire set of nine equations two are redundant and a unique solution can be obtained? Hints: What implicit equations exist that we have not written down? In what order would you solve the equations?

Can you show that making one or more component mass balances around the combination of systems I and II, or II and III in Example 3.24, or the entire set of three units, will add *no* additional independent mass balances on each unit? Can you substitute one of the indicated alternative mass balances for an independent component mass balances? Yes (as long as the precision of the balance is about the same).

We next look at examples of making and solving material balances for systems composed of multiple units.

EXAMPLE 3.16 Multiple Units in Which No Reaction Occurs

Acetone is used in the manufacture of many chemicals and also as a solvent. In its latter role, many restrictions are placed on the release of acetone vapor to the environment. You are asked to design an acetone recovery system having the flowsheet illustrated in Figure E3.16. All the concentrations shown in Figure E3.16 of both gases and liquids are specified in *weight percent* in this special case to make the calculations simpler. Calculate A, F, W, B, and D per hour.

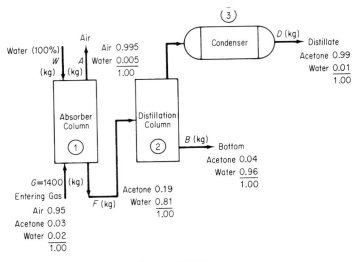

Figure E3.16

Solution

This is a steady-state process without reaction. Three subsystems exist.

     Steps 1, 2, 3 and 4   All the stream compositions are given. All of the unknown stream flows are designated by symbols in the figure.

     Step 5   Pick 1 hr as a basis so that $G = 1400$ kg.

     Steps 6 and 7   We could start the analysis with overall balances, but since the subsystems are connected serially, we will start the analysis with column 1 and then do unit 2, and then unit 3. Three components exist in column 1, and three values of the streams are unknown: W, A, and F; hence a unique solution can be obtained for column 1.

     Steps 7 and 8   The mass balances for column 1 are as follows:

	In		Out		
Air:	1400 (0.95)	=		A(0.995)	(a)
Acetone:	1400 (0.03)	=	F (0.19)		(b)
Water:	1400 (0.02) + W (1.00)	=	F (0.81) + A(0.005)		(c)

The air and acetone are tie components. (Check to make sure that the equations are independent.)

     Step 9   Solve Eqs. (a), (b), and (c) to get

$$
\begin{aligned}
A &= 1336.7 \ \text{kg/hr} \\
F &= 221.05 \ \text{kg/hr} \\
W &= 157.7 \ \text{kg/hr}
\end{aligned}
$$

     Step 10 (Check)   Use the total balance.

$$
\begin{array}{rcl}
G + W & = & A + F \\
1400 & & 1336 \\
\underline{157.7} & & \underline{221.05} \\
1557.7 & \cong & 1557.1
\end{array}
$$

Steps 5 and 6 Applied to Units 2 and 3 Combined

Although we could solve the material balances for column 2 first and then for condenser 3, since we have no information about the stream between column 2 and condenser 3, and are not asked to calculate any values of variables for that stream, we will lump column 2 and condenser 3 into one system for the calculations. Draw a dashed line about units 2 and 3 in Figure E3.16 to designate the system boundary. Two components exist, and two values of the flow streams are unknown, D and B; hence a unique solution exists (if the mass balances are independent, as they are).

Steps 7 and 8   The mass balances are

Acetone:	$221.05\,(0.19)$	$= D(0.99)$	$+$	$B(0.04)$	(d)
Water:	$221.05\,(0.81)$	$= D\,(0.01)$	$+$	$B\,(0.96)$	(e)

Step 9   Solve Eqs. (d) and (e) to get

$$D = 34.91 \text{ kg/hr}$$

$$B = 186.1 \text{ kg/hr}$$

Step 10 (Check)   Use the total balance

$$F = D + B \text{ or } 221.05 \cong 34.91 + 186.1 = 221.01$$

As a matter of interest, what other mass balances could be written for the system and substituted for any one of the Eqs. (a)-(e)? Typical balances would be the overall balances

	In		Out				
Air:	$G(0.95)$	$= A(0.995)$					(f)
Acetone:	$G(0.03)$	$=$		$D(0.99)$	$+$	$B(0.04)$	(g)
Water:	$G(0.02)+ W$	$= A(0.005)$	$+$	$D(0.01)$	$+$	$B(0.96)$	(h)
Total	$G + W$	$= A$	$+$	D	$+$	B	(i)

Equations (f)-(i) do not add any extra information to the problem; the degrees of freedom are still zero. But any of the equations can be substituted for one of Eqs. (a)–(e) as long as you make sure that the resulting set of equations is independent.

EXAMPLE 3.17 Multiple Units in Which a Reaction Occurs

In the face of higher fuel costs and the uncertainty of the supply of a particular fuel, many companies operate two furnaces, one with natural gas and the other with fuel oil. In the RAMAD Corp., each furnace had its own supply of oxygen; the oil furnace used

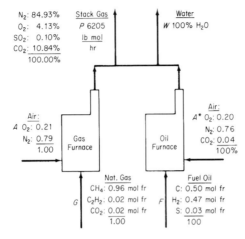

Figure E3.17

a gas stream that analyzed: O_2, 20%; N_2, 76%; and CO_2, 4%, but the stack gases went up a common stack. See Figure E3.17.

During one blizzard, all transportation to the RAMAD Corp. was cut off, and officials were worried about the dwindling reserves of fuel oil because the natural gas supply was being used at its maximum rate possible. The reserve of fuel oil was only 560 bbl. How many hours could the company operate before shutting down if no additional fuel oil was attainable? How many lb mol/hr of natural gas were being consumed? The minimum heating load for the company when translated into the stack gas output was 6205 lb mol/hr of dry stack gas. Analysis of the fuels and stack gas at this time were:

	Natural gas	Fuel oil (API gravity = 24.0)) (Mol %)		Stack gas (Orsat analysis)	
CH_4	96%	C	50	N_2	84.93%
C_2H_2	2%	H_2	47	O_2	4.13%
CO_2	2%	S	3	CO_2	10.84%
				SO_2	0.10%

The molecular weight of the fuel oil was 7.91 lb/lb mol, and its density was 7.578 lb/gal.

Solution

This is a steady state process with reaction. Two subsystems exist. We want to calculate F and G in lb mol/hr and then F in bbl/hr.

Steps 1, 2, 3 and 4   We will use atomic species for the mole balances. The units of all the variables whose values are unknown will be moles. Rather than making balances for each furnace, since we do not have any information about the individual outlet streams of each furnace, we will choose to make overall balances and draw the system boundary about both furnaces.

Step 5

$$\text{Basis: 1 hr, so that } P = 6205 \text{ lb mol}$$

Steps 6 and 7   We have five atomic components in this problem and five streams whose values are unknown, A, G, F, A^*, and W; hence, if the elemental mole balances are independent, we can obtain a unique solution for the problem.

Step 8   The overall mole balances for the elements are

	In			*Out*
H2:	$G(1.94)$	$+$ $F(0.47)$	$=$	$W(1)$
N2:	$A(0.79)$	$+$ $A^*(0.76)$	$=$	$6205(0.8493)$
O2:	$A(0.21)$	$+$ $A^*(0.20 + 0.04)$		$6205(0.0413 + 0.001 + 0.1084)$
		$+$ $G(0.02)$	$=$	$+ W(\tfrac{1}{2})$
S:	$F(0.03)$		$=$	$6205(0.0010)$
C:	$G[(0.96)$	$+$ $(2)(0.02) + 0.02)]$		
		$+$ $F(0.5) + 0.04A^*$	$=$	$6205(0.1084)$

Step 9   Solve the S balance for F (inaccuracy in the SO_2 concentrations will cause some error in F, unfortunately); the sulfur is a tie component. Then solve for the other four balances simultaneously for G.

$$F = 207 \text{ lb mol/h\i}$$
$$G = 498 \text{ lb mol/h\i}$$

Finally, the fuel oil consumption is

$$\frac{207 \text{ mol}}{\text{hr}} \left| \frac{7.91 \text{ lb}}{\text{mol}} \right| \frac{\text{gal}}{7.578 \text{ lb}} \left| \frac{\text{bbl}}{42 \text{ gal}} \right. = 5.14 \text{ bbl/hr}$$

If the fuel oil reserve were only 560 bbl, this amount could last at the most

$$\frac{560 \text{ bbl}}{5.14 \dfrac{\text{bbl}}{\text{hr}}} = 109 \text{ hr}$$

LOOKING BACK

In this section you have seen how systems composed of more than one subsystem can be treated in exactly the same fashion by the 10–step strategy as you treat single subsystems. Whether you use combinations of material balances from several units or lump all of the units into one system, all you have to do is check to see that the number of independent equations you prepared was adequate to solve for the variables whose values are unknown.

Key Ideas

1. Subsystems can be comprised of equipment, junction points of pipes, and/or combinations of subsystems.
2. Material balances can be written for the components of each subsystem and for the total mass flow as well as for the components of the overall system and total mass flow of the overall system.
3. The number of independent equations in a problem cannot be increased by writing overall balances in addition to the component balances for each subsystem, but the latter can be substituted for the former.

Key Terms

Mixing point (p. 196)   Serial connections (p. 195)

Overall process (p. 195)   Splitter (p. 196)

Self-Assessment Test

1. A two-stage separations unit is shown in Figure AT3.5–1. Given the input stream F1 is 1000 lb/hr, calculate the value of F2 and the composition of F2.
2. A simplified process for the production of SO_3 to be used in the manufacture of sulfuric acid is illustrated in Figure AT3.5–2. Sulfur is burned with 100% excess air in the burner, but for the reaction $S + O_2 \rightarrow SO_2$, only 90% conversion of the S to SO_2 is

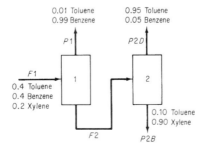

Figure AT3.5–1

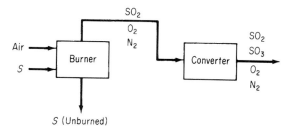

Figure AT3.5–2

achieved. In the converter, the conversion of SO_2 to SO_3 is 95% complete. Calculate the lb of air required per 100 lb of sulfur burned, and the concentrations in mole fraction of the components in the exit gas from the burner and from the converter.

Discussion Problem

1. A number of technologies have been proposed to reduce NO_x emissions from industrial boilers to acceptable levels. A report by the National Research Council has concluded that in most urban areas reductions in NO_x will lead to decreased levels of atmospheric ozone.

The flowsheet for selective noncatalytic reduction (SNCR) (see Figure DP3.5–1) indicates that a reagent, typically ammonia or urea, is used to bond with NO_x to produce inert gases that are released to the atmosphere. The reaction occurs at about 1000°C. SNCR is generally less expensive than a catalytic process. What are some of the possible other advantages and disadvantages of the SNCR process?

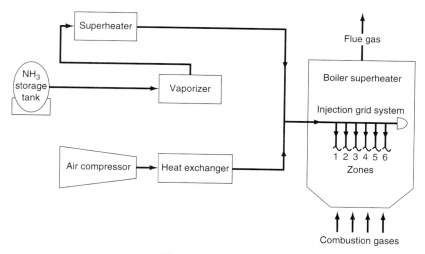

Figure DP3.5–1

3.6 RECYCLE, BYPASS, AND PURGE CALCULATIONS

***Your objectives in studying this
section are to be able to:***

1. Draw a flow diagram for problems involving recycle, bypass, and purge.
2. Apply the 10-step strategy to solve steady-state problems (with and without chemical reaction) involving recycle, and/or bypass, and/or purge streams.
3. Solve problems in which a modest number of interconnected units are involved by making appropriate balances.
4. Use the overall conversion and single-pass (once-through) conversion concepts in solving recycle problems involving reactors.
5. Explain the purpose of a recycle stream, a bypass stream, and a purge stream.

LOOKING AHEAD

In this section we take up material balances involving recycle—instances in which material returns back from downstream of the process and reenters the process again. Cases with and without reaction will be discussed.

MAIN CONCEPTS

Recycling involves returning material (or energy) that leaves a process back to the process for further processing. Recycling can involve a whole city, such as occurs in the recycling of newspapers and cans, or can involve just a single process such as a reactor. A **recycle stream** is a term denoting a process stream that returns material from downstream of a process unit back to the process unit (or to a unit upstream of the unit). For example, examine Figure 3.12, which illustrates various recycle streams designed to make long space missions feasible. All the water and oxygen will have to be recycled so that the starting load in the spacecraft will not be excessive.

Many industrial processes employ recycle streams. In some drying operations, the humidity in the air is controlled by recirculating part of the wet air that

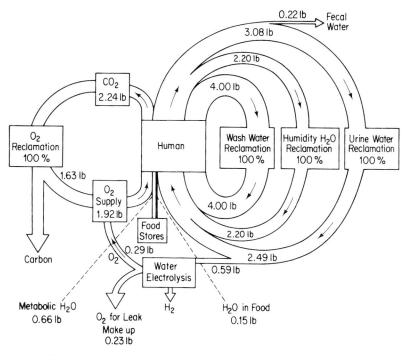

Figure 3.12   Water and oxygen recycle in a space vehicle.

leaves the dryer. In chemical reactions, exit catalyst is returned to the reactor for reuse. Another example of the use of recycling is in fractionating columns where part of the distillate is refluxed back into the column to maintain the quantity of liquid within the column.

The steps in the analysis and solution of material balance problems involving recycle are exactly the same as described in Table 3.1. With a little practice in solving problems involving recycle, you should experience little difficulty in solving recycle problems in general. The **essential point** you should grasp with respect to recycle calculations in this chapter is that the processes such as shown in Figure 3.12 are operating in the *steady state*.

No buildup or depletion of material takes place inside the process or in the recycle stream.

The values of the flows in Figure 3.12 are *constant*. Unsteady-state processes such as startup and shutdown are discussed in Chapter 7.

3.6-1 Recycle in Processes without Chemical Reaction

We first look at processes in which no reactions take place. The strategy listed in Table 3.1 is the strategy to be used in solving recycle problems. You can make component and total material balances for each subsystem as discussed previously in Sect. 3.5, as well as component and total balances for the overall process. Not all of the equations so formulated will be independent, of course. Depending on the information available concerning the amount and composition of each stream, you can determine the amount and composition of the unknowns. If tie components are available, they often simplify the calculations.

Examine Figure 3.13. Material balances can be written for several different systems, four of which are shown by dashed lines in Figure 3.13, namely

1. About the entire process including the recycle stream, as indicated by the dashed lines identified by 1 in Figure 3.13. These balances contain no information about the recycle stream.
2. About the junction point at which the **fresh feed** is combined with the recycle stream (identified by 2 in Figure 3.13).
3. About the process only (identified by 3 in Figure 3.13). These balances contain no information about the recycle stream.
4. About the junction point at which the **gross product** is separated into recycle and **net product** (identified by 4 in Figure 3.13).

In addition, balances can be made about combinations of subsystems, such as the process plus the separator. Only three of the four balances (a)-(d) are independent when made for one component. However, balance 1 will not include the

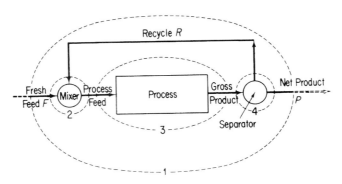

Figure 3.13 Process with recycle (the numbers designate possible system boundaries for the material balances—see the text).

recycle stream, so that the balance will not be directly useful in calculating a value for the recycle R. Balances 2 and 4 do include R. You could write a material balance for the combination of subsystems 2 and 3 or 3 and 4 and include the recycle stream as shown in the following examples.

EXAMPLE 3.18 Recycle without Chemical Reaction

A distillation column separates 10,000 kg/hr of a 50% benzene-50% toluene mixture. The product D recovered from the condenser at the top of the column contains 95% benzene, and the bottom W from the column contains 96% toluene. The vapor stream V entering the condenser from the top of the column is 8000 kg/hr. A portion of the product from the condenser is returned to the column as reflux, and the rest is withdrawn for use elsewhere. Assume that the compositions of the streams at the top of the column (V), the product withdrawn (D), and the reflux (R) are identical because the V stream is condensed completely. Find the ratio of the amount refluxed R to the product withdrawn (D).

Solution

This is a steady-state problem without reaction occurring.

Steps 1, 2, 3, and 4   See Figure E3.18 for the known data, symbols, and other information.

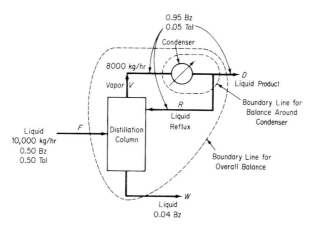

Figure E3.18

Step 5   Select a basis of 1 hr (equal to $F = 10,000$ kg).

Steps 6 and 7   All the compositions are known and three stream flows, D, W, *and* R, are unknown. No tie components are evident in this problem. Two component material balances can be made for the still and two for the condenser. Presumably three of these are independent; hence the problem has a unique solution. We can check as we proceed. A balance around either the distillation column or the condenser would involve the stream R. An overall balance would involve D and W but not R.

Steps 7, 8, and 9   What balances to select to solve for R are somewhat arbitrary. We will choose to use two overall balances first to get D and W, and then use a balance on the condenser to get R. Once D is obtained, R can be obtained by subtraction.

Overall Material balances:
   Total material

$$F = D + W$$
$$10{,}000 = D + W \tag{a}$$

Component (benzene):

$$F\omega_F = D\omega_D + W\omega_W$$
$$10{,}000(0.50) = D(0.95) + W(0.04) \tag{b}$$

Solving (a) and (b) together, we obtain

$$W = 4950 \text{ kg/hr}$$
$$D = 5050 \text{ kg/hr}$$

Balance around the condenser:
   Total material:

$$V = R + D$$
$$8000 = R + 5050 \tag{c}$$
$$R = 2950 \text{ kg/hr}$$
$$\frac{R}{D} = \frac{2950}{5050} = 0.58$$

Would the benzene or toluene balances on the condenser yield additional information to that obtained from the total balance, Eq. (c)? Write the balances down and check to see if they are redundant with Eq. (c).

EXAMPLE 3.19 Recycle without Chemical Reaction

The manufacture of such products as penicillin, tetracycline, vitamins, and other pharmaceuticals, as well as photographic chemicals, dyes, and other fine organic compounds, usually requires separating the suspended solids from their mother liquor by centrifuging, and then drying the wet cake. A closed-loop system (see Figure E3.19a) for centrifuge unloading, drying, conveying, and solvent recovery is comprised of equipment especially designed for handling materials requiring sterile and contamination-free conditions.

    Given the experimental measurements on the pilot plant equipment outlined in Figure E3.19a, what is the lb/hr of the recycle stream R?

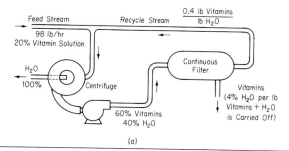

Figure E3.19a

Solution

This is a steady-state problem without reaction and with recycle.

Steps 1, 2, 3 and 4   Figure E3.19a can be simplified with all the known flows and compositions placed on the simplification. Examine Figure E3.19b. We computed the weight fraction of vitamins V in the recycle R from the data given in Fig. E3.19a. On the basis of 1 lb of water, the recycle stream contains (1.0 lb of H_2O + 0.4 lb of V) = 1.4 lb total. The recycle stream composition is

$$\frac{0.4 \text{ lb } V}{1 \text{ lb } H_2O} \left| \frac{1 \text{ lb } H_2O}{1.4 \text{ lb solution}} \right. = 0.286 \text{ lb } V \text{ /lb solution}$$

so that there is 0.714 lb H_2O/lb solution.

Step 5   Pick as a basis 1 hr so that $F = 98$ lb.

Steps 6 and 7   We have four unknown values of variables, W, C, P, and R. We also know nothing about the mass flows of the two components entering the centrifuge after the mixing of F and R. Thus, we need six independent balances. We can make balances on the junction point of F and R. If all of these balances are independent, the problem has a unique solution. We will proceed with the solution assuming the balances are independent.

Steps 7, 8 and 9   Probably the simplest and most effective procedure to solve this problem is to first make overall mass balances to calculate W and P, and then write mass balances about the filter or the junction point to calculate R.

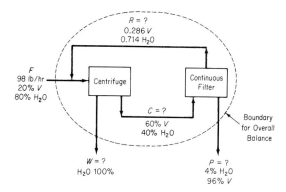

Figure E3.19b

Overall mass balances:

$$V: \qquad 0.20(98) = 0 \qquad\quad + \;\; 0.96P \qquad\qquad \text{(a)}$$

$$H_2O: \qquad 0.80(98) = (1.0)W \;\; + \;\; 0.04P \qquad\qquad \text{(b)}$$

$$\text{Total:} \qquad 98 \qquad\;\; = W \qquad\;\; + \;\; P \qquad\qquad\quad \text{(c)}$$

Observe that V is a tie component so that P can be calculated directly in Eq. (a): $P = 20.4$ lb, and W can be calculated from Eq. (c).

$$W = 98 - 20.4 = 77.6 \text{ lb}$$

Steps 7, 8, and 9 (Continued)   To determine the recycle stream R, we need to make a balance that involves stream R. Either (a) balances around the centrifuge or (b) balances around the filter will do. The latter are easier to formulate since the mixing of R and F does not have to be calculated first.

Total balance on filter:

$$C = R + P$$
$$C = R + 20.4 \qquad\qquad\qquad\qquad \text{(d)}$$

Component V balance on filter:

$$C\omega_C = R\omega_R + P\omega_P$$

$$0.6C = 0.286R + 0.96(20.4) \qquad\qquad\qquad \text{(e)}$$

Solving Eqs. (d) and (e), we obtain $R = 23.4$ lb/hr.

Step 10   Check the value of R using a material balance around the centrifuge.

3.6-2 Recycle in Processes with Chemical Reaction

Now let us turn to recycle problems in which a chemical reaction occurs. Recall from Sec. 1.7 that not all of the limiting reactant necessarily reacts in a process. Do you remember the concept of conversion as discussed in Sec. 1.7? Two different bases for conversion are used in connection with the reactions that occur in a process. Examine Figure 3.13 for the various streams used in the definitions.

1. *Overall fraction conversion*

$$\frac{\text{mass (moles) of reactant in fresh feed - mass (moles) of reactant in output of the overall process (net product)}}{\text{mass (moles) of reactant in fresh feed}}$$

2. *Single-pass ("once-through") fraction conversion*

$$\frac{\text{mass (moles) of reactant fed into the reactor (process feed) - mass (moles) of reactant existing the reactor (gross product)}}{\text{mass (moles) of reactant fed into the reactor}}$$

When the fresh feed consists of more than one material, the conversion must be stated for a single component, usually the limiting reactant, the most expensive reactant, or some similar compound.

Note the distinction between *fresh feed* and *process feed.* The feed to the process itself is made up of two streams, the fresh feed and the recycled material. The gross product leaving the process is separated into two streams, the net product and the material to be recycled. In some cases the recycle stream may have the same composition as the gross product stream, while in other instances the composition may be entirely different depending on how the separation takes place and what happens in the process. Suppose that you are given the data that 30% of a compound A is converted to B on a single pass through the reactor, as illustrated in Figure 3.14, and are asked to calculate the value of R, the recycle, on the basis of 100 moles of fresh feed, F. We will make a balance for A with the reactor as the system.

Recall from Eq. (3.1) that for a specific compound the steady-state material balance for a reactor is (the accumulation term in zero)

$$\left\{\begin{array}{c}\textbf{input}\\\textbf{through}\\\textbf{system}\\\textbf{boundary}\end{array}\right\} - \left\{\begin{array}{c}\textbf{output}\\\textbf{through}\\\textbf{system}\\\textbf{boundary}\end{array}\right\} + \left\{\begin{array}{c}\textbf{generation}\\\textbf{within the}\\\textbf{system}\end{array}\right\} - \left\{\begin{array}{c}\textbf{consumption}\\\textbf{within the}\\\textbf{system}\end{array}\right\} = 0$$

To make a balance for a particular compound such as A, you have to be given, look up, calculate, or experiment to find the extent of the reaction of A to form products. In one type of problem you are given the fraction conversion of A on one pass through the reactor, and asked to calculate R and the other stream flows. The reverse of this problem is to calculate the fraction conversion given the stream flows (or given data to calculate the stream flows).

Let us examine how to apply Eq. (3.1) for a recycle reactor such as shown in Figure 3.14, in which A is converted to B. We want to calculate the value of R. To do so we must make a balance that cuts the recycle stream. A mole balance for the compound A with one system being the mixing point and one being the separator will serve as a start.

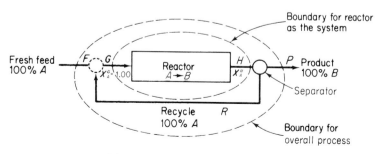

Figure 3.14   Recycle problem.

$$\overset{\text{\textit{System: Mixing Point}}}{\text{Balance on }A:\ \overline{1.00(100) + 1.00\ R = 1.00\ G}}\quad \overset{\text{\textit{System: Separator}}}{\overline{Hx_A = 0(P) + 1.00\ R}}$$

Because we have more unknowns than equations, clearly we need at least one more balance, and this is where the fraction conversion comes in. We will make a balance on the reactor using the fraction conversion in formulating the consumption term.

$$\text{\textit{System\ Reactor}}$$

$$\text{Balance on }A:\ \underset{\textit{In}}{\overline{1.00\ G}}\ -\ \underset{\textit{Out}}{\overline{Hx_A}}\ +\ \underset{\textit{Generation}}{\overline{0}}\ -\ \underset{\textit{Consumption}}{\overline{0.30\ G}}\ =\ \underset{\textit{Accumulation}}{\overline{0}}$$

If we substitute the separator balance into the above relation to eliminate Hx_A and the mixing point balance to eliminate G, we obtain one equation containing just the desired unknown variable R

$$\underset{G}{\underbrace{(100 + R)}}\ -\ \underset{Hx_A}{\underbrace{R}}\ -\ \underset{\text{consumption of }A}{\underbrace{0.30(100 + R)}} = 0$$

the solution for which is

$$R = 233 \text{ mol}$$

Note that all of the A was recycled for simplicity of illustration of the principle in our discussion, but such is rarely the case in general. Nevertheless, Eq. (3.1) still applies as shown in the examples that follow.

EXAMPLE 3.20 Recycle with a Reaction Occurring

Immobilized glucose isomerase is used as a catalyst in producing fructose from glucose in a fixed-bed reactor (water is the solvent). For the system shown in Figure E3.20a, what percent conversion of glucose results on one pass through the reactor when the ratio of the exit stream to the recycle stream in mass units is equal to 8.33? The reaction is

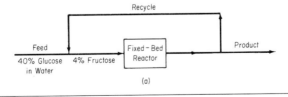

(a)

Figure E3.20a

Solution

We have a steady-state process with a reaction occurring and recycle.

Steps 1, 2, 3, and 4   Figure E3.20b includes all the known and unknown values

of the variables using appropriate notation (*W* stands for water, *G* for glucose, and *F* for fructose in the second position of the mass fraction subscripts). Note that the recycle stream and product stream have the same composition and consequently the same mass symbols are used in the diagram for each stream.

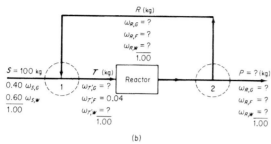

(b)   **Figure E3.20b**

Step 5   Pick as a basis $S = 100$ kg.

Step 6   We have not provided notation for the reactor exit stream and composition as we will not be using these values in our balances. Let f be the fraction conversion for one pass through the reactor. The unknowns are R, F, P, T, $\omega_{R,G}$, $\omega_{R,T}$, $\omega_{R,W}$, $\omega_{T,W}$, $\omega_{T,G}$, and f, for a total of 9.

Step 7   The balances are $\Sigma\omega_{R,\ i} = 1$, $\Sigma\omega_{T,\ i} = 1$, $R = P/8.33$, plus three compound balances each on the mixing point 1, the separator 2, and the reactor. We will assume we can find 9 independent balances among the 12, and proceed. We do not have to solve all of the equations simultaneously. The units are mass (kg).

Steps 7, 8 and 9   We will start with overall balances as they are easy to form and are often decoupled for solution.

> ***Overall balances:***
>
>   *Total:* $P = 100$ kg (How simple!)
>
> Consequently $R = \dfrac{100}{8.33} = 12.0$ kg

Overall no water is generated or consumed, hence

>   *Water:* $100(0.60) = P(\omega_{R,W}) = 100\ \omega_{R,W}$
>
> $\omega_{R,W} = 0.60$

We now have 6 unknowns left for which to solve.

> ***Mixing point I***
>
> No reaction occurs so that compound balances can be used with consumption and generation terms being included:
>
>   *Total:* $100 + 12 = T = 112$
>
>   *Glucose:* $100(0.40) + 12(\omega_{P,\ G}) = 112(\omega_{T,G})$
>
>   *Fructose:* $0 + 12(\omega_{R,\ F}) = 112(0.04)$

or

$$\omega_{R,F} = 0.373$$

Also, because $\omega_{R,F} + \omega_{R,G} + \omega_{R,W} = 1$,

$$\omega_{R,G} = 1 - 0.373 - 0.600 = 0.027$$

and then from the glucose balance,

$$\omega_{T,G} = 0.360$$

Next, rather than make separate balances on the reactor and separator, we will combine the two into one system (and thus avoid having to calculate values associated with the reactor exit stream).

Reactor Plus Separator 2:

Total: $T = 12 + 100 = 112$ (redundant equation)

Glucose:

$T\,\omega_{T,G}$	$-$	Out $(R+P)\,\omega_{R,G}$	$-$	Consumed $fT\omega_{T,G}$	$=$	0
112(0.360)	$-$	112(0.027)	$-$	$f(112)(0.360)$	$=$	0

$$f = 0.93$$

EXAMPLE 3.21 Recycle with a Reaction Occurring

Refined sugar (sucrose) can be converted to glucose and fructose by the inversion process

$$C_{12}H_{22}O_{11} + H_2O \longrightarrow C_6H_{12}O_6 + C_6H_{12}O_6$$

Sucrose   *d*-Glucose   *d*-Fructose

The combined quantity glucose plus fructose is called inversion sugar. If 90% conversion of sucrose occurs on one pass through the reactor, what would be the recycle stream flow per 100 lb fresh feed of sucrose solution entering the process shown in Figure E3.21a? What is the concentration of inversion sugar (*I*) in the recycle stream and in the product stream? The concentrations of components in the recycle stream and product stream are the same.

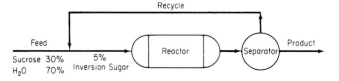

Figure E3.21a

Solution

This problem is the inverse of the previous example.

Steps 1, 2, 3, and 4   First we need to enter the concentrations and stream flows on the diagram. See Figure E3.21b. (W stands for water, S for sucrose, and I for inversion sugar in the mass fraction subscripts.)

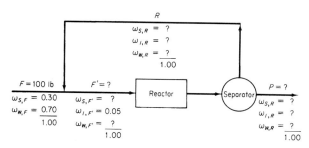

Figure E3.21b

Step 5   Basis: $F = 100$ lb

Step 6   The unknowns are $R, P, F', \omega_{S,F'}, \omega_{W,F'}, \omega_{S,R}, \omega_{I,R}, \omega_{W,R}$, a total of 8.

Step 7   We have as balances $\Sigma\omega_{i,F'} = 1$ and $\Sigma\omega_{i,R} = 1$ plus, as in the previous example, 3 component balances each for the mixing point, the reactor, and the separator so that we should be able to find 8 independent equations among the lot.

Steps 8 and 9   We will again start with the overall balances.

Overall balance

Only the total balance among the overall balances is directly useful for the moment because the S and I balances involve the generation and consumption terms in Eq. (3.1)

$$\text{Total: } P = 100 \text{ lb}$$

Next we make balances on the mixing point (where F and R join). The units are lb.

Mixing Point:

Total:	$100 + R$	$= F'$	(a)
Sucrose:	$100(0.30) + R\omega_{S,R}$	$= F'\omega_{S,F'}$	(b)
Inversion:	$0 + R\omega_{I,R}$	$= F'(0.05)$	(c)

Finally, we make balances on the system of the reactor plus the separator for the reasons stated in the previous example.

Reactor plus Separator

$$\underset{\text{In}}{\underline{F'\omega_{S,F'}}} \quad \underset{\text{Out}}{\underline{-(R+100)\,\omega_{S,R}}} \quad \underset{\text{Consumed}}{\underline{-F'\omega_{S,F'}(0.90)}} = 0 \qquad \text{(d)}$$

We now have remaining 7 unknowns and 6 (hopefully independent) equations. We need one more equation. We could try a total balance as being the simplest

$$F' = R + P$$

but clearly this equation is redundant with Eq. (a). Suppose we set up a water balance. To make the water balance we need to calculate the pounds of water consumed in the reaction per pound of sucrose consumed in the reaction.

1 mol of sucrose uses 1 mole of water

$$\frac{1 \text{ mol } W}{1 \text{ mol } S} \; \bigg| \; \frac{1 \text{ mol } S}{342.35 \text{ lb } S} \; \bigg| \; \frac{18 \text{ lb} W}{1 \text{ mol } W} = 0.0526 \; \frac{\text{lb} W}{\text{lb } S}$$

$$\text{Water:} \quad \underbrace{F'\omega_{W,F'}}_{In} \quad \underbrace{- (R + 100)\, \omega_{S,R}}_{Out} \quad \underbrace{- F'\omega_{S,F}(0.90)(0.0526)}_{Consumed} = 0 \quad \text{(e)}$$

If Eqs. (a) – (e) plus the two sum of mass fraction equations are independent, we can solve for R and $\omega_{I,P} = \omega_{I,R}$. A computer program will ease the solution:

$$R = 20.9 \text{ lb}$$
$$\omega_{I,P} = 0.313$$

3.6-3 Bypass and Purge

Two additional commonly encountered types of process streams are shown in Figure 3.15.

1. A **bypass** stream—one that skips one or more stages of the process and goes directly to another downstream stage (Figure 3.15a).

2. A **purge** stream—a stream bled off to remove an accumulation of inerts or unwanted material that might otherwise build up in the recycle stream (Figure 3.15b).

A bypass stream can be used to control the composition of a final exit stream from a unit by mixing the bypass stream and the unit exit stream in suitable proportions to obtain the desired final composition.

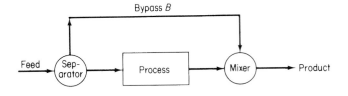

Figure 3.15a   Recycle stream with bypass.

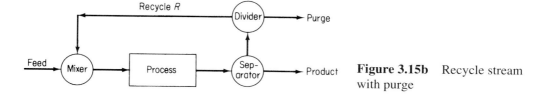

Figure 3.15b   Recycle stream with purge

As an example of the use of a purge stream, consider the production of NH_3. Stream reforming, with a feedstock of natural gas, LPG, or naphtha, is the most widely accepted process for ammonia manufacture. The route includes four major chemical steps:

Reforming:	CH_4	+	H_2O	→	CO	+	$3H_2$
Shift:	CO	+	H_2O	→	CO_2	+	H_2
Methanation:	CO	+	$3H_2$	→	H_2O	+	CH_4
Synthesis	$3H_2$	+	N_2	→	$2NH_3$		

In the final stage, for the fourth reaction, the synthesis gas stream is approximately a 3:1 mixture of hydrogen to nitrogen, with the remainder about 0.9% methane and 0.3% argon.

Compressors step up the gas pressure from atmospheric to about 3000 psi—the high pressure that is needed to favor the synthesis equilibrium. Once pressurized and mixed with recycle gas, the stream enters the synthesis converter, where ammonia is catalytically formed at 400 to 500°C. The NH_3 is recovered as a liquid via refrigeration, and the unreacted syngas is recycled.

In the synthesis step, however, some of the gas stream must be purged to prevent buildup of argon and methane. But purging causes a significant loss of hydrogen that could be used for additional ammonia manufacture, a loss that process designers seek to minimize.

Do you understand why the recycle process without a purge stream will cause an impurity to build up even though the recycle rate is constant? The purge rate is adjusted so that the amount of purged material remains below an acceptable specified economic level or so that the

$$\left\{ \begin{matrix} \text{rate of} \\ \text{accumulation} \end{matrix} \right\} = 0 = \left\{ \begin{matrix} \text{rate of entering material} \\ \text{and / or production} \end{matrix} \right\} - \left\{ \begin{matrix} \text{rate of purge} \\ \text{and / or loss} \end{matrix} \right\}$$

Calculations for bypass and purge streams introduce no new principles or techniques beyond those presented so far. Two examples will make that clear.

EXAMPLE 3.22 Bypass Calculations

In the feedstock preparation section of a plant manufacturing natural gasoline, isopentane is removed from butane-free gasoline. Assume for purposes of simplification that the process and components are as shown in Figure E3.22. What fraction of the butane-free gasoline is passed through the isopentane tower? Detailed steps will not be listed in the analysis and solution of this problem. The process is in the steady state and no reaction occurs.

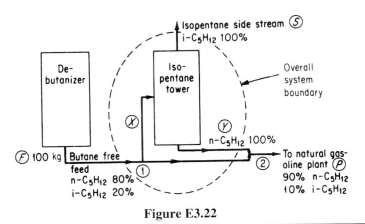

Figure E3.22

Solution

By examining the flow diagram you can see that part of the butane-free gasoline bypasses the isopentane tower and proceeds to the next stage in the natural gasoline plant. All the compositions (the streams are liquid) are known. What kind of balances can we write for this process? We can write the following:

$$\text{Basis: } 100 \text{ kg feed}$$

(a) *Overall balances* (each stream is designates by the letter F, S, or P with units being kg)

Total material balance:

$$\frac{In}{100} = \frac{Out}{S+P} \tag{a}$$

Component balance ($n\text{-}C_5$)(*tie component*)

$$\frac{In}{100(0.80)} = \frac{Out}{S(0) + P(0.90)} \tag{b}$$

Consequently,

$$P = 100\left(\frac{0.80}{0.90}\right) = 88.9 \text{ kg}$$

$$S = 100 - 88.9 = 11.1 \text{ kg}$$

The overall balances will not tell us the fraction of the feed going to the isopentane tower; for this calculation we need another balance.

(b) *Balance around isopentane tower:* Let x = kg of butane-free gas going to the isopentane tower and y be the kg of the n-C_5H_{12} stream leaving the isopentane tower.

Total material balance:

$$\frac{In}{x} = \frac{Out}{11.1 + y} \tag{c}$$

Component (n-C_5) (a tie component):

$$x(0.80) = y \tag{d}$$

Consequently, combining (c) and (d)

$$x = 55.5 \text{ kg, or the desired fraction is } 0.55$$

Another approach to this problem would be to make a balance at mixing points 1 and 2. Although there are no pieces of equipment at those points, you can see that streams enter and leave the junctions.

(c) *Balance around mixing point 2:*

$$\text{material into junction} = \text{material out}$$

Total material:

$$(100 - x) + y = 88.9 \tag{e}$$

Component (iso-C_5):

$$(100 - x)(0.20) + 0 = 88.9(0.10) \tag{f}$$

Equation (f) avoids the use of y. Solving yields

$$x = 55.5 \text{ kg as before}$$

EXAMPLE 3.23 Purge

Considerable interest exists in the conversion of coal into more convenient liquid products for subsequent production of chemicals. Two of the main gases that can be generated under suitable conditions from insitu coal combustion in the presence of steam (as occurs naturally in the presence of groundwater) are H_2 and CO. After cleanup, these two gases can be combined to yield methanol according to the follow-ing equation

$$CO + 2H_2 \longrightarrow CH_3OH$$

Figure E3.23 illustrates a steady-state process. All compositions are in the mole frac-tions or percent. The stream flows are in moles.

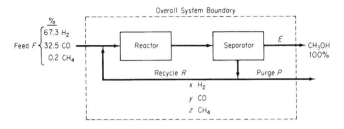

Figure E3.23

You will note in Figure E3.23 that some CH_4 enters the process, but the CH_4 does not participate in the reaction. A purge stream is used to maintain the CH_4 concentration in the exit to the separator at no more than 3.2 mol%. The once-through conversion of the CO in the reactor is 18%.

Compute the moles of recycle, CH_3OH, and purge per mole of feed, and also compute the purge gas composition.

Solution

Steps 1, 2, 3, and 4 All of the known information has been placed on the dia-gram. The process is in the steady state with reaction. The purge and recycle streams have the same composition (implied by the splitter in the figure). The mole fraction of the components in the purge stream have been designated as x, y, and z for H_2, CO, and CH_4, respectively.

Step 5 Select a convenient basis:

$$F = 100 \text{ mol}$$

Step 6 The variables whose values are unknown are x, y, z, E, P, and R. We will ignore the stream between the reactor and separator as no questions are asked about it.

Step 7 Because the problem is presented in terms of moles, making an overall mass balance is not convenient. We will use element balances instead. We can make three independent element balances for the overall process: H, C, and O balances. A CO mole balance on the reactor plus separator will provide one additional balance. How can we obtain fifth and sixth balances so that the system of equations is determi-nate? One piece of information given in the problem statement that we have not used is the information about the upper limit on the CH_4 concentration in purge stream. This limit can be expressed as $z \le 0.032$. Let us assume that purge stream contains the max-imum allowed CH_4 and make

$$z = 0.032 \tag{a}$$

Another piece of information is the implicit balance in the recycle stream

$$x + y + z = 1 \qquad\qquad (b)$$

Steps 7 and 8   The overall element balances are (in moles):

H2: $67.3 + 0.2(2)$ $= E(2) + P(x + 2z)$ (c)
C: $32.5 + 0.2$ $= E(1) + P(y + z)$ (d)
O: 32.5 $= E(1) + P(y)$ (e)

For a system composed of the reactor plus the separator (chosen to avoid calculating the unknown information about the direct output of the reactor), the CO balance is

$$\begin{array}{ccccc} \underline{\textit{In}} & & \underline{\textit{Out}} & & \underline{\textit{Consumed}} \\ \text{CO:} \quad 32.5 + Ry & - & y(R + P) & = & (32.5 + Ry)(0.18) \end{array} \qquad (f)$$

Step 9   Equation (a) can be substituted into Eqs. (b)-(f) and the resulting five equations solved by successive substitution or by using a computer program. The resulting values obtained are (in moles)

E	CH_3OH	31.25
P	purge	6.25
R	recycle	705
x	H_2	0.768
y	CO	0.200
z	CH_4	0.032

Step 10   Check to see that each of the balances (b)–(f) is satisfied.

Until now we have discussed material balances of a rather simple order of complexity. If you try to visualize all the calculations that might be involved in even a moderate-size plant, as illustrated in Figure 3.5, the stepwise or simultaneous solution of material balances for each phase of the entire plant may seem to be a staggering task. However, the task can be eased considerably by the use of computer codes as discussed in Sec. 2.2. Keep in mind that a plant can be described by a number of individual, interlocking material balances each of which, however tedious they are to set up and solve, can be set down according to the principles and techniques discussed in this chapter. In application there is always the problem of collecting suitable information and evaluating its accuracy, but this matter calls for detailed familiarity with any specific process and is not a suitable topic for discussion here. We can merely remark that some of the problems you will encounter have such conflicting data or so little useful data that the ability to perceive what kind of data are needed is the most important attribute that you can bring to bear in their solution.

LOOKING BACK

From the examples presented in this section you should have concluded that problems involving recycle, purge, and bypass are no different from the viewpoint of how they are analyzed than any of the problems solved in earlier sections of this chapter. The one different factor brought out in Sec. 3.6-2 is that recycle for a reactor usually involves information about the fraction conversion of a reactant, and thus the full form of the material balance, Eq. (3.1), must be employed along with or in lieu of balances on the elements.

Key Concepts

1. The general material balance, Eq. (3.1) applies for processes that involve recycle, purge, and bypass as it does for other processes.
2. The new key feature of material balances with recycle in a process in which a reaction takes place is that specification of the fraction conversion of a reactant is the added bit of information useful in constructing the consumption term of Eq. (3.1).
3. Purge is used to maintain a concentration of a minor component of a process stream below some set-point so that it does not accumulate in the process.

Key Terms

Bypass (p. 218)   Overall fraction conversion (p. 212)
Feed to the process (p. 213)   Process feed (p. 208)
Fresh feed (p. 213)   Purge (p. 218)
Gross product (p. 208)   Recycle (p. 206)
Net product (p. 208)   Recycle stream (p. 206)
Once-through conversion (p. 212)   Single pass conversion (p. 212)

Self-Assessment Test

1. Explain what recycle and bypassing involve by means of words and also by a diagram.
2. Repeat for the term "purge."
3. If the components in the feed to a process appear in stoichiometric quantities and the subsequent separation process is complete so that all of the unreacted reactants are recycled, what is the ratio of reactants in the recycle stream?
4. A material containing 75% water and 25% solid is fed to a granulator at a rate of 4000 kg/hr. The feed is premixed in the granulator with recycled product from a dryer which follows the granulator (to reduce the water concentration of the overall material in the granulator to 50% water, 50% solid). The product that leaves the dryer is 16.7% water. In the dryer, air is passed over the solid being dried. The air entering the dryer contains 3% water by weight (mass), and the air leaving the dryer contains 6% water by weight (mass).
   (a) What is the recycle ratio to the granulator?
   (b) What is the rate of air flow to the dryer on a dry basis?

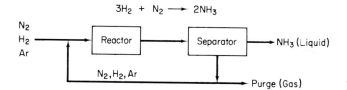

Figure AT3.6–5

5. In the famous Haber process to manufacture ammonia, the reaction is carried out at pressures of 800 to 1000 atm and at 500 to 600°C using a suitable catalyst. Only a small fraction of the material entering the reactor reacts on one pass, so recycle is needed. Also, because the nitrogen is obtained from the air, it contains almost 1% rare gases (chiefly argon) that do not react. The rare gases would continue to build up in the recycle until their effect on the reaction equilibirum would become adverse, so that a small purge stream is used.

      As shown in Figure AT3.6–5, the fresh feed of gas composed of 75.16% H_2, 24.57% N_2, and 0.27% Ar is mixed with the recycled gas and enters the reactor with a composition of 79.52% H_2. The gas stream leaving the ammonia separator contains 80.01% H_2 and no ammonia. The product ammonia contains no dissolved gases. Per 100 moles of fresh feed:

(a) How many moles are recycled and purged?

(b) What is the percent conversion of hydrogen per pass?

6. Ethyl ether is made by the dehydration of ethyl alcohol in the presence of sulfuric acid at 140°C:

$$2C_2H_5OH \longrightarrow C_2H_5OC_2H_5 + H_2O$$

A simplified process diagram is shown in Figure AT3.6–6. If 87% conversion of the alcohol fed to the reactor occurs per pass in the reactor, calculate:

(a) Kilograms per hour of fresh feed

(b) Kilograms per hour of recycle

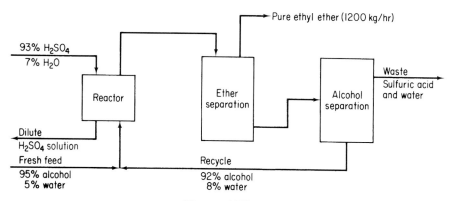

Figure AT3.6–6

Thought Problem

1. Centrifugal pumps cannot run dry, and must have a minimum fluid flow to operate properly—to avoid cavitation, and subsequent mechanical damage to the pump. A storage tank is to be setup to provide liquid flow to a process, but sometimes the demand will drop below the minimum flow rate (10-15% of the rated capacity of the pump). What equipment setup would you recommend be implemented so that the pump is not damaged by the low flows? Draw a picture of the layout so that the minimum flow can go through the pump no matter what the level of liquid is in the feed tank and no matter what the outlet pressure and demand may be.

Discussion Question

1. Because of limitations in supply as well as economics, many industries reuse their water over and over again. For example, recirculation occurs in cooling towers, boilers, powdered coal transport, multistage evaporation, humidifiers, and many devices to wash agricultural products.

   Write a brief report discussing one of these processes, and include in the report a description of the process, a simplified flowsheet, problems with recycling, the extent of purge, and, if you can find the information, the savings made by recycling.

S U P P L E M E N T A R Y   R E F E R E N C E S

General

BHATT, B. I. and S. M., VORA. *Stoichiometry*, 2nd ed. New Delhi: Tate McGraw Hill, 1984.

FLATHMAN, E., D., E. JERGER, and H. EXNER (eds.). *Bioremediation: Field Experience.* Boca Raton, FL: Lewis Publishers, 1993.

FELDER, R. M., and R. W. ROUSSEAU. *Elementary Principles of Chemical Processes*, 2nd ed., New York: Wiley, 1986.

GANAPATHY, V. *Basic Programs for Steam Plant Engineers*. New York: Marcel Dekker, 1986.

HUSAIN, A. *Chemical Process Simulation*. New Delhi: Halstead (Wiley), 1986.

KURTZ, W. A., T. M. WEBB, and P. J. MCNAMEE. *The Carbon Budget of the Canadian Forest Sector*. Edmonton: Forestry Canada, 1993.

LAGREGA, M. D., L. BUCKINGHAM, and J. C. EVANS. *Hazardous Waste Management*. New York: McGraw-Hill, Inc., 1994.

LUYBEN, W. L., and L. A. WENZEL. *Chemical Process Analysis: Mass and Energy Balances*. Intl. Ser. in Phys. & Chem. Engin. Sci. Englewood Cliffs, NJ: Prentice Hall, 1988.

MYERS, A. L., and W. D. SEIDER. *Introduction to Chemical Engineering and Computer Calculations*. Englewood Cliffs, NJ: Prentice-Hall, 1976.

National Technical Information Service. *Flue Gases: Detection, Sampling, Analysis*, PB 86-871290/GAR. Springfield, VA: NTIS, 1986.

RAO, Y. K. *Stoichiometry and Thermodynamics of Metallurgical Processes*. New York: Cambridge University Press,1985.

REKLAITIS, G. V. *Introduction to Material and Energy Balances*. New York: Wiley, 1983.

RUSSELL, T. W. F., and M. M. DENN. *Introduction to Chemical Engineering Analysis*. New York: Wiley, 1972.

PROBLEMS

Section 3.1

3.1. Examine the flow sheet in Figure P3.1 (adapted from *Hydrocarbon Processing*, November 1974, p. 159) for the atmospheric distillation and pyrolysis of all atmospheric distillates for fuels and petrochemicals. Does the mass in equal the mass out? Give one or two reasons why the mass does or does not balance. Note: T/A is metric tons/year.

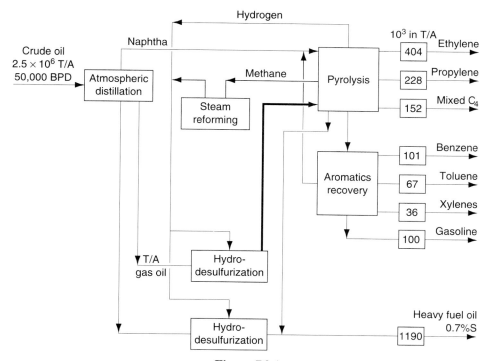

Figure P3.1

3.2. State whether the following processes represent open or closed systems.
   (a) The global carbon cycle of the earth.
   (b) The carbon cycle for a forest.
   (c) An outboard motor for a boat.
   (d) Your home air conditioner with respect to the coolant.

3.3. Explain why the total moles entering a process may not be equal the total moles leaving.

3.4. Give an example of:
   (a) An unsteady-state process
   (b) A steady-state process
   Draw a picture or explain the process in not more than three sentences. Any type of process you can think of will be acceptable—a chemical engineering process is not required.

3.5. Examine Figure P3.5. (adapted from *Environ. Sci. Technol.*, **27** (1993)p. 1975). What would be a good system to designate for this bioremediation process? Sketch the process and draw the system boundary. Is your system open or closed? Is it steady state or unsteady state?

3.6. The danger from a particular extremely hazardous substance depends primarily on the product of its dose of exposure and its hazard potential—a risk-index number, somewhat like the cold chill factor used in weather reports. Dose is determined primarily by the rate of release, proximity of the source to targets, topography of the area, and meteorological conditions. The dose is modified by physicochemical properties like flash point; boiling point, density relative to air; particle size (for nonvolatile solids); and further, by any subsequent biodegradation, hydrolysis, and photolysis.

   Hazards to human health and life can be approximated from potency for

A system for treating soil above the water table (bioventing).

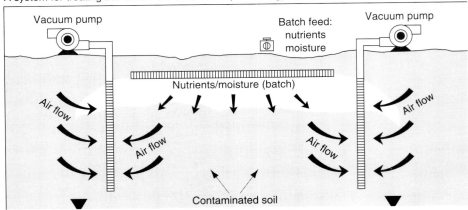

Figure P3.5

carcinogens, IDLH (immediately dangerous to life and health), and/or TLVs (threshold limit values) for noncarcinogenic toxicants. The Environmental Protection Agency–mandated R form, with a valid "risk-index," may be adequate to address chronic health effects to nearby communities from routine industrial chemical emissions in quantities too small to initiate acute problems to average individuals.

One hundred pounds of methyl isocyanate, MIC (boiling point 37 to 39°C, formula weight 57) will quickly vaporize to form a mile-long 30-ft × 200-ft plume that exceeds the ambient concentration that is immediately dangerous to life and health (IDLH, 47 mg per cubic meter). Normal atmospheric moisture will not hydrolyze it fast enough. Continuing research on Bhopal victims indicates that glutathione-transported carbamylation reactions may lead to widespread acute and chronic toxic effects from even a one-time MIC exposure.

If the cross section of the plume is indeed rectangular, check to determine if the concentration of the plume exceeds 47 mg/m³. What assumptions are needed?

3.7. Read each one of the following scenarios. State what the system is. Draw the picture. Classify each as belonging to one or more of the following: open system, closed system, steady-state process, unsteady-state process.
(a) You fill your car radiator with coolant.
(b) You drain your car radiator.
(c) You overfill the car radiator and the coolant runs on the ground.
(d) The radiator is full and the water pump circulates water to and from the engine while the engine is running.

3.8. One of the most common commercial methods for the production of pure silicon that is to be used for the manufacture of semiconductors is the Siemens process (see Figure P3.8) of chemical vapor deposition (CVD). A chamber contains a heated silicon rod, and a mixture of high purity trichlorosilane mixed with high purity hydrogen that is passed over the rod. Pure silicon (EGS-electronic grade silicon) deposits on the rod as a polycrystalline solid. (Single crystals of Si are later made by subsequently melting the EGS and drawing a single crystal from the melt.) The reaction is: $H_2(g) + SiHCl_3(g) \rightarrow Si(s) + 3HCl(g)$.

The rod initially has a mass of 1,460g, and the mole fraction of H_2 in the exit gas is 0.223. The mole fraction of H_2 in the feed to the reactor is 0.580, and the feed enters at the rate of 6.22 kg mol/hr. What will be the mass of the rod at the end of 20 minutes?

3.9. Examine Figure P3.9. Is the material balance satisfactory?

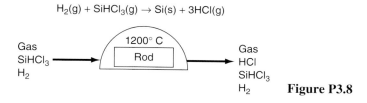

$H_2(g) + SiHCl_3(g) \rightarrow Si(s) + 3HCl(g)$

Gas
SiHCl₃
H₂
1200° C
Rod
Gas
HCl
SiHCl₃
H₂   **Figure P3.8**

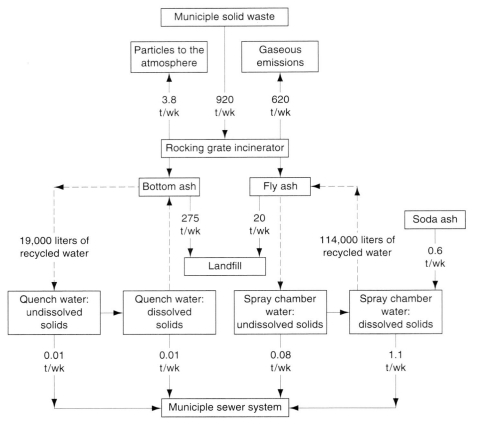

Figure P3.9

3.10. Examine the processes in Figure P3.10. Each box represents a system. For each, state whether:

(a) The process is in the

(1) steady state, (2) unsteady state, or (3) unknown condition

(b) The system is

(1) closed, (2) open, (3) neither, or (4) both

The wavy line represents the initial fluid level when the flows begin. In case (c), the tank stays full.

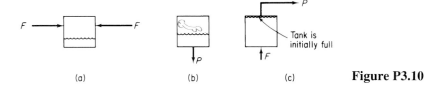

Figure P3.10

Section 3.2

3.11. Are the following equations independent? Do they have a unique solution? Explain your answers.

(a) $x_1 + 2x_2 = 1$
$\ x_1 + 2x_2 = 3$

(b) $(x_1-1)^2 + (x_2-1)^2 = 0$
$\ x_1 + x_2 = 1$

Does the following set of equations have a unique solution for x, y, and z?

$$x + 2y - 3z = -1$$
$$3x - y + 2z = 7$$
$$5x + 3y - 4z = 2$$

3.12. Determine the rank of the following matrices:

(a) $\begin{bmatrix} 1 & 3 & 3 & 2 \\ 2 & 6 & 9 & 5 \\ -1 & -3 & 3 & 0 \end{bmatrix}$ **(b)** $\begin{bmatrix} 1 & 2 & 3 \\ 2 & 3 & 4 \\ 3 & 5 & 7 \end{bmatrix}$ **(c)** $\begin{bmatrix} 1 & 0 & 2 \\ 3 & 0 & 4 \\ 2 & -5 & 1 \end{bmatrix}$

3.13. Do the following sets of equations have a unique solution?

(a) $u + v + w = 0$
$\ u + 2v + 3w = 0$
$\ 3u + 5v + 7w = 1$

(b) $u + w = 0$
$\ 5u + 4v + 9w = 0$
$\ 2u + 4v + 6w = 0$

3.14. For one process your assistant has prepared four material balances

$$0.25\ NaCl + 0.35\ KCl + 0.55\ H_2O = 0.30$$
$$0.35\ NaCl + 0.20\ KCl + 0.40\ H_2O = 0.30$$
$$0.40\ NaCl + 0.45\ KCl + 0.05\ H_2O = 0.40$$
$$1.00\ NaCl + 1.00\ KCl + 1.00\ H_2O = 1.00$$

He says that since the four equations exceed the number of unknowns, three, no solution exists. Is he correct? Explain briefly whether it is possible to achieve a unique solution.

3.15. A cylinder containing CH_4, C_2H_6, and N_2 has to be prepared containing a mole ratio of CH_4 to C_2H_6 of 1.5 to 1. Available are (1) a cylinder of a N_2-CH_4 mixture (80% N_2, 20% CH_4), (2) a cylinder of 90% N_2 and 10% C_2H_6, and (3) a cylinder of pure N_2. List the total number and names (or symbols) of the unknown quantities and concentrations, and list by name the balances that you would use to solve this problem (do not solve the balances). What is the residual number of degrees of freedom that have to be specified to make the problem completely determined (i.e., have a unique solution)?

3.16. For the process shown in Figure P3.16, how many material balance equations can be written? Write them. How many independent material balance equations are there in the set?

3.17. Examine the process in Figure P3.17. No chemical reaction takes place, and x stands for mole fraction. How many stream values are unknown? How many concentrations? Can this problem be solved uniquely for the unknowns?

3.18. You have been asked to check out the process shown in Figure P3.18. What will be the minimum number of measurements to make in order to compute the value of each of the stream flow rates and stream concentrations? Explain your answer.

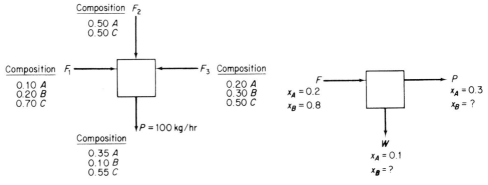

Figure P3.16   **Figure P3.17**

Can any arbitrary set of five be used; that is, can you measure just the three flow rates and two concentrations? Can you measure just three concentrations in stream F and two concentrations in stream W? No chemical reaction takes place and x is the mole fraction of component A, B, or C.

3.19. In preparing 2.50 moles of a mixture of three gases, SO_2, H_2S, and CS_2, gases from three tanks are combined into a fourth tank. The tanks have the following compositions (mole fractions):

	Tanks			Combined Mixture
Gas	1	2	3	4
SO_2	0.23	0.20	0.54	0.25
H_2S	0.36	0.33	0.27	0.23
CS_2	0.41	0.47	0.19	0.52

In the right-hand column is listed the supposed composition obtained by analysis of the mixture. Does the set of three mole balances for the three compounds have a solution for the number of moles taken from each of the three tanks and used to make up the mixture? If so, what does the solution mean?

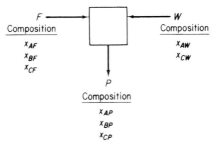

Figure P3.18

3.20. A problem is posed as follows. It is desired to mix three LPG (Liquefied Petroleum Gas) streams denoted by A, B, and C, in certain proportions so that the final mixture will meet certain vapor pressure specifications. These specifications will be met by a stream of composition D as indicated below. Calculate the proportions in which stream A, B, and C must be mixed to give a product with a composition of D. The values are liquid volume %, but the volumes are additive for these compounds.

Component	Stream A	B	C	D
C_2	5.0			1.4
C_3	90.0	10.0		31.2
iso-C_4	5.0	85.0	8.0	53.4
n-C_4		5.0	80.0	12.6
iso-C_5^+			12.0	1.4
	100.0	100.0	100.0	100.0

The subscripts on the C's represent the number of carbons, and the + sign on C_5^+ indicates all compounds of higher molecular weight as well as iso-C_5.

Does this problem have a unique solution?

3.21. Effluent from a fertilizer plant is processed by the system shown in Figure P3.21. How many additional concentration and stream flow measurements must be made to completely specify the problem (so that a unique solution exists)? Does only one unique set of specifications exist?

3.22. A drier takes in wet timber (20.1% water) and reduces the water content to 8.6% water. You want to determine the kg of water removed per kg of timber that enter the process. Draw a picture of the process, put the data on the figure, pick a basis, determine the number of variables whose values are unknown, and the number of independent equations that can be written for the process. Is a unique solution possible?

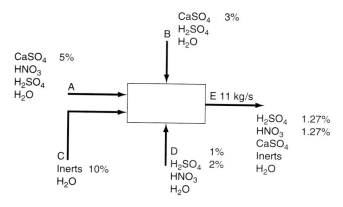

Figure P3.21

Section 3.3

3.23. Three gaseous mixtures, A, B, and C, with the compositions listed in the table are blended into a single mixture.

Gas	A	B	C
CH_4	25	25	60
C_2H_6	35	30	25
C_3H_8	40	45	15
Total	100	100	100

A new analyst reports that the composition of the mixture is 25% CH_4, 25% C_2H_6, and 50% C_3H_8. Without making any detailed calculations, explain how you know the analysis is incorrect.

3.24. A lacquer plant must deliver 1000 lb of an 8% nitrocellulose solution. They have in stock a 5.5% solution. How much dry nitrocellulose must be dissolved in the solution to fill the order?

3.25. A manufacturer of briquets has a contract to make briquets for barbecuing that are guaranteed to not contain over 10% moisture or 10% ash. The basic material they use has the analysis: moisture 12.4%, volatile material 16.6%, carbon 57.5%, and ash 13.5%. To meet the specifications (at their limits) they plan to mix with the base material a certain amount of petroleum coke that has the analysis: volatile material 8.2%, carbon 88.7%, and moisture 3.1%. How much petroleum coke must be added per 100 lb of the base material?

3.26. Sludge is wet solids that result from the processing in municipal sewage systems. The sludge has to be dried before it can be composted or otherwise handled. If a sludge containing 70% water and 30% solids is passed through a drier, and the resulting product contains 25% water, how much water is evaporated per ton of sludge sent to the drier.

3.27. A gas containing 80% CH_4 and 20% He is sent through a quartz diffusion tube (see Figure P3.27) to recover the helium. Twenty percent by weight of the original gas is recovered, and its composition is 50% He. Calculate the composition of the waste gas if 100 kg moles of gas are processed per minute. The initial gas pressure

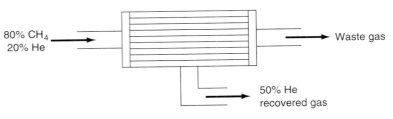

80% CH_4
20% He   Waste gas

50% He
recovered gas

Figure P3.27

is 120 kPa, and the final gas pressure is 115 kPa. The barometer reads 740 mm Hg. The temperature of the process is 22°C.

3.28. To prepare a solution of 50.0% sulfuric acid, a dilute waste acid containing 28.0% H_2SO_4 is fortified with a purchased acid containing 96.0% H_2SO_4. How many kilograms of the purchased acid must be bought for each 100 kg of dilute acid?

3.29. In many fermentations, the maximum amount of cell mass must be obtained. However, the amount of mass that can be made is ultimately limited by the cell volume. Cells occupy a finite volume and have a rigid shape so that they cannot be packed beyond a certain limit. There will always be some water remaining in the interstices between the adjacent cells, which represents the void volume that at best can be as low as 40% of the fermenter volume. Calculate the maximum cell mass on a dry basis per L of the fermenter that can be obtained if the wet cell density is 1.1 g/cm³. Note that cells themselves consist of about 75% water and 25% solids, and cell mass is reported as dry weight in the fermentation industry.

3.30. A dairy produces casein that when wet contains 23.7% moisture. They sell this for $8.00/100 lb. They also dry this casein to produce a product containing 10% moisture. Their drying costs are $0.80/100 lb water removed. What should be the selling price of the dried casein to maintain the same margin of profit?

3.31. A polymer blend is to be formed from the three compounds whose compositions and approximate formulas are listed in the table. Determine the percentages of each compound A, B, and C to be introduced into the mixture to achieve the desired composition.

	Compound (%)			
Composition	A	B	C	Desired mixture
$(CH_4)_x$	25	35	55	30
$(C_2H_6)_x$	35	20	40	30
$(C_3H_8)_x$	40	45	5	40
Total	100	100	100	100

How would you decide to blend compounds A, B, C, and D [$(CH_4)_x = 10\%$, $(C_2H_6)_x = 30\%$, $(C_3H_8)_x = 60\%$] to achieve the desired mixture?

3.32. In a gas-separation plant, the feed-to-butane splitter has the following constituents:

Component	Mole %
C_3	1.9
i-C_4	51.5
n-C_4	46.0
C_5^+	0.6
Total	100.0

The flow rate is 5804 kg mol/day. If the overhead and bottoms streams from the butane splitter have the following compositions, what are the flow rates of the overhead and bottoms streams in kg mol/day?

	Mole %	
Component	Overhead	Bottoms
C_3	3.4	—
i-C_4	95.7	1.1
n-C_4	0.9	97.6
C_5^+	—	1.3
Total	100.0	100.0

3.33. The organic fraction in the wastewater is measured in terms of the biological oxygen demand (BOD) material, namely the amount of dissolved oxygen required to biodegrade the organic contents. If the dissolved oxygen (DO) concentration in a body of water drops too low, the fish in the stream or lake may die. The Environmental Protection Agency has set the minimum summer levels for lakes at 5 mg/L of DO.

(a) If a stream is flowing at 0.3 m³/s and has an initial BOD of 5 mg/L before reaching the discharge point of a sewage treatment plant, and the plant discharges 3.785 ML/day of wastewater, with a concentration of 0.15 g/L of BOD, what will be the BOD concentration immediately below the discharge point of the plant?

(b) The plant reports a discharge of 15.8 ML/day having a BOD of 72.09 mg/L. If the EPA measures the flow of the stream before the discharge point at 530 ML/day with 3 mg/L of BOD, and measures the downstream concentration of 5 mg/L of BOD, is the report correct?

3.34. In the manufacture of vinyl acetate, some unreacted acetic acid and other compounds are discharged to the sewer. Your company has been cited as discharging more than the specified limit of several pollutants. It is not possible to measure the discharge directly because no flow-measuring devices are in place, but you can take samples of liquid at different places in the sewer line, and measure the concentration of potassium chloride. At one manhole, the concentration is 0.105%. You introduce a solution of 400 g of 1L of H_2O at a steady rate of 1L per minute over 1/2 hr at a manhole 50 ft downstream, and at 1200 ft downstream measure the average steady-state concentration of KCI as 0.281%. What is the flow rate of fluid in the sewer in kg/min?

3.35. Ammonia is a gas for which reliable analytical methods are available to determine its concentration in other gases. To measure flow in a natural gas pipeline, pure ammonia gas is injected into the pipeline at a constant rate of 72.3 kg/min for 12 min. Five miles downstream from the injection point, the steady-state ammonia concentration is found to be 0.382 weight percent. The gas upstream from the point of ammonia injection contains no measurable ammonia. How many kilograms of natural gas are flowing through the pipelines per hour?

3.36. Water pollution in the Hudson River has claimed considerable recent attention, especially pollution from sewage outlets and industrial wastes. To determine accurately how much effluent enters the river is quite difficult because to catch and weigh the material is impossible, weirs are hard to construct, and so on. One suggestion that has been offered is to add a tracer of Br ion to a given sewage stream, let it mix well, and sample the sewage stream after it mixes. On one test of the proposal you add ten pounds of NaBr per hour for 24 hours to a sewage stream with essentially no Br in it. Somewhat downstream of the introduction point a sampling of the sewage stream shows 0.012% NaBr. The sewage density is 60.3 lb/ft^3 and river water density is 62.4 lb/ft^3. What is the flow rate of the sewage in lb/min?

3.37. The solubility of manganous sulfate at 20°C is 62.9g/100g H_2O. How much $MnSO_4 \cdot 5H_2O$ must be dissolved in 100 kg of water to give a saturated solution at 20°C?

3.38. Crystals of $Na_2CO_3 \cdot 10H_2O$ are dropped into a saturated solution of Na_2CO_3 in water at 100°C. What percent of the Na_2CO_3 in the $Na_2CO_3 \cdot 10H_2O$ is recovered in the precipitated solid? The precipitated solid is $Na_2CO_3 \cdot H_2O$. Data at 100°C: The saturated solution is 31.2% Na_2CO_3; the molecular weight of Na_2CO_3 is 106.

3.39. A water solution contains 60% $Na_2S_2O_2$ together with 1% soluble impurity. Upon cooling to 10°C, $Na_2S_2O_2 \cdot 5H_2O$ crystallizes out. The solubility of this hydrate is 1.4lb $Na_2S_2O_2 \cdot 5H_2O$/lb free water. The crystals removed carry as adhering solution 0.06lb solution/lb crystals. These are dried to remove the remaining water (but not the water of hydration). The final dry $Na_2S_2O_2 \cdot 5H_2O$ crystals must not contain more than 0.1% impurity. To meet this specification, the original solution, before cooling, is further diluted with water. On the basis of 100lb of the original solution, calculate:
(a) The amount of water added before cooling.
(b) The percentage recovery of the $Na_2S_2O_2$ in the dried hydrated crystals.

3.40. Suppose that 100L/min are drawn from a fermentation tank and passed through an extraction tank in which the fermentation product (in the aqueous phase) is mixed with an organic solvent, and then the aqueous phase is separated from the organic phase. The concentration of the desired enzyene (3-hydroxybutyrate dehydrogenase) in the aqueous feed to the extraction tank is 10.2 g/L. The pure organic extraction solvent runs into the extraction tank at the rate of 9.7L/min. If the ratio of the enzyme in the exit product stream (the organic phase) from the extraction tank to the concentration of the enzyme in the exit waste stream (the aqueous phase) from the tank is D = 18.5 (g/L organic)/(g/L aqueous), what is the fraction recovery of the enzyme and the amount recovered per min? Assume negligible miscibility between the aqueous and organic liguids in each other. And ignore any change in density on removal or addition of the enzyme to either stream.

3.41. Consider the following process for recovering NH_3 from a gas stream composed of N_2 and NH_3 (see Figure P3.41).

Flowing upward through the process is the gas stream, which can contain NH_3 and N_2 but *not* solvent S, and flowing downward through the device is a liquid stream which can contain NH_3 and liquid S but *not* N_2.

The weight fraction of NH_3 in the gas stream *A* leaving the process is related to the weight fraction of NH_3 in the liquid stream *B* leaving the process by the following empirical relationship: $\omega^A_{NH3} = 2\omega^B_{NH3}.$

Given the data shown in Figure P3.41, calculate the flow rates and compositions of streams *A* and *B*. Can you solve the problem by knowing the relation $x^A_{NH_3} = 2x^B_{NH_3}$, (mole fractions) instead of the mass fraction relation?

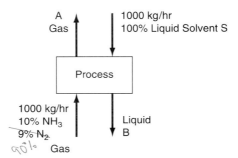

Figure P3.41

3.42. An aqueous etching solution containing 8.8% KI is to be prepared to etch gold in printed circuit boards. The desired solution is to be formed by combining a strong solution (12% KI and 3% I_2 in H_2O) with a weak solution (2.5% KI and 0.625% I_2).

(a) What should be the value of *R*, the ratio of the weights of the strong to the weak solution, to make up the desired etching solution? What will be the concentration of I_2 in the final solution?

(b) Note that you cannot independently vary the concentration of both KI and I_2 in the final mixture simply by varying the value of *R*. Derive a relationship between the weight fraction of KI and the weight fraction of I_2 in the mixture to illustrate this point.

Section 3.4

3.43. A synthesis gas analyzing 6.4% CO_2, 0.2% O_2, 40.0% CO, and 50.8% H_2, (the balance is N_2), is burned with 40% dry excess air. What is the composition of the flue gas?

3.44. Hydrogen-free carbon in the form of coke is burned:
   (a) With complete combustion using theoretical air
   (b) With complete combustion using 50% excess air
   (c) Using 50% excess air but with 10% of the carbon burning to CO only.
   In each case calculate the gas analysis that will be found by testing the flue gases with an Orsat apparatus.

3.45. Thirty pounds of coal (analysis 80% C and 20% H ignoring the ash) are burned with 600 lb of air, yielding a gas having an Orsat analysis in which the ratio of CO_2 to CO is 3 to 2. What is the percent excess air?

3.46. A gas containing only CH_4 and N_2 is burned with air yielding a flue gas that has an Orsat analysis of CO_2: 8.7%, CO: 1.0%, O_2: 3.8%, and N_2: 86.5%. Calculate the percent excess air used in combustion and the composition of the CH_4–N_2 mixture.

3.47. A natural gas consisting entirely of methane (CH_4) is burned with an oxygen enriched air of composition 40% O_2 and 60% N_2. The Orsat analysis of the product gas as reported by the laboratory is CO_2: 20.2%, O_2: 4.1%, and N_2: 75.7%. Can the reported analysis be correct? Show all calculations.

3.48. Dry coke composed of 4% inert solids (ash), 90% carbon, and 6% hydrogen is burned in a furnace with dry air. The solid refuse left after combustion contains 10% carbon and 90% inert ash (and no hydrogen). The inert ash content does not enter into the reaction.

   The Orsat analysis of the flue-gas gives 13.9% CO_2, 0.8% CO, 4.3% O_2, and 81.0% N_2. Calculate the percent of excess air based on complete combustion of the coke.

3.49. A gas with the following composition is burned with 50% excess air in a furnace. What is the composition of the flue gas by percent?

$$CH_4:\ 60\%;\ C_2H_6:\ 20\%;\ CO:\ 5\%,\ O_2:\ 5\%;\ N_2:10\%$$

3.50. A flare is used to convert unburned gases to innocuous products such as CO_2 and H_2O. If a gas of the following composition (in percent) is burned in the flare—CH_4: 70%, C_3H_8: 5%, CO: 15%, O_2: 5%, N_2: 5%—and the flue gas contains 7.73% CO_2, 12.35% H_2O, and the balance is O_2 and N_2, what was the percent excess air used?

3.51. Examine the reactor in Figure P3.51. Your boss says something has gone wrong with the yield of CH_3O, and it is up to you to find out what the problem is. You start by making material balances (naturally!). What is your recommendation as to the probable difficulty? Show all calculations on which your recommendation is based.

3.52. One of the products of sewage treatment is sludge. After microorganisms grow in the activated sludge process to remove nutrients and organic material, a substantial amount of wet sludge is produced. This sludge must be dewatered, one of the most expensive parts of most treatment plant operations.

   How to dispose of the dewatered sludge is a major problem. Some organiza-

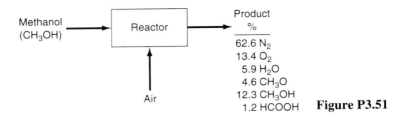

Figure P3.51

tions sell dried sludge for fertilizer, some spread the sludge on farmland, and in some places it is burned. To burn a dried sludge, fuel oil is mixed with it, and the mixture is burned in a furnace with air. If you collect the following analysis for the sludge and for the product gas

Sludge (%)		Product Gas (%)	
S	32	SO_2	1.52
C	40	CO_2	10.14
H_2	4	O_2	4.65
O_2	24	N_2	81.67
		CO	2.02

(a) Determine the weight percent of carbon and hydrogen in the fuel oil.

(b) Determine the ratio of pounds of dry sludge to pounds of fuel oil in the mixture fed to the furnace.

3.53. Ethane is initially mixed with oxygen to obtain a gas containing 80% C_2H_6 and 20% O_2 that is then burned in an engine with 200% excess air. Eighty percent of the ethane goes to CO_2, 10% goes to CO, and 10% remains unburned. Calculate the composition of the exhaust gas on a wet basis.

3.54. In a handbook you find the following expression for the approximate percent excess air in combustion based on the percent CO_2 in the dry flue gas analysis and a parameter K

$$\text{percent excess air} = \left(\frac{K}{\%CO_2} - 1\right)100$$

To test this relation, use it to calculate for the $\%CO_2 = 12.1$ the percent excess air for the combustion of (a) coke (all C), K = 20.5, and (b) natural gas (all CH_4), K = 12.5. Then check the accuracy of the prediction by calculating the dry flue gas analysis based on the percent excess air determined from the relation. How much relative difference is there?

Another relation is

$$\text{percent excess air} = F\left(\frac{\%CO_2}{21 - \%CO_2}\right)$$

where % CO_2 is the percent CO_2 in the dry flue gas. Repeat the calculations for this relation given that

(a) F = 100 for coke and

(b) F = 90 for natural gas.

Use as the percent oxygen the percentage calculated for the first formula.

3.55. In order to neutralize the acid in a waste stream (composed of H_2SO_4 and H_2O), dry ground limestone (composition 95% $CaCO_3$ and 5% inerts) is mixed in. The dried sludge collected from the process is only partly analyzed by firing it in a furnace which results in only CO_2 being driven off. By weight the CO_2 represents 10% of the dry sludge. What percent of the pure $CaCO_3$ in the limestone did not react in the neutralization?

3.56. A specially prepared gas composed of 80.0% CH_4 and 20.0% N_2 is burned with air in a boiler. The Orsat analysis of the exit flue gas is 1.2% CO_2, 4.9% O_2, and 93.9% N_2. To reduce the concentration of CO_2 exiting from the stack, much of the CO_2 from the boiler is absorbed before the combustion gases enter the stack. Assume complete combustion takes place. Calculate the percent excess air, and the ratio of the moles of the gases exiting from the stack to the moles of the special feed gas.

3.57. A gas with the following composition, CH_4: 60%; C_2H_6: 20%; CO: 5%; O_2: 5%; N_2: 10%, is burned with 50% excess air in furnace. You are asked to calculate the number of moles of each component in the flue gases.

The solution prepared by one student was as follows:

Basis: 100 mol gas

50% xs air: Calculate O_2 and N_2 in the air Reqd. O_2 (moles)

$CH_4 + 2O_2 \rightarrow CO_2 + 2H_2O$	60(2)	= 120
$C_2H_6 + 3\,\tfrac{1}{2}\,O_2 \rightarrow 2CO_2 + 3H_2O$	20(3.5)	= 70
$CO + \tfrac{1}{2}\,O_2 \rightarrow CO_2$	5($\tfrac{1}{2}$)	= 2.5
		= 192.5
	xsO_2: 192.5(.5)	= 96.25
	Total:	= 288.75
	N_2 in with O_2 $\dfrac{288.75}{} \Big| \dfrac{0.79}{0.21}$	= 1086

Material balances

	In		*Out*			
C:	60 + 2(20) +5	=	n_{CO_2}		n_{CO_2}	= 105
H:	60(4) + 20(6)	=	$n_{H_2O}(2)$		n_{H_2O}	= 180
N2:	10 + 1086	=	n_{N_2}		n_{N_2}	= 1096
O2:	(5 1/2)+ 5 + 288.75	=	$n_{CO_2} + (n_{H_2O}/2) + n_{O_2}$		n_{O_2}	= 100.25

Is the solution correct? Explain.

3.58. A waste stream from a plant is being disposed of by burning in a flare with air. The waste gas has the composition

CH_4: 30%, CO_2: 10%, CO: 8%, H_2: 10%, O_2: 2%, H_2S: 2%, H_2O: 2%, N_2: 36%

The Orsat analysis of the exit gas shows 0.3% SO_2 along with CO_2, O_2, and N_2. Calculate the percent excess air and the complete Orsat analysis. How much error do you think is in the Orsat analysis you calculate?

3.59. A number of proposed trials for toxic waste disposal involve injection of small amounts of the waste into a lime kiln in which $CaCO_3$ is decomposed ("burned") to CaO and CO_2 at high temperature. The QuickLime Co. has a possible kiln for use on waste disposal, and you are asked to establish a base case for operation so that any differences in operation with the addition of waste can be isolated.

The kiln uses natural gas (80.1% CH_4, 9.8% C_2H_6, and 10.1 % N_2) and air to burn completely the limestone, which has a composition of 2.1% moisture, 96.7% $CaCO_3$, and the balance inert material. The Orsat analysis of the exit gases from the kiln is CO_2 17.7%, CO 1.0%, O_2 2.4%, N_2 78.9%. Calculate the kg of lime (CaO) produced per 100 kg mol of natural gas used.

3.60. The products and byproducts from coal combustion can create environmental problems if the combustion process is not carried out properly. A fuel analyzing 74% C, 14% H, and 12% ash is burned to yield a flue gas containing 12.4% CO_2, 1.2% CO, 5.7% O_2, and 80.7% N_2 on a dry basis. Your boss asks you to determine:
  (a) The lb of coal fired per 100 lb mol of flue gas
  (b) The percent excess air used
  (c) The lb of air used per lb of coal
  (d) Will the calculations be valid? Explain.

3.61. In the anaerobic fermentation of grain, the yeast *Saccharomyces cerevisiae* digests glucose from plants to form the products ethanol and propenoic acid by the following overall reactions:

$$\text{Reaction 1: } C_6H_{12}O_6 \;\rightarrow\; 2C_2H_5OH + 2CO_2$$
$$\text{Reaction 2: } C_6H_{12}O_6 \;\rightarrow\; 2C_2H_3CO_2H + 2H_2O$$

In a batch process, a tank is charged with 4000 kg of a 12% glucose/water solution. After fermentation 120 kg of carbon dioxide are produced together with 90 kg of unreacted glucose. What are the weight percents of ethyl alcohol and propenoic acid remaining in the broth? Assume that none of the glucose is assimilated into the bacteria.

3.62. Semiconductor microchip processing often involves chemical vapor deposition (CVD) of thin layers. The material being deposited needs to have certain desirable properties. For instance, to overlay on aluminum or other bases, a phosphorus pentoxide-doped silicon dioxide coating is deposited as passivation (protective) coating by the simultaneous reactions

$$\text{Reaction 1: } SiH_4 + O_2 \rightarrow SiO_2 + 2H_2$$
$$\text{Reaction 2: } 4PH_3 + 5O_2 \rightarrow 2P_2O_5 + 6H_2$$

Determine the relative masses of SiH_4 and PH_3 required to deposit a film of 5% by weight of phosphorus (P) in the protective coating.

3.63. A power company operates one of its boilers on natural gas and another on oil. The analyses of the fuels show 96% CH_4, 2% C_2H_2, and 2% CO_2 for the natural gas and $C_nH_{1.8n}$ for the oil. The flue gases from both groups enter the same stack, and an Orsat analysis of this combined flue gas shows 10.0% CO_2, 0.63% CO, and 4.55% O_2. What percentage of the total carbon burned comes from the oil?

3.64. Printed circuit boards (PCBs) are used in the electronic industry to both connect and hold components in place. In production, 0.03 in. of copper foil is laminated to an insulating plastic board. A circuit pattern made of a chemically resistant polymer is then printed on the board. Next, the unwanted copper is chemically etched away by using selected reagents. If copper is treated with $Cu(NH_3)_4Cl_2$ (cupric ammonium chloride) and NH_4OH (ammonium hydroxide), the products are water and $Cu(NH_3)_4Cl$ (cuprous ammonium chloride). Once the copper is dissolved, the polymer is removed by solvents leaving the printed circuit ready for further processing. If a single-sided board 4 in. by 8 in. is to have 75% of the copper layer removed using the reagents above, how many grams of each reagent will be consumed? Data: The density of copper is 8.96 g/cm^3.

3.65. The thermal destruction of hazardous wastes involves the controlled exposure of waste to high temperatures (usually 900°C or greater) in an oxidizing environment. Types of thermal destruction equipment include high-temperature boilers, cement kilns, and industrial furnaces in which hazardous waste is burned as fuel. In a properly designed system, primary fuel (100% combustible material) is mixed with waste to produce a feed for the boiler.

  (a) Sand containing 30% by weight of 4,4′-dichlorobiphenyl [an example of a polychorinated biphenyl (PCB)] is to be cleaned by combustion with excess hexane to produce a feed that is 60% combustible by weight. To decontaminate 8 tons of such contaminated sand, how many pounds of hexane would be required?

  (b) Write the two reactions that would take place under ideal conditions if the mixture of hexane and the contaminated sand were fed to the thermal oxidation process to produce the most environmentally satisfactory products. How would you suggest treating the exhaust from the burner? Explain.

  (c) The incinerator is supplied with an oxygen-enriched airstream containing 40% O_2 and 60% N_2 to promote high-temperature operation. The exit gas is found to have a $x_{CO_2} = 0.1654$ and $x_{O_2} = 0.1220$. Use this information and the data about the feed composition above to find: (1) the complete exit gas concentrations and (2) the % excess O_2 used in the reaction.

3.66. The Clean Air Act requires automobile manufacturers to warrant their control systems as satisfying the emission standards for 50,000 mi. It requires owners to have their engine control systems serviced exactly according to manufacturers' specifications and to always use the correct gasoline. In testing an engine exhaust having a known Orsat analysis of 16.2% CO_2, 4.8% O_2, and 79% N_2 at the outlet, you find to your surprise that at the end of the muffler the Orsat analysis is 13.1% CO_2. Can this discrepancy be caused by an air leak into the muffler? (Assume that the analyses are satisfactory.) If so, compute the moles of air leaking in per mole of exhaust gas leaving the engine.

3.67. A low-grade pyrite containing 32% S is mixed with 10 lb of pure sulfur per 100 lb of pyrites so the mixture will burn readily, forming a burner gas that analyzes (Orsat) 13.4% SO_2, 2.7% O_2, and 83.9% N_2. No sulfur is left in the cinder. Calculate the percentage of the sulfur fired that burned to SO_3. (The SO_3 is not detected by the Orsat analysis.)

Section 3.5

3.68. The diagram in Figure P3.68 represents a typical but simplified distillation column. Streams 3 and 6 consist of steam and water, and do not come in contact with the fluids in the column that contains two components. Write the total and component material balances for the three sections of the column. How many independent equations would these balances represent?

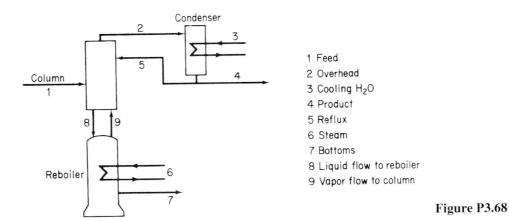

1 Feed
2 Overhead
3 Cooling H_2O
4 Product
5 Reflux
6 Steam
7 Bottoms
8 Liquid flow to reboiler
9 Vapor flow to column

Figure P3.68

3.69. A distillation process is shown in Figure P3.69. You are asked to solve for all the values of the stream flows and compositions. How many unknowns are there in the system? How many independent material balance equations can you write? Explain each answer and show all details whereby you reached your decision. For each stream, the only components that occur are labeled below the stream.

3.70. Metallurgical-grade silicon is purified to electronic grade for use in the semiconductor industry by chemically separating it from its impurities. The Si metal reacts in varying degrees with hydrogen chloride gas at 300°C to form several polychlo-

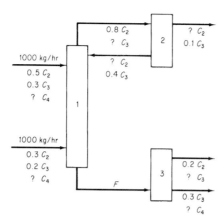

Figure P3.69

rinated silanes. Trichlorosilane is liquid at room temperature and is easily separated by fractional distillation from the other gases. If 100 kg of silicon is reacted as shown in Figure P3.70, how much trichlorosilane is produced?

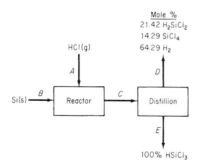

Figure P3.70

3.71. Several streams are mixed as shown in Figure P3.71. Calculate the flows of each stream in kg/s.

3.72. Cryogenic purification of hydrogen takes place as shown in Figure P3.72. No mass is exchanged in the heat exchangers. Try to calculate the composition of stream A. Can you do it? Explain your results. Next, assume you measure the moles of H_2 in stream B as 0.3 mol. Can you now solve for the composition of A? Finally, if you measure the H_2 in B as 0.3 mol and the CH_4 as 6.8 mol, is the problem solvable? Is the problem now overspecified, exactly specified, or underspecified?

3.73. A furnace burns fuel gas of the following composition: 70% Methane (CH_4), 20% Hydrogen (H_2) and 10% Ethane (C_2H_6) with excess air. An oxygen probe placed at the exit of the furnace reads 2% oxygen in the exit gases. The gases are passed then through a long duct to a heat exchanger. At the entrance to the heat exchanger

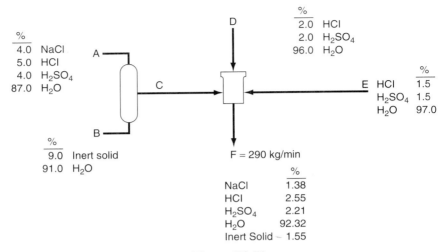

Figure P3.71

the Orsat analysis of the gas reads 6%. Is the discrepancy due to the fact that the first analysis is on a wet basis and the second analysis on a dry basis (no water condenses in the duct), or due to an air leak in the duct? If the former, give the Orsat analysis of the exit gas from the furnace. If the latter, calculate the amount of air that leaks into the duct per 100 mole of fuel gas burned.

3.74. In 1988, the U.S. Chemical Manufacturers Association (CMA) embarked upon an

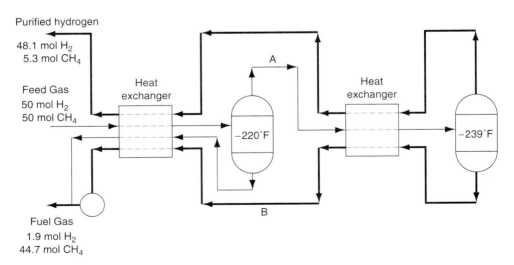

Figure P3.72

ambitious and comprehensive environmental improvement effort—the Responsible Care initiative. Responsible Care commited all of the 185 members of the CMA to ensure continual improvement in the areas of health, safety, and environmental quality, as well as in eliciting and responding to public concerns about their products and operations. The real core of Responsible Care lies in its six Codes of Management Practices. These codes require member companies to establish specific goals and timetables for improving their performance along the whole lifecycle of their chemical products—from research through production, waste management, product transporation, use and, ultimately, disposal. One of the codes, Pollution Prevention, addresses a problem cited by the public as one of its major concerns regarding the industry—the handling of hazardous wastes. The code requires companies to establish specific goals for reducing their generation of all wastes, hazardous or otherwise. One of the best ways to reduce or eliminate hazardous waste is through source reduction. Generally, this means using different raw materials or redesigning the production process to eliminate the generation of hazardous byproducts. As an example, consider the following countercurrent extraction process (Figure P3.74) to recover xylene from a stream that contains 10% xylene and 90% solids by weight.

The stream from which xylene is to be extracted enters Unit 2 at a flow rate of 2000 kg/hr. To provide a solvent for the extraction, pure benzene is fed to Unit 1 at a flow rate of 1000 kg/hr. Assume that the liquid flowing with the solids leaving a unit has the same composition as the clear liquid stream exiting from the same unit. Determine the benzene and xylene concentrations in all of the streams. What is the percent recovery of the xylene entering the process at Unit 2?

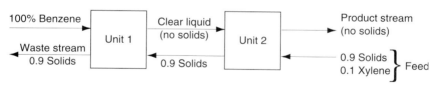

Figure P3.74

3.75. Figure P3.75 shows a three-stage separation process. The ratio of P_3/D_3 is 3, the ratio of P_2/D_2 is 1, and the ratio of A to B in stream P_2 is 4 to 1. Calculate the composition and percent of each component in stream E.

Hint: Although the problem comprises connected units, application of the standard strategy of problem solving will enable you to solve it without solving an excessive number of equations simultaneously.

3.76. A simplified flowsheet for the manufacture of sugar is shown in Figure P3.76. Sugarcane is fed to a mill where a syrup is squeezed out, and the resulting "bagasse" contains 80% pulp. The syrup (E) containing finely divided pieces of

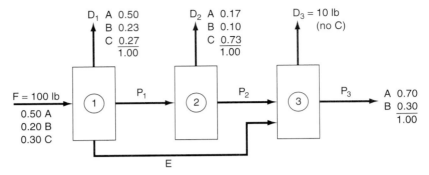

Figure P3.75

pulp is fed to a screen which removes all the pulp and produces a clear syrup (*H*) containing 15% sugar and 85% water. The evaporator makes a "heavy" syrup and the crystallizer produces 1000 lb/hr of sugar crystals.

(a) Find the water removed in the evaporator, lb/hr

(b) Find the mass fractions of the components in the waste stream *G*

(c) Find the rate of feed of cane to the unit, lb/hr

(d) Of the sugar fed in the cane, what percentage is lost with the bagasse?

(e) Is this an efficient operation? Explain why or why not.

3.77. Sodium hydroxide is usually produced from common salt by electrolysis. The essential elements of the system are shown in Figure P3.77.

(a) What is the percent conversion of salt to sodium hydroxide?

(b) How much chlorine gas is produced per pound of product?

(c) Per pound of product, how much water must be evaporated in the evaporator?

3.78. The flowsheet shown in Figure P3.78 represents the process for the production of titanium dioxide (TiO_2) used by Canadian Titanium Pigments at Varennis, Quebec. Sorel slag of the following analysis:

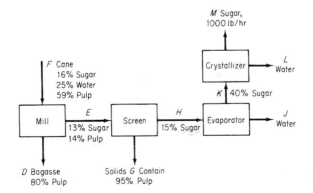

Figure P3.76

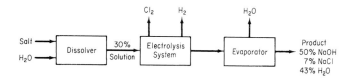

Figure P3.77

	Wt %
TiO_2	70
Fe	8
Inert silicates	22

is fed to a digester and reacted with H_2SO_4, which enters as 67% by weight H_2SO_4 in a water solution. The reactions in the digester are as follows:

$$TiO_2 + H_2SO_4 \rightarrow TiOSO_4 + H_2O \tag{1}$$

$$Fe + \tfrac{1}{2}O_2 + H_2SO_4 \rightarrow FeSO_4 + H_2O \tag{2}$$

Both reactions are complete. The theoretically required amount of H_2SO_4 for the Sorel slag is fed. Pure oxygen is fed in the theoretical amount for all the Fe in the Sorel slag. Scrap iron (pure Fe) is added to the digester to reduce the formation of ferric sulfate to negligible amounts. Thirty-six pounds of scrap iron are added per pound of Sorel slag.

The products of the digester are sent to the clarifier, where all the inert silicates and unreacted Fe are removed. The solution of $TiOSO_4$ and $FeSO_4$ from the clarifier is cooled, crystallizing the $FeSO_4$, which is completely removed by a filter. The product $TiOSO_4$ solution from the filter is evaporated down to a slurry that is 82% by weight $TiOSO_4$.

The slurry is sent to a dryer from which a product of pure hydrate, $TiOSO_4 \cdot H_2O$, is obtained. The hydrate crystals are sent to a direct-fired rotary kiln, where the pure TiO_2 is produced according to the following reaction:

$$TiOSO_4 \cdot H_2O \rightarrow TiO_2 + H_2SO_4 \tag{3}$$

Reaction (3) is complete.

On the basis of 100 lb of Sorel slag feed, calculate:

(a) The pounds of water removed by the evaporator.

(b) The exit lb of H_2O per lb dry air from the dryer if the air enters having 0.036 moles H_2O per mole dry air and the air rate is 18 lb mol of dry air per 100 lb of Sorel slag.

(c) The pounds of product TiO_2 produced.

3.79. In a paper machine system (see Figure P3.79), a stock suspension of cellulose

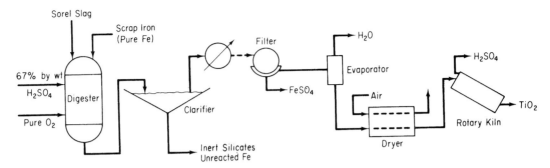

Figure P3.78

fibers in water enters the continuous refining system at 4% consistency (4 lb of fiber per 100 lb of mixture). A dry filler clay is added in the refining system, and this slurry then goes to the paper machine headbox. The headbox distributes the suspension of cellulose fibers and filler clay uniformly over a wire belt in the form of a thin sheet. This wet sheet becomes the desired paper after it is dried. In the headbox, the slurry is diluted with a recycle stream of white water that passed through the wire belt during the formation of the sheet, to a fiber consistency of 0.5%. Analysis shows that the filler consistency is also 0.5% at this point. The damp sheet leaving the press section and entering the drying section, where water is removed by the application of heat, contains 33.3% solids consisting of 90% fiber and 10% filler. Any excess white water from the machine, not needed for headbox dilution, is discarded.

Assuming no fiber loss through the wire and at the presses, calculate:

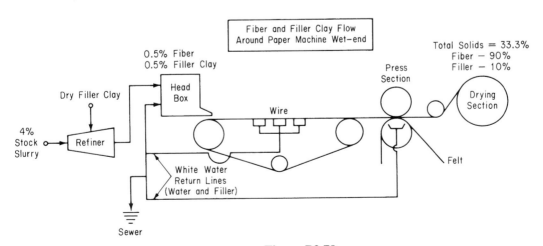

Figure P3.79

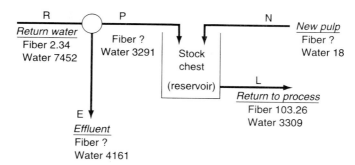

Figure P3.80

(a) The overall percent filler retention on the wet sheet
(b) The percent filler retention per pass (on the basis of the total filler that can be retained).

3.80. In a tissue paper machine, after the water is removed from the paper at various stages it returned to carry more pulp (85% fiber) through the system. See Figure P3.80. Find the unknown fiber values (all values in the figure are in kg) in kg.

Section 3.6

3.81. Examine Figure P3.81. What is the quantity of the recycle stream in kg/hr?

3.82. Seawater is to be desalinized by reverse osmosis using the scheme indicated in Figure P3.82. Use the data given in the figure to determine:
(a) The rate of waste brine removal *(B)*
(b) The rate of desalinized water (called potable water) production *(D)*
(c) The fraction of the brine leaving the osmosis cell (which acts in essence as a separator) that is recycled

3.83. A plating plant has a waste stream containing zinc and nickel in quantities in excess of that allowed to be discharged into the sewer. The proposed process to be used as a first step in reducing the concentration of Zn and Ni is shown in Figure P3.83. Each stream contains water. The concentrations of several of the streams

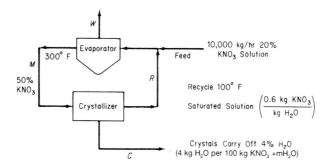

Figure P3.81

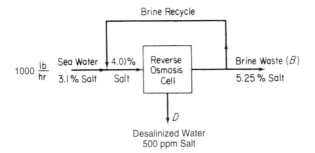

Figure P3.82

are listed in the table. What is the flow (in L/hr) of the recycle stream R_{32} if the feed is 1 L/hr?

Stream	Concentration; g/L	
	Zn	Ni
F	100	10.0
P_0	190.1	17.02
P_2	3.50	2.19
R_{32}	4.35	2.36
W	0	0
D	0.10	1.00

3.84. Ultrafiltration is a method for cleaning up input and output streams from a number of industrial processes. The lure of the technology is its simplicity, merely putting a membrane across a stream to sieve out physically undesirable oil, dirt, metal particles, polymers, and the like. The trick, of course, is coming up with the right

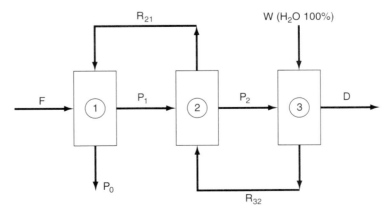

Figure P3.83

membrane. The screening material has to meet a formidable set of conditions. It has to be very thin (less than 1 micron), highly porous, yet strong enough to hold up month after month under severe stresses of liquid flow, pH, particle abrasion, temperature, and other plant operating characteristics.

A commercial system consists of standard modules made up of bundles of porous carbon tubes coated on the inside with a series of proprietary inorganic compositions. A standard module is 6 inches in diameter and contains 151 tubes each 4 feet long with a total working area of 37.5 sq. ft and daily production of 2,000 to 5,000 gallons of filtrate. Optimum tube diameter is about 0.25 inch. A system probably will last at least two to three years before the tubes need replacing from too much residue buildup over the membrane. A periodic automatic chemical cleanout of the tube bundles is part of the system's normal operation. On passing through the filter, the exit stream concentration of oil plus dirt is increased by a factor of 20 over the entering stream.

Calculate the recycle rate in gallons per day (g.p.d.) for the set up shown in Figure P3.84, and calculate the concentration of oil plus dirt in the stream going to the process. The circled values in Figure P3.84 are the known concentration of oil plus dirt.

3.85. Toluene reacts with H_2 to form benzene (B), but a side reaction occurs in which a by-product diphenyl (D) is formed:

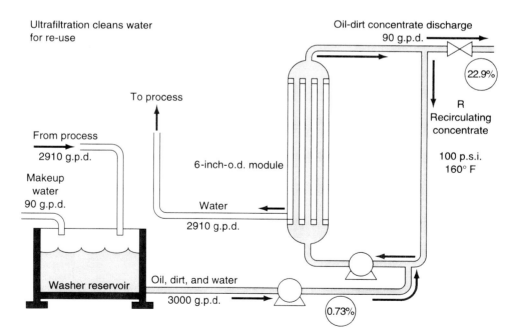

Figure P3.84

$$\underset{\text{Toluene}}{C_7H_8} + \underset{\text{hydrogen}}{H_2} \rightarrow \underset{\text{benzene}}{C_6H_6} + \underset{\text{methane}}{CH_4} \tag{a}$$

$$2C_7H_8 + H_2 \rightarrow \underset{\text{diphenyl}}{C_{12}H_{10}} + 2CH_4 \tag{b}$$

The process is shown in Figure P3.85. Hydrogen is added to the gas recycle stream to make the ratio of H_2 to CH_4 1 to 1 before the gas enters the mixer. The ratio of H_2 to toluene entering the reactor at G is $4H_2$ to 1 toluene. The conversion of toluene to benzene on one pass through the reactor is 80%, and the conversion of toluene to the by-product diphenyl is 8% on the same pass.

Calculate the moles of R_G and R_L per hour.

Data:

Compound:	H_2	CH_4	C_2H_6	C_7H_8	$C_{12}H_{10}$
M W:	2	16	78	92	154

3.86. To save energy, stack gas from a furnace is used to dry rice. The flow sheet and known data are shown in Figure P3.86. What is the amount of recycle gas (in lb mol) per 100 lb of P if the concentration of water in the gas stream entering the dryer is 5.20%?

3.87. This problem is based on the data of G.F. Payne, "Bioseparations of Traditional Fermentation Products" in *Chemical Engineering Problems in Biotechnology,* ed. M.L. Schuler. American Institute of Chemical Engineers, New York: 1989. Examine Figure P3.87. Three separation schemes are proposed to separate the desired fermentation products from the rest of the solution. Ten liters/min of a broth containing 100g/L of undesirable product is to be separated so that the concentration in the exit wastestream is reduced to (not more than) 0.1 g/L. Which of the three

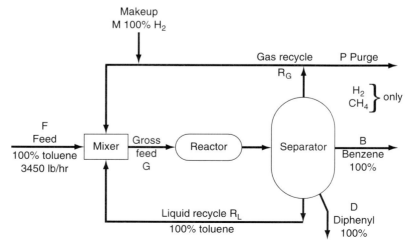

Figure P3.85

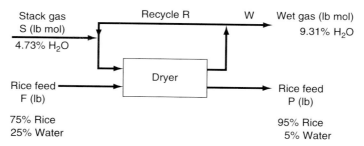

Figure P3.86

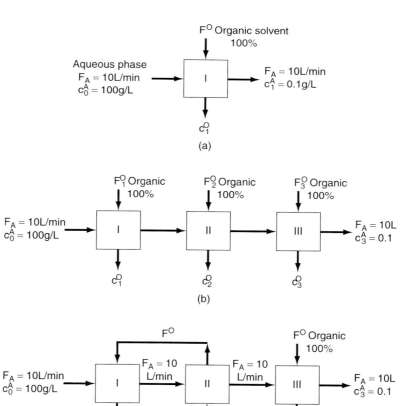

Figure P3.87

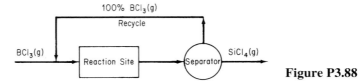

Figure P3.88

flowsheets requires the least fresh pure organic solvent? Ignore any possible density changes in the solutions. Use equal values of the organic solvent in (b), i.e., $F_1^\circ + F_2^\circ + F_3^\circ = F^\circ$. The relation between the concentration of the undesirable material in the aqueous phase and that in the organic phase in 10 to 1 that is, $c^A/c^o = 10$ in the outlet streams of each unit.

3.88. Boron trichloride (BCl_3) gas can be fed into a gas stream and used for doping silicon. The simplest reaction (not the only one) is

$$4\,BCl_3 + 3Si \;\rightarrow\; 3SiCl_4 + 4B$$

If all the BCl_3 not reacted is recycled (see Figure P3.88), what is the mole ratio of recycle to $SiCl_4$ exiting the separator? The conversion on one pass through the reactor is 87% and 1 mole per hour of BCl_3 is fed to the reactor.

3.89. It is proposed to produce ethylene oxide $((CH_2)_2O)$ by the oxidation of ethane (C_2H_6) in the gas phase

$$C_2H_6 + O_2 \;\rightarrow\; (CH_2)_2O + H_2O$$

The ratio of the air to the C_2H_6 in the gross feed into the reactor is 10 to 1, and conversion of C_2H_6 on one pass through the reactor is 18%. The unreacted ethane is separated from the reactor products and recycled as shown in Figure P3.89. What is the ratio of the recycle stream to the feed stream, and what is the composition of the outlet gas from the reactor?

3.90. Acetylene (C_2H_2), ethylene (C_2H_4), and added water are fed as a gas into a reactor in which on one pass 60% of the ethylene and some of the acetylene reacts to form ethanol (C_2H_5OH) and acetic acid ($C_2H_4O_2$). The mole ratio of C_2H_2 to C_2H_4 in the

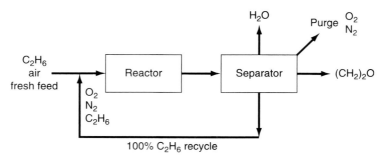

Figure P3.89

gross feed to the reactor is 1.105 to 1, and there is also 0.4 kg of water per kg of ethylene in the feed to the reactor.

The products of the reaction from the reactor are separated into two streams: (1) liquid ethanol (70%), water, and acetic acid, and (2) a recycle stream composed only of gaseous ethylene and acetylene that is combined with the fresh feed, which itself is composed of 52.5% acetylene and the rest ethylene. The combination of recycle, fresh feed, and water is the stream introduced into the reactor.

Compute the ratio of the moles of recycle stream flow to the moles of fresh stream flow, and the mole fractions of C_2H_4 and C_2H_2 in the recycle stream.

3.91. The process shown in Figure P3.91 is the dehydrogenation of propane (C_3H_8) to propylene (C_3H_6) according to the reaction

$$C_3H_8 \rightarrow C_3H_6 + H_2$$

The conversion of propane to propylene based on the *total* propane feed into the reactor at F_2 is 40%. The product flow rate F_5 is 50 kg. mol/hr.

(a) Calculate all the six flow rates F_1 to F_6 in kg mol/hr.

(b) What is the percent conversion of propane in the reactor based on the fresh propane fed to the process (F_1).

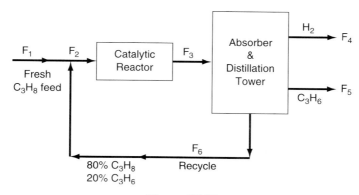

Figure P3.91

3.92. *d*-Glucose and *d*-Fructose have the same chemical formula ($C_6H_{12}O_6$) but different properties. Glucose is converted to fructose as shown in Figure P3.92, but only 60% is converted on one pass through the converter vessel so that unconverted material is recycled. Calculate the flow of the recycle stream per kg 100% glucose fed to the converter. Ignore the solvent water used to carry the glucose and fructose.

3.93. Natural gas (CH_4) is burned in a furnace using 15% excess air based on the complete combustion of CH_4. One of the concerns is that the exit concentration of NO (from the combustion of N_2) is about 415 ppm. To lower the NO concentration in the stack gas to 50 ppm it suggested that the system be redesigned to recycle a

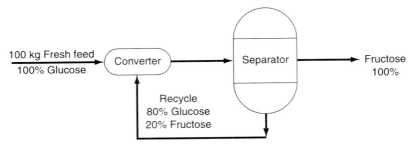

Figure P3.92

portion of the stack gas back through the furnace. You are asked to calculate the amount of recycle required. Will the scheme work? Ignore the effect of temperature on the conversion of N_2 to NO, that is, assume the conversion factor is constant.

3.94. Sulfur dioxide may be converted to SO_3, which has many uses including the production of H_2SO_4 and sulphonation of detergent. A gas stream having the composition shown in Figure P3.94 is to be passed through a two-stage converter. The fraction conversion of the SO_2 to SO_3 (on one pass though) in the first stage is 0.75 and in the second stage 0.65. To boost the overall conversion to 0.95, some of the exit gas from stage 2 is recycled back to the inlet of stage 2. How much must be recycled per 100 moles of inlet gas? Ignore the effect of temperature on the conversion.

3.95. Many chemical processes generate emissions of volatile compounds that need to be controlled. In the process shown in Figure P3.95, the exhaust of CO is eliminated by its separation from the reactor effluent and recycling of 100% of the CO generated in the reactor together with some reactant back to the reactor feed.

Although the product is proprietary, information is provided that the feed stream contains 40% reactant, 50% inert, and 10% CO, and that on reaction 2 moles of reactant yield 2.5 moles of product. Conversion of reactant to product is only 73% on one pass through the reactor, and 90% overall. Calculate the ratio of moles of recycle to moles of product.

3.96. Nitroglycerine, a widely used high explosive, when mixed with wood flour, is

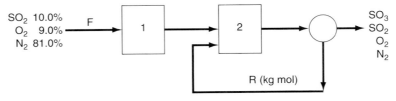

Figure P3.94

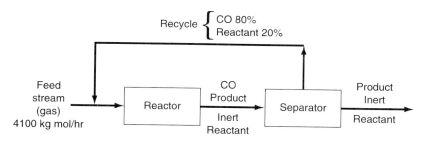

Figure P3.95

called "dynamite." It is made by mixing high-purity glycerine (99.9 + percent pure) with nitration acid, which contains 50.00 percent H_2SO_4, 43.00 percent HNO_3, and 7.00 percent water by weight. The reaction is:

$$C_3H_8O_3 + 3\ HNO_3 + (H_2SO_4)\ \rightarrow\ C_3H_5O_3(NO_2)_3 + 3\ H_2O + (H_2SO_4)$$

The sulfuric acid does not take part in the reaction, but is present to "catch" the water formed. Conversion of the glycerine in the nitrator is complete, and there are no side reactions, so all of the glycerine fed to the nitrator forms nitroglycerine. The mixed acid entering the nitrator (Stream G) contains 20.00 percent excess HNO_3 to assure that all the glycerine reacts.

Figure P3.96. is a process flow diagram.

After nitration, the mixture of nitroglycerine and spent acid (HNO_3, H_2SO_4, and water) goes to a separator (a settling tank). The nitroglycerine is insoluble in the spent acid, and its density is less, so it rises to the top. It is carefully drawn off as product stream P and sent to wash tanks for purification. The spent acid is withdrawn from the bottom of the separator and sent to an acid recovery tank, where the HNO_3 and H_2SO_4 are separated. The H_2SO_4-H_2O mixture is Stream W, and is

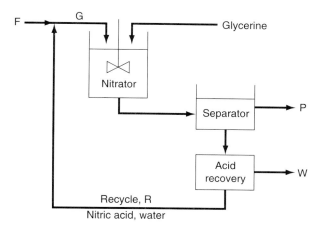

Figure P3.96

concentrated and sold for industrial purposes. The recycle stream to the nitrator is a 70.00% by weight solution of HNO_3 in water. In the above diagram, product stream P is 96.50% nitroglycerine and 3.50% water by weight.

To summarize:

Stream F = 50.00 wt% H_2SO_4, 43.00% HNO_3, 7.00% H_2O
Stream G contains 20.00 percent excess nitric acid
Stream P = 96.50 wt% nitroglycerine, 3.50 wt% water
Stream R = 70.00 wt% nitric acid, 30.00% water.

(a) If 1.000×10^3 kg of glycerine per hour are fed to the nitrator, how many kg per hour of stream P result?

(b) How many kg per hour are in the recycle stream?

(c) How many kg of fresh feed, stream F, are fed per hour?

(d) Stream W is how many kg per hour? What is its analysis in weight percent?

Molecular weights: glycerine = 92.11, nitroglycerine = 227.09, nitric acid = 63.01, sulfuric acid = 98.08, and water = 18.02.

3.97. Alkyl halides are used as an alkylating agent in various chemical transformations. The alkyl halide ethyl chloride can be prepared by the following chemical reaction:

$$2C_2H_6 + Cl_2 \quad \rightarrow \quad 2C_2H_5Cl + H_2$$

In the reaction process shown in Figure P3.97, fresh ethane and chlorine gas and recycled ethane are combined and fed into the reactor. A test shows that if 100% excess chlorine is mixed with ethane, a single-pass optimal conversion of 60% results, and of the ethane that reacts, all is converted to products and none goes into undesired products. Calculate:

(a) The fresh feed concentrations required for operation

(b) The moles of C_2H_5Cl produced in P per mole of C_2H_6 in the fresh feed F_1.

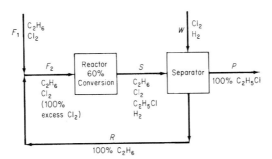

Figure P3.97

3.98. A process for methanol synthesis is shown in Figure P3.98. The pertinent chemical reactions involved are

$CH_4 + 2H_2O \rightarrow CO_2 + 4H_2$	(main reformer reactionn)	(a)
$CH_4 + H_2O \rightarrow CO + 3H_2$	(reformer side reaction)	(b)
$2CO + O_2 \rightarrow 2CO_2$	(CO converter reaction)	(c)
$CO_2 + 3H_2O \rightarrow CH_3OH + H_2$		(d)

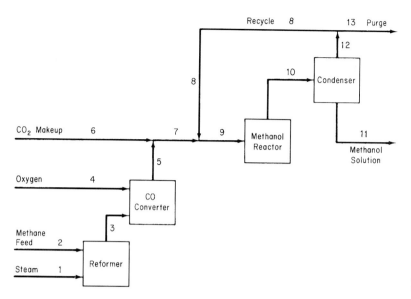

Figure P3.98

Ten percent excess steam, based on reaction (a), is fed to the reformer, and conversion of methane is 100%, with a 90% yield of CO_2. Conversion in the methanol reactor is 55% on one pass through the reactor.

A stoichiometric quantity of oxygen is fed to the CO converter, and the CO is completely converted to CO_2. Additional makeup CO_2 is then introduced to establish a 3-to-1 ratio of H_2 to CO_2 in the feed stream to the methanol reactor.

The methanol reactor effluent is cooled to condense all the methanol and water, with the noncondensible gases recycled to the methanol reactor feed. The H_2/CO_2 ratio in the recycle stream is also 3-to-1.

Because the methane feed contains 1% nitrogen as an impurity, a portion of the recycle stream must be purged as shown in Figure P3.98 to prevent the accumulation of nitrogen in the system. The purge stream analyzes 5% nitrogen.

On the basis of 100 mol of methane feed (including the N_2), calculate:

(a) How many moles of H_2 are lost in the purge

(b) How many moles of makeup CO_2 are required

(c) The recycle to purge ratio in mol/mol

(d) How much methanol solution (in kg) of what strength (weight percent) is produced.

GASES, VAPORS, LIQUIDS, AND SOLIDS

4

In planning and decision making for modern technology, engineers and scientists must know with reasonable accuracy the properties of the fluids and solids with which they deal. By **property** we mean any measurable characteristic of a substance, such as pressure, volume, or temperature, or a characteristic that can be calculated or deduced, such as internal energy, to be discussed in Chapter 5. For example, if you are engaged in the design of equipment—such as calculating the volume required for a process vessel—you need to know the specific volume or density of the gas or liquid that will be in the vessel as a function of temperature and pressure.

Clearly, it is not possible to have reliable, detailed experimental data at hand for the properties of all of the useful pure compounds and mixtures that exist in the world. Consequently, in the absence of experimental information, we estimate (predict) properties based on empirical correlations. Thus, the foundation of the estimation methods ranges from quite theoretical to completely empirical, and their reliability ranges from excellent to terrible.

Before starting our examination of physical properties, we need to mention two concepts: **state** and **equilibrium.** A system will *possess a unique set of properties,* such as temperature, pressure, density, and so on, at a given time, *and thus is said to be in a particular state.* A change in the state of the system results in a change in at least one of its properties.

By equilibrium we mean *a state in which there is no tendency toward spontaneous change.* Another way to say the same thing is to say that equilibrium is a state in which all the rates of attaining and departing from the state are balanced. When a system is in equilibrium with another system, or the surroundings, it will not change its state unless the other system, or the surroundings, also changes.

What are solids, liquids, vapors, and gases? Rather than define the states of matter, a task that is not easy to accomplish with precision in the brief space we have here, let us instead characterize the states in terms of two quantities, flow and structure of the molecules:

State	Flow of collections of molecules	Structure of collections of molecules
Perfect gas	Extensive	None
Real gas	Extensive	Almost none
Liquid	Short distance	Related structure
Liquid crystal	Some	Some crystal structure
Amorphous solid	Little	Little
Real crystal	Almost none	Highly structured
Perfect crystal	None	Completely structured

At any temperature and pressure, a pure compound can exist as a gas, liquid, or solid, and at certain specific values of T and p, mixtures of phases exist, such as when water boils or freezes. Thus a compound (or a mixture of compounds) may consist of one or more phases. A **phase** is defined as a completely homogeneous and uniform state of matter. Liquid water would be a phase; ice would be another phase. Two immiscible liquids in the same container, such as mercury and water, would represent two different phases because the liquids have different properties.

In this chapter we first discuss ideal and real gas relationships. You will learn about methods of expressing the p-V-T properties of real gases by means of equations of state and, alternatively, by compressibility factors. Next we introduce the concepts of vaporization, condensation, and vapor pressure, and illustrate how material balances are made for saturated and partially saturated gases.

4.1 IDEAL GAS LAW CALCULATIONS

> ### *Your objectives in studying this section are to be able to:*
>
> 1. Write down the ideal gas law, and define all its variables and parameters and their associated dimensions.
> 2. Calculate the values and units of the ideal gas law constant R in any set of units from the standard conditions.

3. Convert gas volumes to moles (and mass), and vice versa.
4. Calculate one variable, p, V, T, or n, from the given set values of the other three variables.
5. Calculate the specific gravity of a gas even if the reference condition is not clearly specified.
6. Calculate the density of a gas given its specific gravity.
7. Define and use partial pressure in gas calculations.
8. Show that under certain assumptions the volume fraction equals the mole fraction in a gas.
9. Solve material balances involving gases.

LOOKING AHEAD

In this section we explain how the ideal (perfect) gas law can be used to calculate the pressure, temperature, volume, or number of moles in a quantity of gas. We also discuss how to calculate the specific gravity and density of a gas.

MAIN CONCEPTS

4.1-1 The Ideal Gas Law

Under conditions such that the average distance between the molecules in a substance is great enough to neglect the effect of the intermolecular forces and the volume of the molecules themselves, a gas can be termed an **ideal gas.** More properly, an ideal gas is an imaginary gas that obeys exactly the following relationship

$$pV = nRT \tag{4.1}$$

where p = **absolute pressure** of the gas
V = total volume occupied by the gas
n = number of moles of the gas
R = ideal gas constant in appropriate units
T = **absolute temperature** of the gas

Sometimes the ideal gas law is written as

$$p\hat{V} = RT \tag{4.1a}$$

where $\hat{V}$ is the specific volume (volume per mole or mass) of the gas. Figure 4.1 illustrates the surface generated by Eq. (4.1a) in terms of the three properties p, $\hat{V}$, and T. Eq. (4.1) can be applied to a pure component or a mixture.

Several arbitrarily specified standard states (usually known as *standard conditions,* or S.C.) of temperature and pressure have been selected by custom. See Table 4.1 for the most common ones. The fact that a substance cannot exist as a gas at 0°C and 1 atm is immaterial. Thus, as we see later, water vapor at 0°C cannot exist at a pressure greater than its saturation pressure of 0.61 kPa (0.18 in. Hg) without condensation occurring. However, the imaginary volume at standard conditions can be calculated and is just as useful a quantity in the calculation of volume-mole-relationships as though it could exist. In what follows, the symbol V will stand for total volume and the symbol $\hat{V}$ for volume per mole, or per unit mass.

Because the SI, universal scientific, and American engineering standard are identical, you can use the values in Table 4.1 with their units to change from one system of units to another. Knowing the standard conditions also makes it easy for you to work with mixtures of units from different systems.

The next example illustrates how the standard conditions can be employed to convert mass or moles to volume. Do you see how to convert volume to moles or mass?

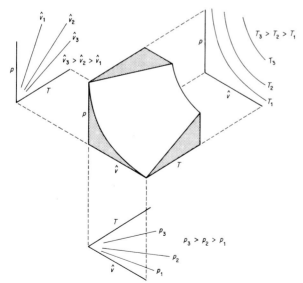

Figure 4.1   Representation of the ideal gas law in three dimensions as a surface.

TABLE 4.1   Common Standard Conditions for the Ideal Gas

System	T	p	$\hat{V}$
SI	273.15K	101.325 kPa	22.415m^3/kg mol
Universal scientific	0.0°C	760 mm Hg	22.415 liters/g mol
Natural gas industry	60.0°C	14.696 psia	379.4 ft^3/lb mol
		(15.0°C)	(101.325 kPa)
American engineering	32°F	1 atm	359.05 ft^3/lb mol

EXAMPLE 4.1 Use of Standard Conditions

Calculate the volume, in cubic meters, occupied by 40 kg of CO_2 at standard conditions.

Solution

Basis: 40 kg of CO_2

$$\frac{40 \text{ kg CO}_2}{} \left| \frac{1 \text{ kg mol CO}_2}{44 \text{ kg CO}_2} \right| \frac{22.42 \text{ m}^3 \text{ CO}_2}{1 \text{ kg mol } \text{CO}_2} = 20.4 \text{ m}^3 \text{ CO}_2 \text{ at S.C.}$$

Notice in this problem how the information that 22.42 m^3 at S.C. = 1 kg mol is applied to transform a known number of moles into an equivalent number of cubic meters.

   Incidentally, whenever you use cubic measure for volume, you must establish the conditions of temperature and pressure at which the cubic measure for volume exists, since the term "m^3" or "ft^3," standing alone, is really not any particular *quantity* of material.

   You can apply the ideal gas law, Eq. (4.1), directly by introducing values for three of the four quantities, n, p, T, and V, solving for the fourth. To do so, you need to look up or calculate R in the proper units. Inside the front cover you will find selected values of R for different combinations of units. Example 4.2 illustrates how to calculate the value of R in any set of units you want from the values of p, T, and $\hat{V}$ at standard conditions.

EXAMPLE 4.2 Calculation of *R*

Find the value for the universal gas constant R for the following combination of units: For 1 g mol of ideal gas when the pressure in atm, the volume in cm^3, and the temperature in K.

Solution

At standard conditions we will use the approximate values

$$p = 1 \text{ atm}$$
$$\hat{V} = 22,415 \text{ cm}^3/\text{g mol}$$
$$T = 273.15\text{K}$$

$$R = \frac{p\hat{V}}{T} = \frac{1 \text{ atm}}{273.15 \text{ K}} \left| \frac{22,415 \text{ cm}^3}{1 \text{ g mol}} \right. = 82.06 \frac{(\text{cm}^3)(\text{atm})}{(\text{K})(\text{g mol})}$$

In many processes going from an initial state to a final state, you can use the ratio of the ideal gas laws in the respective states and eliminate R as follows (the subscript 1 designates the initial state, and the subscript 2 designates the final state)

$$\frac{p_1 V_1}{p_2 V_2} = \frac{n_1 R T_1}{n_2 R T_2}$$

or

$$\left(\frac{p_1}{p_2} \right) \left(\frac{V_1}{V_2} \right) = \left(\frac{n_1}{n_2} \right) \left(\frac{T_1}{T_2} \right) \tag{4.2}$$

Note how Eq. (4.2) involves ratios of the same variable. This result of the application of the ideal gas law has the convenient feature that the pressures may be expressed in any system of units you choose, such as kPa, in. Hg, mm Hg, atm, and so on, as long as the same units are used for both conditions of pressure (do not forget that the pressure must be *absolute* pressure in both cases). Similarly, the ratio of the *absolute* temperatures and the volumes gives ratios that are dimensionless. Notice how the ideal gas constant R is eliminated in taking the ratio of the initial to the final state.

Let us see how we can apply the gas law both in the form of Eq. (4.2) and Eq. (4.1) to problems.

EXAMPLE 4.3 Application of the Ideal Gas Law

Calculate the volume occupied by 88 lb of CO_2 at a pressure of 32.2 ft of water and at 15°C.

Solution

Examine Figure E4.3. To use Eq. (4.2) the initial volume has to be calculated as

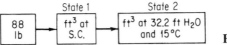

Figure E4.3

shown in Example 4.1. Then the final volume can be calculated via Eq. (4.2) in which both R and $(n_1)/(n_2)$ cancel out:

$$V_2 = V_1 \left(\frac{p_1}{p_2}\right)\left(\frac{T_2}{T_1}\right)$$

Assume that the given pressure is absolute pressure.

At S.C. (state 1)	*At state 2*
$p = 33.91$ ft H_2O	$p = 32.2$ ft H_2O
$T = 273$ K	$T = 273 + 15 = 288$ K

Basis: 88 lb of CO_2

$$\frac{88 \text{ lb } CO_2}{} \left| \frac{1 \text{ lb mol } CO_2}{44 \text{ lb } CO_2} \right| \frac{359 \text{ ft}^3}{1 \text{ lb mol}} \left| \frac{288}{273} \right| \frac{33.91}{32.2} = \frac{798 \text{ ft}^3 \ CO_2}{\text{at 32.2 ft } H_2O \text{ and } 15°C}$$

Calculation of V_1

You can mentally check your calculations by saying to yourself: The temperature goes up from 0°C at S.C. to 15°C at the final state, hence the volume must increase from S.C., and the temperature ratio must be greater than unity. Similarly, you can say: The pressure goes down from S.C. to the final state, so that the volume must increase from S.C., hence the pressure ratio must be greater than unity.

The same result can be obtained by using Eq. (4.1). First, the value of R must be obtained in the same units as the variables p, $\hat{V}$, and T. Look it up or calculate the value from p, $\hat{V}$, and T at S.C.

$$R = \frac{p\hat{V}}{T}$$

At S.C.,

$p = 33.91$ ft H_2O \qquad $V = 359$ ft^3/lb mol \qquad $T = 273$K

$$R = \frac{33.91}{} \left| \frac{359}{273} \right. = 44.59 \ \frac{(\text{ft } H_2O)(\text{ft}^3)}{(\text{lb mol})(K)}$$

Now, using Eq. (4.1), insert the given values, and perform the necessary calculations.

Basis: 88 lb of CO_2

$$V = \frac{nRT}{p} = \frac{88 \text{ lb } CO_2}{\dfrac{44 \text{ lb } CO_2}{\text{lb mol } CO_2}} \left| \frac{44.59 \text{ (ft } H_2O)(ft^3)}{(\text{lb mol})(K)} \right| \frac{288 \text{ K}}{32.2 \text{ ft } H_2O}$$

$$= 798 \text{ ft}^3 CO_2 \text{ at } 32.2 \text{ ft } H_2O \text{ and } 15°C$$

If you will inspect both solutions closely, you will observe that in both cases the same numbers appear and that both are identical.

The **density of a gas** is defined as the mass per unit volume and can be expressed in kilograms per cubic meter, pounds per cubic foot, grams per liter, or other units. Inasmuch as the mass contained in a unit volume varies with the temperature and pressure, as we have previously mentioned, you should always be careful to specify these two conditions. If not otherwise specified, the densities are presumed to be at S.C.

EXAMPLE 4.4 Calculation of Gas Density

What is the density of N_2 at 27°C and 100 k Pa in SI units?

Solution

Basis: 1 m³ of N_2 at 27°C and 100 kPa

$$\frac{1 \text{ m}^3}{} \left| \frac{273 \text{ K}}{300 \text{ K}} \right| \frac{100 \text{ kPa}}{101.3 \text{ kPa}} \left| \frac{1 \text{ kg mol}}{22.4 \text{ m}^3} \right| \frac{28 \text{ kg}}{1 \text{ kg mol}} = 1.123 \text{ kg}$$

density = 1.123 kg/m³ of N_2 at 27°C (300 K) and 100 kPa

The **specific gravity of a gas** is usually defined as the ratio of the density of the gas at a desired temperature and pressure to that of air (or any specified reference gas) at a certain temperature and pressure. The use of specific gravity occasionally may be confusing because of the manner in which the values of specific gravity are reported in the literature. You must be very careful in using literature values of the specific gravity to ascertain that the conditions of temperature and pressure are known both for the gas in question and for the reference gas. Thus, this question is not well posed: *What is the specific gravity of methane?* Actually, this question may have the same answer as the question: How many grapes are in a bunch? Unfortunately, occasionally one may see this question and the best possible answer is

$$\text{sp. gr.} = \frac{\text{density of methane at S.C.}}{\text{density of air at S.C.}}$$

by which the temperature and pressure of the methane and reference air are clearly specified.

EXAMPLE 4.5 Specific Gravity of a Gas

What is the specific gravity of N_2 at 80° F and 745 mm Hg compared to air at 80°F and 745 mm Hg?

Solution

First you must obtain the density of the N_2 and the air at their respective conditions of temperature and pressure, and then calculate the specific gravity by taking a ratio of their densities. Example 4.4 covers the calculation of the density of a gas, and therefore, to save space, no units will appear in the intermediate calculation here:

Basis: 1 ft³ of air at 80°F and 745 mm Hg

$$\frac{1}{}\ \left|\ \frac{492}{540}\ \right|\ \frac{745}{760}\ \left|\ \frac{}{359}\ \right|\ \frac{29}{}\ = 0.0721\ \text{lb/ft}^3\ \text{at 80°F and 745 mm Hg}$$

Basis: 1 ft³ of N_2 at 80°F and 745 mm Hg

$$\frac{1}{}\ \left|\ \frac{492}{540}\ \right|\ \frac{745}{760}\ \left|\ \frac{}{359}\ \right|\ \frac{28}{}\ = 0.0697\ \text{lb/ft}^3\ \text{at 80°F and 745 mm Hg}$$

$$(\text{sp. gr.})_{N_2} = \frac{0.0697}{0.0721} = 0.967\ \frac{\text{lb } N_2/\text{ft}^3\ \text{at 80°F, 745 mm Hg}}{\text{lb air/ft}^3\ \text{air at 80°F, 745 mm Hg}}$$

Did you note from Example 4.5 that for gases at the *same* temperature and pressure, the specific gravity is just the ratio of the respective molecular weights? Let A be one gas and B be another.

$$p\hat{V} = RT \qquad\qquad \text{or} \qquad\qquad p\,\frac{1}{\rho} = RT$$

Thus

$$\text{sp. gr.} = \frac{\rho_A}{\rho_B}\left(\frac{\text{mol. wt.}_A}{\text{mol. wt.}_B}\right)\left(\frac{T_B}{T_A}\right) \tag{4.3}$$

4.1-2 Ideal Gas Mixtures and Partial Pressure

Frequently, as an engineer, you will want to make calculations for mixtures of gases instead of individual gases. You can use the ideal gas law (under the proper assumptions, of course, for a mixture of gases by interpreting p as the total pressure of the mixture, V as the volume occupied by the mixture, n as the total number of moles of all components in the mixture, and T as the temperature of the mixture.

Engineers use a fictious but useful quantity called the **partial pressure** in many of their calculations involving gases. The partial pressure of gas i defined by Dalton, p_i, namely the pressure that would be exerted by a single component in a gaseous mixture if it existed by itself in the *same volume* as occupied by the mixture and at the *same temperature* of the mixture is

$$p_i V_{\text{total}} = n_i R T_{\text{total}} \tag{4.4}$$

where p_i is the partial pressure of component i. If you divide Eq. (4.4) by Eq. (4.1), you find that

$$\frac{p_i V_{\text{total}}}{p_{\text{total}} V_{\text{total}}} = \frac{n_i R T_{\text{total}}}{n_{\text{total}} R T_{\text{total}}}$$

or

$$p_i = p_{\text{total}} \frac{n_i}{n_{\text{total}}} = p_{\text{total}} \, y_i \tag{4.5}$$

where y_i is the mole fraction of component i. Can you show that Dalton's law of the summation of partial pressures is true using Eq. (4.5)?

$$p_1 + p_2 + \ldots + p_n = p_t \tag{4.6}$$

Although you cannot measure the partial pressure directly with an instrument, you can calculate the value from Eqs. (4.5) and/or (4.6). To illustrate the significance of Eq. (4.4) and the meaning of partial pressure, suppose that you carried out the following experiment with ideal gases. Two tanks of 1.50 m³ volume, one containing gas A at 300 kPa and the other gas B at 400 kPa (both gases being at the same temperature of 20°C), are connected together. All the gas in B is forced into tank A isothermally. Now that you have a 1.50-m³ tank of $A + B$ at 700 kPa and 20°C for this mixture, you could say that gas A exerts a partial pressure of 300 kPa and gas B exerts a partial pressure of 400 kPa. Of course you cannot put a pressure gauge on the tank and check this conclusion, because the pressure gauge will read only the total pressure. These partial pressures are hypothetical pressures that the individual gases would exert and are equivalent to the pressures they actually would have if they were put into the same volume at the same temperature all by themselves. If the total pressure of the gaseous mixture is known as well as the mole fraction of a component, the partial pressure of the component can be calculated via Eq. (4.5):

$$p_A = 700\left(\frac{3}{7}\right) = 300 \text{ kPa}$$

EXAMPLE 4.6 Calculation of Partial Pressures from a Gas Analysis

A flue gas analyzes 14.0% CO_2, 6.0% O_2, and 80.0% N_2. It is at 400°F and 765.0 mm Hg pressure. Calculate the partial pressure of each component.

Solution

Use Eq. (4.5): $p_i = p_t y_i$.

Basis: 1.00 kg (or lb) mol flue gas

Component	kg (or lb) mol	p (mm Hg)
CO_2	0.140	107.1
O_2	0.060	45.9
N_2	0.800	612.0
Total	1.000	765.0

On the basis of 1.00 mole of flue gas, the mole fraction y of each component, when multiplied by the total pressure, gives the partial pressure of that component.

4.1-3 Material Balances Involving Gases

Now that you have had a chance to practice applying the ideal gas law to simple problems, we turn our attention back to material balances. The only difference between the subject matter of Chapter 3 and this section is that here the amount of material flow can be specified in terms of p, V, and T rather than solely mass or moles. For example, the basis for a problem, or the quantity to be solved for, might be a volume of gas at a given temperature and pressure rather than a mass of gas. The next two examples illustrate two of the same types of the problems you have encountered before but now involve gases.

EXAMPLE 4.7 Material Balance with Combustion

A gas produced by gasifying wood chips analyzes 6.4% CO_2, 0.1% O_2, 39% CO, 51.8% H_2, 0.6% CH_4, and 2.1% N_2. It enters the combustion chamber at 90°F and a pressure of 35.0 in. Hg, and is burned with 40% excess air (dry) which is at 70°F and the atmospheric pressure of 29.4 in Hg; 10% of the CO remains unburned. How many cubic feet of air are supplied per cubic foot of entering gas? How many cubic feet of product are produced per cubic foot of entering gas if the exit gas is at 29.4 in. Hg and 400°F?

Solution

This is a steady-state problem with reaction in which you can directly add and subtract quantities; you do not have to solve simultaneous equations. The system is the gases in combustion chamber.

Steps 1, 2, 3, and 4   Figure E4.7 illustrates the process and notation. With 40% excess air, certainly all of the CO, H_2, and CH_4 should burn to CO_2 and H_2O; apparently, for some unknown reason, not all of the CO burns to CO_2. The product gases are shown in the figure.

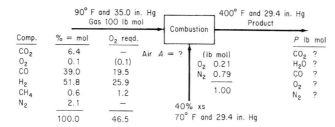

Figure E4.7

Step 5   You could take 1 ft^3 at 90°F and 35.0 in. Hg as the basis, but it is just as easy to take 1 lb mol because then % = lb mol. At the end of the problem you can convert lb mol to ft$_3$.

Basis: 100 lb mol synthesis gas

Step 4 (continued)   The entering air can be calculated from the specified 40% excess air; the reactions for complete combustion are

$$CO + \tfrac{1}{2}O_2 \rightarrow CO_2$$

$$H_2 + \tfrac{1}{2} O_2 \rightarrow H_2O$$

$$CH_4 + 2O_2 \rightarrow CO_2 + 2H_2O$$

The moles of oxygen required are listed in Figure E4.7. The excess oxygen is

$$\text{Excess } O_2\text{: } 0.4(46.5) = 18.6$$

$$\text{Total } O_2 = 46.5 + 18.6 = 65.10$$

$$N_2 \text{ in is } 65.10 \left(\frac{79}{21} \right) = 244.9$$

Total moles of air in are 244.9 + 65.10 = 310 lb mol.

Steps 6 and 7 Five unknowns exist, the five products. You can make four element balances plus know the fraction of the entering CO that exits in P. Hence the problem has a unique solution.

Steps 8 and 9 We make element balances in moles to calculate the unknown quantities.

	In		*Out*
N2:	2.1 + 244.9	=	n_{N_2}
C:	6.4 + 39.0 + 0.6	=	$n_{CO_2} + 0.10(39.0)$
H2:	51.8 + 0.6(2)	=	n_{H_2O}
O2:	6.4 + 0.1 + 0.5(39) + 65.1	=	$n_{O_2} + n_{CO_2} + 0.5(n_{H_2O}) + n_{CO})$

The solution of these equations is

$$n_{N_2} = 247.0 \qquad n_{CO} = 3.9 \qquad n_{CO_2} = 42.1 \qquad n_{H_2O} = 53.0 \qquad n_{O_2} = 20.55$$

The total moles exiting sum up to be 366.55 lb mol.

Finally, we can convert the lb mol of air and product into the volumes requested:

$$T_{gas} = 90 + 460 = 550°R$$

$$T_{air} = 70 + 460 = 530°R$$

$$T_{product} = 400 + 460 = 860°R$$

ft^3 of gas : $\dfrac{100 \text{ lb mol entering gas}}{} \left| \dfrac{359 \text{ ft}^3 \text{ at SC}}{1 \text{ lb mol}} \right| \dfrac{550°R}{492°R} \left| \dfrac{29.92 \text{ in. Hg}}{35.0 \text{ in Hg}} \right. = 343 \times 10^2$

ft^3 of air : $\dfrac{310 \text{ lb mol air}}{} \left| \dfrac{359 \text{ ft}^3 \text{ at SC}}{1 \text{ lb mol}} \right| \dfrac{530°R}{492°R} \left| \dfrac{29.92 \text{ in. Hg}}{29.4 \text{ in. Hg}} \right. = 1220 \times 10^2$

ft^3 of product: $\dfrac{366.55 \text{ lb mol } P}{} \left| \dfrac{359 \text{ ft}^3 \text{ at SC}}{1 \text{ lb mol}} \right| \dfrac{860°R}{492°R} \left| \dfrac{29.92 \text{ in. Hg}}{35.0 \text{ in. Hg}} \right. = 2340 \times 10^2$

The answers to the questions are

$$\frac{1220 \times 10^2}{343 \times 10^2} = 3.56 \frac{\text{ft}^3 \text{ air at } 530° \text{ R and } 29.4 \text{ in. Hg}}{\text{ft}^3 \text{ gas at } 550° \text{ R and } 35.0 \text{ in. Hg}}$$

$$\frac{2340 \times 10^2}{343 \times 10^2} = 6.82 \frac{\text{ft}^3 \text{ product at } 860° \text{ R and } 29.4 \text{ in. Hg}}{\text{ft}^3 \text{ gas at } 550° \text{ R and } 35.0 \text{ in. Hg}}$$

EXAMPLE 4.8 Material Balance without Reaction

Gas at 15°C and 105 kPa is flowing through an irregular duct. To determine the rate of flow of the gas, CO_2 from a tank was passed into the gas stream. The gas analyzed 1.2% CO_2 by volume before, and 3.4% CO_2 by volume after, addition. The CO_2 left the tank, was passed through a rotameter, and was found to flow at the rate of 0.0917 m³/min at 7°C and 131 kPa. What was the rate of flow of the entering gas in the duct in cubic meters per minute?

Solution

This is a steady-state problem without reaction.
Steps 1, 2, 3, and 4

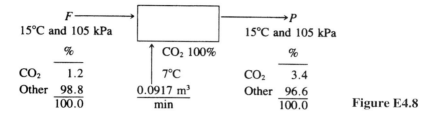

	F ⟶			⟶ P	
	15°C and 105 kPa	CO₂ 100%		15°C and 105 kPa	
	%	↑ 7°C		%	
CO_2	1.2	0.0917 m³	CO_2	3.4	
Other	98.8	——————	Other	96.6	
	100.0	min		100.0	**Figure E4.8**

Assume that the entering and exit gases are at the same temperature and pressure.

Step 5   Take as a basis 1 min ≡ 0.0917 m³ of CO_2 at 7°C and 131 kPa. The gas analysis is in volume percent, which is the same as mole percent. We could convert the 0.0917 m³ to moles and solve the problem in terms of moles, but there is no need to do so because we just as easily convert the known flow rate to 15°C and 105 kPa, and solve the problem using m³ all at the same conditions. We could similarly convert all of the data to 7°C and 131 kPa, but more calculations would be required to get the answer than for 15°C and 105 kPa.

$$\frac{0.0917 \text{ m}^3}{} \left| \frac{273 + 15}{273 + 7} \right| \frac{131}{105} = 0.1177 \text{ m}^3 \text{ at} \atop 15°C \text{ and } 105 \text{ kPa}$$

Steps 6 and 7   We do not know F and P, but can make two independent component balances, hence the problem has a unique solution.
Steps 7, 8, and 9

"Other" balance (m³ at 15°C and 105 kPa): $F(0.988) = P(0.966)$ (a)

CO_2 balance (m³ at 15°C and 105 kPa): $F(0.012) + 0.1177 = P(0.034)$ (b)

Total balance (m³) at 15°C and 105 kPa): $F + 0.1177 = P$ (c)

Note that "other" is a tie component. Solution of Eqs. (a) and (c) gives

$$F = 5.17 \text{ m}^3/\text{min at } 15°C \text{ and } 105 \text{ kPa}$$

Step 10 (Check)　Use the redundant equation:

By Eq. (b): $5.17\,(0.012) + 0.1177 = 0.180 \overset{?}{=} 5.17(0.988/0.966)(0.034) = 0.180$

The equation checks out to a satisfactory degree of precision.

LOOKING BACK

We have completed our review of the ideal gas law as applied to pure components and gas mixtures. In making material balances, process measurements are frequently made as volumetric flows rather than molar or mass flows, and the molar flows are calculated from the volumetric flows rather than the reverse.

Key Ideas

1. The p, V, and T properties of a gas, either a pure component or a mixture, at moderate pressures and temperatures can be represented by the ideal gas law $pV = nRT$. This relation represents a surface in the three dimensions of p, $\hat{V}$, and T.
2. The values of p, $\hat{V}$, and T at standard conditions can be used to convert volume to moles.
3. Gas specific gravity is the ratio of the density of a gas at a specified T and p to the density of a reference gas, usually air, at a specified T and p.
4. The partial pressure of one gas in a mixture can be calculated by multiplying the total pressure by the mole fraction of the component.
5. Material balances in which data is given as volumetric quantities of gas are solved by exactly the same procedure as in Table 3.1.

Key Terms

Density (p. 269)　　　　　　　Phase (p. 263)
Equilibrium (p. 262)　　　　　Properties (p. 262)
Gas constant (p. 264)　　　　Specific gravity (p. 269)
Ideal gas (p. 264)　　　　　　Standard conditions (SC) (p. 266)
Ideal gas law (p. 264)　　　　State (p. 262)
Partial pressure (p. 271)

Self-Assessment Test

1. Write down the ideal gas law.
2. What are the dimensions of T, p, V, n, and R?
3. List the standard conditions for a gas in the SI and American Engineering systems of units.
4. Calculate the volume in ft^3 of 10 lb mol of an ideal gas at 68°F and 30 psia.

5. A steel cylinder of volume 2 m^3 contains methane gas (CH_4) at 50°C and 250 kPa absolute. How many kilograms of methane are in the cylinder?

6. What is the value of the ideal gas constant R to use if the pressure is to be expressed in atm, the temperature in kelvin, the volume in cubic feet, and the quantity of material in pound moles?

7. Twenty-two kilograms per hour of CH_4 are flowing in a gas pipeline at 30°C and 920 mm Hg. What is the volumetric flow rate of the CH_4 in m^3 per hour?

8. What is the density of a gas that has a molecular weight of 0.123 kg/kg mol at 300 K and 1000 kPa?

9. What is the specific gravity of CH_4 at 70°F and 2 atm compared to air at S.C.?

10. A gas has the following composition at 120°F and 13.8 psia.

Component	Mol %
N_2	2
CH_4	79
C_2H_6	19

(a) What is the partial pressure of each component?
(b) What is the volume fraction of each component?

11. (a) If the C_2H_6 were removed from the gas in problem 10, what would be the subsequent pressure in the vessel?
(b) What would be the subsequent partial pressure of the N_2?

12. A furnace is fired with 1000 ft^3 per hour at 60°F and 1 atm of a natural gas having the following volumetric analysis: CH_4: 80%, C_2H_6: 16%, O_2: 2%, CO_2: 1%, N_2: 1%. The exit flue gas temperature is 800°F and the pressure is 760 mm Hg absolute; 15% excess air is used and combustion is complete. Calculate the:
(a) Volume of CO_2 produced per hour
(b) Volume of H_2O vapor produced per hour
(c) Volume of N_2 produced per hour
(d) Total volume of flue gas produced per hour

13. A flue gas contains 60% N_2 and is mixed with air., If the resulting mixture flows at a rate of 250,000 ft^3/hr and contains 70% N_2,what is the flow rate of the flue gas? State all your assumptions concerning the temperatures and pressures of the gas streams.

Thought Problems

1. A candle is placed vertically in a soup plate, and the soup plate filled with water. Then the candle is lit. An inverted water glass is carefully placed over the candle. The candle soon goes out, and the water rises inside the glass. It is often said that this shows how much oxygen in the air has been used up. Is this conclusion correct?

2. A scientific supply house markets aerosol-type cans containing compressed helium for

filling balloons, doing demonstrations, and the like. On the label there appears the notice: "Because the can contains helium, it quite naturally feels empty. It is actually lighter full than empty." Is this statement correct? If so, why? If not, why not?

3. Some reviewers of books and articles have suggested that the gas constant R in the ideal gas equation be forced to take the value of unity (1). What would this step require as far as using the ideal gas equation?

4. In an article by P. Hickman in *Physics Teacher*, **25**, (1987): 430 it was suggested that the density of air be determined using marshmallows as follows:
   1. Put the marshmallows in a vacuum so they expand and release trapped air. Repressurize the vessel and compress the marshmallows to a fraction of their original volume.
   2. Weigh the marshmallows on a balance before and after the vacuum treatment. The difference should be the mass of air trapped inside.
   3. Measure their volume before and after the vacuum treatment. The difference should be the volume of the air trapped inside.

      Is this a sound way of getting the density of air?

Discussion Questions

1. Three identical glasses are arranged as in Figure DQ4.1–1. Hollow stirrers are also needed. Glasses A and B are completely filled with water (by submerging them jointly in a bucket or sink and joining the mouths before removing), and C is empty. Glasses A and B are carefully placed on a few hollow stirrers as shown in the figure. How can most of the water in glass A be transferred to glass C without ever touching or moving the glasses or their supporting stirrers?

2. An employee was cleaning a cylindrical vessel that contained CS_2 in which solid residues had built up on the stirrer. The vessel had been pumped out and blanketed with nitrogen. The manhole cover was removed and the solid residue removed from the stirrer with a scraper rod. The employee went to lunch, leaving the manhole cover

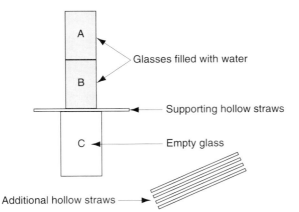

Figure DQ 4.1–1

off, and when returning to complete the job, started a flash fire with a spark from striking the stirrer with the scraper. What might be some of the causes of the accident?

3. Sea breezes provide welcome relief from the summer heat for residents who live close to the shore. No matter what part of the world—the coast of California, Australia where sea breezes can be very strong, even along the shores of the Great Lakes—the daily pattern in summertime is the same. The sea breeze, a wind blowing from sea to land begins to develop three or four hours after sunrise and reaches its peak intensity by midafternoon. It may penetrate inland as much as 60 or 70 km. The sea breeze dies out in the evening and three or four hours after sunset may be replaced by a land breeze, blowing from land to the sea. The land breeze, much weaker than the sea breeze, reaches its peak intensity just before sunrise. What causes these breezes?

4. A distillation column reboiler in a room, as shown in Figure DQ 4.1–4, had been cleaned, but the manhole cover was not securely fastened on startup. As a result benzene vapor escaped from the manhole and one operator died by asphyxiation. How could this accident occur?

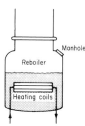

Figure DQ 4.1–4

4.2 REAL GAS RELATIONSHIPS

We have said in Section 4.1 that at room temperature and pressure many gases can be assumed to act as ideal gases. However, for some gases under normal conditions, and for most gases under conditions of high pressure, values of predictions of the gas properties that you might obtain using the ideal gas law would be at wide variance with the experimental evidence. You might wonder exactly how the behavior of **real gases** compares with that calculated from the ideal gas laws. In Figure 4.2 you can see how the $(p\hat{V})$ product of several gases deviates from that predicted by the ideal gas laws as the pressure increases substantially. Thus, it is clear that we need some way of computing the p-V-T properties of a gas that is not ideal, i.e., a real gas.

Physical properties and/or the equations used to predict physical properties can be stored on spreadsheets. No programming experience is required, and simple calculations are easy to carry out using the spreadsheet software. Numerical data such as tables can be typed into cells, where they are easily seen. Equations

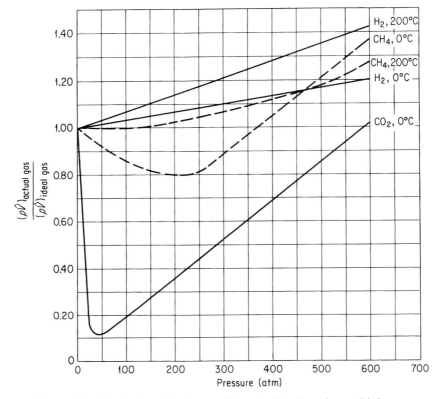

Figure 4.2 Deviation of real gases from the ideal gas law at high pressures.

can also be typed into cells and introduced as needed. Labels and units as well as remarks can be added as desired.

In this section we discuss three methods of getting or predicting real gas properties in lieu of having experimental data:

(1) Compressibility charts

(2) Equations of state

(3) Estimated properties

Even if experimental data are available, the other three techniques still may be quite useful for certain types of calculations. Keep in mind that under conditions such that part of the gas liquefies, the gas laws apply only to the gas-phase portion of the system—you cannot extend these real gas laws into the liquid region any more than you can apply the ideal gas laws to a liquid.

4.2-1 Critical State, Reduced Parameters, and Compressibility

Your objectives in studying this section are to be able to:

1. State the law of corresponding states.
2. Define the critical state.
3. Calculate the reduced temperature, reduced pressure, and reduced volume, and use any two of these parameters to obtain the compressibility factor, z, from the compressibility charts.
4. Use compressibility factors and appropriate charts to predict the p-V-T behavior of a gas, or given the required data, find compressibility factors.
5. Calculate the ideal critical volume and ideal reduced volume, and be able to use the V_{r_i} parameter in the compressibility charts.

LOOKING AHEAD

In this section we explain how the critical properties of gases can be used to facilitate a graphical method of solving for p, V, n, and T for single- and multi-component real gases.

MAIN CONCEPTS

In the attempt to devise some truly universal gas law that predicts well at high pressures, the idea of corresponding states was developed. Early experimenters found that at the critical point all substances are in approximately the same state of molecular dispersion. Consequently, it was felt that their thermodynamic and physical properties should be similar. The **law of corresponding states** expresses the idea that in the critical state all substances should behave alike.

What does the **critical state** mean? You can find many definitions, but the one most suitable for general use with pure component systems as well as with mixtures of gases is the following:

> **The critical state for the gas-liquid transition is the set of physical conditions at which the density and other properties of the liquid and vapor become identical.**

This point, for a pure component (only), is the highest temperature at which liquid and vapor can exist in equilibrium.

Refer to Figure 4.3 in which the states of water are illustrated. As the temperature increases, the density of the liquid and vapor approach each other until finally at 374.14°C, the values are the same. At the critical state, if you watch a liquid held at or above the critical temperature and expand its volume, you cannot tell when the liquid becomes a vapor, because no interface is formed between the phases—no liquid surface can be seen. This phenomenon occurs at such a high pressure and temperature for water that it is outside your everyday experience.

You can find experimental values of the critical temperature (T_c) and the critical pressure (p_c) for various compounds in Appendix D. If you cannot find a desired critical value in this text or in a handbook, you can always consult Reid, Prausnitz, and Poling (1987), (see references at the end of this chapter), which describe and evaluate methods of estimating critical constants for various compounds.

The gas-liquid transition described above is only one of several possible transitions exhibiting a critical point. Critical phenomena are observed in liquids and solids as well.

A **supercritical fluid**, that is, a compound in a state above the critical point, combines some of the properties of both gases and liquids. Supercritical fluids are used to replace the void left by solvents such as trichloroethylene and methylene chloride, the emissions from which, and contact with, have been severely limited. For example, in coffee decaffeination, removal of cholesterol from egg yolk with CO_2, the production of vanilla extract, and the destruction of undesirable organic compounds all take place using supercritical water. Supercritical-water ox-

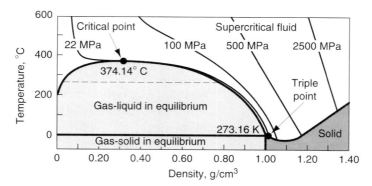

Figure 4.3   The regions of existence of solid, liquid, gaseous, and supercritical water. At the triple point solid, liquid, and gas are all in equilibrium.

idation has been shown to destroy 99.99999% of all of the major types of toxic agents in the U.S. chemical weapons stockpile.

Another set of terms with which you should become familiar are the **reduced parameters**. These are *corrected*, or *normalized*, conditions of temperature, pressure, and volume, normalized by their respective critical conditions, as follows:

$$T_r = \frac{T}{T_c}$$

$$p_r = \frac{p}{p_c}$$

$$V_r = \frac{\hat{V}}{\hat{V}_c}$$

Now how can the ideas presented above be used? One common way is to modify the ideal gas law by inserting an adjustable coefficient z, the **compressibility factor**, a factor that compensates for the nonideality of the gas. Thus, the ideal gas law becomes a real gas law, a **generalized equation of state**.

$$pV = znRT \tag{4.7}$$

One way to look at z is to consider it to be a factor that makes Eq. 4.7 an equality. If the compressibility factor is plotted for a given temperature against the pressure for different gases, you obtain plots as shown in Figure 44(b). However, if the compressibility is plotted against the reduced pressure as a function of the reduced temperature, then for most gases the compressibility values at the same reduced temperature and reduced pressure fall at about the same point, as illustrated in Figure 4.4 (b).

This outcome permits the use of what is called a generalized compressibility factor, and Figure 4.5 shows the **generalized compressibility charts** or z-factor chart prepared by Nelson and Obert.[1] These charts are based on 30 gases. Figures 4.5(b) and (c) represent z for 26 gases (excluding H_2, He, NH_3, H_2O) with a maximum deviation of 1%, and H_2 and H_2O within a deviation of 1.5%. Figure 4.5(d) is for 26 gases and is accurate to 2.5%, while Figure 4.5(e) is for 9 gases and errors can be as high as 5%. For H_2 and He only, corrections to the actual critical constants are used to give pseudocritical constants

$$T'_c = T_c + 8 \text{ K}$$

$$p'_c = p_c + 8 \text{ atm}$$

---

[1]L. C. Nelson and E. F. Obert. *Chem. Eng.*, 61 (7) (1954): 203–208. Figures 4.5(b) and (c) include data reported by P. E. Liley. *Chem. Eng.*, (July 20, 1987): 123.

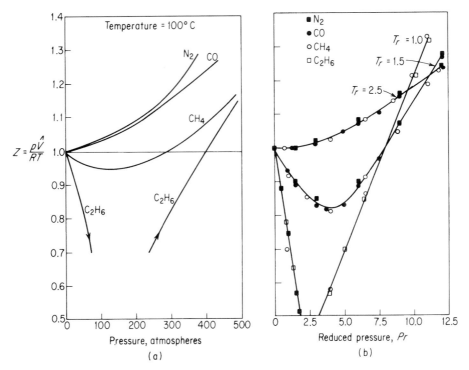

Figure 4.4 (a) Compressibility factor as a function of temperature and pressure; (b) compressibility as a function of reduced temperature and reduced pressure.

which enable you to use Figures 4.5(a-e) for these two gases as well with minimum error.

Figure 4.5(f) is a unique chart that, by having several parameters plotted simultaneously on it, helps you avoid trial-and-error solutions or graphical solutions of real gas problems. One of these helpful parameters is the ideal reduced volume defined as

$$V_{r_i} = \frac{\hat{V}}{\hat{V}_{c_i}}$$

V_{c_i} the ideal critical volume (not the experimental value of the critical volume), or

$$\hat{V}_{c_i} = \frac{RT_c}{p_c}$$

Both V_{r_i} and $\hat{V}_{c_i}$ are easy to calculate since T_c and p_c are presumed known. The development of the generalized compressibility chart is of considerable practical

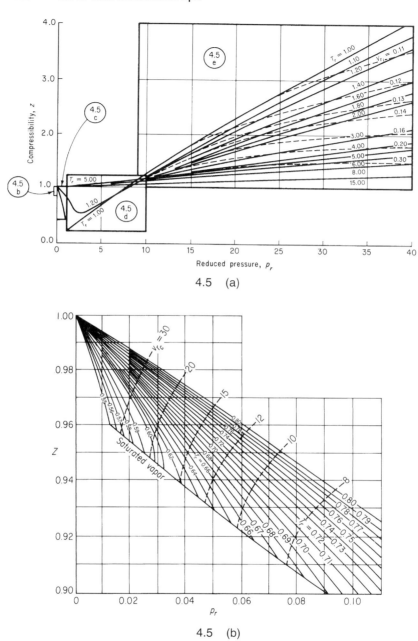

Figure 4.5 (a) Generalized compressibility chart showing the respective portions of the subsequent expanded charts. (b) Generalized compressibility chart, very low reduced pressure. (c) Generalized compressibility chart, low pressure. (d) Generalized compressibility chart, medium pressures. (e) Generalized compressibility chart, high pressures.

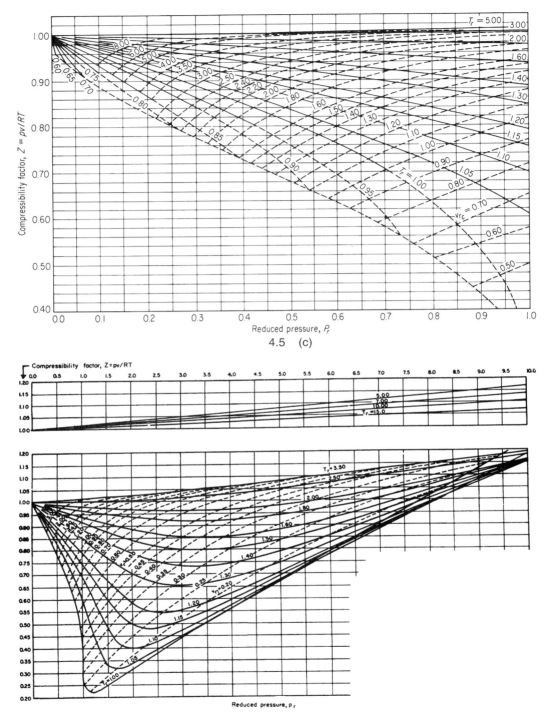

4.5 (c)

4.5 (d)

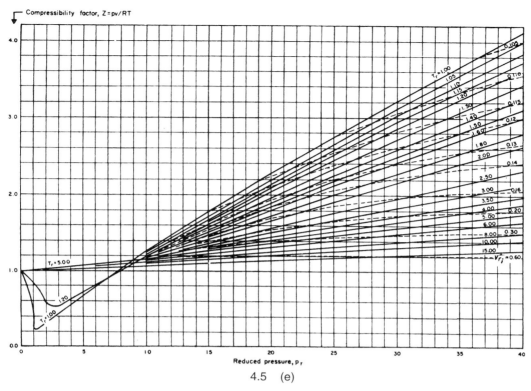

4.5 (e)

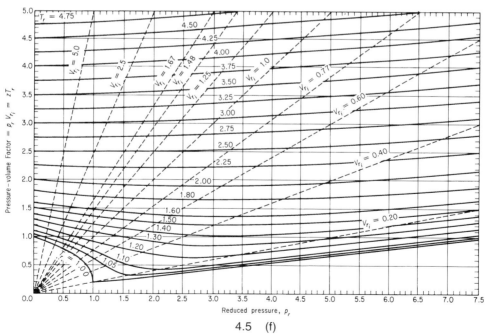

4.5 (f)

as well as pedagogical value because it enables engineering calculations to be made with considerable ease and also permits the development of thermodynamic functions for gases for which no experimental data are available. All you need to know to use these charts are the critical temperature and the critical pressure for a pure substance (or the pseudovalues for a mixture, as we shall see later). The value $z = 1$ represents ideality, and the value $z = 0.27$ is the compressibility factor at the critical point.

By adding another physical parameter of the gas besides T_r and p_r to help correlate data for z, you can gain some additional accuracy, but you would have to employ a *set* of tables or charts rather than a single table or chart it in your calculations. Because the increased precision that may accompany the use of a third parameter in calculating z is not necessary for our purposes, and because the presentation of z values so as to include the third parameter is considerably more cumbersome, we shall not show three-parameter charts or tables here.

EXAMPLE 4.9 Use of the Compressibility Factor

In spreading liquid ammonia fertilizer, the charges for the amount of NH_3 are based on the time involved plus the pounds of NH_3 injected into the soil. After the liquid has been spread, there is still some ammonia left in the source tank (volume = 120 ft^3), but in the form of a gas. Suppose that your weight tally, which is obtained by difference, shows a net weight of 125 lb of NH_3 left in the tank at 292 psig. Because the tank is sitting in the sun, the temperature in the tank is 125°F.

Your boss complains that his calculations show that the specific volume of the gas is 1.20 ft^3/lb, and hence that there are only 100 lb of NH_3 in the tank. Could he be correct? See Figure E4.9.

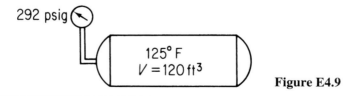

Figure E4.9

Solution

Basis: 1 lb of NH_3

Apparently, your boss used the ideal gas law in getting his figure of 1.20 ft^3/lb of NH_3 gas:

$$R = 10.73 \frac{(\text{psia})(\text{ft}^3)}{(\text{lb mol})(°R)}$$

$$p = 292 + 14.7 = 306.7 \text{ psia}$$

$$T = 125°F + 460 = 585°R$$

$$n = \frac{1 \text{ lb}}{17 \text{ lb/lb mol}}$$

$$\hat{V} = \frac{RT}{p} = \frac{\frac{1}{17}(10.73)(585)}{306.7} = 1.20 \text{ ft}^3/\text{lb}$$

However, he should have used the compressibility factor, because NH_3 does not behave as an ideal gas under the observed conditions of temperature and pressure. Let us again compute the mass of gas in the tank this time using

$$pV = znRT$$

We know all of the values of the variables in the equation except z. The additional information needed (taken from Appendix D) is

$$T_c = 405.5K \simeq 729.9R$$

$$p_c = 111.3 \text{ atm} \simeq 1636 \text{ psia}$$

Then, since z is a function of T_r and p_r,

$$T_r = \frac{T}{T_c} = \frac{585° \text{R}}{729.9° \text{R}} = 0.801$$

$$p_r = \frac{p}{p_c} = \frac{306.7 \text{ psia}}{1636 \text{ psia}} = 0.187$$

From the Nelson and Obert chart, Figure 4.5(c), you can read $z \simeq 0.855$. Now $\hat{V}$ can be calculated as

$$\hat{V} = \frac{1.20 \text{ ft}^3 \text{ ideal}}{\text{lb}} \left| \frac{0.855}{1} \right. = 1.03 \text{ ft}^3/\text{lb } NH_3$$

$$\frac{1 \text{ lb } NH_3}{1.03 \text{ ft}^3} \left| \frac{120 \text{ ft}^3}{} \right. = 117 \text{ lb } NH_3$$

Certainly 117 lb is a more realistic figure than 100 lb, and it is easily possible to be in error by 8 lb if the residual weight of NH_3 in the tank is determined by difference. As a matter of interest you might look up the specific volume of NH_3 at the conditions in the tank in a handbook. You would find that $\hat{V} = 0.973$ ft³/lb, and hence the compressibility factor calculation yielded a volume with an error of only about 4%.

EXAMPLE 4.10 Use of the Compressibility Factor

Liquid oxygen is used in the steel industry, in the chemical industry, in hospitals, as rocket fuel, and for wastewater treatment as well as many other applications. In a hospital a tank of 0.0284-m^3 volume is filled with 3.500 kg of liquid O_2 that vaporized at $-25°C$. Will the pressure in the tank exceed the safety limit of the tank (10^4 kPa)?

Solution

$$\text{Basis: 3.500 kg } O_2$$

We know from Appendix D that

$$T_c = 154.4 \text{ K}$$

$$p_c = 49.7 \text{ atm} \rightarrow 5,035 \text{ kPa}$$

However, this problem cannot be worked exactly the same way as the preceding problem because we do not know the pressure of the O_2 in the tank. Thus, we need to use the other parameter, V_{r_i}, that is available on the Nelson and Obert charts. We first calculate

$$\hat{V} \text{ (molal volume)} = \frac{0.0284 \text{ m}^3}{3.500 \text{ kg}} \left| \frac{32 \text{ kg}}{1 \text{ kg mol}} \right. = 0.260 \text{ m}^3/\text{kg mol}$$

Note that the *molal volume must* be used in calculating V_{r_i}. Since $\hat{V}_{c_i}$ is a volume per mole.

$$\hat{V}_{c_i} = \frac{RT_c}{p_c} = \frac{8.313 \text{ (m}^3\text{) (kPa)}}{\text{(kg mol)(K)}} \left| \frac{154.4 \text{ K}}{5,035 \text{ kPa}} \right. = 0.255 \frac{\text{m}^3}{\text{kg mol}}$$

Then

$$V_{r_i} = \frac{\hat{V}}{\hat{V}_{c_i}} = \frac{0.260}{0.255} = 1.02$$

Now we know two parameters, V_{r_i} and,

$$T_r = \frac{248 \text{ K}}{154.4 \text{ K}} = 1.61$$

From the Nelson and Obert chart [Figure 4.5(f)],

$$p_r = 1.43$$

Then

$$p = p_r p_c$$

$$= 1.43 \,(5,035) = 7200 \text{ kPa}$$

The pressure of 100 atm will not be exceeded. Even at room temperature the pressure will be less than 10^4 kPa.

LOOKING BACK

In this section we showed how the ideal gas equation can be generalized to apply to real gases by inserting the compressibility factor which is a function of the reduced properties of the gas.

Key Ideas

1. The critical state is the state in which gas and liquid properties merge to become the same.

2. The equation $pV = znRT$ can be used to predict real gas properties.

3. The compressibility factor z is a function of the reduced parameters T_r, p_r, and V_{r_i}. Given two of these four variables, the other two can be determined from the compressibility charts.

4. Use of z may not result in as high an accuracy in predicting gas properties as that obtained from equations of state described in the next section, but for instructional purposes, and for many engineering calculations, the technique is more convenient to use and usually has adequate accuracy.

Key Terms

Compressibility factor (p. 283)   Generalized compressibility charts (p. 283)
Corresponding states (p. 281)   Ideal critical volume (p. 284)
Critical state (p. 281)   Reduced parameters (p. 283)
Generalized equation of state (p. 283)   Supercritical fluid (p. 282)

Self-Assessment Test

1. What is the ideal critical volume? What is the advantage of using V_{c_i}?

2. In a proposed low pollution vehicle burning H_2—O_2, the gases are to be stored in tanks at 2000 psia. The vehicle has to operate from -40 to $130°F$.
   (a) Is the ideal gas law a sufficiently good approximation for the design of these tanks?
   (b) A practical operating range requires that 3 lb_m of hydrogen be stored. How large must the hydrogen tank be if the pressure is not to exceed 2000 psia?
   (c) The H_2/O_2 ratio is 2 on a molar basis. How large must the oxygen tank be?

3. A carbon dioxide fire extinguisher has a volume of 40 L and is to be charged to a pressure of 20 atm at a storage temperature of $20°C$. Determine the mass in kilograms of CO_2 at 1 atm.

4. Calculate the pressure of 4.00 g mol CO_2 contained in a 6.25×10^{-3} m^3 fire extinguisher at $25°C$.

Discussion Question

1. In the early 1970s it was believed that geopressured brines might contain appreciable potential for electric power production because the thermal and kinetic energy of the

waters represents approximately 50 percent of the total energy in the fluids. However, field tests have indicated that temperatures at 15,000 feet are usually in the range of 275 to 300°F, making electricity production marginal.

However, the brines are saturated with natural gas. Ten test wells have been tested in coastal Texas and Louisiana at an average cost of $2 million each. Test results have been mixed. Brine salinities have ranged from 13,000 to 191,000 ppm. Most tests indicate gas at or near solution gas levels. Temperatures have ranged from 237 to 307°F and formation pressures from 11,050 to 13,700 psia. Flow rates have ranged from 13,000 to 29,000 barrels per day of water. Because natural gas content is proportional to temperature and pressure but inversely proportional to salinity, the natural gas content in design wells ranges from 19 to 50 standard cubic feet per barrel. All of the wells have indicated some carbon dioxide (CO_2) in the gases produced. Content seems to be correlative with temperature, ranging from 4 to 9 percent of total gas content. The CO_2 has the potential to cause scaling problems, requiring use of inhibitors. Reservoir limits are very large. Transient pressure testing on two wells indicates no barriers to an outer limit of about 4 miles.

Is recovery of natural gas from such wells a viable source of energy? Prepare a brief report that includes economic, engineering, and environmental considerations for the proposed gas production.

4.2-2 Equations of State

Your objectives in studying this section are to be able to:

1. Cite two reasons for using equations of state to predict p-V-T properties of gases.
2. Solve an equation of state given the values of the coefficients in the equation and values of three of the four variables for either p, V, n, or T.
3. Convert the coefficients in an equation of state from one set of units to another.

LOOKING AHEAD

In this section we describe how empirical equations fit from experimental data can be used to solve for p, V, n, and T for single component real gases.

MAIN CONCEPTS

We now examine another way of predicting p, V, n, and T for real gases (either pure components or mixtures), namely by using **equations of state**. The simplest example of an equation of state is the ideal gas law itself. Equations of state are formulated by collecting experimental data and calculating the coefficients in a proposed equation by statistical fitting. Table 4.2 lists a few of the commonly used equations of state from among the hundreds that have been proposed that involve two or more coefficients. In the supplementary references at the end of this chapter, you will find numerous sources of information about equations of state

TABLE 4.2   Equations of State (for 1 Mole)*

Van der Waals:

$$\left(p + \frac{a}{\hat{V}^2}\right)(\hat{V} - b) = RT$$

$$a = \left(\frac{27}{64}\right)\frac{R^2 T_c^2}{p_c}$$

$$b = \left(\frac{1}{8}\right)\frac{RT_c}{p_c}$$

Peng-Robinson:

$$p = \frac{RT}{\hat{V} - b} - \frac{a\alpha}{\hat{V}(\hat{V} + b) + b(\hat{V} - b)}$$

$$a = 0.45724\left(\frac{R^2 T_c^2}{p_c}\right)$$

$$b = 0.07780\left(\frac{RT_c}{p_c}\right)$$

$$\alpha = [1 + \kappa(1 - T_r^{1/2})]^2$$

$$\kappa = 0.37464 + 1.54226\,\omega - 0.26992\,\omega^2$$

ω = acentric factor

Soave-Redlich-Kwong

$$p = \frac{RT}{\hat{V} - b} - \frac{a'\lambda}{\hat{V}(\hat{V} + b)}$$

$$a' = \frac{0.42748\,R^2 T_c^2}{p_c}$$

$$b = \frac{0.08664\,RT_c}{p_c}$$

$$\lambda = [1 + \kappa(1 - T1/2\,r]^2$$

$$\kappa = (0.480 + 1.574\omega - 0.176\omega^2$$

Redlich-Kwong:

$$p = \frac{RT}{(\hat{V} - b)} - \frac{a}{T^{1/2}\hat{V}(\hat{V} + b)}$$

$$a = 0.42748\frac{R^2 T_c^{2.5}}{p_c}$$

$$b = 0.08664\frac{RT_c}{p_c}$$

*T_c and p_c are explained in Sec. 4.1-1; $\hat{V}$ is the specific volume.

and values of their respective coefficients for various compounds (and sometimes for mixtures). Computer databases are also good sources of the same type of information in a convenient form.

In Table 4.2 two of the equations require the use of the Pitzer **acentric factor,** which is defined as being equal to $- \ln p_{rs} - 1$, where p_{rs} is the value of the reduced vapor pressure at $T_r = 0.70$. Table 4.3 lists a few values of ω for common gases.

Table 4.4 lists some of the values of the coefficients in the Van der Waals' and Redlich-Kwong equations.

An equation of state to be effective must represent experimental data for the p-V-T properties in the gas region with reasonable precision. Cubic equations of state such as Redlich-Kwong, Soave-Redlich-Kwong, and Peng-Robinson listed in Table 4.2 can have an accuracy of one to two percent over a large range of conditions for many compounds. Figure 4.6 compares Van der Waals' equation and the Redlich-Kwong equation with experimental data. Other classical equations of state are formulated as a power series (the **virial** form) with p being a function of $1/\hat{V}$ with 3 to 6 coefficients. Computer databases offer several choices for most compounds. Such relations may be (but may not be) more accurate than cubic equations of state. Equations of state in databases may have as many as 30 or 40 coefficients to achieve high accuracy (see for example the AIChE DIPPR reports). Keep in mind you must know the region of validity of any equation of state and not extrapolate outside that region, particularly not into the liquid re-

TABLE 4.3   Values of the Pitzer* Acentric Factor

Compound	Acentric Factor	Compound	Acentric Factor
Acetone	0.309	Hydrogen sulfide	0.100
Benzene	0.212	Methane	0.008
Ammonia	0.250	Methanol	0.559
Argon	0.000	n-butane	0.193
Carbon dioxide	0.225	n-pentane	0.251
Carbon monoxide	0.049	Nitric oxide	0.607
Chlorine	0.073	Nitrogen	0.040
Ethane	0.098	Oxygen	0.021
Ethanol	0.635	Propane	0.152
Ethylene	0.085	Propylene	0.148
Freon-12	0.176	Sulfur dioxide	0.251
Hydrogen	−0.220	Water vapor	0.344

*Pitzer, K. S., J. *Am. Chem. Soc.* 77 (1955): 3427.

TABLE 4.4   Constants for the Van der Waals and Redlich-Kwong Equations

	van der Waals		Redlich-Kwong	
	$a*$ $\left[atm\left(\dfrac{cm^3}{g\ mol} \right)^2 \right]$	$b^\dagger$ $\left(\dfrac{cm^3}{g\ mol} \right)$	$a^\ddagger$ $\left[(atm)(K^{1/2})\left(\dfrac{cm^3}{g\ mol} \right)^2 \right]$	$b^\dagger$ $\left(\dfrac{cm^3}{g\ mol} \right)$
Air	1.33×10^6	36.6	15.65×10^6	25.3
Ammonia	4.19×10^6	37.3	85.00×10^6	25.7
Carbon dioxide	3.60×10^6	42.8	63.81×10^6	29.7
Ethane	5.50×10^6	65.1	97.42×10^6	45.1
Ethylene	4.48×10^6	57.2	76.92×10^6	39.9
Hydrogen	0.246×10^6	26.6	1.439×10^6	18.5
Methane	2.25×10^6	42.8	31.59×10^6	29.6
Nitrogen	1.347×10^6	38.6	15.34×10^6	26.8
Oxygen	1.36×10^6	31.9	17.12×10^6	22.1
Propane	9.24×10^6	90.7	180.5×10^6	62.7
Water vapor	5.48×10^6	30.6	140.9×10^6	21.1

*To convert to psia (ft³/lb mol)², multiply table value by 3.776×10^{-3}.
†To convert to ft³/lb mol, multiply table value by 1.60×10^{-2}.
‡To convert to psia (°R)$^{1/2}$(ft³/lb mol)², multiply table value by 5.067×10^{-3}.

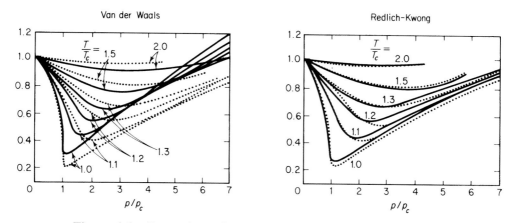

Figure 4.6 Comparison of experimental values (dots) with predicted values (solid lines) for two equations of state. T_c and p_c are explained in Sec. 4.2–1.

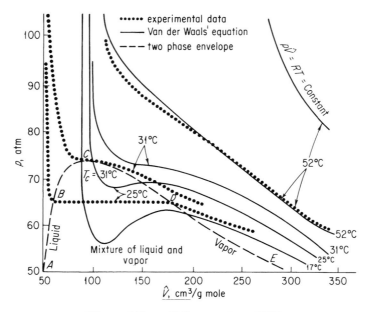

Figure 4.7   *p-V-T* properties of CO_2

gion! For example, examine Figure 4.7, which shows how well van der Waals'
equation predicts the properties of CO_2 in comparison with experimental data.
(Note how far the ideal gas law departs from the experimental data even at 52°C.)
The region under the dashed line is the two-phase region of a mixture of liquid
and vapor.

One feature of the cubic equations of state that merits comment is as follows. In solving for n or V, you must solve a cubic equation that might have more
than one real root as indicated in Figure 4.8. For example, van der Waals equation
can easily be solved explicitly for p as follows:

$$p = \frac{nRT}{V - nb} - \frac{n^2 a}{V^2}$$

However, if you want to solve for V (or n), you can see that the equation becomes
cubic in V (or n):

$$f(V) = V^3 - \left(nb + \frac{nRT}{p}\right)V^2 + \frac{n^2 a}{p}V - \frac{n^3 ab}{p} = 0$$

and can have multiple roots (see Figure 4.8). We want a *positive real root*. In the
disk in the back of the book you will find computer programs that can solve non-

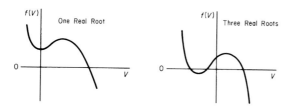

Figure 4.8 Graphs showing the existence of one or three real roots of a cubic equation.

linear equations and can be used to solve for V if you have a *reasonable initial guess* for V, say from the ideal gas law. Refer to Appendix L for details of the methods of solving nonlinear equations. An alternative technique for finding the root of a function of a single variable is to plot the nonlinear function by substituting p and T into the equation of state, and finding at what value of V the value of the left-hand side crosses the zero axis as V changes.

EXAMPLE 4.11 Application of Van der Waals' Equation

A cylinder 0.150 m³ in volume containing 22.7 kg of propane C_3H_8 stands in the hot sun. A pressure gauge shows that the pressure is 4790 kPa gauge. What is the temperature of the propane in the cylinder? Use van der Waals equation.

Solution

Basis: 22.7 kg of propane

The van der Waals constants obtained from any suitable handbook or Table 4.4 are

$$a = 9.24 \times 10^6 \text{ atm} \left(\frac{\text{cm}^3}{\text{g mol}} \right)^2$$

$$b = 90.7 \frac{\text{cm}^3}{\text{g mol}}$$

$$p + \left(\frac{n^2 a}{V^2} \right) (V - nb) = nRT$$

All the additional information you need is as follows:

$$p = \frac{(4790 + 101)\text{kPa}}{} \left| \frac{1 \text{ atm}}{101.3 \text{ kPa}} \right. = 48.3 \text{ atm abs}$$

$$R \text{ in the proper units is} = \frac{82.06(\text{cm}^3)(\text{atm})}{(\text{g mol})(\text{K})}$$

$$n = \frac{22.7}{44 \text{ kg/kg mol}} = 0.516 \text{ kg mol propane}$$

$$\left[48.3 + \frac{(0.516 \times 10^3)^2 (9.24 \times 10^6)}{(0.150 \times 10^6)^2} \right] [0.150 \times 10^6$$

$$- (0.516 \times 10^3)(90.7)] = (0.516 \times 10^3)(82.06)(T_K)$$

$$T = 384 \text{ K}$$

EXAMPLE 4.12 Solution of Van der Waals' Equation for V

Given the values in a vessel of

$p = 679.7 \text{ psia}$

$a = 3.49 \times 10^4 \text{ psia} \left(\frac{\text{ft}^3}{\text{lb mol}} \right)^2$

$n = 1.136 \text{ lb mol}$

$R = 10.73 \dfrac{(\text{psia})(\text{ft}^3)}{(\text{lb mol})(°R)}$

$T = 683°R$

$b = 1.45 \dfrac{\text{ft}^3}{\text{lb mol}}$

solve for the volume of the vessel.

Solution

Write van der Waals' equation as a cubic equation in one unknown variable, V.

$$f(V) = V^3 - \frac{pnb + nRT}{p} V^2 + \frac{n^2 a}{p} V - \frac{n^3 ab}{p} = 0 \qquad \text{(a)}$$

Let us apply Newton's method (refer to Appendix L) to obtain the desired root:

$$V_{k+1} = V_k - \frac{f(V_k)}{f'(V_k)} \qquad \text{(b)}$$

where $f'(V_k)$ is the derivative of $f(V)$ with respect to V evaluated at V_k:

$$f'(V) = 3V^2 - \frac{2(pnb + nRT)}{p} V + \frac{n^2 a}{p} \qquad \text{(c)}$$

You can obtain a reasonably close approximation to V (or n) in many cases from the ideal gas law, useful at least for the first trial in which $k = 0$ in Eq. (b).

$$V_0 = \frac{nRT}{p} = \frac{1.136 \text{ lb mol} \mid 10.73 \text{ (psia)(ft}^3) \mid 683°R}{(\text{lb mol})(°R) \mid 679.7 \text{ psia}}$$

= 12.26 ft^3 at 679.7 psia and 683°R

The second and subsequent estimates of V will be calculated using Eq. (b):

$$V_1 = V_0 - \frac{f(V_0)}{f'(V_0)}$$

$$f(V_0) = (12.26)^3 - \frac{(679.7)(1.137)(1.45) + (1.137)(10.73)(683)}{679.7}(12.26)^2$$

$$+ \frac{(1.137)^2(3.49 \times 10^4)}{679.7}(12.26) - \frac{(1.137)^3(3.49 \times 10^4)(1.45)}{679.7} = 738.3$$

$$f'(V_0) = 3(12.26)^2 - \frac{2[(679.7)(1.137)(1.45) + (1.137)(10.73)(6831)]}{679.7}(12.26)$$

$$+ \frac{(1.137)^2(3.49 \times 10^4)}{679.7} = 216.7$$

$$V_1 = 12.26 - \frac{738.3}{216.7} = 8.85$$

On the next iteration

$$V_2 = V_1 - \frac{f(V_1)}{f'(V_1)}$$

and so on until the change in V_k from one iteration to the next is sufficiently small. The computer program on the disk in back of the book will execute this technique for you. The final solution is 5.0 ft^3 at 679.7 psia and 683°R.

ADDITIONAL DETAILS

The **group contribution method** has been successful in estimating p-V-T properties of pure components (as well as other thermodynamic properties). As indicated by the name, the idea is that compounds can be constituted from combinations of functional groups, the contribution of each group to a property can be tabulated, and the group contributions can be correlated and/or summed to give the desired property of the compound. The assumption is that a group such as $-CH_3$, or $-OH$, behaves identically no matter what the molecule may be in which it appears. This assumption is not quite true, so that any group contribution method yields approximate properties. Probably the most widely used group contribution method is UNIFAC,[2] which forms a part of many computer databases.

[2] A. Fredenslund, J. Gmehling, and P. Rasmussen. *Vapor-Liquid Equilibria Using UNIFAC.* Amsterdam: Elsevier, 1977; D. Tiegs, J. Gmehling, P. Rasmussen, and A. Fredenslund. *Ind. Eng. Chem. Res., 26* (1987):159.

LOOKING BACK

We reviewed how gas properties can be represented by equations of state. In spite of their complications, equations of state are important for several reasons. They permit a concise summary of a large mass of experimental data and also permit accurate interpolation between experimental data points. They provide a continuous function to facilitate calculation of physical properties involving differentiation and integration. Finally, they provide a point of departure for the treatment of thermodynamic properties of mixtures.

Key Ideas

1. Real gas relations should be used for gases in general unless you know that that ideal gas equation applies.
2. Numerous equations exist that relate p, V, T, and n.
3. The accuracy of a particular equation depends on its functional formulation, the gas itself, and the region of application.

Key Terms

Acentric factor (p. 294)   Redlich-Kwong (p. 293)
Cubic equation of state (p. 297)   Soave-Redlich-Kwong (p. 293)
Equation of state (p. 292)   Van der Waals (p. 293)
Group contribution method (p. 299) Virial form (p. 294)
Peng-Robinson (p. 293)

Self-Assessment Test

1. Equations of state for gases are often used to predict p-V-T properties of a gas. Cite two reasons why.
2. What are the units of a and b in the SI system for the Redlich-Kwong equation?
3. You measure that 0.00220 lb mol of a certain gas occupies a volume of 0.95 ft^3 at 1 atm and 32°F. If the equation of state for this gas is $pV = nRT(1 + bp)$, where b is a constant, find the volume at 2 atm and 71°F.
4. Calculate the temperature of 2 g mol of a gas using van der Waals' equation with $a = 1.35 \times 10^{-6} \text{m}^6(\text{atm})(\text{g mol}^{-2})$, $b = 0.0322 \times 10^{-3}\text{m}^3)(\text{g mol}^{-1})$ if the pressure is 100 kPa and the volume is 0.0515 m^3.
5. Calculate the pressure of 10 kg mol of ethane in a 4.86 m^3 vessel at 300 K using two equations of state: (a) ideal gas and (b) Soave-Redlich-Kwong. Compare with your answer the observed value of 34.0 atm.
6. The van der Waals constants for a gas are $a = 2.31 \times 10^6$ (atm)(cm^3/g mol)^2 and $b = 44.9$ cm^3/g mol. Find the volume per kilogram mole if the gas is at 90 atm and 373 K.

Thought Problems

1. Municipal sludge is being converted to sterile ash and readily biodegradable liquid effluent in a mile-deep well at Longmont, Colorado. The nation's first deep well wet-air oxidation sludge destruction process reduces chemical oxygen demand (COD) by up to 68% and destroys all living organisms, according to Howard C. Delaney, superintendent of Longmont's wastewater treatment plant. The VerTech Treatment System is suspended in the conventionally drilled, concrete-encased well. Tubes of various diameters are concentrically fitted within the well to create annular spaces for the two-phase flow. Oxidation takes place at the bottom of the well. Why is it advantageous to oxidize the sludge at the bottom of a well rather than in a pond at ground level?

2. From the *Oil and Gas Journal*, p. 55, December 2, 1985.

    > Gas trapped in a subsea blowout preventer (BOP) can be a serious problem in a deepwater drilling. Amoco used nitrogen to confirm a safe method to remove gas trapped in a subsea BOP. The gas can be recovered up the choke line by displacing the kill mud with water and then allowing the gas to expand against the lower hydrostatic gradient of the water. The test was conducted in the Gulf of Mexico from a semisubmersible working in 1,015 ft. of water.

    What is the difficulty with a gas bubble trapped in the BOP?

Discussion Questions

1. What are some of the tests that you might apply to see how an equation of state fits *p-V-T* data?

2. What factors in a real gas cause the gas to behave in a nonideal manner? How are these factors taken into account in the equations of state listed in Table 4.2?

3. A separations column (a depropanizer) was designed to operate at 1900 kPa on a corrosive stream that had a high fouling factor. The column was protected by a relief valve set at 2100 kPa. The pressure in the column was controlled by the heat transferred to the bottom of the column (via the reboiler). A single pressure transmitter signaled from the same tap both the state of the high pressure alarm and the control for the heat to the reboiler. See Figure DQ4.3. What do you think might go wrong with this arrangement?

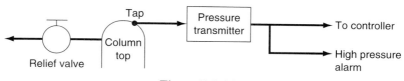

Figure DQ4.3

4.2-3 Gaseous Mixtures

> ### Your objectives in studying this section are to be able to:
>
> 1. Use Kay's method of pseudocritical values to calculate the pseudo-reduced values and predict p, V, T, and n via the compressibility factor.
> 2. Suggest a method of using equations of state to predict gas properties.

LOOKING AHEAD

In this section we describe ways to estimate p-V-T values for gas mixtures.

MAIN CONCEPTS

So far we have discussed only how to predict p-V-T properties of single component real gases. In this section the question is: How can we predict p-V-T properties with reasonable accuracy for real gaseous mixtures? We treated ideal gas mixtures in Sec. 4.1-2, and saw, as we shall see later (in Chapter 5 for the thermodynamic properties), that for ideal mixtures the properties of the individual components can be weighed and added together to give the desired property of the mixture. But this technique does not prove to be as satisfactory for real gases. The most desirable technique would be to develop methods of calculating p-V-T properties for mixtures based solely on the properties of the pure components. Possible ways of doing this for real gases are discussed below.

If you plan to use the generalized equation of state, $pV = znRT$, should you calculate z for each component, and then compute an average z by some type of weighting such as by mole fractions in the mixture? This is one possibility. It turns out that a way to get more accurate answers is to use Kay's method.

In Kay's method, **pseudocritical** values for mixtures of gases are calculated on the assumption that each component in the mixture contributes to the pseudo-critical value in the same proportion as the number of moles of that component. Thus, the pseudocritical values are computed as follows:

$$p_c' = p_{c_A} y_A + p_{c_B} y_B + \cdots \tag{4.8a}$$

$$T_c' = T_{c_A} y_A + T_{c_B} y_B + \cdots \tag{4.8b}$$

where y_i is the mole fraction, p'_c is the pseudocritical pressure and T'_c is the pseudocritical temperature. (It has also been found convenient in some problems to calculate similarly a weighted pseudo-ideal-critical volume V''_{c_i}.) You can see that these are linearly weighted mole average pseudocritical properties. Figure 4.9 compares the true critical points of a gaseous mixture of CO_2 and SO_2 with the pseudocritical points. The respective pseudoreduced variables are

$$p'_r = \frac{p}{p'_c}$$

$$T'_r = \frac{T}{T'_c}$$

Kay's method is known as a two-parameter rule since only p_c and T_c for each component are involved in the calculation of z. If a third parameter such as z_c, the Pitzer acentric factor, or $\hat{V}_{c_i}$ is included in the determination of the compressibility factor, then we would have a three-parameter rule. Other pseudocritical methods better provide accuracy in predicting p-V-T properties than Kay's method, but Kay's method can suffice for engineering work and it is easy to use.

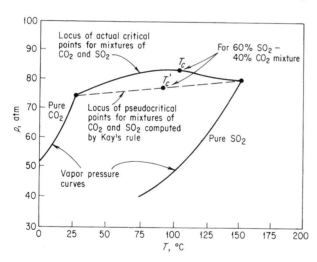

Figure 4.9 Critical and pseudo-critical points for a mixture of CO_2 and SO_2

EXAMPLE 4.13 Calculation of p-V-T Properties for Real Gas Mixture

A gaseous mixture has the following composition (in mole percent):

Methane, CH_4	20
Ethylene, C_2H_4	30
Nitrogen, N_2	50

at 90 atm pressure and 100°C. Compare the volume per mole as computed by the methods of:

(a) the perfect gas law

(b) the pseudoreduced technique (Kay's method)

Solution

Basis: 1 g mol of gas mixture

Additional data needed are:

Component	T_c (K)	p_c (atm)
CH_4	191	45.8
C_2H_4	283	50.5
N_2	126	33.5

$$R = 82.06 \frac{(cm^3)(atm)}{(g\ mol)(K)}$$

(a) Perfect gas law:

$$\hat{V} = \frac{RT}{p} = \frac{1(82.06)(373)}{90} = 340\ cm^3 \text{ at 90 atm and 373 K}$$

(b) According to Kay's method, we first calculate the pseudocritical values for the mixture

$$p'_c = p_{c_A} y_A + p_{c_B} y_B + p_{c_C} y_C = (45.8)(0.2) + (50.5)(0.3) + (33.5)(0.5)$$

$$= 41.2\ atm$$

$$T'_c = T_{c_A} y_A + T_{c_B} y_B + T_{c_C} y_C = (191)(0.2) + (283)(0.3) + (126)(0.5)$$

$$= 186\ K$$

Then we calculate the pseudo-reduced values for the mixture

$$p'_r = \frac{p}{p'_c} = \frac{90}{41.2} = 2.18, \qquad T'_r = \frac{T}{T'_c} = \frac{373}{186} = 2.01$$

With the aid of these two parameters we can find from Figure 4.5d that $z = 0.965$. Thus

$$\hat{V} = \frac{zRT}{p} = \frac{0.965(1)(82.06)(373)}{90} = 328\ cm^3 \text{ at 90 atm and 373 K}$$

In instances in which the temperature or pressure of a gas mixture is un-known, to avoid a trial-and-error solution using the generalized compressibility charts, you can compute the pseudocritical ideal volume and a **pseudoreduced ideal reduced volume** V_{r_i} thus

$$\hat{V}'_{c_i} = \frac{RT'_c}{p'_c} \quad \text{and} \quad \hat{V}'_{r_i} = \frac{\hat{V}}{\hat{V}_{c_i}}$$

$\hat{V}'_{r_i}$ can be used in lieu of p'_r or T'_r in the compressibility charts.

If you plan to use a specific equation of state such as one of those listed in Table 4.2, you have numerous choices, no one of which will consistently give the best results. You could calculate average values for the coefficients in the equation, you could calculate average compressibility factors by using pure component coefficients in the equation for each gas, and so on. Consult the references at the end of this chapter for further information.

LOOKING BACK

In this section we described how to apply Kay's method (the pseudocritical method) to predict p, V, n, and T in gas mixtures, and mentioned how equations of state might be used.

Key Concepts

1. Kay's pseudocritical method will provide reasonable estimates of the variables in $pV = znRT$ based on pure component properties, although other methods of greater accuracy (and of greater complexity) are available in the literature. All these methods begin to break down near the true critical point of the mixture and for highly polar compounds.
2. Equations of state can also be used to predict p, V, n, and T if suitable weighting of the properties of pure components is used.

Key Terms

Pseudocritical (p. 302)   Pseudoreduced ideal reduced volume (p. 305)

Self-Assessment Test

1. One pound mole of a mixture containing 0.400 lb mol of N_2 and 0.600 lb mol C_2H_4 at 50°C occupies a volume of 1.44 ft³. What is the pressure in the vessel? Compute your answer by Kay's method.

Discussion Question

1. Would it make more sense from a theoretical viewpoint to use a mean compressibility

$$z_m = z_A y_A + z_B y_B + \ldots$$

in applying the compressibility charts via $pV = znRT$, or to use the z for each component, calculate the property desired, and average the properties by the mole fraction y_i?

4.3 VAPOR PRESSURE AND LIQUIDS

> ### Your objectives in studying this section are to be able to:
>
> 1. Define and explain a number of important terms listed as "Key Terms" at the end of this section.
> 2. Calculate the vapor pressure of a substance from an equation that relates the vapor pressure to the temperature, such as the Antoine equation, given values for the coefficients in the equation, and look up the vapor pressure in reference books.
> 3. Calculate the temperature of a substance from a vapor-pressure equation given the values for the coefficients in the equation, and the vapor pressure.
> 4. Estimate the vapor pressure of a compound via a Cox chart.

LOOKING AHEAD

In this section we review the concept of vapor and liquid, and in particular discuss the prediction of vapor pressure as a function of temperature. In addition, a number of new terms are introduced with which you should become familiar.

MAIN CONCEPTS

The terms *vapor* and *gas* are used very loosely. A gas that exists below its critical temperature is usually called a vapor because it can condense. If you continually compress a pure gas at constant temperature, provided that the temperature is below the critical temperature, some pressure is eventually reached at which the gas starts to condense into a liquid (or a solid). Further compression does not increase the pressure but merely increases the fraction of gas that condenses. A reversal of the procedure just described will cause the liquid to be transformed into the gaseous state again (i.e., vaporize). From now on, the word **vapor** will be reserved to describe a gas below its critical point in a process in which the phase change is of primary interest, while the word **gas** or **noncondensable gas** will be used to describe a gas above the critical point or a gas in a process in which it cannot condense.

If the vapor and liquid of a pure component are in **equilibrium**, then the **equilibrium pressure is called the vapor pressure**. At a given temperature there is only one pressure at which the liquid and vapor phases of a pure substance may exist in equilibrium. Either phase alone may exist, of course, over a wide range of conditions.

The concept of vapor pressure can be grasped most easily by inspecting Figure 4.10, which in three dimensions shows the $p, \hat{V}, T$ surface for water. Vapor pressure is represented by the two-dimensional projection, a curve, of the three-dimensional surface into the p - T plane. The view is directly across the liquid-vapor region (a slight expansion of the projection in the lower left has been introduced for clarity). Figure 4.11 is an additional expansion of a segment of the region in Figure 4.10. For each temperature you can read the corresponding pressure at which water vapor and water liquid exist in equilibrium. You have encountered this condition of

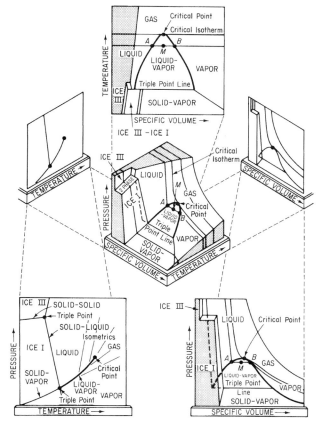

Figure 4.10 p-$\hat{V}$-T surface and projections for H_2O.

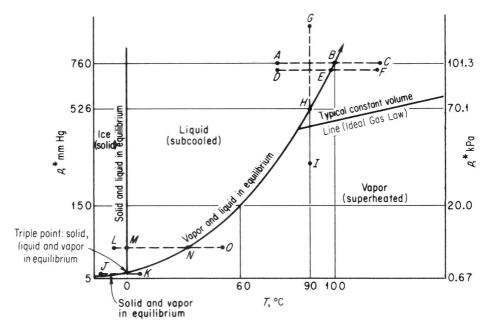

Figure 4.11   Vapor pressure curve for water. The vapor pressure curve terminates at the critical point as shown in Figure 4.10.

equilibrium many times—for example, in boiling. Any substance has an infinite number of boiling points, but by custom we say the **"normal" boiling point** is the temperature at which boiling takes place under a pressure of l atm (101.3 kPa, 760 mm Hg). Unless another pressure is specified, 1 atm is assumed and the term **boiling point** is taken to mean the "normal boiling point." A piston exerting a force of 101.3 kPa could just as well take the place of the atmosphere, as shown in Figure 4.12. For example, you know that at 100°C water will boil (**vaporize**) and the pressure will be 101.3 kPa or 1 atm (point *B*s) Suppose that you heat water starting at 77°C in a container as in Figure 4.12—what happens? We assume that the water vapor is at all times in equilibrium with the liquid water. This is a constant-pressure process. As the temperature rises and the confining pressure stays constant, nothing particularly noticeable occurs until 100°C is reached, at which time the water begins to boil, that is, **evaporate**. If the water evaporates, the vapor further pushes back the piston and the water will completely change from liquid into vapor. If you heated the water at constant pressure after it had all evaporated at point *B*, you could apply the gas laws in the temperature region *B-C* (and at higher temperatures). A reversal of this process from the temperature *C* would cause the vapor to **condense** at *B* to form a liquid. The temperature at point *B* is called the **dew point**.

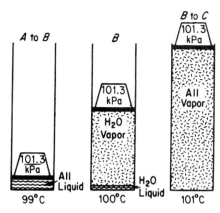

Figure 4.12 Transformation of liquid water into vapor at constant pressure. The 101.3 kPa exerted by the piston includes the force of the atmosphere above the piston.

Suppose that you went to the top of Pikes Peak and repeated the experiment in the open air—what would happen then? Everything would be the same (points *D-E-F*) with the exception of the temperature at which the water would begin to boil, or condense. Since the pressure of the atmosphere at the top of Pikes Peak is lower than 101.3 kPa, the water would start to displace the air, or boil, at a lower temperature. You can see that (1) at any given temperature water exerts its vapor pressure (at equilibrium); (2) as the temperature goes up, the vapor pressure goes up; and (3) it makes no difference whether water vaporizes into air, into a cylinder closed by a piston, or into an evacuated cylinder—at any temperature it still exerts the same vapor pressure as long as the water is in equilibrium with its vapor.

A pure compound can change from a liquid to a vapor, or the reverse, via a constant temperature process as well as a constant pressure process. A process of **vaporization** or **condensation at constant temperature** is illustrated by the lines *G-H-I* or *I-H-G*, respectively, in Figure 4.11. Water would vaporize or condense at constant temperature as the pressure reached point *H* on the vapor-pressure curve (also look at Figure 4.13).

Figure 4.11 also shows the *p-T* conditions at which ice (in its common form) and water vapor are in equilibrium. When the solid passes directly into the vapor phase without first melting to become a liquid (line *J-K* as opposed to line *L-M-N-O*) it is said to **sublime**. Iodine crystals do this at room temperature; water sublimes only below 0°C, as when the frost disappears in the winter when the thermometer reads −6°C.

A term commonly applied to the vapor-liquid portion of the vapor-pressure curve is the word **saturated, meaning the same thing as vapor and liquid in equilibrium** with each other. If a gas is just ready to start to condense its first drop of liquid, the gas is called a **saturated gas**; if a liquid is just about to vapor-

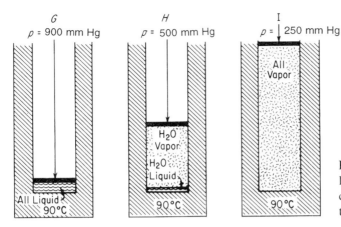

Figure 4.13 Transformation of liquid water into water vapor at a constant temperature by changing the imposed pressure.

ize, it is called a **saturated liquid.** These two conditions are also known as the **dew point** and **bubble point**, respectively.

The region to the right of the vapor-pressure curve in Figure 4.11 is called the **superheated** region and the one to the left of the vapor-pressure curve is called the **subcooled** region. The temperatures in the superheated region, if measured as the difference (*O-N*) between the actual temperature of the superheated vapor and the saturation temperature for the same pressure, are called degrees of superheat. For example, steam at 500°F and 100 psia (the saturation temperature for 100 psia is 327.8°F) has $(500 - 327.8) = 172.2°F$ of superheat. Another new term you will find used frequently is the word **quality.** A *wet* vapor consists of saturated vapor and saturated liquid in equilibrium. The mass fraction of vapor is known as the quality.

EXAMPLE 4.14 Vapor-Liquid Properties of Water

For each of the conditions of temperature and pressure listed below for water, state whether the water is a solid phase, liquid phase (superheated), or is a saturated mixture, and if the latter, calculate the quality. Use the steam tables (inside the back cover) to assist in the calculations.

State	p (kPa)	T (K)	$\hat{V}$ (m³/kg)
1	2000	475	—
2	1000	500	0.2206
3	101.3	200	—
4	245.6	400	0.505

Solution

State 1: liquid   State 3: solid
State 2: vapor   State 4: satuated vapor and liquid

You can calculate the properties of a mixture of vapor and liquid in equilibrium (for a single component) from the individual properties of the saturated vapor and saturated liquid. At 400 K and 245.6 kPa with the specific volume of a wet steam mixture being 0.505 m³/kg, what is the quality of the steam? From the steam tables the specific volumes of the saturated liquid and vapor are

$$\hat{V}_1 = 0.001067 \text{ m}^3/\text{kg} \qquad \hat{V}_g = 0.7308 \text{ m}^3/\text{kg}$$

Basis: 1 kg of wet steam mixture

Let x = mass fraction vapor.

$$\frac{0.001067 \text{ m}^3 \, | \, (1-x) \text{ kg liquid}}{1 \text{ kg liquid}} + \frac{0.7308 \text{ m}^3 \, | \, x \text{ kg vapor}}{1 \text{ kg vapor}} = 0.505 \text{ m}^3$$

$$0.001067 - 0.001067x + 0.7308x = 0.505$$

$$x = 0.69$$

Other properties of wet mixtures can be treated in the same manner.

4.3-1 Change of Vapor Pressure with Temperature

You can see from Figure 4.11 that the function of p^* versus T is not a linear function (except as an approximation over a very small temperature range). Many functional forms have been proposed to predict p^* from T (see the references at the end of the chapter), and the reverse, some with numerous coefficients. We will use the Antoine equation in this book—it has sufficient accuracy for our needs:

$$\ln (p^*) = A - \frac{B}{C + T} \tag{4.9}$$

where A, B, C = constants for each substance
$\qquad T$ = temperature, K

Refer to Appendix G for the values of A, B, and C for various compounds. Yaws[3] prepared a set of values for A, B, and C for approximatelly 700 compounds that can be found on the disk in the back of this book.

You can estimate the values of A, B, and C in Eq. (4.9) from experimental

---

[3] Yaws, C. L. and H. C. Yang. "To Estimate Vapor Pressure Easily." *Hydrocarbon Processing*. (October, 1989): 65.

data by using a statistical program such as the one found on the disk in the back of this book. With just three experimental values you have to solve three nonlinear equations by another code, also found on the disk in the back of the book.

EXAMPLE 4.15 Vaporization of Metals for Thin Film Deposition

Three methods of providing vaporized metals for thin film deposition are evaporation from a boat or a filament, and transfer via an electronic beam. Figure E4.15 illustrates evaporation from a boat placed in a vacuum chamber.

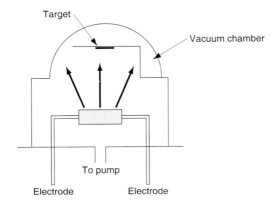

Target

Vacuum chamber

To pump

Electrode   Electrode   **Figure E4.15**

The boat made of tungsten has a negligible vapor pressure at 972°C, the operating temperature for the vaporization of aluminum (which melts at 660°C and fills the boat). The approximate rate of evaporation is given in g/(m^2)(s) by

$$m = 0.437 \frac{p^*(MW)^{1/2}}{T^{1/2}}$$

where p^* is the pressure in kPa and T is the temperature in K.
What is the vaporization rate for Al at 972°C in g/(cm^2)(s)?

Solution

We have to calculate p^* for Al at 972°C. The Antoine equation is suitable if data are known for the vapor pressure of Al. Considerable variation exists in the data for Al at high temperatures, but we will use $A = 8.779$, $B = 1.615 \times 10^4$, and $C = 0$ with p^* in mm Hg and T in K.

$$\ln p^*_{972°C} = 8.799 - \frac{1.615 \times 10^4}{972 + 273} = 0.0154 \text{ mmHg (0.00201 kPa)}$$

$$m = 0.437 \frac{(0.00201)(26.98)^{1/2}}{(972 + 273)^{1/2}} = 1.3 \times 10^{-4} \text{ g/(cm}^2)(s)$$

Another way to relate vapor pressure to temperature is by a graphical technique. The curvature illustrated in Figure 4.11 for p^* versus T can be straightened out by a special plot known as a **Cox chart,**[4] which is prepared as follows (refer to Figure 4.14).

1. Mark on the horizontal scale values of log p^* so as to cover the desired range of p^*.
2. Next draw a straight line on the plot at a suitable angle, say 45°, that covers the range of p^*.
3. To calibrate the vertical axis in common integers such as 25, 50, 100, 200 degrees, and so on, you use a **reference substance**, namely water. For the first integer, say 100°F, you look up the vapor pressure of water in the steam tables, or calculate it from the Antoine equation, to get 0.9487 psia. Locate this value on the horizontal axis, and proceed vertically until you hit the straight line. Then proceed horizontally until you hit the vertical axis. Mark the scale there as 100°F.

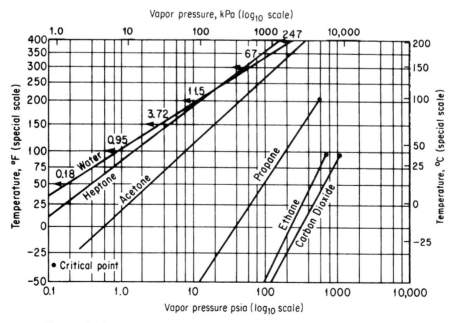

Figure 4.14   Cox chart. The vapor pressure of compounds other than water can be observed to fall on straight lines.

[4]E. R. Cox. *Ind. Eng. Chem., 15*, (1923): 15

4. Pick the next temperature, say 200°F.

5. Continue as in 3 and 4 until the vertical scale is established over the desired range for the temperature.

What proves useful about the Cox chart is that the vapor pressures of other substances plotted on this specially prepared set of coordinates will yield straight lines over extensive temperature ranges, and thus facilitate the extrapolation and interpolation of vapor-pressure data. It has been found that lines so constructed for closely related compounds, such as hydrocarbons, all meet at a common point. Since straight lines can be obtained in a Cox chart only *two sets* of vapor-pressure data are needed to provide complete information about the vapor pressure of a substance over a considerable temperature range.

EXAMPLE 4.16 Extrapolation of Vapor-Pressure Data

The control of existing solvents is described in the *Federal Register, 36*, (158), August 14, 1971, under Title 42, Chapter 4, Appendix 4.0, Control of Organic Compound Emissions. Chlorinated solvents and many other solvents used in industrial finishing and processing, dry-cleaning plants, metal degreasing, printing operations, and so forth, can be recycled and reused by the introduction of carbon adsorption equipment. To predict the size of the adsorber, you first need to know the vapor pressure of the compound being adsorbed at the process conditions. The vapor pressure of chlorobenzene is 400 mm Hg at 110°C and 5 atm at 205°C. Estimate the vapor pressure at 245°C and at the critical point (359°C).

Solution

The vapor pressures will be estimated by use of a Cox chart. The temperature scale (vertical) and vapor pressure scale (horizontal) are constructed as described above in connection with Figure 4.14. Vapor pressures of water from 3.72 to 3094 psia corresponding to 150F to 700F are on the horizontal logarithmic scale and the respective temperature marked on the vertical scale as shown in Figure E4.16.

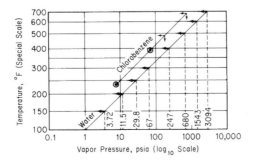

Figure E4.16

Next we convert the two vapor pressures of chlorobenzene into psia.

$$\frac{400 \text{ mm Hg}}{} \left| \frac{14.7 \text{ psia}}{760 \text{ mm Hg}} \right. = 7.74 \text{ psia} \qquad 110°C = 230°F$$

$$\frac{5 \text{ atm Hg}}{} \left| \frac{14.7 \text{ psia}}{1 \text{ atm Hg}} \right. = 73.5 \text{ psia} \qquad 205°C = 401°F$$

and plot these two points on the graph paper. Examine the encircled dots. Next we draw a straight line between the encircled points and extrapolate to 471°F (245°C) and 678°F (359°C). At these two temperatures, we can read off the estimated vapor pressures.

	471°F (245°C)	678°F (359°C)
Estimated:	150 psia	700 psia
Experimental:	147 psia	666 psia

Experimental values are given for comparison.

4.3-2 Liquid Properties

Considerable experimental data are available for liquid densities of pure compounds as a function of temperature and pressure. Refer to the sources listed in the supplementary reference at the end of this chapter for data and correlation formulas. Wooley[5] gives formulas (and a computer program) to estimate the liquid density given T_c, p, and ω, and/or any liquid density reference point. Often making density a linear function of temperature that is independent of pressure provides sufficiently accurate predictions

$$\rho = \rho_o - \beta (T - T_o) \tag{4.10}$$

where ρ_o and β are constants, and ρ is the density of the liquid. Most liquids can be considered to be incompressible, that is, their density is independent of pressure.

As to liquid mixtures, it is even more difficult to predict the p-V-T properties of liquid mixtures than of real gas mixtures. Probably more experimental data (especially at low temperatures) are available than for gases, but less is known about the estimation of the p-V-T properties of liquid mixtures. For compounds with like molecular structures, such as hydrocarbons of similar molecular weight, called **ideal liquids**, the density of a liquid mixture can be approximated by assuming that the weighted specific volumes are additive:

[5]R. J. Wooley. *Chem. Eng.* (March 31, 1986): 109.

$$\hat{V}_{average} = x_1 \hat{V}_1 + x_2 \hat{V}_2 + \cdots x_n \hat{V}_n = \sum_{i=1}^{n} x_i \hat{V}_i \qquad (4.11)$$

where n is the number of components in the mixture. For **nonideal liquids** or solids dissolved in liquids, experimental data or estimation techniques described in many of the references listed at the end of this chapter must be employed, as the specific volumes are not additive.

ADDITIONAL DETAILS

To meet emission standards, refiners are formulating gasoline and diesel fuel in new ways. The rules on emissions are related to the vapor pressure of a fuel, but the vapor pressure is specified in terms of the **Reid Vapor Pressure (RVP),** which is determined at 100°F in a bomb that permits partial vaporization. For a pure component the RVP is the true vapor pressure, but for a mixture (as are most fuels) the RVP is lower than the true vapor pressure of the mixture (by roughly 10% for gasoline). Refer to Vazquez-Esparragoza et al.[6] for specific details about estimating the RVP.

LOOKING BACK

In this section we explained (1) the meaning of a number of new terms, (2) what vapor pressure is, and (3) how to predict vapor pressure as a function of temperature.

Key Ideas

1. If the vapor and liquid of a pure component are in equilibrium at a particular temperature, the pressure established is known as the vapor pressure and is unique.
2. Vapor condenses into a liquid, or liquid vaporizes into a vapor as the pressure or temperature is changed when the vapor and liquid exist at equilibrium.
3. The vapor pressure of a liquid can be predicted as a function of temperature via the Antoine equation.

Key Terms

Antoine equation (p. 311)
Bubble point (p. 310)
Condense (p. 308)
Cox chart (p. 313)

Degrees of superheat (p. 310)
Dew point (p. 308)
Equilibrium (p. 307)
Gas (p. 306)

[6]J.J. Vazquez-Esparragoza, G. A. Iglesias-Silva, M. W. Hlavinka, and J. Bulin, "How to Estimate RVP of Blends." *Hydrocarbon Processing,* (August, 1992): 135.

Ideal liquids (p. 315)
Noncondensable gas (p. 306)
Nonideal liquids (p. 315)
Normal boiling point (p. 308)
Quality (p. 310)
Reference substance (p. 313)
Reid vapor pressure (p. 316)
Saturated (p. 309)

Subcooled (p. 310)
Sublime (p. 309)
Superheated (p. 310)
Triple point (p. 307)
Vapor (p. 306)
Vapor pressure (p. 307)
Vaporize (p. 308)

Self-Assessment Test

1. Draw a p-T chart for water. Label the following clearly: vapor-pressure curve, dew-point curve, saturated region, superheated region, subcooled region, and triple point. Show where evaporation, condensation, and sublimation take place by arrows.

2. Describe the state and pressure conditions of water initially at 20°F as the temperature is increased to 250°F in a fixed volume.

3. Look in Appendix J at the diagram for CO_2.
   (a) At what pressure is CO_2 solid in equilibrium with CO_2 liquid and vapor?
   (b) If the solid is placed in the atmosphere, what happens?

4. Use the Antoine equation to calculate the vapor pressure of ethanol at 50°C, and compare with the experimental value.

5. Determine the normal boiling point for benzene from the Antoine equation.

6. Prepare a Cox chart from which the vapor pressure of toluene can be predicted over the temperature range −20 to 140°C.

Thought Problems

1. In the startup of a process, Dowtherm, an organic liquid with a very low vapor pressure, was being heated from room temperature to 335°F. The operator suddenly noticed that the gauge pressure was not the expected 15 psig but instead was 125 psig. Fortunately, a relief valve in the exit line ruptured into a vent (expansion) tank so that a serious accident was avoided. Why was the pressure in the exit line so high?

2. A cylinder containing butadiene exploded in a research laboratory, killing one employee. The cylinder had been used to supply butadiene to a pilot plant. When butadiene gas was required, heat was supplied to the cylinder to raise the pressure of the butadiene in the tank. The maximum temperature that could be achieved in the tank on subsequent tests with a like tank was 160°C. At 152°C, the critical temperature for butadiene, the pressure in 628 $lb_f/in.^2$, less than one-half of the pressure required to rupture the tank by hydraulic test. Why did the tank explode?

3. **"Careless Campers Contaminate Mountain Water"** was a recent headline in the newspaper. The article went on:

Beware! There are little monsters loose in those seemingly clean, pristine mountain streams. Their name: *Giardia*, and of specific interest to humans and pigs, *Giardia lamblia*. *Giardia* is a beet-shaped organism with no less than eight flagella. It is of concern to anyone who happens to slurp any down because it attaches itself by means of a sucking organism to the intestinal mucous membranes. The result is severe diarrhea, bordering on dysentery.

The incidence of *Giardia* in the wilderness areas of New Mexico and Colorado has vastly increased over the past five years. The disease it causes, giardiasis, is contracted by drinking water containing the organism. Unfortunately for all backpackers, horse packers, and day hikers, many of the lakes and streams are already tainted. No problem, you say—just drop in a chemical purification tablet and let it do the job? While chemical purification such as Halazone, iodine, or chlorine may kill many bacteria, the hale and hearty giardia goes unscathed.

What steps would you take to avoid the problem other than carrying a portable water supply with you? You need at least 210°F to kill the organisms by boiling water.

Discussion Questions

1. The following description of a waste disposal system appeared in Chemical Engineering, June 1993, p. 23.

   The first commercial application of VerTech Deep Shaft, a wet oxidation process that takes place in a reactor suspended in a borehole about 1,250 m deep, has been commissioned to treat 25,000 m.t./yr of sewage sludge at a municipal wastewater treatment plant in Apeldoorn, The Netherlands. The reactor consists of three steel tubes suspended in a 1,250-m drilled shaft of 95 cm dia. A 5% sludge solution is pumped down the inner 19.5-cm-dia. open-ended tube, while the oxidized sludge and dissolved gases come up through the annular space between the first tube and the closed second tube (34 cm dia.). About 400 m down, pure oxygen is sparged into the sludge to promote oxidation of the suspended organic matter. The static head of the 1,200-m column prevents boiling at about 280°C and 100 bars. The alternatives: incineration, composting, drying, or conventional wet oxidation have per m.t. costs of Dutch guilders (DGL) 790, DGL 790, DGL 700 and DGL 600, respectively, versus DGL 580 for the VerTech process.

   Check the consistency of the data, and offer an opinion as to the problems that might occur if the process operated on a 20% sludge solution in water.

2. The term BLEVE (boiling-liquid expanding-vapor explosion) refers to an explosion that occurs when the pressure on a liquid is suddenly reduced substantially below its vapor pressure. Explain how such an explosion can occur. Might it occur if a tank car ruptures in an accident even if no fire occurs? Give one or two other examples of vessel failures that might cause a BLEVE.

3. Many distillation columns are designed to withstand a pressure of 25 or 50 psig. The reboiler at the bottom of the column is where the heat used to vaporize the fluid in the

column is introduced. What would you recommend as to the type of heat source among these three: (a) steam (heat exchanger), (b) fired heater (analogous to a boiler), and (c) hot oil (heat exchanger)?

4.4 SATURATION

Your objectives in studying this section are to be able to:

1. Define saturated gas.
2. Calculate the partial pressure of the components of a saturated ideal gas given combinations of the temperature, pressure, volume and/or number of moles present, or calculate the number of moles of vapor.
3. Determine the condensation temperature (dew point) of the vapor in a saturated gas given the pressure, volume, and/or number of moles.

LOOKING AHEAD

In this section we explain how to determine the composition and dew point for a saturated vapor mixed with a noncondensable gas.

MAIN CONCEPTS

How can you predict the conditions of a mixture of a *pure vapor* (which can condense) *and a noncondensable gas* at equilibrium? A mixture containing a vapor behaves somewhat differently than does a pure component by itself. A typical example with which you are quite familiar is that of water vapor in air. It condenses, rains, and freezes, and the reverse, depending on the temperature.

When any pure gas (or a gaseous mixture) comes in contact with a liquid, the gas will acquire molecules from the liquid. If contact is maintained for a considerable length of time, vaporization continues until equilibrium is attained, at which time the *partial pressure of the vapor in the gas will equal the vapor pressure* of the liquid at the temperature of the system. Regardless of the duration of contact between the liquid and gas, when equilibrium is reached no more net liquid will vaporize into the gas phase. The gas is then said to be **saturated** with the particular vapor at the given temperature. We also say that the gas mixture is at its dew point. **Dew point for the mixture of pure vapor and noncondensable gas means the temperature at which the vapor just starts to condense when cooled at constant pressure.**

Now, what do these concepts mean with respect to the quantitative measurement of gas-vapor conditions? Suppose that you inject liquid water at 65°C into a cylinder of air at the same temperature, and keep the system at a constant temperature of 65°C. Suppose further that the pressure at the top of the cylinder is maintained at 101.3 kPa (1 atm). What happens to the volume of the cylinder as a function of time? Figure 4.15 shows that the volume of the air plus the water vapor increases until the air is saturated with water vapor, after which stage the volume remains constant. Figure 4.16(a) indicates how the partial pressure of the water vapor increases with time until it reaches its vapor pressure of 24.9 kPa (187 mm Hg). Why does the partial pressure of the air decrease?

Next, suppose that you carry out a similar experiment, but maintain the volume constant and let the total pressure vary in the cylinder. Will the pressure go up or down with respect to time? What will be the asymptotic value of the partial pressure of the water vapor? The air? Look at Figure 4.16(b) to see if your answers to these questions were correct.

Finally, is it possible to have the water evaporate into air and saturate the air, and yet maintain both a constant temperature, volume, and pressure in the cylinder? (*Hint:* What would happen if you let some of the gas-vapor mixture escape from the cylinder?)

Assuming that the ideal gas law applies to both air and water vapor with excellent precision, we can say that the following relations hold *at saturation:*

$$\frac{p_{air}V}{p_{H_2O}V} = \frac{n_{air}RT}{n_{H_2O}RT}$$

or

$$\frac{p_{air}}{p_{H_2O}} = \frac{n_{air}}{n_{H_2O}} = \frac{p_{air}}{p_{total} - p_{air}}$$

in a volume V at temperature T.

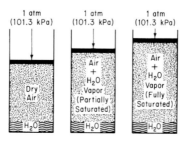

Figure 4.15   Evaporation of water at constant pressure and temperature of 65°C.

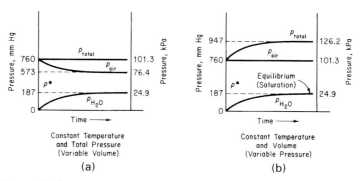

Figure 4.16   Change of partial and total pressures during the vaporization of water into air at constant temperature: (a) constant temperature and total pressure (variable volume); (b) constant temperature and volume (variable pressure).

EXAMPLE 4.17 Saturation

What is the minimum number of cubic meters of dry air at 20°C and 100 kP necessary to evaporate 6.0 kg of ethyl alcohol if the total pressure remains constant at 100 kPa and the temperature remains 20°C? Assume that the air is blown through the alcohol to evaporate it in such a way that the exit pressure of the air-alcohol mixture is at 100 kPa.

Solution

See Figure E4.17. Assume that the process is isothermal. The additional data needed are

$$p^*_{\text{alcohol}} \text{ at } 20°C = 5.93 \text{ kPa}$$

$$\text{mol. wt. ethyl alcohol} = 46.07$$

Figure E4.17

The minimum volume of air calls for a saturated mixture; any condition less than saturated would require more air.

Basis: 6.0 kg of alcohol

The ratio of moles of ethyl alcohol to moles of air in the final gaseous mixture is the same as the ratio of the partial pressures of these two substances. Since we know the moles of alcohol, we can find the number of moles of air.

$$\frac{p^*_{\text{alcohol}}}{p_{\text{air}}} = \frac{n_{\text{alcohol}}}{n_{\text{air}}}$$

and once we know the number of moles of air we can apply the ideal gas law. Since
$p^*_{alcohol} = 5.93$ kPa

$$p_{air} = p_{total} - p^*_{alcohol} = (100 - 5.93) \text{ kPa} = 94.07 \text{ kPa}$$

$$\frac{6.0 \text{ kg alcohol}}{} \left| \frac{1 \text{ kg mol alcohol}}{46.07 \text{ kg alcohol}} \right| \frac{94.07 \text{ kg mol air}}{5.93 \text{ kg mol alcohol}} = 2.07 \text{ kg mol air}$$

$$V_{air} = \frac{2.07 \text{ kg mol air}}{} \left| \frac{8.314 \text{ (kPa)(m}^3)}{\text{(kg mol)(K)}} \right| \frac{293 \text{ K}}{100 \text{ kPa}} = 50.3 \text{ m}^3 \text{ at } 20°C \text{ and } 100 \text{ kPa}$$

Another way to view this problem is to say that the final volume contains

V m^3 of alcohol at 5.93 kPa and 20°C
V m^3 of air at 94.07 kPa and 20°C
V m^3 of air plus alcohol at 100 kPa and 20°C

Thus, the volume could be calculated from the information about the alcohol

$$V_{alcohol} = \frac{\left(\dfrac{6}{46.07}\right) \left| 8.314 \right| 293}{5.93} = 53.5 \text{ m}^3 \text{ at } 20°C \text{ and } 5.93 \text{ kPa}$$

$$= V_{air} \text{ at } 94.07 \text{ kPa and } 20°C$$

$$V_{air} = \frac{53.5 \text{ m}^3}{} \left| \frac{94.07}{100} \right. = 50.3 \text{ m}^3 \text{ at } 100 \text{ kPa and } 20°C$$

EXAMPLE 4.18 Smokestack Emission and Pollution

A local pollution-solutions group has reported the Simtron Co. boiler plant as being an air polluter and has provided as proof photographs of heavy smokestack emissions on 20 different days. As the chief engineer for the Simtron Co., you know that your plant is not a source of pollution because you burn natural gas (essentially methane) and your boiler plant is operating correctly. Your boss believes the pollution-solutions group has made an error in identifying the stack—it must belong to the company next door that burns coal. Is he correct? Is the pollution-solutions group correct? See Figure E4.18a.

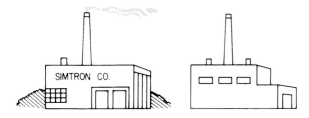

Figure E4.18a

Solution

Methane (CH_4) contains 2 kg mol of H_2 per kilogram mole of C, while coal contains 71 kg of C per 5.6 kg of H_2 in 100 kg of coal. The coal analysis is equivalent to

$$\frac{71 \text{ kg C}}{} \left| \frac{1 \text{ kg mol C}}{12 \text{ kg C}} \right. = 5.92 \text{ kg mol C}$$

$$\frac{5.6 \text{ kg H}_2}{} \left| \frac{1 \text{ kg mol H}_2}{2.016 \text{ kg H}_2} \right. = 2.78 \text{ kg mol H}_2$$

or a ratio of $2.78/5.92 = 0.47$ kg mol of H_2/kg mol of C. Suppose that each fuel burns with 40% excess air and that combustion is complete. We can compute the mole fraction of water vapor in each stack gas.

   Steps 1, 2, 3, and 4   The process is shown in Figure E4.18b.

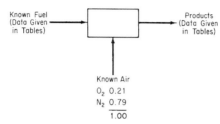

Figure E4.18b

   Step 5

Basis: 1 kg mol C

   Steps 6 and 7   The combustion problem is a standard type of problem having a unique solution in which both the fuel and air flows are given, and the product flows are calculated directly.
   Steps 7, 8, and 9   Tables will make the analysis and calculations compact.
Natural Gas

$$CH_4 + 2O_2 \rightarrow CO_2 + 2H_2O$$

Components	kg mol	Composition of combustion gases (kg mol)			
		CO_2	H_2O	Excess O_2	N_2
C	1.0	1.0			
H_2	2.0		2.0		
Air				0.80	10.5
Total		1.0	2.0	0.80	10.5

Required O_2: 2
Excess O_2: $2(0.40) = 0.80$
N_2: $(2.80)(79/21) = 10.5$

The total kilogram moles of gas produced are 14.3 and the mole fraction H_2O is

$$\frac{2.0}{14.3} = 0.14$$

Coal

$$C + O_2 \rightarrow CO_2 \qquad H_2 + \tfrac{1}{2}O_2 \rightarrow H_2O$$

Components	kg mol	Composition of combustion gases (kg mol)			
		CO_2	H_2O	Excess O_2	N_2
C	1	1			
H_2	0.47		0.47		
Air				0.49	6.5
Total		$\overline{1}$	$\overline{0.47}$	$\overline{0.49}$	$\overline{6.5}$

Required O_2: $1 + 0.47(1/2) = 1.24$
Excess O_2 $(1.24)(0.40) = 0.49$
N_2 $1.40(79/21)[1 + 0.47(1/2) = 6.50$

The total kilogram moles of gas produced are 8.46 and the mole fraction H_2O is

$$\frac{0.47}{8.46} = 0.056$$

If the barometric pressure is, say, 100 kPa, the stack gas would become saturated and water vapor would start to condense at $p^*_{H_2O}$:

	Natural gas	*Coal*
Partial pressure:	$100(0.14) = 14$ kPa	$100(0.056) = 5.6$ kPa
Equivalent temperature:	52.5°C	35°C

Thus, the stack will emit condensed water vapor at the top at higher ambient temperatures for boilers burning natural gas than for those burning coal. The public, unfortunately, sometimes concludes that all the emissions they perceive are pollution. Natural gas could appear to the public to be a greater pollutant than either oil or coal when, in fact, the emissions are just water vapor. The sulfur content of coal and oil can be released as sulfur dioxide to the atmosphere, and the polluting capacities of coal and oil are much greater than natural gas when all three are being burned properly. The sulfur contents as delivered to the consumers are as follows: natural gas, $4 \times 10^{-4}\%$ (as added mercaptans); number 6 fuel oil, up to 2.6%; and coal, from 0.5 to 5%. In addition, coal may release particulate matter into the stack plume. By mixing the stack gas with air, and by convective mixing above the stack, the mole fraction water vapor is reduced, and hence the condensation temperature is reduced. However, for equivalent dilution, the coal-burning plant will always have a lower condensation temperature. What steps would you take to resolve the questions that were originally posed?

LOOKING BACK

In this section we explained what saturation means, and showed how the vapor pressure of a compound mixed with a noncondensable gas reaches saturated conditions which are a function of the temperature and total pressure on the system.

Key Ideas

1. The partial pressure of a vapor mixed with a noncondensable gas at equilibrium with the liquid phase will be its vapor pressure.
2. The dew point is the temperature at which the vapor in such a mixture will just start to condense.

Key Terms

Dew point (p. 320)   Saturated (p. 319)
Noncondensable gas (p. 319)   Vapor (p. 319)

Self-Assessment Test

1. What does the term "saturated gas" mean?
2. If a container with a volumetric ratio of air to liquid water of 5 is heated to 60°C and equilibrium is reached, will there still be liquid water present? at 125°C?
3. A mixture of air and benzene contains 10 mole % benzene at 43°C and 105 kPa pressure. At what temperature does the first liquid form? What is the liquid?
4. The dew point of water in atmospheric air is 82°F. What is the mole fraction of water vapor in the air if the barometric pressure is 750 mm Hg?
5. Ten pounds of $KClO_3$ is completely decomposed and the oxygen evolved collected over water at 80°F. The barometer reads 29.7 in. Hg. What weight of saturated oxygen is obtained?
6. If a gas is saturated with water vapor, describe the state of the water vapor and the air if it is:
   (a) Heated at constant pressure
   (b) Cooled at constant pressure
   (c) Expanded at constant temperature
   (d) Compressed at constant temperature

Thought Problems

1. Water was drained from the bottom of a gasoline tank into a sewer and shortly thereafter a flash fire occurred in the sewer. The operator took special care to make sure that none of the gasoline entered the sewer. What caused the sewer fire?
2. To reduce problems of condensation associated with continuous monitoring of stack gases, a special probe and flow controller were developed to dilute the flue gas with

outside air in a controlled ratio (such as 10 to 1). Does this seem like a sound idea? What problems might occur with continuous operation of the probe?

3. Why is it important to know the concentration of water in the air entering a boiler?

Discussion Question

1. Whenever fossil fuels containing sulfur are burned in heaters or boilers, sulfur dioxide, carbon dioxide and water vapor are formed analogously, when municipal solid wastes are incinerated HCl and HBr form as well as sulfur dioxide. These acid gases form quite corrosive solutions if the water vapor in the flue gas condenses.

How would you go about estimating the dew point of a flue gas such as one that contained 8% CO_2, 12% H_2O, 73% N_2, 0.02% SO_2, 0.015% HCl, 6% O_2, and 0.01% HBr? What protective measures might be used to avoid corrosion of the heater or stack surfaces? Will the dew point alone be adequate information to alleviate corrosion? *Hint:* Compounds such as ferric chloride are very hydroscopic even at high temperatures.

2. Expired breath is at body temperature, that is, 37°C, and essentially saturated. Condensation of moisture in respiratory equipment from the expired breath occurs as the breath cools. Such a high humidity level has several implications in developing space suits, diving equipment, oxygen masks in hospitals, and so on. What might they be?

4.5 VAPOR-LIQUID EQUILIBRIA FOR MULTICOMPONENT SYSTEMS

Your objectives in studying this section are to be able to

1. Use Raoult's law and Henry's law to predict the partial pressure of a solute and a solvent.
2. Use the relationship $K_i = y_i / x_i$ to calculate any one of the variables, given the other two.
3. Calculate the composition of multicomponent systems at equilibrium between the liquid and vapor phases.

LOOKING AHEAD

In this section we show how to calculate the partial pressures and mole fractions of solutes and solvents in multicomponent mixtures in which the gas and liquid phases are at equilibrium.

MAIN CONCEPTS

In a two-phase vapor-liquid mixture at equilibrium, a component in one phase is in equilibrium with the same component in the other phase. The equilibrium relationship depends on the temperature, pressure, and composition of the mixture. Figure 4.17 illustrates two cases, one at constant pressure and the other at constant temperature. At the pairs of points A and B, and C and D, the respective pure components exert their respective vapor pressures at the equilibrium temperature. In between the pairs of points, as the overall composition of the mixture changes, two phases exist, each having a different composition for the same component as indicated by the dashed lines.

First we consider two cases in which linear ("ideal") equations exist to relate the mole fraction of one component in the vapor phase to the mole fraction of the same component in the liquid phase.

Henry's law.   Used primarily for a component whose mole fraction approaches zero, such as a dilute gas dissolved in as liquid:

$$p_i = H_i x_i \tag{4.12}$$

where p_i is the pressure in the gas phase of the dilute component at equilibrium at some temperature, and H_i is the *Henry's law constant*. Note that in the limit where $x_i \equiv 0$, $p_i \equiv 0$. Values of H_i can be found in several handbooks.

Calculation of the partial pressure of a gas in the gas phase that in equilibrium the gas dissolved the liquid phase when Henry's law applies is quite simple. Take for example CO_2 dissolved in water at 40°C for which the value of H is 69,600 atm/mol fraction. (The large value of H shows that $CO_2(g)$ is only sparing soluble in water.) For example, if $x_{CO_2} = 4.2 \times 10^{-6}$, the partial pressure of the CO_2 in the gas phase is

$$p_{CO_2} = 69,600(4.2 \times 10^{-6}) = 0.29 \text{ atm}$$

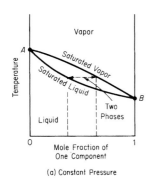

(a) Constant Pressure

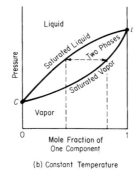

(b) Constant Temperature

Figure 4.17   Vapor-liquid equilibria for a binary mixture. The dashed lines show the equilibrium compositions (a) when the total pressure is constant and (b) when the temperature is constant over the composition range.

Note that

$$y_i = \frac{p_i}{p_{tot}} = \frac{H_i x_i}{p_{tot}}$$

and since H_i is roughly independent of p_{tot}, the higher the total pressure, the larger x_i. Carbonated beverages utilize this relation by increasing the pressure in the bottle or can.

Raoult's law.   Used primarily for a component whose mole fraction approaches unity or for solutions of components quite similar in chemical nature, such as straight chain hydrocarbons. Let the subscript i denote the component, p_i be the partial pressure of component i in the gas phase, y_i be the gas-phase mole fraction, and x_i be the liquid-phase mole fraction. Then:

$$p_i = p_i^* x_i \tag{4.13}$$

Note that in the limit where $x_i \equiv 1$, $p_i \equiv p_i^*$. Often an **equilibrium constant** K_i is defined using Eq. (4.13) as follows by assuming that Dalton's law applies to the gas phase ($p_i = p_{tot} y_i$):

$$K_i = \frac{y_i}{x_i} = \frac{p_i^*}{p_{tot}} \tag{4.14}$$

Equation (4.14) gives reasonable estimates of K_i values at low pressures for components well below their critical temperatures, but yields values too large for components above their critical temperatures, at high pressures, and/or for polar compounds. Nevertheless, Eq. (4.14) can be adapted to nonideal mixtures if K_i is made a function of temperature, pressure, and composition so that relations for K_i can be fit by experimental data and used directly, or in the form of charts, for design calculations as explained in some of the references at the end of this chapter. A useful approximate relation for K_i was recommended by Sandler[7] if $T_{c,i}/T > 1.2$:

$$K_i = \frac{p_{c,i}^{[7.224 - 7.534/T_{r,i} - 2.598 \ln T_{r,i}]}}{p_{tot}} \tag{4.15}$$

Typical problems you may be asked to solve that involve the use of the equilibrium coefficient K_i are:

(1) Calculate the bubble point temperature of a liquid mixture given the total pressure and liquid composition.

[7]S. I. Sandler, in *Foundations of Computer Aided Design*, Vol. 2, p. 83, R. H. S. Mah and W. D. Seider, eds., New York: American Institute of Chemical Engineers, 1981.

(2) Calculate the dew point temperature of a vapor mixture given the total pressure and vapor composition.

(3) Calculate the related equilibrium vapor-liquid compositions over the range of mole fractions from 0 to 1 as a function of temperature given the total pressure.

Analogous problems occur with respect to calculating the total pressure given a fixed temperature. Another useful calculation is:

(4) Calculate the composition of the vapor and liquid streams, and their respective quantities, when a liquid of given composition is partially vaporized (flashed) at a given temperature and pressure (the temperature must lie between the bubble and dew point temperatures of the feed).

To calculate the **bubble point temperature** (given the total pressure and liquid composition), you can write Eq. (4.14) as $y_i = K_i x_i$ and you know that $\Sigma y_i = 1$ in the vapor phase.

$$1 = \sum_{i=1}^{n} K_i x_i \tag{4.15}$$

in which the K_i's are functions of solely the temperature and n is the number of components. Because each of the K_i's increases with temperature, Eq. (4.16) has only one positive root. You can employ Newton's method to get the root (see Appendix L) if you can express each K_i as an explicit function of temperature. For an ideal solution, Eq. (4.16) becomes

$$P_{tot} = \sum_{i=1}^{n} p_i^* x_i \tag{4.16}$$

and you might use Antoine's equation for p_i^*. Once the bubble point temperature is determined, the vapor composition can be calculated from

$$y_i = \frac{p_i^* x_i}{P_{tot}} \tag{4.17}$$

To calculate the **dew point temperature** (given the total pressure and vapor composition), you can write Eq. (4.14) as $x_i = y_i/K_i$, and you know $\Sigma x_i = 1$ in the liquid phase. Consequently, you want to solve the equation

$$1 = \sum_{i=1}^{n} \frac{y_i}{K_i} \tag{4.18}$$

in which the K's are function of temperature as explained for the bubble point temperature calculation. For an ideal solution,

$$1 = p_{tot} \sum_{i=1}^{n} \frac{y_i}{p_i^*} \tag{4.19}$$

To calculate the amount of the respective vapor and liquid phases that evolve at equilibrium when a liquid of known composition flashes (**flash vaporization**) at a known temperature and pressure, you must use Eq. (4.14) together with a material balance. Figure 4.18 illustrates the steady-state process. A mole balance for component i gives

$$Fx_{F_i} = Lx_i + Vy_i \tag{4.20}$$

where F is the moles of liquid to be flashed, L is the moles of liquid at equilibrium, and V is the moles of vapor at equilibrium. Introduction of $y_i = K_i x_i$ into Eq. (4.20) gives

$$Fx_{F_i} = L\left(\frac{y_i}{K_i}\right) + Vy_i$$

$$y_i = \frac{Fx_{F_i}}{\dfrac{L}{K_i} + (F - L)} = \frac{x_{F_i}}{1 - \dfrac{L}{F}\left(1 - \dfrac{1}{K_i}\right)} \tag{4.21}$$

where L/F is the fraction of liquid formed on vaporization. Consequently, since $\Sigma y_i = 1$, after summing the y_i's you want to solve the following equation

$$1 = \sum_{i=1}^{n} \frac{x_{F_i}}{1 - \dfrac{L}{F}\left(1 - \dfrac{1}{K_i}\right)} \tag{4.22}$$

for L/F (>0). Numerous computer programs are available to solve the flash vaporization problem.

Using the above equations, you can prepare figures such as Figure 4.17 for binary mixtures.

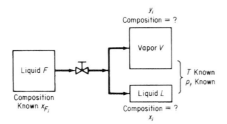

Figure 4.18 Flash vaporization with V and L in equilibrium

The equations above are not the only formulations in which vapor-liquid calculations are conducted. For example, for the bubble point calculation, an equation equivalent to Eq. (4.16) would be

$$\ln \sum_{i=1}^{n} K_i x_i = 0$$

and an equation equivalent to Eq. (4.22) would be

$$\sum_{i=1}^{n} \frac{x_{F_i} (K_i - 1)}{(K_i - 1)(V/F) + 1} = 0$$

In selecting a particular form of the equation to be used for your equilibria calculations, you must select a method of solving the equation that has desirable convergence characteristics. Convergency to the solution should

(1) Lead to the desired root if the equation has multiple roots
(2) Be stable, that is, approach the desired root asymptotically rather than by oscillating
(3) Be rapid, and not become slower as the solution is approached.

EXAMPLE 4.19 Vapor-Liquid Equilibrium Calculation

Suppose that a liquid mixture of 4.0% n-hexane in n-octane is vaporized. What is the composition of the first vapor formed if the total pressure is 1.00 atm?

Solution

Examine Figure 4.17a. The mixture can be treated as an ideal mixture because the components are quite similar. As an intermediate step, you must calculate the bubble point temperature with the aid of Eq. (4.17). You have to look up the coefficients of the Antoine equation to obtain the vapor pressures of the two components:

$$\ln (p^*) = A - \frac{B}{C + T}$$

where p^* is in mm Hg and T is in K:

	A	B	C
n-hexane (C_6):	15.8737	2697.55	−48.784
n-octane (C_8):	15.9798	3127.60	−63.633

Basis: 1 kg mol of liquid

We need to solve the following equation to get the bubble point temperature using one of the techniques described in Sec. L.2:

$$760 = \exp\left(15.8737 - \frac{2697.55}{-48.784 + T}\right)0.040 + \exp\left(15.9787 - \frac{3127.60}{-63.633 + T}\right)0.960$$

The solution is $T = 393.3$K, where the vapor pressure of hexane is 3114 mm Hg and the vapor pressure of octane is 661 mm Hg.

$$y_{C_6} = \frac{p_{C_6}^*}{p_{tot}}x_{C_6} = \frac{3114}{760}(0.040) = 0.164$$

$$y_{C_8} = 1 - 0.164 = 0.836$$

EXAMPLE 4.20 Flash Calculation

Calculate the fraction of liquid that will exist at equilibrium at 150°F and 50 psia when the liquid concentrations of the solution to be vaporized are as follows:

Component	Initial liquid mole fraction	K
C_2	0.0079	16.20
C_3	0.1321	5.2
i-C_4	0.0849	2.6
n-C_4	0.2690	1.98
i-C_5	0.0589	0.91
n-C_5	0.1321	0.72
C_6	0.3151	0.28
Total	1.0000	

The K values come from the *Engineering Data Book* of the Gas Processors Supply Association (1980).

Solution

We want to solve Eq. (4.22) for L/F. Introduce the values of x_i and K_i into Eq. (4.22). Start with an initial guess of $L/F = 1.0$. Successive stages of the iteration by Newton's method give

Stage	L/F
1	1.0
2	0.8565
3	0.6567
4	0.5102
5	0.4573
6	0.4511 which is sufficiently accurate

ADDITIONAL DETAILS

Liquid-liquid equilibria are treated essentially the same way as nonideal vapor-liquid equilibria with activity-coefficient corrections in both phases. For liquid-solid equilibria, engineers rely almost entirely on recorded experimental data, as prediction of the liquid-phase composition is quite difficult. Refer to the references at the end of this chapter for more information.

You will find the Gibb's **phase rule** a useful guide in establishing how many properties, such as pressure and temperature, have to be specified to definitely fix all the remaining properties and number of phases that can coexist for any physical system. **The rule can be applied only to systems in equilibrium.** The Gibbs phase rule states that

$$F = C - \mathcal{P} + 2 \tag{4.23}$$

where F = number of degrees of freedom (i.e., the number of independent properties that have to be specified to determine all the intensive properties of each phase of the system of interest)

C = number of components in the system; for circumstances involving chemical reactions, C is *not* identical to the number of chemical compounds in the system but is equal to the number of chemical compounds less the number of independent-reaction and other equilibrium relationships among these compounds.

$\mathcal{P}$ = number of phases that can exist in the system; a phase is a homogeneous quantity of material such as a gas, a pure liquid, a solution, or a homogeneous solid

Variables of the kind with which the phase rule is concerned are called **phase-rule variables**, and they are **intensive** properties of the system. **By this we mean properties that do not depend on the quantity of material present.** If you think about the properties we have employed so far in this book, you have the feeling that pressure and temperature are independent of the amount of material present. So is concentration, but what about volume? The total volume of a system is called an **extensive variable because it does depend on how much material you have; the specific volume, on the other hand, the cubic meters per kilogram, for example, is an intensive property** because it is independent of the amount of material present. You should remember that that the specific (per unit mass) values are intensive properties; the total quantities are extensive properties.

An example will clarify the use of these terms in the phase rule. You will remember for a pure gas that we had to specify three of the four variables in the ideal gas equation $pV = nRT$ in order to be able to determine the remaining one unknown. You might conclude that $F = 3$. If we apply the phase rule, for a single phase $\mathcal{P} = 1$, and for a pure gas $C = 1$, so that

$$F = C - \mathcal{P} + 2 = 1 - 1 + 2 = 2 \quad \text{variables to be specified}$$

How can we reconcile this apparent paradox with our previous statement? Since the phase rule is concerned with intensive properties only, the following are phase-rule variables in the ideal gas law:

$$\left.\begin{array}{l} p \\ \hat{V} \text{ (specific molar volume)} \\ T \end{array}\right\} \quad \text{3 intensive properties}$$

Thus the ideal gas law would be written

$$p\hat{V} = RT$$

and written in this form you can see that by specifying two intensive variables ($F = 2$), the third can be calculated.

An **invariant** system is one in which no variation of conditions is possible without one phase disappearing. An example with which you may be familiar is the ice-water-water vapor system, which exists at only one temperature (0.0°C) and pressure (0.611 Pa):

$$F = C - \mathcal{P} + 2 = 1 - 3 + 2 = 0$$

With all three phases present, none of the physical conditions can be varied without the loss of one phase. As a corollary, if the three phases are present, the temperature, the specific volume, and so on, must always be fixed at the same values. This phenomenon is useful in calibrating thermometers and other instruments.

A discussion of the term C in the phase rule is beyond the scope of this book; consult one of the references at the end of this chapter for details.

EXAMPLE 4.21 Application of the Phase Rule

Calculate the number of degrees of freedom (how many additional intensive variables must be specified to fix the system) from the phase rule for the following materials at equilibrium:

(a) Pure liquid benzene

(b) A mixture of ice and water only

(c) A mixture of liquid benzene, benzene vapor, and helium gas

(d) A mixture of salt and water designed to achieve a specific vapor pressure.

What variables might be specified in each case?

Solution

$$F = C - P + 2$$

(a) $C = 1$, $P = 1$, hence $F = 1 - 1 + 2 = 2$. The temperature and pressure might be specified in the range in which benzene remains a liquid.

(b) $C = 1$, $P = 2$, hence $F = 1 - 2 + 2 = 1$. Once either the temperature or the pressure is specified, the other intensive variables are fixed.

(c) $C = 2$, $P = 2$, hence $F = 2 - 2 + 2 = 2$. A pair from temperature, pressure, or mole fraction can be specified.

(d) $C = 2$, $P = 2$, hence $F = 2 - 2 + 2 = 2$. Since a particular pressure is to be achieved, you would adjust the salt concentration and the temperature of the solution.

Note in (a) and (b) it would be unlikely that a vapor phase would not exist in practice, increasing P by 1 and reducing F by one.

LOOKING BACK

In this section we described how to calculate the partial pressure of the components in a vapor-liquid mixture at equilibrium using Henry's law, Raoult's law, or the equilibrium coefficient K_i. We also showed how the phase rule applies to systems in equilibrium.

Key Ideas

1. The partial pressure in the gas phase of a dilute solute in a mixture at equilibrium can be often calculated using Henry's law.

2. The partial pressure in the gas phase of a solvent or of compounds of similar chemical nature at equilibrium can often be calculated via Raoult's law.

3. Vapor-liquid compositions can be determined using an equilibrium coefficient (that itself may be a function of temperature, pressure, and composition).

4. The phase rule relates the number of degrees of freedom in a system at equilibrium to the number of phases and components present.

Key Terms

Bubble point temperature (p. 329)
Degrees of freedom (p. 333)
Dewpoint temperature (p. 329)
Equilibrium constant (p. 328)
Extensive variable (p. 334)
Flash vaporization (p. 330)

Henry's law (p. 327)
Intensive variable (p. 333)
Invariant (p. 334)
Phase rule variables (p. 333)
Raoult's law (p. 328)

Self-Assessment Test

1. Calculate (a) the pressure at the dew point for the following mixture at 100°F and (b) the liquid composition.

Component	Mole fraction	K values at psia of		
		190	200	210
C_2H_6	0.2180	3.22	3.07	2.92
C_3H_8	0.6650	1.005	0.973	0.920
i-C_4H_{10}	0.1073	0.45	0.43	0.41
n-C_4H_{10}	0.0097	0.315	0.305	0.295
Total	1.0000			

2. Is the critical point a single phase? If not, what phases are present? Repeat for the triple point (for water).

3. A vessel contains air: $N_2(g)$, $O_2(g)$, and $Ar(g)$.
   (a) How many phases, components, and degrees of freedom are there according to the phase rule?
   (b) Repeat for a vessel one-third filled with liquid ethanol and two-thirds filled with N_2 plus ethanol vapor.

Thought Problems

1. The fluid in a large tank caught on fire 40 minutes after the start of a blending operation in which one grade of naphtha was being added to another. The fire was soon put out and the naphtha was moved to another tank. The next day blending was resumed in the second tank; 40 minutes later another fire started. Can you explain the reason for this sequence of events? What might be done to prevent such accidents?

2. CO_2 can be used to clean optical or semiconductor surfaces and remove particles or organic contaminants. A bottle of CO_2 at 4000 kPa is attached to a jet that sprays onto the optical surface. Two precautions must be taken with this technique. The surface must be heated to about 30–35°C to minimize moisture condensation, and a CO_2 source with no residual heavy hydrocarbons (lubricants) must be employed to mini-

mize recontamination in critical cleaning applications. Describe the physical conditions of the CO_2 as it hits the optical surface. Is it gas, liquid, or solid? How does the decontamination take place?

3. The advertisement reads "Solid dry ice blocks in 60 seconds right in your own lab! Now you can have dry ice available to you at any time, day or night, with this small, safe, efficient machine and readily available CO_2 cylinders. No batteries or electrical energy are required." How is it possible to make dry ice in 60 seconds without a compressor?

4. An inventor is trying to sell a machine that transforms water vapor into liquid water without ever condensing the water vapor. You are asked to explain if such a process is technically possible. What is your answer?

5. Examine the statements below:
   (a) "The vapor pressure of gasoline is about 14 psia at 130°F."
   (b) "The vapor pressure of the system, water-furfural diacetate, is 101 kPa at 99.96°C."
   Are the statements correct? If not, correct them. Assume the numerical values are correct.

6. To maintain safe loading of hydrocarbon fluids, one of the many objectives of the Coast Guard is to prevent underpressuring of the tank(s) of the vessel being loaded. Overpressuring is easy to understand—to much fluid is pumped into a tank. How can underpressuring occur?

Discussion Questions

1. Gasoline tanks that have leaked have posed a problem in cleaning up the soil at the leak site. To avoid digging up the soil around the gasoline, which is located 5-10 m deep, it has been suggested that high pressure stream be injected underneath the gasoline site via wells to drive the trapped gasoline into a central extraction well which, under vacuum, would extract the gasoline. How might you design an experiment to test the concept of removal? What kinds of soils might be hard to treat? Why do you think steam was used for injection rather than water?

2. How to meet increasingly severe federal and state regulations for gasoline, oxygenated fuels, and low-sulfur diesel fuel represents a real challenge. The table below shows some typical values for gasoline components prior to the implementation of the regulations in the State of California, and the limits afterwards.

Fuel parameter	Former (typical gasoline)	Current (limit for refineries)
Sulfur (ppmw)	150	40
Benzene (vol %)	2	1
Olefins (vol %)	9.9	6
Oxygen (wt %)	0	2.2
Boiling point for 90% of the gasoline (°F)	330	300

Read some of the chemical engineering literature, and prepare a brief report on some of the feasible and economic ways that have been proposed or used to meet the new standards. Will enforcing emission standards on old automobiles (or junking them) be an effective technique of reducing emissions relative to modifying the gasoline? What about control of evaporative emissions from the fuel tank. What about degradation or malfunction of emission controls?

2. The EPA negotiated an agreement on reformulated gasoline that included a waiver permitting the use of higher vapor pressure gasoline with added ethanol than gasoline without ethanol. Considerable argument occurred because the ethanol-gasoline fuel leads to more volatile organic compounds finding their way into the atmosphere. Supposedly blending 10% ethanol into gasoline increases the vapor pressure of the mixture over ethanol free gasoline by 1 psi/measured as Reid Vapor Pressure (RVP). Tests show that the evaporation of hydrocarbons increases by 50%. What would be the vapor pressure of a 10% ethanol-gasoline mixture versus the vapor pressure of gasoline alone at 25°C, and indicate whether the reported 50% increase in vaporization of hydrocarbons from the fuel seems reasonable. (Note the aromatics in the gasoline are not more than 25% and the benzene is not more than 1% by volume.) What other factors must be take into account in blending gasoline?

4.6 PARTIAL SATURATION AND HUMIDITY

***Your objectives in studying this
section are to be able to:***

1. Define relative saturation (humidity), molal saturation (humidity), absolute saturation (humidity), and humidity by formulas involving partial pressures of the gas components.
2. Given the value of the partial saturation in one form, calculate the corresponding values in the other three forms as well as the dew point.

LOOKING AHEAD

In Sec. 4.4 we dealt with a mixture of a noncondensable gas and a vapor that saturated the gas. Here we examine cases in which the noncondensable gas is less than saturated.

MAIN CONCEPTS

Often, the contact time required in a process for equilibrium (or saturation) to be attained between the gas and liquid is too long, and the gas is not completely saturated with the vapor. Then the vapor is not in equilibrium with a liquid phase, and the partial pressure of the vapor is less than the vapor pressure of the liquid at the given temperature. This condition is called **partial saturation**. What you have is simply a mixture of two or more gases that obey the gas laws. What distinguishes this case from the previous examples for gas mixtures is that under suitable conditions it is possible to condense part of one of the gaseous components. Examine Figure 4.19. You can see how the partial pressure of the water vapor in a gaseous mixture at constant volume obeys the ideal gas laws as the temperature drops until saturation is reached, at which time the water vapor starts to condense. Until these conditions are achieved, you can confidently apply the gas laws to the mixture.

Several ways exist to express the concentration of a vapor in a mixture with a noncondensable gas. You sometimes encounter mass or mole fraction (or percent), but more frequently one of the following:

(1) Relative saturation (relative humidity)

(2) Molal saturation (molal humidity)

(3) "Absolute" saturation ("absolute" humidity) or percent saturation (percent humidity)

(4) Humidity

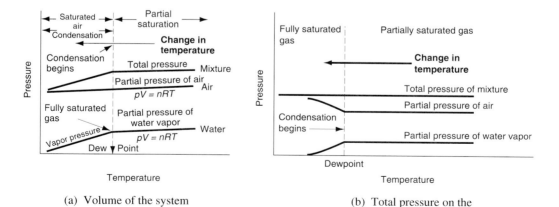

(a) Volume of the system remains constant

(b) Total pressure on the system is constant.

Figure 4.19   Transformation of a partially saturated water vapor-air mixture into a saturated mixture as the temperature is lowered.

When the vapor is water vapor and the gas is air, the special term **humidity** applies. For other gases or vapors, the term **saturation** is used.

Relative Saturation. Relative saturation is defined as

$$RS = \frac{p_{vapor}}{p_{satd}} = \text{relative saturation} \tag{4.24}$$

where p_{vapor} = partial pressure of the vapor in the gas mixture
p_{satd} = partial pressure of the vapor in the gas mixture if the gas were saturated at the given temperature of the mixture (i.e., the vapor pressure of the vapor component)

Then, for brevity, if the subscript 1 denotes the vapor,

$$RS = \frac{p_1}{p_1^*} = \frac{p_1/p_t}{p_1^*/p_t} = \frac{V_1/V_t}{V_{satd}/V_t} = \frac{n_t}{n_{satd}} = \frac{\text{mass}_1}{\text{mass}_{satd}}$$

You can see that relative saturation, in effect, represents the fractional approach to total saturation, as shown in Figure 4.20. If you listen to the radio or TV and hear the announcer say that the temperature is 25°C (77°F) and the *relative humidity* is 60%, he or she implies that

$$\frac{p_{H_2O}}{p_{H_2O}^*}(100) = \% \, RH = 60$$

with both the p_{H_2O} and the $p_{H_2O}^*$ being measured at 25°C. Zero percent relative saturation means no vapor in the gas. What does 100% relative saturation mean? It means that the partial pressure of the vapor in the gas is the same as the vapor pressure of the substance that is the vapor.

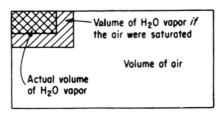

Figure 4.20   Partially saturated gas with the water and air separated conceptually.

EXAMPLE 4.22 Application of Relative Humidity

The weather report on the radio this morning was that the temperature this afternoon would reach 94°F, the relative humidity would be 43%, the barometer 29.67 in. Hg, partly cloudy to clear, with the wind from SSE at 8 mi/hr. How many pounds of water vapor would be in 1 mi^3 of afternoon air? What would be the dew point of this air?

Solution

The vapor pressure of water at 94°F is 1.61 in. Hg. We can calculate the partial pressure of the water vapor in the air from the given percent relative humidity.

$$p_w = (1.61 \text{ in. Hg})(0.43) = 0.692 \text{ in. Hg}$$

$$(p_{air} = p_t - p_w = 29.67 - 0.692 = 28.98 \text{ in. Hg})$$

Basis: 1 mi^3 water vapor at 94°F and 0.692 in. Hg

$$\frac{1 \text{ mi}^3}{} \left|\left(\frac{5280 \text{ ft}}{1 \text{ mi}}\right)^3\right| \frac{492°R}{555°R} \left|\frac{0.692 \text{ in. Hg}}{29.92 \text{ in. Hg}}\right| \frac{1 \text{ lb mol}}{359 \text{ ft}^3} \left|\frac{18 \text{ lb H}_2\text{O}}{1 \text{ lb mol}}\right| = 1.52 \times 10^8 \text{ lb H}_2\text{O}$$

Now the dew point is the temperature at which the water vapor in the air will first condense on cooling at *constant pressure and composition*. As the gas is cooled you can see from Eq. (4.24) that the relative humidity increases, since the partial pressure of the water vapor is constant while the vapor pressure of water decreases with temperature. When the percent relative humidity reaches 100%

$$100 \, \frac{p_{\text{H}_2\text{O}}}{p_{\text{H}_2\text{O}}^*} = 100\% \qquad \text{or} \qquad p_{\text{H}_2\text{O}} = p_{\text{H}_2\text{O}}^*$$

the water vapor will start to condense. Look at Figure 4.19. This means that at the dew point the vapor pressure of water will be 0.692 in. Hg. From the steam tables you can see that this corresponds to a temperature of about 68-69°F.

Molal Saturation.   Another way to express vapor concentration in a gas is to use the ratio of the moles of vapor to the moles of vapor-free gas:

$$\frac{n_{\text{vapor}}}{n_{\text{vapor-free gas}}} = \textbf{molal saturation} \tag{4.25}$$

If subscripts 1 and 2 represent the vapor and the dry gas, respectively, then for a binary system,

$$p_1 + p_2 = p_{tot}$$

$$n_1 + n_2 = n_{tot}$$

$$\frac{n_1}{n_2} = \frac{p_1}{p_2} = \frac{V_1}{V_2} = \frac{n_1}{n_{tot} - n_1} = \frac{p_1}{p_{tot} - p_1} = \frac{V_1}{V_{tot} - V_1} \tag{4.26}$$

Humidity. The special term **humidity** ($\mathcal{H}$) refers to the mass of water vapor per mass of bone-dry air, and is used in connection with the humidity charts in Sec. 5.7. By multiplying Eq. (4.25) by the appropriate molecular weights, you can find the mass of vapor per mass of dry gas:

$$\mathcal{H} = \frac{(n_{vapor})(\text{mol. wt.}_{vapor})}{(n_{dry\ gas})(\text{mol. wt.}_{dry\ gas})} = \frac{\text{mass}_{vapor}}{\text{mass}_{dry\ gas}} \tag{4.27}$$

"Absolute" Saturation (Humidity); Percentage Saturation (Humidity). **"Absolute" saturation** is defined as the ratio of the moles of vapor per mole of *vapor-free* gas to the moles of vapor *that would be present* per mole of *vapor-free* gas *if the mixture were completely saturated* at the existing temperature and total pressure:

$$\mathcal{AS} = \text{"absolute saturation"} = \frac{\left(\dfrac{\text{moles vapor}}{\text{moles vapor - free gas}}\right)_{actual}}{\left(\dfrac{\text{moles vapor}}{\text{moles vapor - free gas}}\right)_{saturated}} \tag{4.28}$$

Using the subscripts 1 for vapor and 2 for vapor-free gas,

$$\text{percent absolute saturation} = \frac{\left(\dfrac{n_1}{n_2}\right)_{actual}}{\left(\dfrac{n_1}{n_2}\right)_{saturated}}(100) = \frac{\left(\dfrac{p_1}{p_2}\right)_{actual}}{\left(\dfrac{p_1}{p_2}\right)_{saturated}}(100)$$

Since p_1 saturated $= p_1^*$ and $p_{tot} = p_1 + p_2$,

$$\text{percent absolute saturation} = 100 \ \frac{\dfrac{p_1}{p_{total} - p_1}}{\dfrac{p_1^*}{p_{total} - p_1^*}} = \frac{p_1}{p_1^*}\left(\frac{p_{total} - p_1^*}{p_{total} - p_1^*}\right)100 \tag{4.29}$$

Now you will recall that $p_1/p_1^* = $ relative saturation. Therefore,

$$\text{percent absolute saturation} = (\text{relative saturation})\left(\frac{p_{total} - p_1^*}{p_{total} - p_1}\right)100 \tag{4.30}$$

Percent absolute saturation is always less than relative saturation except at saturated conditions (or at zero percent saturation) when percent absolute saturation = percent relative saturation.

Dew Point.   As a partially saturated gas cools either at constant volume, as in Figure 4.19, or at constant total pressure, the noncondensable gas eventually becomes saturated with vapor, and the vapor starts to condense. The temperature at which condensation commences is the **dew point**. If the process takes place at constant total pressure, the mole fraction and partial pressure of the vapor remain constant until condensation starts.

EXAMPLE 4.23 Partial Saturation

The percent absolute humidity of air at 30°C (86°F) and a total pressure of 750 mm Hg (100 kPa) is 20%. Calculate (a) the percent relative humidity, (b) the humidity, and (c) the partial pressure of the water vapor in the air. What is the dew point of the air?

Solution

Data from the steam tables are

$$p_{H_2O}^* \text{ at } 86°F = 1.253 \text{ in. Hg } = 31.8 \text{ mm Hg} = 4.242 \text{ kPa}$$

To get the relative humidity, $p_{H_2O}/p_{H_2O}^*$, we need to find the partial pressure of the water vapor in the air. This may be obtained from

$$100(\mathcal{AH}) = 20 = \cfrac{\cfrac{p_{H_2O}}{p_{tot}-p_{H_2O}}}{\cfrac{p_{H_2O}^*}{p_{tot}-p_{H_2O}^*}} \; 100 = \cfrac{\cfrac{p_{H_2O}}{750-p_{H_2O}}}{\cfrac{31.8}{750-31.8}} \; 100$$

This equation can be solved for p_{H_2O}:

(c)
$$p_{H_2O} = 6.58 \text{ mm Hg}$$

(a)
$$\% \; \mathcal{RH} = 100 \frac{6.58}{31.8} = 20.7\%$$

(b)
$$\mathcal{H} = \frac{(MW_{H_2O})(n_{H_2O})}{(MW_{air})(n_{air})} = \frac{(18)(p_{H_2O})}{(29)(p_{air})} = \frac{18(6.58)}{29(750-6.58)} = 0.0055$$

The units are lb H_2O/lb air or kg H_2O/kg air, etc.

(c) The dew point is the temperature at which the water vapor in the air would first commence to condense, when cooled at constant total pressure, because the gas becomes completely saturated. This would be at the vapor pressure of 6.58 mm Hg, or about 5.1°C (41°F).

LOOKING BACK

In this section we explained a number of new terms all of which are used to identify partially saturated conditions in a noncondensable gas.

Key Ideas

1. The vapor in a partially saturated gas will condense when saturation is reached (the dew point).
2. Several different measures exist for partial saturation all of which are equivalent.

Key Terms

Absolute humidity (p. 342)   Partial saturation (p. 339)
Absolute saturation (p. 342)   Percent saturation (p. 340)
Dew point (p. 343)   Relative humidity (p. 340)
Humidity (p. 342)   Relative saturation (p. 340)
Molal saturation (p. 341)

Self-Assessment Test

1. A mixture of air and benzene is found to have a 50% relative saturation at 27°C and an absolute pressure of 110 kPa. What is the mole fraction of benzene in the air?
2. A TV announcer says that the dew point is 92°F. If you compress the air to 110°F and 2 psig, what is the percent absolute humidity?
3. Nine hundred forty-seven cubic feet of wet air at 70°F and 29.2 in. Hg are dehydrated. If 0.94 lb of H_2O are removed, what was the relative humidity of the wet air?

Thought Problems

1. A stirred tank containing liquid CS_2 solvent had to be cleaned because of solid residue that had accumulated in the stirrer. To avoid a possible fire and explosion, the CS_2 was pumped out and the tank blanketed with nitrogen. Then the manhole cover was removed from the top of the tank, and a worker started to remove the solid from the stirrer rod with a scraper. At this point the maintenance worker left for lunch. When he returned to complete the job, a spark caused by the scraper striking the stirrer rod started a flash fire. How could the fire have occurred despite the preventive measure of using a N_2 blanket?

2. *From:*   Marine Board of Investigation

    To: Commandant (MVI)

    Subj.:   SS *V. A. FOGG*, O. N. 244971; sinking with loss of life in Gulf of Mexico on 1 February 1972

Findings of Fact

At 1240 on February 1, 1972, the tankship *V. A. FOGG* departed Freeport, Texas, en route to the Gulf of Mexico to clean cargo tanks that carried benzene residue. The vessel was due to arrive in Galveston, Texas, at 0200, on February 2. At approximately 1545, February 1, the *V. A. FOGG* suffered multiple explosions and sank. All 39 persons aboard died as a result of this casualty. Three bodies were recovered; two of the bodies were identified and one remained unidentified. The other persons were missing and presumed dead.

You are asked your opinion of the most probable cause of the incident. What is your explanation?

3. On two different days the temperature and barometric pressure are the same. On day 1 the humidity is high, on day 2 the humidity is low. On which day is the air the most dense? Justify your answer with arguments using Dalton's law and the ideal gas law. Why does the air feel "heavier" on day 1 than on day 2?

Discussion Question

1. The news recently described a proposed new propellant system for the generation of aerosol sprays. It is based on Polygas liquid—a mixture of carbon dioxide dissolved in an acetone "carrier"—absorbed under pressure by a microporous polymer pellet.

   Each Polygas pellet (about 0.25 g) can hold up to 10 times its weight of Polygas liquid. Among the benefits claimed for Polygas are absence of damage inflicted on the environment at large, and on stratospheric ozone in particular; a negligible volatile organic compound release; nonflammability; and elimination of the potential for solvent abuse by misguided individuals.

   The system is designed to work with conventional barrier-pack aerosol technology. When the button of the aerosol can is pressed, the pressure in the compartment containing the pellet loaded with Polygas liquid drops, causing carbon dioxide to be released from the pellet. The released gas, in turn, compresses a bag or activates a piston, forcing product through the spray nozzle in the usual way. The carbon dioxide, confined within its sealed compartment, stays in the can, while the acetone carrier remains absorbed within the microstructure of the polymer pellet.

   What problems do you anticipate occurring with the proposed propellant versus the use of hydrocarbons under pressure such as ethylene, propane, or butane, or carbon dioxide by itself?

2. Toward the end of football season, the weather often turns out to be cold, snowy, and rainy. One report from Pittsburgh said that 20 groundskeepers worked around the clock in 18-hour shifts, fighting to prevent snow, gusty winds, and temperatures that dropped below zero from wrecking the delicate artificial playing surface at Three Rivers.

   Although the average football fan considers artificial turf fairly immune from the effects of the elements, at least compared to grass fields, nothing could be further from the truth. Artificial turf can become a swamp.

   To combat the elements, the grounds crew brought a 3-million BTU heating unit and a large propane tank onto the field and situated it beneath the 18-gauge, herculite tarpaulins that cover the field.

   After anchoring the tarps with weights and ropes and poles, the crew switched on the heater, blowing the tarps up into a tent of warm air that stood about 15 feet high in the middle of the field and raised the surface temperature to 30 or 35°F. The crew then began a constant vigil to prevent winds from ripping the tarps of the field.

   What would the relative humidity be beneath the tarp? *Hint:* What happens to the combustion products? Would providing a vent to the tarp be better than sealing it to the field from the viewpoint of water removal from the field?

4.7 MATERIAL BALANCES INVOLVING CONDENSATION AND VAPORIZATION

> **Your objectives in studying this section are to be able to:**
>
> **1.** Solve material balance problems involving vaporization and condensation.

LOOKING AHEAD

In this section we illustrate the solution of material balance problems involving partial saturation, condensation and vaporization. No new principles are involved.

MAIN CONCEPTS

In the examples below we apply the 10-step strategy of Table 3.1 to material balance problems in which the compositions are given via partial pressures rather than directly as mole fractions. Do you remember the drying problems in Chapter 3? The top third of Figure 4.21 represents the drying of a solid. The

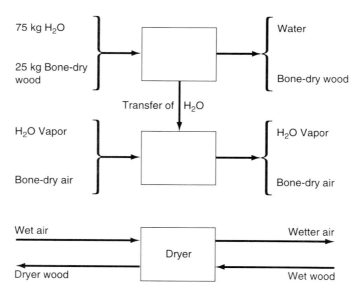

Figure 4.21   Operation of a dryer showing the transfer of water that occurs inside the process.

middle third shows the transfer of water from the solid into the air. In practice, the bottom third shows what the overall process is like in operation. You solve for the unknown variables using the given information, or information you look up in data bases, and material balances. (Humidity and saturation problems that include the use of energy balances and humidity charts are discussed in Chapter 5.)

In connection with the examples that follow, we again stress that if you know the dew point of the water in a gas, you automatically know the partial pressure of the water vapor in the gas. When a partially saturated gas is cooled at constant pressure, as in the cooling of air containing some water vapor at atmospheric pressure, the volume of the mixture may change slightly, but the **mole fractions of the gas and water vapor remain constant until the dew point is reached**. At this point water begins to condense; the gas remains saturated as the temperature is lowered. All that happens is that more water goes from the vapor into the liquid phase. **At the time the cooling is stopped, the gas is still saturated, and at its new dew point.**

The material balance problems that follow are solved using the 10-step strategy presented in Table 3.1.

EXAMPLE 4.24 Dehydration

To avoid deterioration of drugs in a container, you remove all (0.93 kg) of the H_2O from the moist air in the container at 15°C and 98.6 kPa by absorption in silica gel. The same air measures 1000 m³ at 20°C and 108.0 kPa when dry. What was the relative humidity of the moist air?

Solution

> **Steps 1, 2, 3, and 4:**   Figure E4.24 contains the known data.
> **Step 5**   Either the W or the A stream can serve as the basis.

Basis: 1000 m³ bone-dry air (BDA) at 20°C and 108.0 kPa

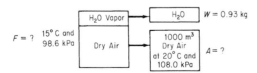

Figure E4.24

Steps 3 and 4   We first need to calculate the amounts (in kg mol) of water vapor and dry air in the original air. $W = 0.93$ kg or

$$\frac{0.93 \text{ kg } H_2O \mid 1 \text{ kg mol } H_2O}{18 \text{ kg } H_2O} = 0.0517 \text{ kg mol } H_2O$$

As for the dry air

$$\frac{1000 \text{ m}^3 \text{ BDA}}{} \left| \frac{273 \text{ K}}{293 \text{ K}} \right| \frac{108.0 \text{ kPa}}{101.3 \text{ kPa}} \left| \frac{1 \text{ kg mol}}{22.4 \text{ m}^3} \right. = 44.35 \text{ kg mol BDA}$$

Steps 5, 6, 7, 8, and 9   All of the water and all of the air in the original moist air are in the removed H_2O (0.0517 kg mol) and the BDA (44.35 kg mol), so that balances are trivial, and the problem has a unique solution.

To get the relative humidity of the moist air, we must calculate the partial pressure of the water vapor in the moist air

$$\frac{p_{H_2O}}{p_{tot}} = \frac{n_{H_2O}}{n_{tot}} = \frac{0.0517}{0.0517 + 44.35}$$

$$p_{H_2O} = 0.1147 \text{ kPa}$$

The vapor pressure at 15°C for water is 1.70 kPa, hence the fractional relative humidity of the original air was

$$\frac{0.1147}{1.70} = 0.067$$

EXAMPLE 4.25 Humidification

To condition the air in an office building in the winter, one thousand cubic meters of moist air at 101 kPa and 22°C and with a dew point of 11°C enters the process. The air leaves the process at 98 kPa with a dew point of 58°C. How many kilograms of water vapor are added to each kilogram of wet air entering the process?

Solution

Steps 1, 2, 3, and 4   The known data appear in Figure E4.25. Additional data needed are

Dew point temp. (°C)	$p^*_{H_2O}$ (mm Hg)	$p^*_{H_2O}$ (kPa)
11	9.84	1.31[†]
58	136.1	18.14[†]

[†]These values give the partial pressures of the water vapor in the initial and final gas mixtures.

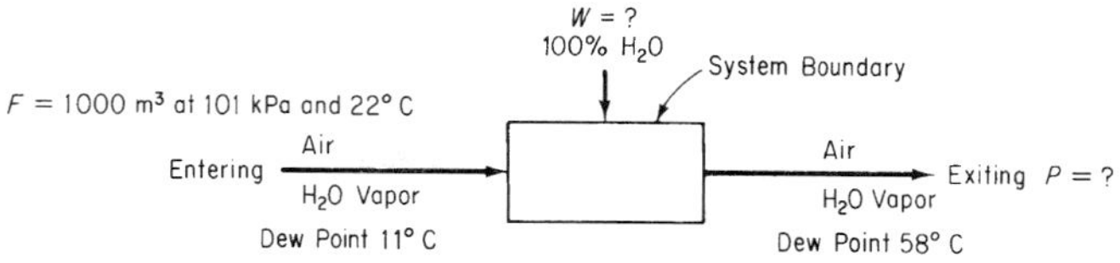

$W = ?$
100% H_2O
System Boundary

$F = 1000$ m³ at 101 kPa and 22° C

Air
Entering
H_2O Vapor
Dew Point 11° C

Air
H_2O Vapor
Dew Point 58° C
Exiting $P = ?$

Figure E4.25

The partial pressures of the water vapor are the pressures at the dew point in each case, and the dry air has a partial pressure which is the difference between the total pressure and the partial pressure of the water vapor.

Let the subscript W stand for the water vapor and DA stand for dry air:

$$\text{In: } p_{DA} = p_{tot} - p_w = 101 - 1.31 = 99.69 \text{ kPa}$$

$$\text{Out: } p_{DA} = 98 - 18.14 = 79.86 \text{ kPa}$$

Step 5   Basis is 1000 m³ at 101 kPa and 22°C. Other bases could be selected such as 101 kg mol of moist entering air or 98 kg mol of moist exit air.

Steps 6 and 7   We have two unknowns, W and P, and can make both an air and a water balance, so that the problem has a unique solution.

Steps 7, 8, and 9

$$\frac{1000 \text{ m}^3}{} \left| \frac{101 \text{ kPa}}{101.3 \text{ kPa}} \right| \frac{273 \text{ K}}{295 \text{ K}} \left| \frac{1 \text{ kg mol}}{22.4 \text{ m}^3} \right. = 41.19 \text{ kg mol wet air}$$

Dry air is a tie component.

$$DA \text{ balance: } 41.19 \left(\frac{99.69}{101} \right) = P \left(\frac{79.86}{98} \right)$$

$$P = 49.87 \text{ kg mol}$$

$$Total \text{ balance}: F + W = P$$

$$W = 49.87 - 41.19 = 8.68 \text{ kg mol } H_2O$$

Step 10   Check using the water balance.

$$Water \text{ balance} : 41.19 \left(\frac{1.31}{101} \right) + W = 49.87 \left(\frac{18.1}{98} \right)$$

$$W = 8.68 \text{ kg mol } H_2O$$

Step 9 (Continued) To calculate the kg of wet air entering

Component	kg mol	Mol. wt.	kg
Dry air	$41.19 \left(\dfrac{99.69}{101} \right)$	29	1179.0
H_2O	$41.19 \left(\dfrac{1.31}{101} \right)$	18	9.6
Total	41.19		1188.6

$$\text{Water added} = \frac{(8.68)(18)}{1188.6} = 0.131 \frac{\text{kg water}}{\text{kg wet air in}}$$

EXAMPLE 4.26 Material Balance Involving Condensation

Soil contaminated with polyaromatric hydrocarbons can be treated with hot air and steam to drive out the contaminants. If 30.0 m³ of air at 100°C and 98.6 kPa with a dew point of 30°C are introduced into the soil, and in the soil the gas cools to 14°C at a pressure of 109.1 kPa, what fraction of the water in the gas at 100°C condenses out in the soil?

Solution

Assume the system at 14°C is at equilibrium, and select a fixed volume of initial gas as the system.

   Steps 1, 2, 3, and 4   Some of the data has been placed in Figure E4.26

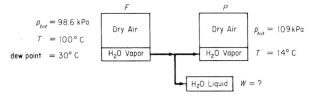

Figure E4.26

In addition we need to get data for the vapor pressure of water at 30°C and 14°C:

$$\text{at } 30°C \quad p^* = 4.24 \text{ kPa}$$

$$\text{at } 14°C \quad p^* = 1.60 \text{ kPa}$$

By subtracting the vapor pressures from the respective total pressures we can calculate the partial pressures of the air, and hence the required compositions of the entering and exit gas streams. Keep in mind that in the transition from 100°C to 30°C (where the water vapor starts to condense) the partial pressure of the water vapor remains constant because the total pressure remains constant.

The needed compositions are (in kPa)

	F	P
Air:	$98.6 - 4.24 = 94.36$	$109.1 - 1.60 = 107.5$
H_2O:	4.24	1.60
Total:	98.6	109.1

Steps 6 and 7   We have 2 unknowns, P and W, and two independent balances can be made, air and water.

Steps 8 and 9   Calculate F (in moles) first from the basis:

$$n = \frac{98.6 \text{ kPa}}{\dfrac{8.314 \text{ (kPa)(m}^3)}{\text{(kg mol)(K)}}} \left| \frac{30.0 \text{ m}^3}{373 \text{ K}} = 0.954 \text{ kg mol}$$

The material balances are

Air: $$0.954 \left(\frac{94.36}{98.6} \right) = P \left(\frac{107.50}{109.1} \right)$$

H_2O: $$0.954 \left(\frac{4.24}{98.6} \right) = P \left(\frac{1.60}{109.1} \right) + W$$

$$P = 0.927 \text{ kg mol} \qquad\qquad W = 0.0270 \text{ kg mol}$$

The initial amount of water was

$$0.954 \left(\frac{4.24}{98.6} \right) = 0.0410 \text{ kg mol}$$

and the fraction condensed was

$$\frac{0.0270}{0.0410} = 0.658$$

Step 10   A check can be made using a balance on the total moles

$$F = P + W \quad \text{or} \quad 0.954 = 0.927 + 0.0270 = 0.954$$

EXAMPLE 4.27 Vaporization of the Products of Ammonia Synthesis

After ammonia is produced from N_2 and H_2, the products are cooled and separated in a partial condenser operated at $-28°F$ and 2000 psia (the reaction pressure). Figure E4.27 illustrates the separation process and gives the known data. Calculate the

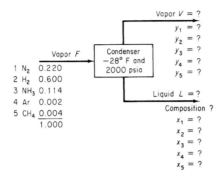

Figure E4.27

composition of the vapor and liquid streams. The equilibrium constants K_i are

N_2:	66.67	Ar:	100	NH_3:	0.015
H_2:	50	CH_4:	33.33		

Solution

Steps 1, 2, 3, and 4   Figure E4.27 contains the known data and symbols for the unknown quantities.

Step 5

$$\text{Basis: 10 lb mol} = F$$

Steps 6 and 7   There are 12 variables whose values are unknown. You can make five component balances, invoke two sum of mole fraction relations for streams L and V, and write five equilibrium relations $y_i = K_i x_i$, hence the problem has a unique solution.

Step 8   Component material balances:

$$\begin{array}{lll} N_2: & 100(0.220) = Vy_1 + Lx_1 & \text{(a)} \\ H_2: & 100(0.660) = Vy_2 + Lx_2 & \text{(b)} \\ NH_3: & 100(0.114) = Vy_3 + Lx_3 & \text{(c)} \\ Ar: & 100(0.002) = Vy_4 + Lx_4 & \text{(d)} \\ CH_4 & 100(0.004) = Vy_5 + Lx_5 & \text{(e)} \end{array}$$

Summation of mole fractions:

$$\begin{array}{lll} V: & y_1 + y_2 + y_3 + y_4 + y_5 = 1 & \text{(f)} \\ L: & x_1 + x_2 + x_3 + x_4 + x_5 = 1 & \text{(g)} \end{array}$$

Equilibrium relations:

$$N_2: \qquad y_1 = 66.67x_1 \tag{h}$$
$$H_2: \qquad y_2 = 50x_2 \tag{i}$$
$$NH_3: \qquad y_3 = 0.015x_3 \tag{j}$$
$$Ar: \qquad y_4 = 100x_4 \tag{k}$$
$$CH_4: \qquad y_5 = 33.33x_5 \tag{l}$$

Step 9   One method of solution is just to solve all 12 equations simultaneously (the component material balances are nonlinear) using one of the computer codes on the disk at the back of this book. By the following algebraic manipulations, the number of variables and equations can be reduced. Rearrange Eqs. (a)–(e). For example, for Eq. (a),

$$x_1 = \frac{100(0.220) - Vy_1}{L} \tag{m}$$

and

$$y_1 = \frac{100(0.220) - Lx_1}{V} \tag{n}$$

Substitute $y_i = K_i x_i$ into the set of equations corresponding to Eq. (m) and solve for x_i; for example,

$$x_1 = \frac{100}{V} \frac{0.220}{(L/V) + 66.67} \tag{o}$$

Introduce all the equations corresponding to Eq. (o) into Eq. (g) to get a single equation in which the unknowns are L and V.

$$\sum_{i=1}^{5} x_i \, (L, \, V) = 1 \tag{p}$$

A similar procedure applied with respect to Eqs. (n) and (f) gives another equation in which L and V are unknowns; for example,

$$y_1 = \frac{100}{V} \frac{66.7(0.220)}{66.7 + (L/V)}$$

Insert all of the y_i's into Eq. (f) to get

$$\sum_{i=1}^{5} y_i \, (L, \, V) = 1 \tag{q}$$

The overall material balance $10 = V + L$ can be combined with Eq. (p) and/or (q) to get a single equation in one unknown variable, V or L.

The solution is

| Stream | lb mol | Concentration, mole fraction | | | | |
		N_2	H_2	NH_3	Ar	CH_4
V	89.73	0.2448	0.7339	0.0147	0.0022	0.0044
L	10.27	0.0037	0.0147	0.9815	0.0000	0.0001

You could use Eq. (4.22) to solve this problem without the detailed analysis above.

We have gone over a number of examples of condensation and vaporization, and you have seen how a given amount of air at atmospheric pressure can hold only a certain maximum amount of water vapor. This amount depends on the temperature of the air, and any decrease in the temperature will lower the water-bearing capacity of the air.

An increase in pressure also will accomplish the same effect. If a pound of saturated air at 75°F is isothermally compressed (with a reduction in volume, of course), liquid water will be deposited out of the air just like water being squeezed out of a wet sponge (Figure 4.22). The vapor pressure of water is 0.43 psia at 75°F.

For example, if a pound of saturated air at 75°F and 1 atm is compressed isothermally to 4 atm (58.8 psia), almost three-fourths of the original content of water vapor now will be in the form of liquid, and the air still has a dew point of

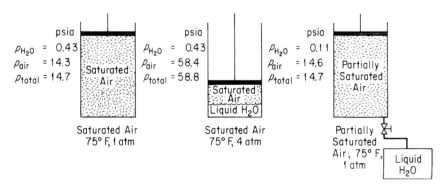

Figure 4.22   Effect of an increase of pressure on saturated air, removal of condensed water, and a return to the initial pressure at constant temperature.

75°F. Remove the liquid water, expand the air isothermally back to 1 atm, and you will find that the dew point has been lowered to about 36°F. Mathematically (1 = state at 1 atm, 4 = state at 4 atm) with $z = 1.00$ for both components:

For saturated air at 75°F and 4 atm:

$$\left(\frac{n_{H_2O}}{n_{air}}\right)_4 = \left(\frac{p^*_{H_2O}}{p_{air}}\right)_4 = \frac{0.43}{58.4}$$

For the same air saturated at 75°F and 1 atm: H_2O

$$\left(\frac{n_{H_2O}}{n_{air}}\right)_1 = \left(\frac{p^*_{H_2O}}{p_{air}}\right)_1 = \frac{0.43}{14.3}$$

Since the air is the tie element in the process, the material balance gives

$$\left(\frac{n_4}{n_1}\right)_{H_2O} = \frac{\dfrac{0.43}{58.4}}{\dfrac{0.43}{14.3}} = \frac{14.3}{58.4} = 0.245$$

24.5% of the original water will remain as vapor after compression.

After the air-water vapor mixture is returned to a total pressure of 1 atm, the following two familiar equations now apply on the basis of 58.4 mol of air and 0.43 mol of H_2O at 4 atm and 75°F:

$$p_{H2O} + p_{air} = 14.7$$

$$\frac{p_{H2O}}{p_{air}} = \frac{n_{H2O}}{n_{air}} = \frac{0.43}{58.4} = 0.00736$$

From these two relations you can find that

$$p_{H_2O} = 0.108 \text{ psia}$$

$$p_{air} = \underline{14.6}$$

$$p_{total} = 14.7 \text{ psia}$$

The pressure of the water vapor represents a dew point of about 36°F, and a percent relative humidity of

$$(100)\frac{p_{H_2O}}{p^*_{H_2O}} = \frac{0.108}{0.43}(100) = 25\%$$

LOOKING BACK

In this section we illustrated how material balances are formulated and solved when information about the compositions of streams is rendered in terms of partial pressures.

Key Ideas

1. The method of analysis for and solution of material balances involving vaporization, condensation, humidification and related problems is the same as outlined in Table 3.1.
2. Compositions of streams can be equally well expressed in terms of partial pressures as mole fractions.
3. As the vapor in a saturated gas condenses, the gas remains saturated.

Self-Assessment Test

1. Gas from a synthetic gas plant analyzes (on a dry basis) 4.8% CO_2 0.2% O_2 25.4% CO, 0.4% C_4, 12.2% H_2, 3.6% CH_4 and 53.4% N_2. The coal used analyzes 70.0% C, 6.5% H, 16.0% O, and 7.5% ash. The entering air for combustion has a partial pressure of water equal to 2.67 kPa. The barometer reads 101 kPa. Records show that 465 kg of steam is supplied to the combustion vessel per metric ton of coal fired. Calculate the dew point of the exit gas.
2. A liquid solution of pharmaceutical material to be dried is sprayed into a stream of hot gas. The water evaporated from the solution leaves with the exit gases. The solid is recovered by means of cyclone separators. Operating data are:

Inlet air:	10,000 ft³/hr, 600°F, 780 mm Hg, humidity of 0.00505 lb H_2O/lb dry air
Inlet Solution:	300 lb/hr, 15% solids, 70°F
Outlet air:	200°F, 760 mm Hg, dew point 100°F
Outlet solid:	130°F

    Calculate the composition of the outlet solid—it is not entirely dry.
3. A gas leaves a solvent recovery system saturated with benzene at 50°C and 750 mm Hg. The gas analyzes, on a benzene-free basis, 15% CO, 4% O_2, and the remainder nitrogen. This mixture is compressed to 3 atm and is subsequently cooled to 20°C. Calculate the percent benzene condensed in the process. What is the relative saturation of the final gas?

Thought Problem

1. A large fermentation tank fitted with a 2-in. open vent was sterilized for 30 minutes by blowing in live steam at 35 psia. After the steam supply was shut off, a cold liquid substrate was quickly added to the tank at which point the tank collapsed inward. What happened to cause the tank to collapse?

SUPPLEMENTARY REFERENCES

General

American Petroleum Institute, Division of Refining. Technical Data Book—Petroleum Refining, 3d ed. New York: API, 1976.

ANGUS, S., B. ARMSTRONG, and K. M. DE REUCK, eds. *International Thermodynamic Tables of the Fluid State*. IUPAC Chemical Data Series, v. 8, Pergamon Press, 1985 and periodically thereafter.

BENEDEK, P., and F. OLTI. *Computer Aided Chemical Thermodynamics of Gases and Liquids: Theory, Models and Programs*. New York: Wiley, 1985.

Compressed Gas Association, Inc. *Handbook of Compressed Gases*. New York: Van Nostrand Reinhold, 1990.

DEAN, J. A., ed. *Lange's Handbook of Chemistry*, 14th ed. New York: McGraw-Hill, 1992.

DAUBERT, T. E. and R. P. DANNERS, eds. *Data Compilation Tables of Properties of Pure Compounds*, New York: AIChE, 1985 and supplements.

DEPARTMENT OF LABOR, OCCUPATIONAL SAFETY, AND HEALTH ADMINISTRATION. *Process Safety Management of Highly Hazardous Chemicals Compliance Guidelines and Enforcement Procedures*. Washington, DC: OSHA Publishing, 1992.

EDMISTER, WAYNE C. *Applied Hydrocarbon Thermodynamics*, 2nd ed. Vol. 1, 1984; Vol. 2, 1988. Houston: Gulf Publishing Co.

REID, R. C., J. M. PRAUSNITZ, and B. E. POLING. *Properties of Gases and Liquids*, 4th ed., New York: McGraw-Hill, 1987.

SELOVER, T. B., ed. *National Standard Reference Data Service of the U.S.S.R.* New York: Hemisphere Publishing Corp., 1987.

Saturation and Vapor-Equilibria

AMMAR, M. N., and H. RENON. "The Isothermal Flash Problem: New Methods of Split Phase Calculations," *AIChE J., 33* (1987): 926.

ANTUNES, C., and D. TASSIOS. "Modified UNIFAC Model for the Prediction of Henry's Constants," *Ind. Eng. Chem. Process Des. Dev., 22* (1983): 457.

ARIT, W., et al. *Liquid-Liquid Equilibria Data Collection*, v. V. West Germany: Dechema, 1980 to date.

CARROLL, J. J. "What Is Henry's Law?" *Chem. Eng. Progr., 87* (Sept. 1991): 48.

Fluid Phase Equilibria. Amsterdam: Elsevier Science Publ. vol. 1–date (1972 to date).

FREDENSLUND, A., J. GMEHLING, and P. RASMUSSEUM. *Vapor-Liquid Equilibria Using UNIFAC*. Amsterdam: Elsevier, 1977.

GMEHLING, J. et al. *Vapor-Liquid Equilibria Data Collection*, 13 parts. Frankfurt, West Germany: Dechema, 1991.

HERSKOWITZ, M. "Estimation of Fluid Properties and Phase Equilibria." *Chem. Eng. Educ., 19* (3) (Summer 1985): 148.

KERTES, A. S., ed. in chief. *Solubility Data Series,* Oxford: Pergamon Press, continuing series.

KNAPP, H., et al. *Vapor-Liquid Equilibria for Mixtures of Low Boiling Substances*, 3 parts. Frankfurt, West Germany: Dechema, 1992 to date.

KNAPP, H., et al. *Solid-Liquid Equilibria Data Collection.* Frankfurt, West Germany: Dechema, 1987.

PRAUSNITZ, J. M., et al. *Computer Calculations for Multicomponent Vapor-Liquid and Liquid-Liquid Equilibria.* Englewood Cliffs, NJ: Prentice-Hall, 1980.

SCHWARTZENTRUBER, J., H. RENON, and S. WATANASIRI. "K-values for Non-ideal Systems: An Easier Way," *Chem. Eng. Progr,* (Mar. 1990): 118.

SHAH, N., and C. L. YAWS. "Densities of Liquids," *Chem. Eng.*, (Oct. 25, 1976): 131.

THOMPSON, G. H., K. R. BROBST, and R. W. HANKINSON. "An Improved Correlation for Densities of Compressed Liquids and Liquid Mixtures," *AIChE J., 28* (1982): 671.

Equations of State

American Society of Heating, Refrigeration, and Air Conditioning Engineers. *ASHRAE Handbook of Fundamentals.* Atlanta, GA: ASHRAE, 1989.

CHAO, K. C., and R. L. ROBINSON. *Equations of State in Engineering and Research.* Washington, DC: American Chemical Society, 1979.

ELIEZER, S., et al. *An Introduction to Equations of State: Theory and Applications.* Cambridge: Cambridge University Press, 1986.

ELLIOTT, J. R., and T. E. DAUBERT. "Evaluation of an Equation of State Method for Calculating the Critical Properties of Mixtures." *Ind. Eng. Chem. Res, 26* (1987): 1689.

GIBBONS, R. M. "Industrial Use of Equations of State." In *Chemical Thermodynamics in Industry*, ed. T. I. Barry. Oxford: Blackwell Scientific, 1985.

MATHIAS, P. M., and M. S. BENSON. "Computational Aspects of Equations of State." *AIChE J., 32* (1986): 2087.

Vapor Pressure and Liquids

AMBROSE, D. "The Correlation and Estimation of Vapor Pressures." In *Proceedings NPL Conference: Chemical Thermodynamic Data on Fluids and Fluid Mixtures*, September 1978 (p. 193). Guildford, UK: IPC Science & Technology Press, 1979.

BOUBLIK, T., V. FRIED, and E. HALA. *The Vapor Pressure of Pure Substances*, 2nd rev. ed. New York: Elsevier, 1984.

KUDELA, L., and M. J. SAMPSON. "Understanding Sublimation Technology." *Chem. Eng.*, *93* (June 23, 1986).

KUSS, E. *PVT Data and Equations of State for Liquids.* Frankfurt, West Germany: Dechema, 1988.

OHE, S. *Computer Aided Data Book of Vapor-Liquid Equilibrium.* Tokyo: Data Book Publishing Company, 1980.

Thermodynamics Research Center. *TRC Chemistry and Engineering Database-Vapor Pressure.* College Station, TX: Texas A&M University, 1993.

YAWS, C. L. *Handbook of Vapor Pressure* (several volumes). Houston, TX: Gulf Publ. Co., 1994.

YAWS, C. L. and H.-C. YANG. "To Estimate Vapor Pressures Easily." *Hydrocarbon Processing, 68,* (10) (Oct. 1989): 65.

Compressibility and Critical Phenomena

CASTILLO, C. A. "An Alternative Method for the Estimation of Critical Temperatures of Mixtures," *AIChE J., 33* (1987): 1025.

EL-GASSIER, M. M.. "Fortran Program Computes Gas Compression." *Oil Gas J.,* (July 13, 1987): 88.

GOMES, J. F. P. "Program Calculates Critical Properties." *Hydrocarbon Processing, 67* (9) (September, 1988): 110.

KEHAT, E., and I. YANIV. "Route Selection for the Computation of Physical Properties." In *Proc. Chemcomp.* Antwerp: KVI, 1982.

SIMMROCK, K. H., et al. *Critical Data of Pure Substances,* 2 parts, v. 2. Frankfurt, West Germany: Dechema, 1986.

STERBACEK, Z., B. BISKUP, and P. TAUSK. *Calculation of Properties Using Corresponding State Methods.* Amsterdam: Elsevier, 1979.

YAWS, C. L., D. CHEN, H. C. YANG, L. TAN, and D. NICO. "Critical Properties of Chemicals," *Hydrocarbon Processing, 68* (7) (July, 1989): 61.

PROBLEMS

Section 4.1

4.1. Dire warnings have appeared of the global ecological consequences of the increased ultraviolet intensity at ground level as a result of ozone depletion. One possibility for maintaining natural ozone levels until CFCs are no longer a problem is to enrich the ozone layer with man-made ozone. The average amount of ozone above each square centimeter of the Earth's surface is only about 8.1×10^{18} molecules. How thick a layer would this be at standard conditions? Ignore the air. How many liters would this be at standard conditions? If the projected depletion rate is 6% per year, how many kg would have to be replaced per day? Is this idea feasible?

4.2. You are making measurements on an air conditioning duct to test its load capacity. The warm air flowing through the circular duct has a density of 0.0796 pounds per cubic foot. Careful measurements of the velocity of the air in the duct disclose that the average air velocity is 11.3 feet per second. The inside radius of the duct is 18.0 inches.

(a) How many cubic feet of air flow through the duct in one hour?

(b) How many pounds of air pass through the duct in 24 hours?

4.3. Ventilation is an extremely important method of reducing the level of toxic air-borne contaminants in the workplace. Since it is impossible to eliminate absolutely all leakage from a process into the workplace, some method is always needed to remove toxic materials from the air in closed rooms when such materials are present in the process streams. The Occupational Safety and Health Administration (OSHA) has set the permissible exposure limit (PEL) of vinyl chloride (VC, MW = 78) at 1.0 ppm as a maximum time-weighted average (TWA) for an eight-hour workday, because VC is believed to be a human carcinogen. If VC escaped into the air, its concentration must be maintained at or below the PEL. If dilution ventilation were to be used, you can estimate the required air flow rate by assuming complete mixing in the workplace air, and then assume that the volume of air flow through the room will carry VC out with it at the concentration of 1.0 ppm.

If a process loses 10 g/min of VC into the room air, what volumetric flow rate of air will be necessary to maintain the PEL of 1.0 ppm by dilution ventilation? (In practice we must also correct for the fact that complete mixing will not be realized in a room so that you must multiply the calculated air flow rate by a safety factor, say a factor of 10.)

If the safety analysis or economics of ventilation do not demonstrate a safe concentration of VC exists, the process might have to be moved into a hood so that no VC enters the room. If the process is carried out in a hood with an opening of 30 in. wide by 25 in. high, and the "face velocity" (average air velocity through the hood opening) is 100 ft/s, what is the volumetric air flow rate at SC? Which method of treating the pollution problem seems to be the best to you? Explain why dilution ventilation is not recommended for maintaining air quality. What might be a problem with the use of a hood? The problem is adapted with permission from the publication *Safety, Health, and Loss Prevention in Chemical Processes* published by The American Institute of Chemical Engineers, New York (1990).

4.4. One liter of a gas is under a pressure of 780 mm Hg. What will be its volume at standard pressure, the temperature remaining constant?

4.5. A gas occupying a volume of 1 m^3 under standard pressure is expanded to 1.200 m^3, the temperature remaining constant. What is the new pressure?

4.6. At standard conditions a gas that behaves as an ideal gas is placed in a 4.13-L container. By using a piston the pressure is increased to 31.2 psia, and the temperature is increased to 212°F. What is the final volume occupied by the gas?

4.7. An oxygen cylinder used an a standby source of oxygen contains 1.000 ft^3 of O_2 at 70°F and 200 psig. What will be the volume of this O_2 in a dry-gas holder at 90°F and 4.00 in. H_2O above atmospheric? The barometer reads 29.92 in. Hg.

4.8. You have 10 lb of CO_2 in a 20-ft^3 fire extinguisher tank at 30°C. Assuming that the ideal gas law holds, what will the pressure gauge on the tank read in a test to see if the extinguisher is full?

4.9. Divers work as far as 500 ft below the water surface. Assume that the water temperature is 45°F. What is the molar specific volume (ft³/lb mol) for an ideal gas under these conditions?

4.10. A 25-L glass vessel is to contain 1.1 g moles of nitrogen. The vessel can withstand a pressure of only 20 kPa above atmospheric pressure (taking into account a suitable safety factor). What is the maximum temperature to which the N_2 can be raised in the vessel?

4.11. Ventilation is an extremely important method of reducing the level of toxic airborne contaminants in the workplace. Trichloroethylene (TCE) is an excellent solvent for a number of applications, and is especially useful in degreasing. Unfortunately, TCE can lead to a number of harmful health effects, and ventilation is essential. TCE has been shown to be carcinogenic in animal tests. (Carcinogenic means that exposure to the agent might increase the likelihood of the subject getting cancer at some time in the future.) It is also an irritant to the eyes and respiratory tract. Acute exposure causes depression of the central nervous system, producing symptoms of dizziness, tremors, and irregular heartbeat, plus others.

Since the molecular weight of TCE is approximately 131.5, it is much more dense than air. As a first thought, you would not expect to find a high concentration of this material above an open tank because you might assume that the vapor would sink to the floor. If this were so, then we would place the inlet of a local exhaust hood for such a tank near the floor. However, toxic concentrations of many materials are not much more dense than the air itself, so where there can be mixing with the air we may not assume that all the vapors will go to the floor. For the case of trichloroethylene OSHA has established a time-weighted average 8 hr permissible exposure limit (PEL) of 100 ppm. What is the fraction increase in the density of a mixture of TCE in air over that of air if the TCE is at a concentration of 100 ppm and at 25°C? This problem has been adapted from *Safety, Health, and Loss Prevention in Chemical Processes*, New York: American Institute of Chemical Engineers, (1990): 3.

4.12. Benzene can cause chronic adverse blood effects such as anemia and possibly leukemia with chronic exposure. Benzene has a PEL (permissible exposure limit) for an 8-hr exposure of 1.0 ppm. If liquid benzene is evaporating into the air at a rate of 2.5 cm³ of liquid/min, what must the ventilation rate be in volume per minute to keep the concentration below the PEL? The ambient temperature is 68°F and the pressure is 740 mm HG. This problem has been adapted from *Safety, Health, and Loss Prevention in Chemical Processes*, New York: American Institute of Chemical Engineers, (1990): 6.

4.13. In a letter to the editor, the author claimed that the air in an inflated basketball contributes 10% of its total mass. Is this true? Data: A properly inflated (7-9 lb is stamped on the surface of all basketballs) basketball must have a circumference of between 29.5 and 30 in., and weigh between 20 and 21 oz. Assume the shell is 0.5 cm thick.

4.14. Will the lungs of a scuba diver who is 10 m deep in the water rupture if the diver rises to the surface without breathing?

4.15. A recent newspaper report states: "Home meters for fuel gas measure the volume of gas usage based on a standard temperature, usually 60 degrees. But gas contracts when it's cold and expands when warm. East Ohio Gas Co. figures that in chilly Cleveland, the homeowner with an outdoor meter gets more gas than the meter says he does, so that's built into the company's gas rates. The guy who loses is the one with an indoor meter: If his home stays at 60 degrees or over, he'll pay for more gas than he gets. (Several companies make temperature-compensating meters, but they cost more and aren't widely used. Not surprisingly, they are sold mainly to utilities in the North.)" Suppose that the outside temperature drops from 60°F to 10°F. What is the percentage increase in the mass of the gas passed by a noncompensated outdoor meter that operates at constant pressure? Assume that the gas is CH_4.

4.16. In Bhopal, India, a Union Carbide storage tank containing methyl isocyanate (CH_3NCO, a component used for the production of insecticide) leaked, resulting in injury and death to thousands of people. The American Occupational Safety rules specify that workplace conditions are to be limited to concentrations less than 0.02 ppm of this compound. Assume ideal gas behavior. What is this concentration in mg/m^3 at 20°C and atmospheric pressure?

4.17. One of the experiments in the fuel-testing laboratory has been giving some trouble because a particular barometer gives erroneous readings owing to the presence of a small amount of air above the mercury column. At a pressure of 755 mm Hg the barometer reads 748 mm Hg, and at 740 the reading is 736. What will the barometer read when the actual pressure is 760 mm Hg?

4.18. From the known standard conditions, calculate the value of the gas law constant R in the following sets of units:

(a) cal/(g mol)(K)   (d) J/(g mol)(K)
(b) Btu/(lb mol)(°R)   (e) (cm^3)(atm)/(g mol)(K)
(c) (psia)(ft^3)/(lb mol)(°R)   (f) (ft^3)(atm)/(lb mol)(°R)

4.19. A natural gas has the following composition:

CH_4 (methane)   87%
C_2H_6 (ethane)   12%
C_3H_8 (propane)   1%

(a) What is the composition in weight percent?
(b) What is the composition in volume percent?
(c) How many m^3 will be occupied by 80.0 kg of the gas at 9°C and 600 kPa?
(d) What is the density of the gas in kg/m^3 at SC?
(e) What is the specific gravity of this gas at 9°C and 600 kPa referred to air at SC?

4.20. What is the density of propane gas (C_3H_8) in kg per cubic meter at 200 kPa and 40°C? What is the specific gravity of propane?

4.21. What is the specific gravity of propane gas (C_3H_8) at 100°F and 800 mm Hg relative to air at 60°F and 760 mm Hg?

4.22. What is the mass of 1 m³ of H_2 at 5°C and 110 kPa? What is the specific gravity of this H_2 compared to air at 5°C and 110 kPa?

4.23. Gas from the Padna Field, Louisiana, is reported to have the following components and volume percent composition. What is:
   (a) The mole percent of each component in the gas?
   (b) The weight percent of each component in the gas?
   (c) The apparent molecular weight of the gas?
   (d) Its specific gravity?

Component	Percent	Component	Percent
Methane	87.09	Pentanes	0.46
Ethane	4.42	Hexanes	0.29
Propane	1.60	Heptanes	0.06
Isobutane	0.40	Nitrogen	4.76
Normal butane	0.50	Carbon dioxide	0.40
		Total	100.00

4.24. Methane is completely burned with 20% excess air, with 30% of the carbon going to CO. What is the partial pressure of the CO in the stack gas if the barometer reads 740 mm Hg, the temperature of the stack gas is 300°F, and the gas leaves the stack at 250 ft above the ground level?

4.25. A mixture of 15 lb N_2 and 20 lb H_2 is at a pressure of 50 psig and a temperature of 60°F. Determine the following (assuming ideality for the gas mixture):
   (a) The partial pressure of each component
   (b) The specific volume of the mixture
   (c) The density of the mixture

4.26. Three thousand cubic meters per day of a gas mixture containing methane and n-butane at 21°C enters an absorber tower. The partial pressures at these conditions are 103 kPa for methane and 586 kPa for n-butane. In the absorber, 80% of the butane is removed and the remaining gas leaves the tower at 38°C and a total pressure of 550 kPa. What is the volumetric flow rate of gas at the exit? How many moles per day of butane are removed from the gas in this process? Assume ideal behavior.

4.27. A heater burns normal butane (n-C_4H_{10}) using 40.0 percent excess air. Combustion is complete. The flue gas leaves the stack at a pressure of 100 kPa and a temperature of 260°C.
   (a) Calculate the complete flue gas analysis.
   (b) What is the volume of the flue gas in cubic meter per kg mol of n-butane?

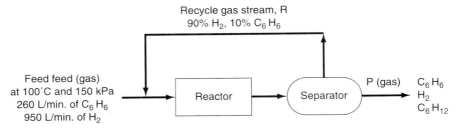

Figure P4.28

4.28. Benzene (C_6H_6) is converted to cyclohexane (C_6H_{12}) by direct reaction with H_2. The fresh feed to the process is 260 L/min of C_6H_6 plus 950 L/min of H_2 at 100°C and 150 kPa. The single pass conversion of H_2 in the reactor is 48% while the overall conversion of H_2 in the process is 75%. The recycle stream contains 90% H_2 and the remainder benzene (no cyclohexane). See Figure P4.28.
    (a) Determine the molar flow rates of H_2, C_6H_6, and C_6H_{12} in the exiting product.
    (b) Determine the volumetric flow rates of the product stream if it exits at 100 kPa and 200°C.
    (c) Determine the molar flow rate of the recycle stream, and the volumetric flow rate if the recycle stream is at 100°C and 100 kPa.

4.29. A flue gas at 790 mm Hg and 200°F has the following composition: CO_2, 12.0%; CO, 0.8%; O_2, 5.2%, and the balance N_2. What is the partial pressure of the CO_2 and the N_2?

4.30. Ammonia is burned in air with a resulting flue gas having the Orsat analysis of 4.1% O_2 and 95.9% N_2. Calculate (a) the m^3 of air used per m^3 of ammonia used, (b) the percent excess air used, and (c) the m^3/min of air at 25°C and 100.0 kPa used if the NH_3 flow rate is 20.7 m^3/min at 40°C and 125.6 kPa.

4.31. The majority of semiconductor chips used in the microelectronics industry are made of silicon doped with trace amounts of materials to enhance conductivity. The silicon initially must contain less than 20 parts per million (ppm) of impurities. Silicon rods are grown by the following chemical deposition reaction of trichlorosilane with hydrogen:

$$HSiCl_3 + H_2 \xrightarrow{\quad 1000°C \quad} 3HCl + Si$$

    Assuming that the ideal gas law applies, what volume of hydrogen at 1000°C and 1 atm must be reacted to increase the diameter of a rod 1 m long from 1 cm to 10 cm? The density of solid silicon is 2.33 g/cm^3.

4.32. The accident at the Three Mile Island nuclear plant began in March 1979 when a pressure relief valve of a water purifier stuck open. A series of operator errors and equipment failures stopped the flow of cooling water to the reactor core. As a result, the water began to boil and the core reached the temperature at which the steam reacted with the zirconium cladding of the fuel rods producing zirconium oxide (ZrO_2) and hydrogen gas (H_2). The hydrogen bubble that formed was estimated to be 28,000 L in volume, and further impeded the flow of cooling water, which was at 250°C and 6900 kPa. From the measurements cited, determine the number of kilograms of zirconium that reacted.

4.33. When a mixture of liquid hydrocarbons (C and H only) are burned in a test engine, the exhaust gas is found to contain 10.0% CO_2 on a dry basis. It is also found that the exhaust gas contains no oxygen or hydrogen (on the dry basis). Careful measurement indicates 173 ft^3 of air of 80°F and 740 mm Hg absolute enter the engine for every pound of fuel used. Calculate the mass ratio of H to C in the fuel.

4.34. A medium-Btu gas analyzes 6.4% CO_2, 0.1% O_2, 39% CO, 51.8% H_2, 0.6% CH_4, and 2.1% N_2. It enters the combustion chamber at 90°F and a pressure of 35.0 in. Hg. It is burned with 40% excess air (dry) at 70°F and an atmospheric pressure of 29.4 in. Hg, but 10% of the CO remains unburned. How many cubic feet of air are supplied per cubic foot of entering gas?

4.35. Methane containing 4.0% N_2 is flowing through a pipeline. To check the flowrate measurement, 2.83 m^3 of N_2 per minute at 22°C and 105.1 kPa are introduced into the pipeline. At some distance down the pipe, by which point complete mixing has taken place, a sample is found to contain 4.82% N_2. What is the flowrate of the pipeline gas (before the addition of N_2) in m^3/min at SC?

4.36. An incinerator produces a dry off gas of the following Orsat composition measured at 60°F and 30 inches of Hg absolute: 4.0% CO_2, 26.0% CO, 2.0% CH_4, 16.0% H_2 and 52.0% N_2. A dry natural gas of the following (Orsat) composition: 80.5% CH_4, 17.8%, C_2H_6, and 1.7% N_2 is used at the rate of 1200 ft^3/min at 60°F and 30 inches of Hg absolute to burn the inineration off gas. The final products of combustion analyze on a dry basis: 12.2% CO_2, 0.7% CO, 2.4% O_2, and 84.7% N_2.

Calculate (a) the rate of flow in ft^3/min of the incerator off gas at 60°F and 30 inches of Hg absolute on a dry basis, and (b) the rate of air flow in ft^3/min, dry, at 80°F and 29.6 inches of Hg absolute.

4.37. A gaseous mixture consisting of 50 mol % hydrogen and 50 mol % acetaldehyde (C_2H_4O) is initially contained in a rigid vessel at a total pressure of 760 mm Hg abs. The formation of ethanol (C_2H_6O) occurs according to

$$C_2H_4O + H_2 \longrightarrow C_2H_6O$$

After a time it was noted that the total pressure in the rigid vessel had dropped to 700 mm Hg abs. Calculate the degree of completion of the reaction using the following assumptions: (1) All reactants and products are in the gaseous state; and (2) the vessel and its contents were at the same temperature when the two pressures were measured.

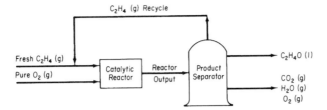

Figure P4.38

4.38. Pure ethylene (C_2H_4) and oxygen are fed to a process for the manufacture of ethylene oxide (C_2H_4O):

$$C_2H_4 + \tfrac{1}{2}O_2 \longrightarrow C_2H_4O$$

Figure P4.38 is the flow diagram for the process. The catalytic reactor operates at 300°C and 1.2 atm. At these conditions, single-pass measurements on the reactor show that 50% of the ethylene entering the reactor is consumed per pass, and of this, 70% is converted to ethylene oxide. The remainder of the ethylene consumed decomposes to form CO_2 and water.

$$C_2H_4 + 3O_2 \longrightarrow 2CO_2 + 2H_2O$$

For a daily production of 10,000 kg of ethylene oxide:
(a) Calculate the m^3/hr of total gas entering the reactor at SC if the ratio of the O_2 (g) fed to fresh C_2H_4 (g) is 3 to 2.
(b) Calculate the recycle ratio, m^3 at 10°C and 100 kPa of C_2H_4 recycled per m^{3}at SC of fresh C_2H_4 fed;
(c) Calculate the m^3 of the mixture of O_2, CO_2 and H_2O leaving the separator per day at 80°C and 100 kPa.

Section 4.2

4.39. You are asked to design a steel tank in which CO_2 will be stored at 290K. The tank is 10.4 m^3 in volume and you want to store 460 kg of CO_2 in it. What pressure will the CO_2 exert? Use the compressibility charts.

4.40. The tank cited in problem 4.39 is constructed and tested, and your boss informs you that you forgot to add a safety factor in the design of the tank. It tests out satisfactorily to 3500 kPa, but you should have added a safety factor of 3 to the design, that is, the tank pressure should not exceed (3500/3) = 1167 kPa, say 1200 kPa. How many kg of CO_2 can be stored in the tank if the safety factor is applied. Use the compressibility charts.

4.41. You have been asked to settle an argument. The argument concerns the maximum allowable working pressure (MWAP) permitted in an A1 gas cylinder. One of your coworkers says that calculating the pressure in a tank via the ideal gas law is best because it gives a conservative (higher) value of the pressure than can actually occur in the tank. The other coworker says that everyone knows the ideal gas law should not be used to calculate real gas pressures as it gives a lower value than the true pressure. Which coworker is correct?

4.42. Safe practices in modern laboratories call for placing gas cylinders in hoods or in utility corridors. In case of leaks, a toxic gas can be properly taken care of. A cylinder of CO is received from the distributor of gases on Friday with a gauge reading of 2000 psig and is placed in the utility corridor. On Monday when you are ready to use the gas you find the gauge reads 1910 psig. The temperature has remained constant at 76°F as the corridor is air conditioned so you conclude that the tank has leaked CO (which does not smell).

   (a) What has been the leak rate from the tank?

   (b) If the tank was placed in a utility corridor whose volume is 1600 ft³, what would be the minimum time that it would take for the CO concentration in the hallway to reach the "Ceiling Threshold Limit Value" (TLV-C) of 100 ppm set by the state Air Pollution Control Commission if the air conditioning did not operate on the weekend?

   (c) In the worst case, what would be the concentration of CO in the corridor if the leak continued from Friday, 3 PM to Monday, 9 AM?

   (d) Why would not either case (b) or (c) occur in practice?

4.43. Levitating solid materials during processing is the best way known to ensure their purity. High-purity materials, which are in great demand in electronics, optics, and other areas, usually are produced by melting a solid. Unfortunately, the containers used to hold the material also tend to contaminate it. And heterogeneous nucleation occurs at the container walls when molten material is cooled. Levitation avoids these problems because the material being processed is not in contact with the container.

   Electromagnetic levitation requires that the sample be electrically conductive, but with a levitation method based on buoyancy, the density of the material is the only limiting factor.

   Suppose that a gas such as argon is to be compressed at room temperature so that silicon (sp gr 2.0) just floats in the gas. What must the pressure of the argon be? If you wanted to use a lower pressure, what different gas might be selected? Is there a limit to the processing temperature for this manufacturing strategy?

4.44. To determine the temperature that occurred in a fire in a warehouse, the arson investigator noticed that the relief valve on a methane storage tank had popped open at 3000 psig, the rated value. Before the fire started, the tank was presumably at ambient conditions, about 80° F. and the gage read 1950 psig. If the volume of the tank was 240 ft³, estimate the temperature during the fire. List any assumptions you make.

4.45. When a scuba diver goes to the dive shop to have her scuba tanks filled with air, the tank is connected to a compressor and filled to about 2100 psia while immersed in a tank of water. (Why immerse the tank in water?—So that the compression of air into the tank will be approximately isothermal.)

Suppose that the tank is filled without inserting it into a water bath, and air at 27°C is compressed very rapidly from 1 atm absolute to the same final pressure. The final temperature would be about 700°C. Compute the fractional increase or decrease in the final quantity of air put in the tank relative to the isothermal case assuming that the air behaves as a pure component real gas with $p_c = 37.2$ atm and $T_c = 132.5$K.

4.46. State whether or not the following gases can be treated as ideal gases in calculations.
   (a) Nitrogen at 100 kPa and 25° C.
   (b) Nitrogen at 10,000 kPa and 25° C.
   (c) Propane at 200 kPa and 25° C.
   (d) Propane at 2000 kPa and 25° C.
   (e) Water at 100 kPa and 25° C.
   (f) Water at 1000 kPa and 25° C.

4.47. A 100-ft^3 tank contains 95.1 lb moles of a nonideal gas at 1250 atm and 440°F. The critical pressure is known to be 50 atm. What is the critical temperature? Use a generalized compressibility chart to obtain your answer.

4.48. A gas is flowing at a rate of 100,000 scfh (standard cubic feet/per hour). What is the actual volumetric gas flow rate if the pressure is 50 atm and the temperature is 600°R? The critical temperature is 40.0°F and the critical pressure is 14.3 atm.

4.49. A steel cylinder contains ethylene (C_2H_4) at 200 psig. The cylinder and gas weigh 222 lb. The supplier refills the cylinder with ethylene until the pressure reaches 1000 psig, at which time the cylinder and gas weigh 250 lb. The temperature is constant at 25°C. Calculate the charge to be made for the ethylene if the ethylene is sold at $0.41 per pound, and what the weight of the cylinder is for use in billing the freight charges. Also find the volume of the empty cylinder in cubic feet.

4.50. A gas has the following composition:

CO_2	10%
CH_4	40%
C_2H_4	50%

It is desired to distribute 33.6 lb of this gas per cylinder. Cylinders are to be designed so that the maximum pressure will not exceed 2400 psig when the temperature is 180°F. Calculate the volume of the cylinder required by Kay's method.

4.51. A gas composed of 20% ethanol and 80% carbon dioxide is at 500 K. What is its pressure if the volume per g mol is 180 cm^3/g mol?

4.52. A sample of natural gas taken at 3500 kPa absolute and 120°C is separated by chromatography at standard conditions. It was found by calculation that the grams of each component in the gas were:

Component	(g)
Methane (CH$_4$)	100
Ethane (C$_2$H$_6$)	240
Propane (C$_3$H$_8$)	150
Nitrogen (N$_2$)	50
Total	540

What was the density of the original gas sample?

4.53. A gaseous mixture has the following composition (in mol %):

C$_2$H$_4$	57
Ar	40
He	3

at 120 atm pressure and 25°C. Compare the experimental volume of 0.14 L/g mol with that computed by Kay's method.

4.54. You are asked to design a steel tank in which CO$_2$ will be stored at 290K. The tank is 10.4 m^3 in volume and you want to store 460 kg of CO$_2$ in it. What pressure will the CO$_2$ exert? Use the Redlich-Kwong equation to calculate the pressure in the tank.

$$p = \frac{RT}{\left(\hat{V} - b\right)} - \frac{a}{\hat{V}\left(\hat{V} + b\right)\left(T^{1/2}\right)}$$

where $a = 0.370$ ($K^{1/2}$)(m^6)(Pa)/(g mol)^2 and $b = 2.97$ x 10^{-5}m^3/g mol.

4.55. The tank cited in problem 4.54 is constructed and tested, and your boss informs you that you forgot to add a safety factor in the design of the tank. It tests out satisfactorily to 3500 kPa, but you should have added a safety factor of 3 to the design, that is, the tank pressure should not exceed (3500/3) = 1167 kPa, say 1200 kPa. How many kg of CO$_2$ can be stored in the tank if the safety factor is applied? Use the Redlich-Kwong equation. *Hint*: Polymath will solve the equation for you.

4.56. A graduate student wants to use van der Waals' equation to express the pressure–volume-temperature relations for a gas. Her project required a reasonable degree of precision in the p-V-T calculations. Therefore, she made the following experimental measurements with her setup to get an idea of how easy the experiment would be:

Temperature, K	Pressure, Atm	Volume, ft^3/lb mol
273.1	200	1.860
273.1	1000	0.741

Determine values for the values of constants a and b to be used in van der Waals' equation that best fit the experimental data.

4.57. The Peng-Robinson equation is listed in Table 4.2. What are the units of a, b, and α in the equation if p is in atm, $\hat{V}$ is in L/g mol, and T is in K?

4.58. The pressure gauge on an O_2 cylinder stored outside at $0°F$ in the winter reads 1375 psia. By weighing the cylinder (whose volume is 6.70 ft^3) you find the net weight, that is, the CO_2, is 63.9 lb. Is the reading on the pressure gauge correct? Use an equation of state to make your calculations.

4.59. Compare the ideal gas, van der Waals, Redlich-Kwong, and Peng-Robinson equations with experimental values for _____ as a function of temperature in 50° increments from $-50°C$ to $200°C$ for pressures from essentially 0 kPa to 3×10^4 kPa. Plot the ratio of p actual divided by p calculated from the respective equation as a function of $\hat{V}$ for the selected temperature (all on the same plot if possible).

4.60. An interesting patent (U.S. 3,718,236) explains how to use CO_2 as the driving gas for aerosol sprays in a can. A plastic pouch is filled with small compartments containing sodium bicarbonate tablets. Citric acid solution is placed in the bottom of the pouch, and a small amount of carbon dioxide is charged under pressure into the pouch as a starter propellant. As the product is CO_2 dispensed, the carbon dioxide expands, rupturing the lowest compartment membrane, thus dropping bicarb tablets into the cirtic acid. That generates more carbon dioxide, giving more pressure in the pouch, which expands and helps push out more product. (The CO_2 does not escape from the can, just the product.)

How many grams of $NaHCO_3$ are needed to generate a residual pressure of 81.0 psig in the can to deliver the very last cm^3 of product if the cylindrical can is 8.10 cm in diameter and 17.0 cm high? Assume the temperature is 25°C.

4.61. Thermal imagers detect the IR radiation from a scene and convert it into a live picture of that scene. To be sensitive enough to resolve small temperature differences, the detectors must be cryogenically cooled to make every object they view appear warm. This is typically accomplished by expanding high-pressure, pure air through a very small diameter orifice close to the detector. For example, ambient air is compressed to approximately 240 atm and a filter removes contaminants (water, carbon dioxide, hydrocarbons, and particulates) that might degrade or prevent air expansion at the cryostat. The system includes a reservoir that contains 0.3 liters of high-pressure air, ready for on demand delivery to the thermal imager. How many moles of air are in the reservoir if it is maintained at $-10°C$?

4.62. Find the molar volume (in cm^3/g mol) of propane at 375 K and 21 atm. Use the Redlich-Kwong and Peng-Robinson equations, and solve for the molar volume using the nonlinear equation solver in the disk in the pocket at the back of this book. The acentric factor for propane to use in the Peng-Robinson equation is 0.1487.

4.63. A 5-L tank of H_2 is left out overnight in Antarctica. You are asked to determine how many g moles of H_2 are in the tank. The pressure gauge reads 39 atm gauge and the temperature is $-50°C$. How many g moles of H_2 are in the tank?

Use the van der Waals and Redlich-Kwong equations of state to solve this problem. (*Hint*: The nonlinear-equation-solving program on the disk in the pocket at the back of this book will make the execution of the calculations quite easy.)

4.64. 4.00 g mol of CO_2 is contained in a 6250-cm^3 vessel at 298.15 K and 14.5 atm. Use the nonlinear equation solver on the disk in the back of the book to solve the Redlich-Kwong equation for the molar volume. Compare the calculated molar volume of the CO_2 in the vessel with the experimental value.

4.65. The fire department is inspecting the fire extinguishers in the chemical engineering building. A No. 2 gas cylinder, weighing 52.27 lb when completely evacuated, is placed on an accurate scale and filled with compressed carbon dioxide gas. When the gas in the cylinder has reached room temperature (54.5°F), the pressure in the cylinder is measured and found to be 338 psig. The capacity of the cylinder is 2.04 ft^3. Use the van der Waals equation of state to estimate what the *scale* will read in lb.

4.66. One of the biggest impediments to the realization of ambitious plans to build coal slurry pipelines has been the lack of water in the right places. Much of the desirable low-sulfur coal in the United States comes from arid or semiarid regions in the West and in the northern Great Plains. Natives of these regions get upset and also litigious when coal companies talk of using thousands of acre-feet of scarce and valuable water every year just to carry powdered coal to the Midwest or East.

One way to get around the problem is to use liquid CO_2 instead of water as the transport medium. One advantage of liquid CO_2 is that it is less viscous than water. Friction in the pipeline would be lower, so less energy would be needed to transport a given amount of coal. Also there is little if any interaction between powdered coal and liquid carbon dioxide. Because of the lower viscosity and nonreactivity of liquid carbon dioxide (compared to water), slurries can carry more coal. That means additional energy savings and also means that a smaller pipelines could provide the same coal throughput.

About 8% of the coal shipped would be needed to produce the carbon dioxide if the CO_2 is discarded at the terminal. To have liquid CO_2 at, say 100°F, would not be possible because the critical temperature of CO_2 is 304.2 K (87.6°F) while the critical pressure is 72.9 atm. However, the fluid that exists at 100°F would serve satisfactorily as a transport medium.

Calculate by the following two ways the ft^3 of CO_2 at 120°F and 1200 psia that would be needed to transport 1 lb of coal (containing 74% carbon). Assume that 0.08 lb of (additional) coal per lb of transported coal is used to make the CO_2.

(a) Compressibility factor

(b) van der Waals equation

4.67. A steel cylinder contains ethylene (C_2H_4) at 10^4 kPa gauge. The weight of the cylinder and gas is 70 kg. Ethylene is removed from the cylinder until the gauge pressure measured falls to one-third of the original reading. The cylinder and gas now weigh 52 kg. The temperature is constant at 25°C. Calculate:

(a) The fraction of the original gas (i.e., at 10^4 kPa) that remains in the cylinder at the lower pressure

(b) The volume of the cylinder in cubic meters.

Section 4.3

4.68. Methanol has been proposed as an alternate fuel for automobile engines. Proponents point out that methanol can be made from many feedstocks such as natural gas, coal, biomass, and garbage, and that it emits 45% less ozone precursor gases than does gasoline. Critics say that methanol combustion emits toxic formaldehyde and that methanol rapidly corrodes automotive parts. Moreover, engines using methanol are hard to start at temperatures below 40°F. Why are engines hard to start? What would you recommend to ameliorate the situation?

4.69. The vapor pressure of formic acid is 20 mm Hg abs at 10.3°C, 60 mm Hg abs at 32.4°C, and 100 mm Hg abs at 43.8°C. Estimate the vapor pressure at 100°C in two ways:

(a) Calculate the constants in the Antoine equation and predict p^* at 100°C;

(b) Prepare a Cox chart and make the same prediction.

The experimental value is about 750 mm Hg abs.

4.70. In a handbook the vapor pressure of solid decaborane ($B_{10}H_{14}$) is given as

$$\log_{10} p^* = 8.3647 - \frac{2642}{T}$$

and of liquid $B_{10}H_{14}$

$$\log_{10} p = 10.3822 - \frac{3392}{T}$$

The handbook also shows the melting point of $B_{10}H_{14}$ is 89.8°C. Can this be correct?

4.71. A storeman recently lost an eye when an old 1-L bottle of 100% formic acid (HCO_2H) exploded as he lifted it off the shelf in the main store of the chemistry department. He was not wearing safety glasses. Concentrated formic acid slowly decomposes to carbon monoxide and water upon prolonged storage, and the gas pressure can be sufficient to rupture sealed glass containers.

Assume that the pressure in the vapor space in the bottle above the formic acid reached 210 kPa, and that the vapor space occupied 10 cm^3. If the bottle was at room temperature (25°C), estimate the fraction of the formic acid that decomposed. Use the results of problem 4.69 to get the vapor pressure of formic acid at 25°C. Assume the water and the carbon monoxide are both in the vapor phase only (the worst case scenario). *Data*: sp gr of formic acid is 1.220 $^{20}\!\!/_4$ and the MW = 43.06. (Note that a design change in manufacturing resulted in vented caps for formic acid bottles).

4.72. Take 10 data points from the steam tables for the vapor pressure of water as a function of temperature from the freezing point to 500K, and fit the following function:

$$p^* = \exp [a + b \ln T + c (\ln T)^2 + d (\ln T)^3]$$

where p is in kPa and T is in K.

4.73. Estimate the vapor pressure of ethyl ether at 40°C given

(a) The melting point is –119.8°C at which temperature the vapor pressure is 0.0027 mm Hg, and the normal boiling point is at 34.6°C;

(b) The experimental values as follows:

p^*(kPa):	2.53	15.0	58.9
T(°C):	–40.0	–10.0	20.0

In each case compare with the experimental value.

4.74. Prepare a Cox chart for:

(a) Acetic acid vapor   (c) Ammonia
(b) Heptane   (d) Ethanol

from 0°C to the critical point (for each substance). Compare the estimated vapor pressure at the critical point with the critical pressure.

4.75. From the following data, estimate the vapor pressure of sulfur dioxide at 100°C; the actual vapor pressure is about 29 atm.

T(°C)	–10	6.3	32.1	55.5
p^* (atm)	1	2	5	10

Section 4.4

4.76. A large chamber contains dry N_2 at 27°C and 101.3 kPa. Water is injected into the chamber. After saturation of the N_2 with water vapor, the temperature in the chamber is 27°C.

(a) What is the pressure inside the chamber after saturation?

(b) How many moles of H_2O per mole of N_2 are present in the saturated mixture?

4.77. The vapor pressure of hexane (C_6H_{14}) at –20°C is 14.1 mm Hg absolute. Dry air at this temperature is saturated with the vapor under a total pressure of 760 mm Hg. What is the percent excess air for combustion?

4.78. In a search for new fumigants, chloropicrin (CCl_3NO_2) has been proposed. To be effective, the concentration of chloropicrin vapor must be 2.0% in air. The easiest way to get this concentration is to saturate air with chloropicrin from a container of liquid. Assume that the pressure on the container is 100 kPa. What temperature should be used to achieve the 2.0% concentration? From a handbook, the vapor pressure data are (T,°C; vapor pressure, mm Hg): 0, 5.7; 10, 10.4; 15, 13.8; 20, 18.3; 25, 23.8; 30, 31.1.

At this temperature and pressure, how many kg of chloropicrin are needed to saturate 100 m^3 of air?

4.79. Suppose that you place in a volume of dry gas that is in a flexible container a quantity of liquid, and allow the system to come to equilibrium at constant temperature and total pressure. Will the volume of the container increase, decrease or stay the same from the initial conditions? Suppose that the container is of a fixed instead of flexible volume, and the temperature is held constant as the liquid vaporizes. Will the pressure increase, decrease or remain the same in the container?

4.80. One way that safety enters into specifications is to specify the composition of a vapor in air that could burn if ignited. If the range of concentration of benzene in air in which ignition could take place is 1.4 to 8.0 percent, what would be the corresponding temperatures for air saturated with benzene in the vapor space of a storage tank? The total pressure in the vapor space is 100 kPa.

4.81. If sufficient water is placed in a dry gas at 15°C and 100 kPa to thoroughly saturate it, what would be the pressure after saturation, the temperature and volume remaining constant?

4.82. In a dry cleaning establishment warm dry air is blown through a revolving drum in which clothes are tumbled until all of the Stoddard solvent is removed. The solvent may be assumed to be n-octane (C_8H_{18}) and have a vapor pressure of 2.36 in. Hg at 120°F. If the air becomes saturated with octane, calculate the:
(a) Pounds of air required to evaporate one pound of octane;
(b) Percent octane by volume in the gases leaving the drum;
(c) ft^3 of inlet air required per lb of octane. The barometer reads 29.66 in. Hg.

4.83. Figure P4.83 shows a typical n-butane loading facility. To prevent explosions either (a) additional butane must be added to the intake lines (a case not shown) to raise the concentration of butane above the upper explosive limit (UEL) of 8.5% butane in air, or (b) air must be added (as shown in the figure) to keep the butane concentration below the lower explosive limit (LEL) of 1.9%. The n-butane gas leaving the water seal is at a concentration of 1.5%, and the exit gas is saturated with water (at 20°C). The pressure of the gas leaving the water seal is 120.0 kPa. How many m^3 of air per minute at 20.0°C and 100.0 kPa must be drawn through the system by the burner if the joint leakage from a single tank car and two trucks is 300cm^3/min at 20.0°C and 100.0 kPa?

Note: The emission standards for loading are 5.7 mg/L in San Francisco and 12 mg/L for ships in Louisiana.

4.84. When people are exposed to certain chemicals at relatively low but toxic concentrations, the toxic effects are only experienced after prolonged exposures. Mercury is such a chemical. Chronic exposure to low concentrations of mercury can cause permanent mental deterioration, anorexia, instability, insomnia, pain and numbness in the hands and feet, and several other symptoms. The level of mercury that can cause these symptoms can be present in the atmosphere without a worker being aware of it because such low concentrations of mercury in the air cannot be seen or smelled.

Federal standards based on the toxicity of various chemicals have been set

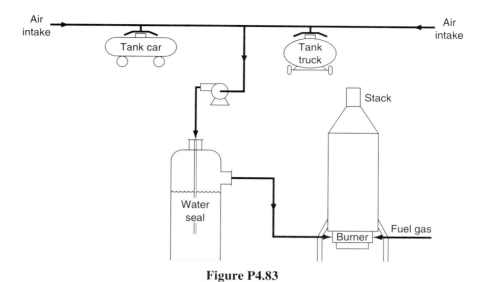

Figure P4.83

for the "Permissible Exposure Limit," or PEL. These limits are set by the Occupational Safety and Health Administration (OSHA). The PEL is the maximum level of exposure permitted in the workplace based on a time weighted average (TWA) exposure. The TWA exposure is the average concentration permitted for exposure day after day without causing adverse effects. It is based on exposure for 8 hours per day for the worker's lifetime.

The present Federal standard (OSHA/PEL) for exposure to mercury in air is 0.1 mg/m^3 as a ceiling value. Workers must be protected from concentrations greater than 0.1 mg/m^3 if they are working in areas where mercury is being used.

Mercury manometers are filled and calibrated in a small store room that has no ventilation. Mercury has been spilled in the storeroom and is not completely cleaned up because the mercury runs into cracks and cracks in the floor covering. What is the maximum mercury concentration that can be reached in the storeroom if the temperature is 20°C? You may assume that the room has no ventilation and that the equilibrium concentration will be reached. Is this level acceptable for worker exposure? *Data*: $p^*_{Hg} = 1.729 \times 10^{-4}$ kPa; the barometer reads 99.5 kPa. This problem has been adapted from the problems in the publication *Safety, Health, and Loss Prevention in Chemical Processes* published by the American Institute of Chemical Engineers, New York (1990) with permission.

4.85. When you fill your gas tank or any closed vessel, the air in the tank rapidly becomes saturated with the vapor of the liquid entering the tank. Consequently, as air leaves the tank and is replaced by liquid, you can often smell the fumes of the liquid around the filling vent such as with gasoline.

Suppose that you are filling a closed five-gallon can with benzene at 75°F. After the air is saturated, what will be the moles of benzene per mole of air expelled from the can? Will this value exceed the OSHA limit for benzene in air (currently 0.1 mg/cm^3)? Should you fill a can in your garage with the door shut in the winter?

4.86. A synthesis gas of the following composition: 4.5% CO_2, 26.0% CO, 13.0% H_2, 0.5% CH_4, and 56.0% N_2 is burned with 10% excess air. The barometer reads 98 kPa. Calculate the dewpoint of the stack gas. To prevent condensation and consequent corrosion, the stack gases must be kept well above their dewpoint.

4.87. CH_4 is completely burned with air. The outlet gases from the burner, which contain no oxygen, are passed through an absorber where some of the water is removed by condensation. The gases leaving the absorber have a nitrogen mole fraction of 0.8335. If the exit gases from the absorber are at 130°F and 20 psia, calculate:

(a) To what temperature must this gas be cooled at constant pressure in order to start condensing more water?

(b) To what pressure must this gas be compressed at constant temperature before more condensation will occur?

Section 4.5

4.88. For any flammable compound to burn in air, a minimum concentration of the flammable gas or vapor must exist at the existing temperature. The minimum concentration at which ignition will occur is called the lower flammable limit (LFL). If the flammable material is normally a liquid, the liquid must be warm enough to provide a vapor-air mixture equal in fuel concentration at least to the LFL concentration. The LFL is determined by experiment using a standard method called a "closed cup flash point" test. The "flash point" of a liquid fuel is thus the liquid temperature at which the concentration of fuel vapor in air is large enough for a flame to flash across the surface of the fuel if an ignition source is present.

The flash point of a liquid mixture in equilibrium with its vapor can be estimated by finding the temperature at which the concentration of the flammable vapors in air satisfy the following relation

$$\Sigma \, (y_i/LFL_i) = 1.0$$

where y_i is the vapor phase mole fraction of component i and LFL_i is the lower flammable limit mole fraction of component i. For ideal solutions, Raoult's law can be used to calculate the values of y_i given x_i. For nonideal solutions, more complex calculations are needed to get the vapor phase mole fractions.

The LFL values are 0.010 for *n*-octane, 0.0080 for *n*-nonane, and 0.0080 for *n*-decane in air. Estimate the flash point of a liquid mixture containing 60 mole percent *n*-octane, 15 mole percent *n*-nonane, and 25 mole percent *n*-decane. To be safe, in filling drums and handling the solution, the temperature should be well below the flash point, and the equipment grounded to discharge static electricity. This problem has been adapted from the problems in the publication *Safety, Health, and Loss Prevention in Chemical Processes* published by the American Institute of Chemical Engineers, New York (1990) with permission.

4.89. Most combustible reactions occur in the gas phase. For any flammable material to burn both fuel and oxidizer must be present, and a minimum concentration of the flammable gas or vapor in the gas phase must also exist. The minimum concentration at which ignition will occur is called the lower flammable limit (LFL). The liquid temperature at which the vapor concentration reaches the LFL can be found experimentally. It is usually measured using a standard method called a "closed cup flash point" test. The "flash point" of a liquid fuel is thus the liquid temperature at which the concentration of fuel vapor in air is large enough for a flame to flash across the surface of the fuel if an ignition source is present.

The flash point and the LFL concentration are closely related through the vapor pressure of the liquid. Thus, if the flash point is known, the LFL concentration can be estimated, and if the LFL concentration is known, the flash point can be estimated. Estimate the flash point (the temperature) of liquid *n*-decane that contains 5.0 mole percent pentane. The LFL for pentane is 1.8% and that for *n*-decane is 0.8%. Assume the propane–*n*-decane mixture is an ideal liquid. Assume the ambient pressure is 100 kPa. This problem has been adapted from *Safety, Health. and Loss Prevention in Chemical Processes*, eds. J.R. Welker and C. Springer, American Institute of Chemical Engineers, New York (1990), with permission.

4.90. Calculate the composition of the liquid that is in equilibrium with the following vapor at 66°C: ethane (10.0%), propane (25.0%), *iso*-butane (30.0%), *n*-butane (25%), and *iso*-pentane (10.0%).

4.91. You are asked to determined the maximum pressure at which steam distillation of naphtha can be carried out at 180°F (the maximum allowable temperature). Steam is injected into the liquid naphtha to vaporize it. If (1) the distillation is carried out at 160°F, (2) the liquid naphtha contains 7.8% (by weight) non-volatile impurities, and (3) if the initial charge to the distillation equipment is 1000 lb of water and 5000 lb of impure naphtha, how much water will be left in the still when the last drop of naphtha is vaporized? *Data*: For naphtha the MW is about 107, and p^* (180°F) = 460 mm Hg, p^*(160°F)= 318 mm Hg.

4.92. Late in the evening of 21 August 1986 a large volume of toxic gas was released from beneath and within Lake Nyos in the Northwest Province of Cameroon. An aerosol of water mixed with toxic gases swept down the valleys to the north of Lake Nyos, leaving more than 1,700 dead and dying people in its wake. The lake had a surface area of 1.48 km^2 and a depth of 200-250 m. It took 4 days to refill the lake, hence it was estimated to have lost about 200,000 tons of water during the gas emission. To the south of the lake and in the small cove immediately to the east of the spillway a wave rose to a height of about 25 m.

The conclusion of investigators studying this incident was that the waters of Lake Nyos were saturated with CO_2 of volcanic origin. Late in the evening of 21 August a pulse of volcanic gas—mainly CO_2 but containing some H_2S—was released above a volcanic vent in the northeast corner of the lake. The stream of bubbles rising to the surface brought up more bottom waters highly charged with CO_2 that gushed out increasing the gas flow and hence the flow of water to the surface much as a warm soda bottle overflows on release of pressure. At the surface, the release of gas transformed the accompanying water into a fine mist and sent a wave of water crashing across the lake. The aerosol of water and CO_2 mixed with a trace of H_2S swept down the valleys to the north of the lake leaving a terrible toll of injury and death in its wake.

If the solution at the bottom of the lake obeyed Henry's law, how much CO_2 was released with the 200,000 metric tons of water, and what would be the volume of CO_2 at SC in cubic meters? At 25°C the Henry's law constant is 1.7×10^3 atm/mol fr.

4.93. A natural gas has the following analysis at 200 psia.

	Mole %
Methane	78.0
Ethane	8.0
Propane	6.0
Butane	5.0
Pentane	3.0

Its temperature must be kept above what value to prevent condensation? If it were cooled, what would be the composition of the liquid that first condenses out of the gas?

4.94. Gasoline is a mixture of numerous compounds. Assume for simplicity that gasoline consists of (in mole percent) 11.0% isobutane, 10.0% isopentane, *n*-hexane 12.0%, *n*-heptane 7.0%, and the remainder less volatile compounds whose vapor pressure can be ignored for this problem. If you put gasoline in a sealed safety can at room temperature (76°F) and the can sits in your yard on a hot summer day (100°F), is it safe, that is, will the can deform or leak? It is tested to 5.0 psig. Estimate the temperature at which 5 psig would be exceeded.

4.95. Examine the statements below:
(a) "The vapor pressure of gasoline is about 14 psia at 130°F."
(b) "The vapor pressure of the system, water-furfural diacetate, is 760 mm Hg at 99.96°C."

Are the statements correct? If not, correct them. Assume the numerical values are correct.

4.96. A mixture of water (15%), dimethyl acetamide (70%), and some inert liquid that has negligible vapor pressure (15%) is to be flashed at 194°F and 1.94 psia. The K values are water: 5.26 and dimethylacetamide: 0.64. How many moles of vapor and liquid are formed at equilibrium, and what are the mole fractions of the vapor and liquid components?

4.97. A mixture of hydrocarbons consisting of an equal number of moles each of propane, normal butane, and normal pentane is to be fractionated in a well insulated column. Figure P4.97 shows the compositions and the overall material balance. The overhead product is completely condensed to a saturated liquid at 100°F. What is the temperature in the top tray of the column where the liquid vaporizes at equilibrium to form the overhead? *Hint:* Assume that the pressure in the condenser is the same as that in the top tray.

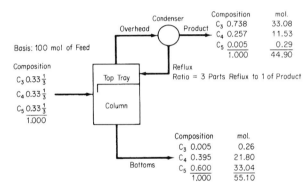

Figure P4.97

4.98. A mixture of 50% benzene and 50% toluene is contained in a cylinder at 19.34 in. Hg absolute. Calculate the temperature range in which a two phase system can exist.

4.99. Calculate the liquid (L) and vapor (V) compositions and the temperature (in °F) for the flash vaporization of the following feed stream (F) at 300 psia for a value of V/F of
(a) 0.3 and
(b) 0.6.

| Component | Mole Faction | b_i values | | |
		b_1	b_2	b_3
n-Hexane	0.05	0.2420	0.2713×10^{-2}	0.0876×10^{-4}
n-Pentane	0.20	0.5087	0.4625×10^{-2}	0.1051×10^{-4}
i-Pentane	0.20	0.5926	0.5159×10^{-2}	0.1094×10^{-4}
n-Butane	0.20	0.2126	0.7618×10^{-2}	0.1064×10^{-4}
i-Butane	0.35	0.2606	0.8650×10^{-2}	0.1017×10^{-4}

The equilibrium constants can be calculated from the relation

$$K_i = b_{1i} + b_{2i}(T - 250) + b_{3i}(T - 250)^2$$

where T is in °F.

4.100. For what temperature range can the following mixture be part liquid and part vapor at 100 psia?

	Mol %
Propane	10
i-Butane	10
n-Butane	40
i-Pentane	10
n-Pentane	30
Total	100

4.101. A mixture of hydrocarbons has the following composition in mol %:

	Mol %
CH_4	1.0
C_2H_6	15.0
C_3H_8	25.0
i-C_4H_{10}	14.0
n-C_4H_{10}	25.0
n-C_5H_{12}	20.0
Total	100.0

The mixture exists at a pressure of 100 psia and is all in the vapor phase.

(a) What is the minimum temperature at which it could be a vapor?

(b) What mole % of the vapor is condensed if the mixture is cooled to 90°F?

(c) Calculate the liquid and vapor compositions in part (b).

(d) For the conditions in part (b), what percent of the total methane present is in the vapor phase? The pentane?

(e) What temperature is required to condense 40 mol % of the total (still at 100 psia)?

(f) What temperature is required to condense 100% of the vapor?

(g) If 100,000 ft³ at S.C. of the original gas is introduced into a separator at 100 psia and 90°F, how many gallons per minute of liquid have to be handled by the pump removing liquid from this separator?

Assume that the liquid specific gravities are the same at 60°F and 90°F. Note the important effect of the 1% CH_4 on the bubble point and the insignificant effect of the pentane. Note also how the opposite is true for the dew point.

4.102. A vessel contains liquid ethanol, ethanol vapor, and N_2 gas. How many phases, components, and degrees of freedom are there according to the phase rule.

4.103. The triple point and the ice point for water differ by 0.0095°C. Why?

4.104. What is the number of degrees of freedom according to the phase rule for each of the following systems:
   (a) Solid iodine in equilibrium with its vapor
   (b) A mixture of liquid water and liquid octane (which is immiscible in water) both in equilibrium with their vapors

4.105. In the decomposition of $CaCO_3$ in a sealed container from which the air was initially pumped out, you generate CO_2 and CaO. If not all of the $CaCO_3$ decomposes at equilibrium, how many degrees of freedom exist for the system according to the Gibbs phase rule?

Section 4.6

4.106. If a gas at 60.0°C and 101.6 kPa abs. has a molal humidity of 0.030, determine:
   (a) the percentage humidity
   (b) the relative humidity
   (c) the dewpoint of the gas (in °C)

4.107. What is the relative humidity of 28.0 m^3 of wet air at 27.0°C that is found to contain 0.636 kg of water vapor?

4.108. Air at 80°F and 1 atm has a dew point of 40°F. What is the relative humidity of this air? If the air is compressed to 2 atm and 58°F, what is the relative humidity of the resulting air?

4.109. If a gas at 140°F and 30 in. Hg abs. has a molal humidity of 0.03 mole of H_2O per mole of dry air, calculate:
   (a) The percentage humidity
   (b) The relative humidity (%)
   (c) The dew point of the gas (°F)

4.110. The Weather Bureau reports a temperature of 90°F, a relative humidity of 85%, and a barometric pressure of 14.696 psia.
   (a) What is the molal humidity?
   (b) What is the humidity (weight basis)?
   (c) What is the percentage "absolute" humidity?
   (d) What is the saturation temperature or the dew point?
   (e) What is the number of degrees of superheat of the water vapor?
   (f) Determine the molal humidity and dew point if the air is heated to 105°F, the pressure remaining steady.
   (g) Determine the molal humidity and dew point if the air is cooled to 60°F, the pressure remaining steady.
   (h) What fraction of the original water is condensed at 60°F?

4.111. The Environmental Protection Agency has promulgated a national ambient air quality standard for hydrocarbons: 160 µg/m^3 is the maximum 3-hr concentration not to be exceeded more than once a year. It was arrived at by considering the role of hydrocarbons in the formation of photochemical smog. Suppose that in an exhaust gas benzene vapor is mixed with air at 25°C such that the partial pressure of the benzene vapor is 2.20 mm Hg. The total pressure is 800 mm Hg. Calculate:

(a) The moles of benzene vapor per mole of gas (total)
(b) The moles of benzene per mole of benzene free gas
(c) The weight of benzene per unit weight of benzene-free gas
(d) The relative saturation
(e) The percent saturation
(f) The micrograms of benzene per cubic meter
(g) The grams of benzene per cubic foot

Does the exhaust gas concentration exceed the national quality standard?

4.112. A constant volume bomb contains air at 66°F and 21.2 psia. One pound of liquid water is introduced into the bomb. The bomb is then heated to a constant temperature of 180°F. After equilibrium is reached, the pressure in the bomb is 33.0 psia. The vapor pressure of water at 180°F is 7.51 psia.

(a) Did all of the water evaporate?
(b) Compute the volume of the bomb in cubic feet.
(c) Compute the humidity of the air in the bomb at the final conditions in pounds of water per pound of air.

4.113. A mixture of ethyl acetate vapor and air has a relative saturation of 50% at 30°C and a total pressure of 740 mm Hg. Calculate (a) the analysis of the vapor and (b) the molal saturation.

4.114. In a gas mixture there are 0.0083 lb mol of water vapor per lb mol of dry CH_4 at a temperature of 80°F and a total pressure of 2 atm.

(a) Calculate the relative saturation of this mixture.
(b) Calculate the percentage saturation of the mixture.
(c) Calculate the temperature to which the mixture must be heated in order that the relative saturation becomes 20%.

Section 4.7

4.115. A drier must remove 200 kg of H_2O per hour from a certain material. Air at 22°C and 50% relative humidity enters the drier and leaves at 72°C and 80% relative humidity. What is the weight (in kg) of bone dry air used per hour? The barometer reads 103.0 kPa.

4.116. One thousand kg (1 metric ton) of a slurry containing 10% by weight of $CaCO_3$ are to be filtered in a rotary vacuum filter. The filter cake from the filter contains 60% water. This cake is then placed into a drier and dried to a moisture content of 9.09 kg H_2O/100kg $CaCO_3$. If the humidity of the air entering the drier is 0.005 kg of water per kg of dry air and the humidity of the air leaving the drier is 0.015 kg of water per kg of dry air, calculate:

(a) the kg of water removed by the filter
(b) the kg of wet air entering the drier

4.117. Methane gas contains CS_2 vapor in an amount such that the relative saturation at 35°C is 85%. To what temperature must the gas mixture be cooled to condense out 60% of the CS_2 by volume? The total pressure is constant at 750 mm Hg and the vapor pressure of CS_2 is given by the following relation:

$$p^* = 15.4\ T + 130$$

where p^* = vapor pressure of CS_2 in mm Hg
        T = temperature in degrees Celcius.

4.118. Recovery of solvents is essential under current regulations—they cannot be vented. Toluene has been evaporated into dry air in a process at 38.0°C and 100.0 kPa. Measurements show a percentage saturation of 50.1 percent. You want to condense 90 percent of the toluene by a combination of cooling and compressing. If the temperature is reduced to 4.0°C, to what pressure (in kPa) must the gas be compressed?

4.119. You are asked to design a silica gel drier capable of removing 500 kg of water per hour from air. The water transferred from the solid to the silica gel remains in the apparatus. Air is to be supplied to the drier at a temperature of 54.0°C, a pressure of 100 kPa, and a dew point of 30.0°C. If the air leaves the drier at a temperature of 32.2°C, a pressure of 100 kPa abs., and a dew point of 7.2°C, calculate the volume of air (in m^3 at the initial conditions) that must be supplied per hour.

4.120. A wet gas at 30°C and 100.0 kPa with a relative humidity of 75.0% was compressed to 275 kPa, and then cooled to 20°C. How many m^3 of the original gas was compressed if 0.341 kg of condensate (water) was removed from the separator that was connected to the cooler?

4.121. An absorber receives a mixture of air containing 12 percent carbon disulfide (CS_2). The absorbing solution is benzene, and the gas exits from the absorber with a CS_2 content of 3 percent and a benzene content of 3 percent (because some of the benzene evaporates). What fraction of the CS_2 was recovered?

4.122. If a liquid with a fairly high vapor pressure at room conditions is stored in a fixed size tank that breathes, that is, has a vent to the atmosphere, because of ambient temperature changes, how much loss per day occurs in g mol/m^3 of fluid under the following conditions, namely the material stored is *n*-octane at the 50°C during the day and at 10°C at night. The space above the octane consists of air and octane vapor that expands and contracts. Ignore changes in the liquid density.

4.123. Thermal pollution is the introduction of waste heat into the environment in such a way as to adversely affect environmental quality. Most thermal pollution results from the discharge of cooling water into the surroundings. It has been suggested that power plants use cooling towers and recycle water rather than dump water into the streams and rivers. In a proposed cooling tower, air enters and passes through baffles over which warm water from the heat exchanger falls. The air enters at a temperature of 80°F and leaves at a temperature of 70°F. The partial pressure of the water vapor in the air entering is 5 mm Hg and the partial pressure of the water vapor in the air leaving the tower is 18 mm Hg. The total pressure is 740 mm Hg. Calculate:

(a) The relative humidity of the air-water vapor mixture entering and of the mixture leaving the tower
(b) The percentage composition by volume of the moist air entering and of that leaving
(c) The percentage composition by weight of the moist air entering and of that leaving
(d) The percent absolute humidity of the moist air entering and leaving
(e) The pounds of water vapor per 1000 ft^3 of mixture both entering and leaving
(f) The pounds of water vapor per 1000 ft^3 of vapor-free air both entering and leaving
(g) The weight of water evaporated if 800,000 ft^3 of air (at 740 mm and 80°F) enters the cooling tower per day

4.124. A drier must evaporate 200 lb/hr of H_2O Air at 70°F and 50% relative humidity enters the drier, leaving at 140°F and 80% relative humidity. What volume of dry air is necessary per hour?

4.125. 1000 ft^3 of air, saturated with H_2O, at 30°C and 740 mm Hg, are cooled to a lower temperature and one-half of the H_2O is condensed out. Calculate:

(a) How many pounds of H_2O are condensed out
(b) The volume of dry air at 30°C and 740 mm Hg

4.126. Moist air at 25°C and 100 kPa with a dew point of 19.5°C is to be dehydrated so that during its passage through a large cold room used for food storage excess ice formation can be avoided on the chilling coils in the room. Two suggestions have been offered: (1) Cool the moist air to below the saturation temperature at 100 kPa, or (2) compress the moist air above the saturation pressure at 25°C. Calculate the saturation temperature for (1) and the total pressure at saturation for (2).

(a) If 60% of the initial water in the entering moist air has to be removed before the air enters the cold room, to what temperature should the air in process (1) be cooled?
(b) What pressure should the moist air in process (2) reach?
(c) Which process appears to be the most satisfactory? Explain.

4.127. To ensure a slow rate of drying and thereby prevent checking of the dried product, the inlet relative humidity to a drier is specified as 70% at 75°F. The air leaving the drier has a relative humidity of 90% at 70°F. If the outside air has a dew point of 40°F, what fraction of the air leaving the drier must be mixed and recycled with the outside air to provide the desired moisture content in the air fed to the drier?

4.128. A hydrocarbon fuel is burned with bone-dry air in a furnace. The stack gas is at 116 kPa and has a dew point of 47°C. The Orsat analysis of the gas shows 10 mol % carbon dioxide; the balance consists of oxygen and nitrogen. What is the ratio of hydrogen to carbon in the hydrocarbon fuel?

4.129.   Hot air that is used to dry pharmaceuticals is recycled in a closed loop to prevent the contamination of the moist material from atmospheric impurities. In the first conditioning step for the air, 5000 kg mol/hr at 105 kPa and 42°C with a 90% relative humidity are fed to a condenser to remove some of the water picked up previously in the dryer. The air exits the condenser at 17°C and 100 kPa containing 91 kg mol/hr of water vapor. Next, the air is heated in a heat exchanger to 90°C, and then goes to the dryer. By the time the air enters the dryer, the pressure of the stream has dropped to 95 kPa and the temperature is 82°C.

   (a) How many moles of water per hour enter the condenser?
   (b) What is the flow rate of the condensate water in kg/hr?
   (c) What is the dew point of the air in the stream exiting the condenser?
   (d) What is the dew point of the air in the stream entering the dryer?

4.130.   A certain gas contains moisture, and you have to remove the moisture by compression and cooling so that the gas will finally contain not more than 1% moisture (by volume). You decide to cool the final gas down to 21°C.

   (a) Determine the minimum final pressure needed.
   (b) If the cost of the compression equipment is

$$\text{cost in } \$ = (\text{pressure in psia})^{1.40}$$

and the cost of the cooling equipment is

$$\text{cost in } \$ = (350 - \text{temp. K})^{1.9}$$

is 21°C the best temperature to use? *Hint*: Look at the list of computer programs on the disk in the back of this book.

4.131.   A flue gas from a furnace leaves at 315°C and has an Orsat analysis of 16.7% CO_2, 4.1% O_2, and 79.2% N_2. It is cooled in a spray cooler and passes under slight suction through a duct to an absorption system at 32.0°C to remove CO_2 for the manufacture of dry ice. The gas at the entrance to the absorber analyzes 14.6% CO_2, 6.2% O_2, and 79.2% N_2, due to air leaking into the system. Calculate the cubic meters of air leaked in per cubic meter of gas to the absorber, both measured at the same temperature and pressure.

4.132.   A wet sewage sludge contains 50% by weight of water. A centrifuging step removes water at a rate of 100 lb/hr. The sludge is dried further by air. Use the data in Figure P4.132 to determine how much moist air (in cubic feet per hour) is required for the process shown.

4.133.   Refer to the process flow diagram (Figure P4.133) for a process that produces maleic anhydride by the partial oxidation of benzene. The moles of O_2 fed to the reactor per mole of pure benzene fed to the reactor is 18.0. All of the maleic acid produced in the reactor is removed with water in the bottom stream from the water scrubber. All of the C_6H_6, O_2, CO_2, and N_2 leaving the reactor leave in the stream from the top of the water scrubber, saturated with H_2O. Originally, the benzene contains trace amounts of a nonvolatile contaminant that would inhibit the reaction. This contaminant is removed by steam distillation in the steam still. The steam still contains liquid phases of both benzene and water

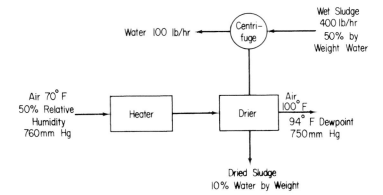

Figure P4.132

(benzene is completely insoluble in water). The benzene phase is 80% by weight, and the water phase is 20% by weight of the total of the two liquid phases in the still. Other process conditions are given in the flow sheet. Use the following vapor-pressure data:

Temperature (°F)	Benzene (psia)	Water (psia)
110	4.045	1.275
120	5.028	1.692
130	6.195	2.223
140	7.570	2.889
150	9.178	3.718
160	11.047	4.741
170	13.205	5.992
180	15.681	7.510
190	18.508	9.339
200	21.715	11.526

The reactions are

$$C_6H_6 + 4\tfrac{1}{2}O_2 \longrightarrow \begin{matrix} CH-C \underset{\displaystyle OH}{\overset{\displaystyle O}{}} \\ \| \\ CH-C \underset{\displaystyle OH}{\overset{\displaystyle O}{}} \end{matrix} + 2CO_2 + H_2O \tag{1}$$

$$C_6H_6 + 7\tfrac{1}{2}O_2 \longrightarrow 6CO_2 + 3H_2O \tag{2}$$

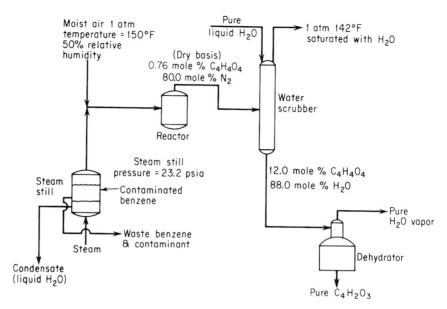

Figure P4.133

Calculate:

(a) The moles of benzene undergoing reaction (2) per mole of benzene feed to the reactor

(b) The pounds of H_2O removed in the top stream from the dehydrator per pound mole of benzene feed to the reactor

(c) The composition (mole percent, wet basis) of the gases leaving the top of the water scrubber

(d) The pounds of pure liquid H_2O added to the top of the water scrubber per pound mole of benzene feed to the reactor

ENERGY BALANCES

5

In this chapter we take up the second prominent topic in this book, energy balances. To provide publicly acceptable, effective, and yet economical conversion of our resources into energy and to properly utilize the energy so generated, you must understand the basic principles underlying the generation, uses, and transformation of energy in its different forms. Figure 5.1 shows the forecast of world energy demand to the year 2020. The answer to questions such as Is thermal pollution inherently necessary? What is the most economic source of fuel? What can be done with waste heat? How much steam at what temperature and pressure are needed to heat a process? and related questions can only arise from an understanding of the treatment of energy transfer by natural processes or machines. As an example, examine Figure 5.2 and try to answer the question: What can be done economically to reduce the loss of energy rejected as heat to the surroundings? Can you offer reasonable suggestions at this stage in your professional life?

In this chapter we discuss energy balances together with the accessory background information needed to understand and apply them correctly. Our main attention will be devoted to heat, work, enthalpy, and internal energy, and energy balances associated with chemical reaction.

5.1 CONCEPTS AND UNITS

Your objectives in studying this section are to be able to:

1. Define or explain the following terms: energy, system, closed system, nonflow system, open system, flow system, surroundings, property, extensive property, intensive prop-

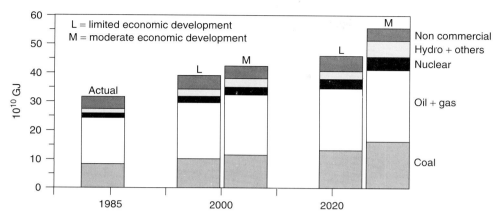

Figure 5.1   Forecast of world energy demand. (Source: Conservation and Studies Committee of the World Energy Conference, Montreal, 1989.)

erty, state, heat, work, kinetic energy, potential energy, internal energy, enthalpy, initial state, final state, point (state) function, state variable, cyclical process, and path function.

2. Select a system suitable for solving a problem, either closed or open, steady or unsteady state, and fix the system boundary.

3. Distinguish among potential, kinetic, and internal energy.

4. Convert energy in one set of units to another set.

5. State the energy balance in words and write the balance in mathematical symbols for closed and open systems.

LOOKING AHEAD

In this section we discuss each term that will be used in the energy balance, and then proceed to formulate the energy balance for both closed and open systems.

MAIN CONCEPTS

To use the energy balance, you have to express the balance as an equation. Each term of the energy balance has to be written in mathematical symbols so that you can simplify the equation as appropriate, and then you can carry out the necessary calculations. The units of each term in the equation must be consistent as is the case with mass balances. In this book we use mainly the joule (J) and the British

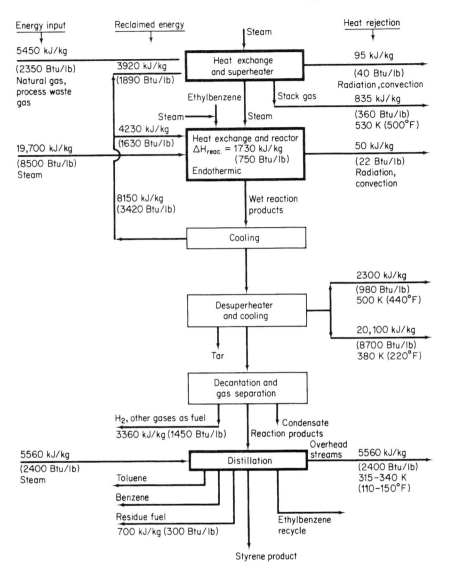

Figure 5.2 Energy balance in styrene production in the United States: 40% conversion of ethylbenzene to products, 90% selectivity to styrene. (From J. T. Reding and B. P. Shepherd. *Energy Consumption: The Chemical Industry*, Report EPA-650/2-75-032a. Washington, DC: Environmental Protection Agency, April 1975.)

thermal unit (Btu), but occasionally other units. Some of the older data sources use the thermochemical calorie (equal to 4.184 J). In the nutrition information listed on foods, the "calorie" is really a kcal, that is, a hamburger reported to contain "500 calories" really contains 500 kcal.

You have to be precise in analyzing the terms in an energy balance, hence we first will review certain terms that have been explained in earlier chapters that occur repeatedly in this chapter; these terms are summarized below with some elaboration in view of their importance.

System. Any arbitrarily specified mass of material or segment of apparatus to which we would like to devote our attention. A system must be defined by surrounding it with a system boundary. The system boundary does not have to coincide with the walls of a vessel. A system enclosed by a boundary through which no transfer of mass occurs is termed a **closed system,** or **nonflow system,** in distinction to an **open system,** or **flow system**, in which the exchange of mass is permitted. All the mass or apparatus external to the defined system is termed the **surroundings**. You should always draw similar boundaries in the solution of your problems, since this step will fix clearly the system and surroundings.

Property. A characteristic of material that can be measured, such as pressure, volume, or temperature—or calculated, if not directly measured, such as certain types of energy. The properties of a system depend on its condition at any given time and not on what has happened to the system in the past.

An **extensive property** (variable, parameter) is one whose value is the sum of the values of each of the subsystems comprising the whole system. For example, a gaseous system can be divided into two subsystems that have volumes or masses different from the original system. Consequently, mass or volume is an extensive property.

An **intensive property** (variable, parameter) is one whose value is not additive and does not vary with the quantity of material in the subsystem. For example, temperature, pressure, density (mass per volume), and so on, do not change in the parts of the system if the system is sliced in half or if the halves are put together.

Two properties are **independent** of each other if at least one variation of state for the system can be found in which one property varies while the other remains fixed. The number of independent intensive properties necessary and sufficient to fix the state of the system can be ascertained from the phase rule (see Sec. 4.5).

State. The given set of properties of material at a given time. The state of a system does not depend on the shape or configuration of the system but only on its intensive properties such as temperature, pressure, and composition.

Now that we have reviewed the concepts of system, property, and state, we can discuss the various types of energy with which we will be involved in this chapter. As you know, energy exists in many different forms. What forms are important? We shall consider here six quantities: work, heat, kinetic energy, potential energy, internal energy, and enthalpy. You probably have encountered many of these terms before. Unfortunately, some of the terms described below are used loosely in our ordinary conversation and writing, and thus have different connotations than those presented below. You may have the impression that you understand the terms from long acquaintance—be sure that you really do. For effective learning, you must feel comfortable with them.

5.1-1 Six Types of Energy

The first two types of energy we discuss are **energy transfer** between the system and surroundings without any accompanying mass transfer.

Work. Work (W) is a term that has wide usage in everyday life (such as "I am going to work"), but has a specialized meaning in connection with energy balances. Work is a form of energy that represents a **transfer** between the system and surroundings. Work cannot be stored. *Work is positive when done on the system*. For work to occur because of a mechanical force, the boundary in a system must move.

$$W = \int_{\text{state 1}}^{\text{state 2}} \mathbf{F} \cdot d\mathbf{s} \tag{5.1}$$

where $\mathbf{F}$ is an external force in the direction of $\mathbf{s}$ acting on the system (or a system force acting on the surroundings). The amount of mechanical work done by or on a system can be difficult to calculate because (a) the displacement may not be easy to define, and (b) the integration of $\mathbf{F} \cdot d\mathbf{s}$ as shown in Eq. (5.1) does not necessarily result in an equal amount of work being done by the system or on the system. In this text, the symbol W refers to the total work done over a period of time, and *not* the rate of work.

Note that unless the process (or path) under which work is carried out is specified from the initial to the final state of the system, you are not able to calculate the value of the work done by integrating Eq. (5.1) In other words, work done in going between the initial and final states can have *any* value, depending on the path taken. **Work is therefore called a path function**, and the value of W depends on the initial state, the path, and the final state of the system as illustrated in the next example.

EXAMPLE 5.1 Calculation of Mechanical Work by a Gas on a Piston

Suppose that an ideal gas at 300 K and 200 kPa is enclosed in a cylinder by a frictionless piston, and the gas slowly forces the piston so that the volume of gas expands from 0.1 to 0.2m³. Examine Figure E5.1a. Calculate the work done by the gas on the piston (the only part of the boundary that moves) if two different paths are used to go from the initial state to the final state:

Path A: the expansion occurs at constant pressure ($p = 200$ kPa)
Path B: the expansion occurs at constant temperature ($T = 300$ K)

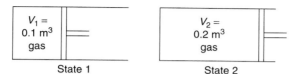

| State 1 | State 2 | **Figure E5.1a** |

Solution

As explained in more detail in Sec. 5.5, the piston must be frictionless and the process be ideal for the following calculations to be valid. The system is the gas. The mechanical work done by the system on the piston is

$$W = -\int_{state\ 1}^{state\ 2} \frac{\mathbf{F}}{A} \cdot A\, d\mathbf{s} = -\int_{V_1}^{V_2} p\, dV$$

because p is exerted normally on the piston face. Note that by definition, work done by the system is *negative*.

Path A

$$W = -p\int_{V_1}^{V_2} dV = -p(V_2 - V_1)$$

$$= -\frac{200 \times 10^3\ \text{Pa}}{} \left| \frac{1\ \dfrac{\text{N}}{\text{m}^2}}{1\ \text{Pa}} \right| 0.1\ \text{m}^3 \left| \frac{1\ \dfrac{\text{J}}{\text{m}}}{1\ \text{N}} \right| = -20\ \text{kJ}$$

Path B

$$W = -\int_{V_1}^{V_2} \frac{nRT}{V}\, dV = -nRT \ln\left(\frac{V_2}{V_1}\right)$$

$$n = \frac{200\ \text{kPa} \left| 0.1\ \text{m}^3 \right|}{\left| 300\ \text{K} \right| 8.314(\text{kPa})(\text{m}^3)} (\text{kg mol})(\text{K}) = 0.00802\ \text{kg mol}$$

$$W = -\frac{0.00802\ \text{kg mol}}{\left| (\text{kg mol})(\text{K}) \right|} \frac{8.314\ \text{kJ}}{} \frac{300\ \text{K}}{} \ln 2 = -20 \ln 2 = -13.86\ \text{kJ}$$

Figure E5.1b shows the two integrals as areas in a $p - V$ plot.

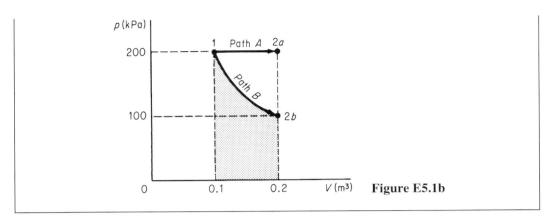

Figure E5.1b

Work done by or to a system can be classified in other categories besides *mechanical work.* For instance, if a voltage is applied to a resistance in a system that results in current flow that in turn increases the internal energy of the system, we would classify the energy transferred to the system under the voltage potential as *"electrical work."* If the system turns the shaft of a motor or compressor, the work done is often called *"shaft work."* And so on.

Heat. In a discussion of *heat* we enter an area in which our everyday use of the term may cause confusion, since we are going to use heat in a very restricted sense when we apply the laws governing energy changes. Heat (Q) is commonly defined as that part of the total energy flow across a system boundary that is caused by a temperature difference between the system and the surroundings. Engineers say "heat" meaning "heat flow." Heat is not stored nor created. *Heat is positive when transferred to the system.* Heat may be transferred by conduction, convection, or radiation. Heat, as is work, is a path function. To evaluate heat transfer *quantitatively,* unless given a priori, you must apply the energy balance that is discussed below, or use an empirical formula to estimate the heat transfer such as (for a steady-state process)

$$\dot{Q} = UA\,\Delta T \qquad\qquad (5.2)$$

where $\dot{Q}$ is the rate of heat transfer, A is the area for heat transfer, ΔT is the effective temperature difference between the system and its surroundings, and U is an empirical coefficient determined from experimental data for the equipment involved. In this text we use the symbol Q to denote the total amount of heat transferred in a time period, and not the rate of heat transfer.

Kinetic energy. Kinetic energy (K) is the energy a system possesses because of its velocity relative to the surroundings at rest. Kinetic energy may be calculated from the relation

$$K = \frac{1}{2}mv^2 \qquad (5.3a)$$

or

$$\hat{K} = \frac{1}{2}v^2 \qquad (5.3b)$$

where the superscript caret (^) refers to the energy per unit mass and not the total kinetic energy, as in Eq. (5.3a).

EXAMPLE 5.2 Calculation of Kinetic Energy

Water is pumped from a storage tank into a tube of 3.00 cm inner diameter at the rate of 0.001 m³/s. See Figure E5.2. What is the specific kinetic energy of the water?

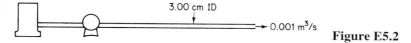

3.00 cm ID

0.001 m³/s

Figure E5.2

Solution

Basis : 0.001 m³/s of water

Assume that $\rho = \dfrac{1000 \text{ kg}}{\text{m}^3}$ $r = \dfrac{1}{2}(3.00) = 1.50$ cm

$$v = \frac{0.001 \text{ m}^3}{\text{s}} \left| \frac{}{\pi(1.50)^2 \text{ cm}^2} \right| \left(\frac{100 \text{ cm}}{1 \text{ m}}\right)^2 = 1.415 \text{ m/s}$$

$$\hat{K} = \frac{1}{2}\left(\frac{1.415}{\text{m/s}}\right)^2 \left| \frac{1 \text{ N}}{1\frac{(\text{kg})(\text{m})}{\text{s}^2}} \right| \frac{1 \text{ J}}{1(\text{N})(\text{m})} = 1.00 \text{ J/kg}$$

Potential energy.   Potential energy (P) is energy the system possesses because of the body force exerted on its mass by a gravitational or electromagnetic field with respect to a reference surface. Potential energy for a gravitational field can be calculated from

$$P = mgh \qquad (5.4a)$$

or

$$\hat{P} = gh \qquad (5.4b)$$

where h is the distance from the reference surface and where the symbol ($\hat{}$) again means potential energy per unit mass. The measurement of h takes place to the center of mass of a system. Thus, if a ball suspended inside a container somehow is permitted to drop from the top of the container to the bottom, and in the process raises the thermal energy of the system slightly, we do not say work is done on the system but instead say that the potential energy of the system is reduced (slightly).

EXAMPLE 5.3 Calculation of Potential Energy

Water is pumped from one reservoir to another 300 ft away, as shown in Figure E5.3. The water level in the second reservoir is 40 ft above the water level of the first reservoir. What is the increase in specific potential energy of the water in Btu/lb$_m$?

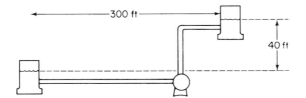

Figure E5.3

Solution

Let the water level in the first reservoir be the reference plane. Then $h = 40$ ft.

$$\hat{P} = \frac{32.2 \text{ ft}}{s^2} \left| \frac{40 \text{ ft}}{} \right| \frac{}{\frac{32.2 \ (lb_m)(ft)}{(lb_f)(s^2)}} \left| \frac{1 \text{ Btu}}{778.2(ft)(lb_f)} \right| = 0.0514 \text{ Btu/lb}_m$$

Internal energy. Internal energy (U) is a macroscopic measure of the molecular, atomic, and subatomic energies, all of which follow definite microscopic conservation rules for dynamic systems. Because no instruments exist with which to measure internal energy directly on a macroscopic scale, internal energy must be calculated from certain other variables that can be measured macroscopically, such as pressure, volume, temperature, and composition.

To calculate the internal energy per unit mass ($\hat{U}$) from the variables that can be measured, we make use of a special property of internal energy, namely that it is an exact differential (because it is a *point* or *state* property, a matter to be described shortly) and, for a pure component, can be expressed in terms of just two intensive variables according to the phase rule for one phase:

$$F = C - \mathcal{P} + 2 = 1 - 1 + 2 = 2$$

Custom dictates the use of temperature and specific volume as the variables. If we say that $(\hat{U})$ is a function of T and V,

$$\hat{U} = \hat{U}(T, V)$$

by taking the total derivative, we find that

$$d\hat{U} = \left(\frac{\partial \hat{U}}{\partial T}\right)_{\hat{V}} dT + \left(\frac{\partial \hat{U}}{\partial \hat{V}}\right)_{T} d\hat{V} \tag{5.5}$$

By definition $(\partial \hat{U}/\partial T)_{\hat{V}}$ is the "heat capacity" at constant volume, given the special symbol C_V. For all practical purposes in this text the term $(\partial \hat{U}/\partial \hat{V})_T$ is so small that the second term on the right-hand side of Eq. (5.5) can be neglected. Consequently, *changes in the internal energy* can be computed by integrating Eq. (5.5) as follows:

$$\hat{U}_2 - \hat{U}_1 = \int_{T_1}^{T_2} C_V dT \tag{5.6}$$

Note that you can only calculate differences in internal energy, or calculate the internal energy relative to a reference state, *but not absolute values* of internal energy.

EXAMPLE 5.4 Internal Energy

An entrance examination for graduate school asked the following two multiple choice questions:

(a) The internal energy of a solid is equal to the
   (1) absolute temperature of the solid
   (2) total kinetic energy of its molecules
   (3) total potential energy of its molecules
   (4) sum of the kinetic and potential energy of its molecules
(b) The internal energy of an object depends on its
   (1) temperature, only
   (2) mass, only
   (3) phase, only
   (4) temperature, mass, and phase

Which answers would you chose?

Solution

Neither of these questions can be answered with the choices given. With respect to question (a), internal energy itself cannot be evaluated—only its change can be. With respect to (b), the specific internal energy *change* depends on temperature, phase, and specific volume while the total internal energy *change* depends also on the mass.

The last class of energy we are going to discuss is enthalpy.

Enthalpy. In applying the energy balance you will encounter a variable which is given the symbol H and the name *enthalpy* (pronounced en'-thal-py). This variable is defined as the combination of two variables which will appear very often in the energy balance:

$$H = U + pV \tag{5.7}$$

where p is the pressure and V is the volume.

To calculate the enthalpy per unit mass, we use the property that the enthalpy is also an exact differential. For a pure substance, the enthalpy for a single phase can be expressed in terms of the temperature and pressure (a more convenient variable than specific volume) alone. If we let

$$\hat{H} = \hat{H}(T, p)$$

by taking the total derivative of $\hat{H}$ we can form an expression corresponding to Eq. (5.5):

$$d\hat{H} = \left(\frac{\partial \hat{H}}{\partial T}\right)_p dT + \left(\frac{\partial \hat{H}}{\partial p}\right)_T dp \tag{5.8}$$

By definition $(\partial \hat{H}/\partial T)_p$ is the heat capacity at constant pressure, and is given the special symbol C_p. For most practical purposes $(\partial \hat{H}/\partial T)_T$ is so small at modest pressures that the second term on the right-hand side of Eq. (5.8) can be neglected. Changes in enthalpy can then be calculated by integration of Eq. (5.8) as follows:

$$\hat{H}_2 - \hat{H}_1 = \int_{T_1}^{T_2} C_p\, dT \tag{5.9}$$

However, in processes operating at high pressures, the second term on the right-hand side of Eq. (5.8) cannot necessarily be neglected, but must be evaluated from experimental data. Consult the references at the end of the chapter for details. One property of ideal gases that should be noted is that their enthalpies and internal energies are functions of temperature only, and are not influenced by changes in pressure or specific volume, respectively.

As with internal energy, *enthalpy has no absolute value*; only changes in

enthalpy can be calculated. Often you will use a reference set of conditions (perhaps implicitly) in computing enthalpy changes. For example, the reference conditions used in the steam tables are liquid water at 0°C (32°F) and its vapor pressure. This does not mean that the enthalpy is actually zero under these conditions, but merely that the enthalpy has arbitrarily been assigned a value of zero at these conditions. In computing enthalpy changes, the reference conditions cancel out as can be seen from the following:

initial state of system   *final state of system*

$$\text{enthalpy} = \hat{H}_1 - \hat{H}_{\text{ref}} \qquad \text{enthalpy} = \hat{H}_2 - \hat{H}_{\text{ref}}$$

$$\textit{net enthalpy change} = (\hat{H}_2 - \hat{H}_{\text{ref}}) - (\hat{H}_1 - \hat{H}_{\text{ref}}) = \hat{H}_2 - \hat{H}_1$$

Point, or state, functions.   The variables specific enthalpy and specific internal energy (as well as temperature, pressure, and density) are called *point functions*, or *state variables*, meaning that their values depend *only* on the state of the material (temperature, pressure, phase, and composition), and *not* on how the material reached that state. Figure 5.3 illustrates the concept of a state variable. In proceeding from state 1 to state 2, the actual process conditions of temperature and pressure are shown by the wiggly line. However, you may calculate $\Delta\hat{H}$ by route *A* or *B*, or any other route, and still obtain the same net enthalpy change as for the route shown by the wiggly line. The change of enthalpy depends only on the initial and final states of the system. A process that proceeds first at constant pressure and then at constant temperature from 1 to 2 will yield exactly the same $\Delta\hat{H}$ as one that takes place first at constant temperature and then at constant pressure, as long as the end point is the same. The concept of the point function is the same as that of an airplane passenger who plans to go straight to Chicago from New York but is detoured because of bad weather by way of Cincinnati. His trip costs him the same whatever way he flies, and he eventually arrives at his destination. The fuel consumption of the plane may vary considerably, and in analogous fashion heat (*Q*) or work (*W*), the two "path" functions with which we deal,

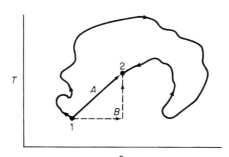

Figure 5.3   Point function

may vary depending on the specific path chosen, while $\Delta \hat{H}$ is the same regardless of path. If the plane were turned back by mechanical problems and landed at New York, the passenger might be irate, but at least could get a refund. Thus $\Delta \hat{H} = 0$ if a cyclical process is involved which goes from state 1 to 2 and back to state 1 again. All the intensive properties we shall work with, such as $\hat{P}$, T, $\hat{U}$, p, $\hat{H}$, and so on, are state variables and exhibit this same feature.

Now that we have discussed all of the terms in the energy balance, we are prepared to write down what the balance is in symbols.

5.1-2 The General Energy Balance

Rather than use the words the "law of the conservation of energy" in this book, we will use "energy balance" to avoid confusion with the colloquial use of the words "energy conservation," that is, reduction of energy waste or increased efficiency of energy utilization. The energy balance is such a fundamental physical principle that we invent new classes of energy to make sure that the equation indeed does balance. Equation (5.10) as written below is a generalization of the results of numerous experiments on relatively simple special cases. We universally believe the equation is valid because we cannot find exceptions to it in practice, taking into account the precision of the measurements.

It is necessary to keep in mind two important points as you read what follows. First, we examine only systems that are homogeneous, not charged, and without surface effects, in order to make the energy balance as simple as possible. Second, the energy balance is developed and applied from the macroscopic viewpoint (overall about the system) rather than from a microscopic viewpoint (i.e., an elemental volume within the system).

The concept of the macroscopic energy balance is similar to the concept of the macroscopic material balance, namely,

$$\left\{ \begin{array}{c} \text{accumulation of} \\ \text{energy within the} \\ \text{system} \end{array} \right\} = \left\{ \begin{array}{c} \text{transfer of energy} \\ \text{into system through} \\ \text{system boundary} \end{array} \right\} - \left\{ \begin{array}{c} \text{transfer of energy out} \\ \text{of system through} \\ \text{system boundary} \end{array} \right\}$$

$$+ \left\{ \begin{array}{c} \text{energy genera-} \\ \text{tion within} \\ \text{system} \end{array} \right\} - \left\{ \begin{array}{c} \text{energy con-} \\ \text{sumption} \\ \text{within system} \end{array} \right\} \qquad (5.10)$$

Equation (5.10) can be applied to a single piece of equipment or to a complex plant such as that shown in Figure 5.2.

While the formulation of the energy balance in words as outlined in Eq. (5.10) is easily understood and rigorous, you will discover in later courses that to express each term of Eq. (5.10) in mathematical notation may require certain simplifications to be introduced, a discussion of which is beyond our scope here, but they have a quite minor influence on our final balance.

Thus the energy generation and consumption terms will not play a role in this book because they involve processes such as radioactive decay or the slowing down of neutrons in a process.

We will first translate Eq. (5.10) into mathematical symbols for closed (batch) systems, and then extend the equation so that it applies also to open (flow) systems.

Energy Balances for Closed Systems (without Chemical Reaction).

Figure 5.4 illustrates a very general process. We will for the moment ignore the mass flow into and out of the system, and assume that the system is a closed one. Then we can make an inventory of the energy quantities in the system corresponding to the accumulation term in Eq. (5.10), and also take into account the two transfer terms in the equation using the symbols Q and W, and using the notation in Figure 5.4. Examine Figure 5.5. As to the notation, the subscripts t_1 and t_2 refer to the initial and final times for the period over which the accumulation is to be evaluated, with $t_2 > t_1$. The other notation has been explained in Sec. 5.1-1.

As is commonly done, we have split the total energy (E) associated with the mass *in the system* into three categories: internal energy (U), kinetic energy (K),

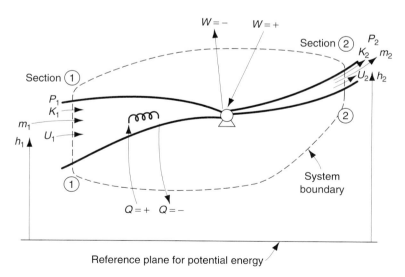

Figure 5.4   General process showing the system boundary and energy transport across the boundary.

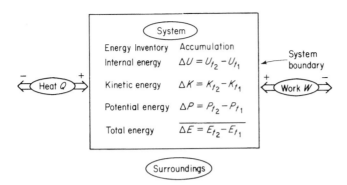

Figure 5.5 Quantities in the energy balance for a closed system.

and potential energy (P). Energy transported across the system boundary can be transferred by two modes: heat (Q) and work (W). Other classes of energy could be split out if significant. All of these terms represent quantities integrated over the time period from t_1 to t_2 such as

$$\Delta U = \int_{U_{t_1}}^{U_{t_2}} dU \qquad Q = \int_{t_1}^{t_2} \dot{Q}\ dt \qquad W = \int_{t_1}^{t_2} \dot{W}\ dt$$

Note that Q and W have been defined as the *net* transfer of heat and work, respectively, between the system and the surroundings. Remember that Q is + when heat is transferred to the system, and W is + when done on the system. The symbol Δ is a difference operator signifying final value minus initial value (in time here).

Given the notation outlined above, Eq. (5.10) in symbols becomes

$$\Delta E = E_{t_2} - E_{t_1} = \Delta U + \Delta P + \Delta K = Q + W \qquad (5.11)$$

We give an alternative formulation of the energy balance in terms of differentials in Chapter 7, where the emphasis is placed on the instantaneous rate of change of energy of a system rather than just the initial and final states of the system.

Now for a word of warning: Be certain you use *consistent units* for all terms; in the American Engineering system the use of foot-pound, for example, and Btu in different places in Eq. (5.11) is a common error for the beginner. Examples of the application of the equation will be found in Sec. 5.3 below.

Energy Balances for Open Systems (without Chemical Reaction). We now extend Eq. (5.11) to include the energy associated with the mass flowing into and out of the system as in Figure 5.4. Table 5.1 lists the notation used in the equation. The overlay caret (^) will mean the energy term per unit

TABLE 5.1   Summary of the Symbols to be Used in the General Energy Balance

Accumulation term		
Type of energy	*At time t_1*	*At time t_2*
Internal	U_{t1}	U_{t2}
Kinetic	K_{t1} E_{t1}	K_{t2} E_{t2}
Potential	P_{t1}	P_{t2}
Mass	m_{t1}	m_{t2}
Energy accompanying mass transport		
Type of energy	*Transport in*	*Transport out*
Internal	U_1	U_2
Kinetic	K_1	K_2
Potential	P_1	P_2
Mass	m_1	m_2
Net heat input to system	Q	
Net work done on system by surroundings	W	
Work to introduce material into system, or work recovered on removing material from system	$m_1(p_1\hat{V}_1)$	$-m_2(p_2\hat{V}_2)$

mass, and m with the subscript t_1 or t_2 denotes mass at t_1 or t_2, respectively, whereas m with the subscript 1 or 2 denotes a steady flow of mass at point 1 or 2, respectively.

In terms of the symbols given in Table 5.1, each of the terms in Eq. (5.10) translates into the following:

> **Accumulation**: $\Delta E = m_{t2}(\hat{U} + \hat{K} + \hat{P})_{t_2} - m_{t1}(\hat{U} + \hat{K} + \hat{P})_{t_1}$
> **Energy transfer in with mass flow**: $(\hat{U}_1 + \hat{K}_1 + \hat{P}_1)m_1$
> **Energy transfer out with mass flow**: $(\hat{U}_2 + \hat{K}_2 + \hat{P}_2)m_2$
> **Net transfer by heat flow in**: Q
> **Net transfer by mechanical or electrical work in**: W
> **Net transfer by work to introduce and remove mass**: $p_1\hat{V}_1 m_1 - p_2\hat{V}_2 m_2$

The terms $p_1\hat{V}_1$ and $p_2\hat{V}_2$ need a little explanation. They represent the so-called "pV work," or "pressure energy" or "flow work," or "flow energy," that is, the work done by the surroundings to put a mass of matter into the system at 1 in Figure 5.4 and the work done by the system on the surroundings as a unit mass leaves the system at 2, respectively. Because the pressures at the entrance and exit to the system are constant for differential displacements of mass, the work done per unit mass by the surroundings on the system adds energy to the system at 1:

$$\hat{W}_1 = \int_0^{\hat{V}_1} p_1 d\hat{V} = p_1(\hat{V}_1 - 0) = p_1 \hat{V}_1$$

where $\hat{V}$ is the volume per unit mass. Similarly, the work done by the fluid on the surroundings as the fluid leaves the system is $W_2 = -p_2 \hat{V}_2$

If we now introduce each of the terms above into the energy balance, we get a somewhat formidable equation

$$\Delta E = (\hat{U}_1 + \hat{K}_1 + \hat{P}_1)m_1 - (\hat{U}_2 + \hat{K}_2 + \hat{P}_2)m_2 + Q + W$$
$$+ p_1 \hat{V}_1 m_1 - p_2 \hat{V}_2 m_2 \tag{5.12}$$

To simplify the notation in Eq. (5.12) we add $p_1\hat{V}_1 m_1$ to $\hat{U}_1 m_1$ and $p_2\hat{V}_2 m_2$ to $\hat{U}_2 m_2$, and then introduce the expression $\Delta U + \Delta pV = \Delta H$ in Eq. (5.12). Thus, we locate the source of the variable called enthalpy in the energy balance. The energy balance can be reduced to a form easier to memorize:

$$\Delta E = E_{t_2} - E_{t_1} = Q + W - \Delta[(H + K + P)] \tag{5.13}$$

In Eq. (5.13) the delta symbol (Δ) has two different meanings:

    (a) In ΔE, Δ means final minus initial in time.
    (b) In ΔH, and so on, Δ means out of the system minus into the system.

Such usage is perhaps initially confusing, but is very common; hence you might as well become accustomed to it. The symbol Δ stands for a change. A more rigorous derivation of Eq. (5.13) from a microscopic energy balance may be found in Slattery.[1]

If there is more than one input and output stream to the system, it becomes convenient to calculate the properties of each stream separately so that Eq. (5.13) would become

$$\Delta E = E_{t_2} - E_{t_1} = \sum_{in} m_i(\hat{H}_i + \hat{K}_i + \hat{P}_i) - \sum_{out} m_o(\hat{H}_o + \hat{K}_o + \hat{P}_o) + Q + W \tag{5.14}$$

Here the subscript (o) designates an output stream and the subscript (i) designates an input stream.

In Sec. 5.3 we will show how to use Eqs. (5.11) and (5.13), but first you need to become familiar with various ways of calculating ΔH and ΔU because they usually comprise terms of large value in the energy balances in chemical engineering processes.

---

[1]J. C. Slattery. *Momentum, Energy, and Mass Transfer in Continua.* New York: McGraw-Hill, 1972.

LOOKING BACK

In this section we first reviewed a number of specialized terms pertinent to making energy balances. Then we described each of the terms employed in energy balances. Finally, we translated the words in the energy balance to mathematical symbols for both open and closed systems.

Key Ideas

1. The energy balance is given in words by Eq. (5.10) and in its most general form (for this book) in symbols by Eqs. (5.13) and (5.14).
2. For a closed system, energy transfer between the system and surroundings only takes place via heat (flow) and work.
3. For an open system, energy transfer accompanies the transfer of mass as well.
4. Work is positive when done on the system. Heat is positive when transferred to the system.
5. Heat and work are path functions. Internal energy and enthalpy are state (point) functions.
6. You cannot calculate absolute values of enthalpy and internal energy, just changes in value.

Key Terms

Batch system (p. 401)	Mechanical work (p. 394)
Closed system (p. 391)	Nonflow system (p. 391)
Electrical work (p. 394)	Open system (p. 391)
Energy (p. 392)	pV work (p. 403)
Energy balance (p. 400)	Path function (p. 392)
Enthalpy (p. 398)	Point function (p. 396)
Extensive property (p. 391)	Point property (p. 399)
Flow system (p. 391)	Potential energy (p. 395)
Flow work (p. 403)	Pressure energy (p. 403)
Heat (p. 394)	Property (p. 391)
Heat capacity at constant p (p. 398)	Shaft work (p. 395)
Heat capacity at constant volume (p. 397)	State (p. 391)
Independent property (p. 391)	State function (p. 396)
Intensive property (p. 391)	State property (p. 399)
Internal energy (p. 396)	System (p. 391)
Kinetic energy (p. 394)	Work (p. 392)

Self-Assessment Test

1. Contrast the following property classifications: extensive-intensive, measurable-unmeasurable, state-path.
2. Define heat and work.

3. Consider the hot-water heater in your house. Classify each case below as an open system, closed system, neither, or both.
   (a) The tank is being filled with cold water.
   (b) Hot water is being drawn from the tank.
   (c) The tank leaks.
   (d) The heater is turned on to heat the water.
   (e) The tank is full and the heater is turned off.

4. The units of potential energy or kinetic energy are (select all the correct expressions):
   (a) $(ft)(lb_f)$
   (b) $(ft)(lb_m)$
   (c) $(ft)(lb_f)/(lb_m)$
   (d) $(ft)(lb_m)/(lb_f)$
   (e) $(ft)(lb_f)/(hr)$
   (f) $(ft)(lb_m)/(hr)$

5. Review the selection of a system and surroundings by reading two or three examples in Secs. 3.3 to 3.6 covering up the solution, and designating the system. Compare with the system shown in the example.

6. Will the kinetic energy per unit mass of an incompressible fluid flowing in a pipe increase, decrease, or remain the same if the pipe diameter is increased at some place in the line?

7. A 100-kg ball initially on the top of a 5-m ladder is dropped and hits the ground. With reference to the ground:
   (a) What is the initial kinetic and potential energy of the ball?
   (b) What is the final kinetic and potential energy of the ball?
   (c) What is the change in kinetic and potential energy for the process?
   (d) If all the initial potential energy were somehow converted to heat, how many calories would this amount to? How many Btu? How many joules?

8. In expanding a balloon, two types of work are done by the air in the balloon (the system). One is stretching the balloon $dW = \sigma dA$, where σ is the surface tension of the balloon. The other is the work of pushing back the atmosphere. If the balloon is spherical and expands slowly from a diameter of 1 m to 1.5 m, what is the work done in pushing back the atmosphere? What assumptions must you make about T and p?

Thought Problems

1. A proposed goal to reduce air pollution from automobiles is to introduce into U.S. domestic gasoline a specified fraction of oxygenated compounds from renewable resources, one of which is ethanol grown from corn. What is your estimate of the fraction of the available U.S. cropland that would be required to replace 10% of the gasoline with alcohol in all of the annual gasoline production of about 1.2×10^{10} gal/yr? Assume 90.0 bu/acre of corn and 2.6 gal ethanol/bu.

2. Another proposal is to supply 10% of the U.S. oil usage by coal liquefaction. What is your estimate of the percentage of the coal now mined in the U.S. that would have to be processed in order to fulfill this proposal? Assume 3.26 bbl of liquid per ton of coal.

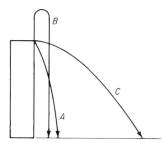

Figure TP5.1–3

3. Three baseballs are thrown from the top corner of a four-story building as shown in Figure TP5.1–3. All have the same initial speed. Which of the following answers would you say best reflects the speed of the balls when they hit the ground?

(**1**) *A* is greatest

(**2**) *B* is greatest

(**3**) *C* is greatest

(**4**) *A* and *C* are the same and greatest

(**5**) *A* and *B* are the same and greatest

(**6**) All have the same speed

4. In one of the explanations of the energy balance, the question was posed "Why do you not burn your hand when you put it into the oven at 300°C, but burn it if you touch the metal tray in the oven?" The answer was

> There are two important reasons. First, the mass of air surrounding the hand is very small (the density of air at 300°C and 1 atm is 0.615 kg/m^3), and hence its heat content is small despite the high oven temperature. Second, the air is a poor conductor and the relatively still hot air will not burn.
>
> The metal oven tray, on the other hand, has a much larger mass and hence contains much more heat than the air. In addition, the metals are better conductors of heat. Therefore, a good conductor coupled to a large reservoir of heat will relay large quantities of heat at a faster rate than blood can take it away from the fingers.

Comment on the answer; tell what aspects are right and are wrong with it.

Discussion Questions

1. You can find suggestions in the popular magazines that electric vehicles be covered with solar cells to reduce use of coal generated electricity. The solar insolation outside the earth's atmosphere is 1 kilowatt per square meter (or square yard, for all practical purposes), yielding 24 kWh per day. Experience has shown that an electric vehicle converted from a conventional car usually requires a minimum of about 30 kWh stored in the battery. If the surfaces of a car referred to are covered with a layer of solar cells 5 ft by 10 ft, or 50 ft^2—about 5 m^2—120 kWh/day will be available from the photovoltaic cells. Carry out an evaluation of this calculation. What deficiencies does it have? Think of the entire transition from solar energy to chemical change in the battery.

2. The following letter appeared in *The Physics Teacher*, Oct. 1990, p. 441:

> The recent article on "The Meaning of Temperature" by R. Baierlein is interesting but I believe it fails in its purpose. Baierlein quotes Eric Rogers' definition that "temperature is

hotness measured in some definite scale." I very much admired Eric Rogers and this definition, coupled with his statement that "heat is something that makes things hotter, or melts solids, or boils away liquids" might be adequate for the layman. But in no way are they acceptable scientific definitions. Similarly, Baierlein's statement that "hotness is the tendency to transfer energy in irregular, microscopic fashion" and his conclusion that "the function of temperature . . . is to tell us about a system's tendency to transfer energy (as heat)" are not only vague and nonoperational, but they are also incorrect or at least misleading.

(a) Would you agree with these statements?

(b) Read the complete letter and the Baierlein's reply in *The Physics Teacher*, and prepare a one page summary of your views on the relation of temperature and heat transfer.

(c) Also read Octave Levenspiel's amusing discussion of thermomometry in *Chemical Engineering Education*, Summer 1975, pp. 102–105 for a different viewpoint.

3. Worldwide interest exists in the possible global warming that occurs as a result of human activities. Prepare a report with tables that lists the sources of and mitigation options that exist for total CO_2, CH_4, CFC, and N_2O emissions in the world. For example, for CH_4 include rice cultivation, enteric fermentation, landfills, and so on. If possible, find data for the values of the annual emissions in Mt/yr. For the mitigation options, list the possible addition approximate costs over current cost to implement the option per 1 ton of emitter.

5.2 CALCULATION OF ENTHALPY CHANGES

Your objectives in studying this section are to be able to:

1. Calculate enthalpy (and internal energy) changes from heat capacity equations, graphs and charts, tables, and computer databases given the initial and final states of the material.

2. Become familiar with the steam tables and their use both in SI and American Engineering units.

3. Ascertain the reference state for enthalpy values in the data source.

4. Fit empirical heat capacity data with a suitable function of temperature by estimating the values of the coefficients in the function.

5. Convert an expression for the heat capacity from one set of units to another.

LOOKING AHEAD

In this section we explain how to look up and/or calculate the enthalpy (and internal energy) changes to be introduced into the energy balance.

MAIN CONCEPTS

Recall from Sec. 4.3 and Figure 4.10 that **phase transitions** occur from the solid to the liquid phase and the liquid to the gas phase, and the reverse. During these transitions large changes in the value of the enthalpy for a substance occur that need to be calculated accurately (so called **latent heat** changes). For a single phase the enthalpy varies as a function of the temperature as illustrated in Figure 5.6. The enthalpy changes that take place in a single phase are often called **"sensible heat"** changes.

   The enthalpy changes for the phase transitions are termed **heat of fusion** (for melting) and **heat of vaporization** (for vaporization). The word "heat" has been carried alone by custom from very old connotations because enthalpy changes have to be calculated from experimental data that frequently require experiments involving heat transfer. Enthalpy of fusion and vaporization are the proper terms but not widely used. **Heat of condensation** is the negative of the heat of vaporization, and the **heat of sublimation** is the enthalpy change from solid directly to vapor. A few values of $\Delta \hat{H}$ for the phases changes are listed in Appendix D.

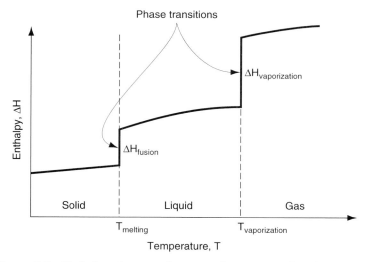

Figure 5.6   Enthalpy changes of a pure substance as a function of temperature. The vertical lines represent the "latent changes" that occur during a phase transition.

Once experimental data has been collected and validated, the information is stored for future use in several ways. We will look at four:

1. Heat capacity equations
2. Tables
3. Enthalpy charts
4. Computer databases

Enthalpies can also be *estimated* by generalized methods based on the theory of corresponding states or additive bond contributions, but we will not discuss these methods. Refer instead to the references at the end of the chapter for information.

5.2-1 Heat Capacity Equations

Eq. (5.9) in Sec. 5.1 shows that the enthalpy for a substance in a single phase (**not for the phase transitions**) can be calculated using the **heat capacity**

$$\Delta \hat{H} \;=\; \int_{T_1}^{T_2} C_p \, dT$$

To give the heat capacity some physical meaning, you can think of C_p as representing the amount of energy required to increase the temperature of a substance by 1 degree, energy that might be provided by heat transfer in certain specialized processes, but can be provided by other means as well. We will discuss C_p as C_v is not used very often.

What are the units of the heat capacity? From the definition of heat capacity in Sec. 5.1 you can see that the units are (energy)/(temperature difference) (mass or moles), hence the term "specific heat" (capacity) is sometimes used for heat capacity. The common units found in engineering practice are (we suppress the Δ symbol for the temperature)

$$\frac{\text{kJ}}{\text{(kg mol)(K)}} \qquad\qquad \frac{\text{Btu}}{\text{(lb mol) (}^\circ\text{F)}}$$

or equivalent values per unit mass.

The heat capacity varies with temperature for solids, liquids, and real gases, but is a continuous function of temperature *only* in the region between the phase transitions; consequently, it is not possible to have a heat capacity equation for a substance that will go from absolute zero up to any desired temperature. What an engineer does is to determine experimentally the heat capacity between the temperatures at which the phase transitions occur, fit the data with an equation, and then determine a new heat capacity equation for the next range of temperatures between the succeeding phase transitions. For the ideal monoatomic gas, the heat capacity at constant pressure is constant even though the temperature varies (see Table 5.2). For typical real gases, examine Figure 5.7.

For ideal gas mixtures, the heat capacity (per mole) of the mixture is the mole weighted average of the heat capacities of the components:

$$C_{p_{\text{avg}}} = \sum_{i=1}^{n} x_i C_{p_i} \tag{5.15}$$

For nonideal mixtures, particularly liquids, you should refer to experimental data or some of the estimation techniques listed in the literature (see the supplementary references at the end of the chapter).

Most of the equations for the heat capacities of solids, liquids, and gases are empirical. We usually express the heat capacity at constant pressure C_p as a function of temperature in a power series, with constants a, b, c, and so on; for example,

$$C_p = a + bT + cT^2 \tag{5.16}$$

where the temperature may be expressed in degrees Celsius, degrees Fahrenheit, degrees Rankine, or degrees kelvin. Since the heat capacity equations are valid only over moderate temperature ranges, it is possible to have equations of different types represent the experimental heat capacity data with almost equal accuracy. The task of fitting heat capacity equations to heat capacity data is greatly simplified by the use of digital computers, which can determine the constants of best fit by means of a standard prepared program and at the same time determine how precise the predicted heat capacities are. Heat capacity information can be

TABLE 5.2   Heat Capacities of Ideal Gases

	Approximate heat capacity,* C_p	
Type of molecule	High temperature (translational, rotational, and virational degrees of freedom)	Room temperature (translational and rotational degrees of freedom only)
Monoatomic	$\frac{5}{2}R$	$\frac{5}{2}R$
Polyatomic, linear	$(3n - \frac{3}{2})R$	$\frac{7}{2}R$
Polyatomic, nonlinear	$(3n - 2)R$	$4R$

*n, number of atoms per molecule; R, gas constant defined in Sec. 4.1.

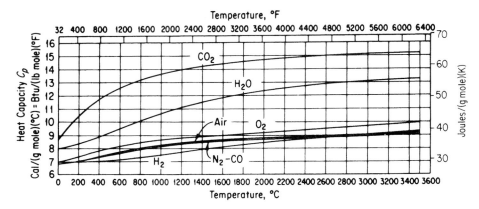

Figure 5.7 Heat capacity curves for the combustion gases.

found in Appendix E and in the disk in the back of the book, obtained from Professor Yaws' which has coefficients for 700 compounds (both gases and liquids). The change of C_p with pressure at high pressures is beyond the scope of our work here. Sources of heat capacity data can be found in several of the references listed at the end of the chapter.

Given the heat capacity equation, say Eq. (5.16) for example, you can calculate the enthalpy change per unit mole or mass by integrating the heat capacity equation with respect to temperature

$$\Delta \hat{H} = \int_{T_1}^{T_2} (a + bT + cT^2)\mathrm{d}T = a(T_2 - T_1) + \frac{b}{2} (T_2^2 - T_1^2) + \frac{c}{3} (T_2^3 - T_1^3) \quad (5.17)$$

If a different functional form of the heat capacity is available, the integration result will have a different form, of course.

EXAMPLE 5.5 Heat Capacity Equation

The heat capacity equation for CO_2 gas is

$$C_p = 2.675 \times 10^4 + 42.27\, T - 1.425 \times 10^{-2} T^2$$

with C_p expressed in J/(kg mol)(ΔK) and T in K. Convert this equation into a form so that the heat capacity will be expressed over the entire temperature range in Btu/(lb mol)(°F) with T in °F.

Solution

Changing a heat capacity equation from one set of units to another is merely a problem in the *conversion of units. Each term in the heat capacity equation must have the same units as the left-hand side of the equation. To avoid confusion in the conversion, you must remember to distinguish between the temperature symbols that represent temperature and the temperature symbols that represent temperature difference even though the same symbol often is used for each concept. In the conversions below we shall distinguish between the temperature and the temperature difference for clarity.*

First, multiply each side of the given equation for C_p by appropriate conversion factors to convert J/(kg mol)(ΔK) to (Btu/lb mol)(Δ°F). Multiply the left-hand side by the factor in the square brackets

$$\frac{C_p \text{ J}}{(\text{kg mol})(\Delta K)} \times \left[\frac{1 \text{ Btu}}{1055 \text{ J}} \middle| \frac{1 \Delta K}{1.8 \Delta °R} \middle| \frac{1 \Delta °R}{1 \Delta °F} \middle| \frac{0.4536 \text{ kg}}{1 \text{ lb}} \right] \to C_p \frac{\text{Btu}}{(\text{lb mol})(\Delta °F)}$$

and multiply right-hand side by the same set of conversion factors.

Next, substitute the relation between the temperature in K and the temperature in °F

$$T_K = \frac{T_{°R}}{1.8} = \frac{T_{°F} + 460}{1.8} \tag{a}$$

into the given equation for C_p where T_K appears.

Finally, carry out the indicated mathematical operations, and consolidate quantities to get

$$C_p = 8.702 \times 10^{-3} + 4.66 \times 10^{-6} T_{°F} - 1.053 \times 10^{-9} T^2_{°F}$$

EXAMPLE 5.6 Fitting Heat Capacity Data

The heat capacity of carbon dioxide gas as a function of temperature has been found by a series of repeated experiments to be as follows:

T(K)	300	400	500	600	700	800
C_p[J/(gmol)(K)]	39.87	45.16	50.72	56.85	63.01	69.52
	39.85	45.23	51.03	56.80	63.09	69.68
	39.90	45.17	50.90	57.02	63.14	69.63

Find the values of the coefficients in the equation

$$C_p = a + bT + cT^2$$

that yield the best fit to the data.

Solution

Use a least squares program (or directly minimize using the minimization code in the disk in the back of this book) to minimize the sum of the squares of the deviations between the predicted values of C_p and the experimental ones (refer to Appendix M):

$$\text{Minimize} \sum_{i=1}^{6} (C_{p_{\text{predicted},i}} - C_{p_{\text{experimental},i}})^2$$

The variables are a, b, and c.

The solution is

$$C_p = 25.47 + 4.367 \times 10^{-2}T - 1.44 \times 10^{-5}T^2$$

EXAMPLE 5.7 Calculation of $\Delta \hat{H}$ for a Gas Mixture Using Heat Capacity Equations

The conversion of solid wastes to innocuous gases can be accomplished in incinerators in an environmentally acceptable fashion. However, the hot exhaust gases often must be cooled or diluted with air. An economic feasibility study indicates that solid municipal waste can be burned to a gas of the following composition (on a dry basis):

CO_2	9.2%
CO	1.5%
O_2	7.3%
N_2	82.0%
	100.0%

What is the enthalpy difference for this gas per lb mol between the bottom and the top of the stack if the temperature at the bottom of the stack is 550°F and the temperature at the top is 200°F? Ignore the water vapor in the gas. You can neglect any energy effects resulting from the mixing of the gaseous components.

Solution

The heat capacity equations are [T in °F; C_p = Btu/(lb mol)(°F)]

N_2: $\quad C_p = 6.895 + 0.7624 \times 10^{-3}T - 0.7009 \times 10^{-7}T^2$

O^2: $\quad C_p = 7.104 + 0.7851 \times 10^{-3}T - 0.5528 \times 10^{-7}T^2$

CO^2: $\quad C_p = 8.448 + 5.757 \times 10^{-3}T - 21.59 \times 10^{-7}T^2 + 3.059 \times 10^{-10}T^3$

CO: $\quad C_p = 6.865 + 0.8024 \times 10^{-3}T - 0.7367 \times 10^{-7}T^2$

Basis: 1.00 lb mol of gas

By multiplying these equations by the respective mole fraction of each component, and then adding them together, you can save time in the integration, but they can be integrated separately, particularly if you use the computer program for calculating ΔH in the disk in the back of this book.

N_2: $\quad 0.82(6.895 + 0.7624 \times 10^{-3}T - 0.7009 \times 10^{-7}T^2)$

O_2: $\quad 0.073(7.104 + 0.7851 \times 10^{-3}T - 0.5528 \times 10^{-7}T^2)$

CO_2: $\quad 0.092(8.448 + 5.757 \times 10^{-3}T - 21.59 \times 10^{-7}T^2 + 3.059 \times 10^{-10}T^3)$

CO: $0.015(6.865 + 0.8024 \times 10^{-3}T - 0.7367 \times 10^{-7}T^2$

$C_{p\text{net}} = 7.053 + 1.2242 \times 10^{-3}T - 2.6124 \times 10^{-7}T^2 + 0.2814 \times 10^{-10}T^3$

$$\Delta H = \int_{550}^{200} (7.053 + 1.2242 \times 10^{-3}T - 2.6124 \times 10^{-7}T^2 + 0.2814 \times 10^{10}\, T^3)dT$$

$$= 7.053[(200) - (550)] + \frac{1.2242 \times 10^{-3}}{2}[(200)^2 - (550)^2]$$

$$- \frac{2.6124 \times 10^{-7}}{3}[(200)^3 - (550)^3] + \frac{0.2814 \times 10^{-10}}{4}[(200)^4 - (550)^4]$$

$$= -2468.6 - 160.7 + 13.8 - 0.633 = -2616 \text{ Btu/lb mol gas}$$

5.2-2 Tables of Enthalpy Values

Tables can accurately cover ranges of physical properties well beyond the range applicable for a single equation. Because the most commonly measured properties are temperature and pressure, tables of enthalpies (and internal energies) for pure compounds usually are organized in columns and rows, with T and p being the independent variables. If the intervals between table entries are close enough, linear interpolation between entries is reasonably accurate.

We will now describe two types of tables to be used in this chapter. The first set is the tables of enthalpies of gases at 1 atm, which can be found in Appendix D. The values of the enthalpies are for the gas phase only. The steam tables represent the second type of tables, namely tables that include enthalpies of phase changes along with enthalpies for a single phase. Remember that enthalpy values are all relative to some reference state. You can calculate enthalpy differences by subtracting the initial enthalpy from the final enthalpy for any two sets of conditions as shown in the following example.

EXAMPLE 5.8 Calculation of the Enthalpy Change of a Gas Using Tabulated Enthalpy Values

Calculate the enthalpy change for 1 kg mol of N_2 gas that is heated at a constant pressure of 100 kPa from 18°C to 1100°C.

Solution

Because 100 kPa is essentially 1 atm, the tables in Appendix D can be used to calculate the enthalpy change.

at 1100°C (1373K): $\Delta \hat{H} = 34{,}715$ kJ/kg mol (by interpolation)
at 18°C (291K): $\Delta \hat{H} = 524$ kJ/kg mol

Basis: 1 kg of N_2

$\Delta \hat{H} = 34{,}715 - 524 = 34{,}191$ kJ/kg mol

In this book you will find two sets of steam tables, one in SI in the foldout in the back pocket, and the other partly in Appendix C and partly in the foldout. To make use of the steam tables you must first locate the region of the physical properties of interest. The tables are organized so that the saturation properties are given separately from the properties of superheated steam and subcooled liquid. In addition, the saturation properties are listed in two ways: (1) The saturation pressure is given at even intervals for easy interpolation, and (2) the saturation temperature is given at even intervals for the same reason.

You can find the region in which a particular state lies by referring to one of the two tables for the saturation properties. If, at the given T or p, the given specific intensive property lies outside the range of properties that can exist for saturated liquid, saturated vapor, or their mixtures, then the state must be in either the superheated or the subcooled region. For example, look at the following brief extract from the steam tables in SI units (from the foldout in the back of this book):

		Specific volume (m³/kg)			Enthalpy (kJ/kg)		
p_{sat}(kPa)	T_{sat}(°C)	v_l	v_{lg}	v_g	h_l	h_{lg}	h_g
101.325	100.0	0.001043	1.672	1.673	419.5	2256.0	2675.6
200.0	120.2	0.001061	0.8846	0.8857	504.7	2201.5	2706.2

For each saturation pressure, the corresponding saturation temperature (boiling point) is given along with the values of specific volume and enthalpy for both saturated liquid and saturated vapor. The volume and enthalpy value in the column designated by the subscript lg is the difference between the values in the saturated vapor and saturated liquid states.

EXAMPLE 5.9 Use of the Steam Tables

Steam is cooled from 640°F and 92 psia to 480°F and 52 psia. What is $\Delta \hat{H}$ in Btu/lb?

Solution

Use the tables for the American Engineering units in the pocket in the back of the book. You must employ double interpolation to get the specific enthalpies, $\Delta \hat{H}$, relative to the reference for the table. Values of $\Delta \hat{H}$ are interpolated first between pressures at fixed temperature, and then between temperatures at fixed (interpolated) pressure.

	T (°F)				T (°F)		
p (psia)	600	700		p	600	640	700
90	1328.7	1378.1		92	1328.6		1378.0
95	1328.4	1377.8				1348.4	

	400	500		p	450	480	500
50	1258.7	1282.6		52	1258.4		1282.4
55	1258.2	1282.2				1272.8	

Note that the steam table values include the effect of pressure on $\Delta \hat{H}$ as well as temperature. An example of the interpolation needed at 600°F is

$$\tfrac{2}{5}(1328.7 - 1328.4) = 0.4(0.3) = 0.12$$

At $p = 92$ psia and $T = 600°F$, $\Delta \hat{H} = 1328.7 - 0.12 = 1328.6$.
The enthalpy change requested is

$$\Delta \hat{H} = 1272.8 - 1348.4 = -75.6 \text{ Btu/lb}$$

EXAMPLE 5.10 Use of the Steam Tables When a Phase Change Is Involved

Four kilograms of water at 27°C and 200 kPa are heated at constant pressure until the volume of the water becomes 1000 times the original value. What is the final temperature of the water?

Solution

The effect of pressure on the volume of liquid water can be neglected (to check refer to the table of the properties of liquid water), hence the initial specific volume is that of saturated liquid water at 300 K, or 0.001004 m³/kg. The final specific volume is

$$0.001004(1000) = 1.004 \text{ m}^3/\text{kg}$$

At 200 kPa, using interpolation between 400 and 450 K which covers the range of the specific volume of 1.004 m³/kg, we find T by solving

$$\Delta T \left(\frac{\Delta V}{V_2 - V_1} \right) + T_1 = 0.9024 \, \frac{\text{m}^3}{\text{kg}} + \frac{(1.025 - 0.9024) \, \text{m}^3/\text{kg}}{(450 - 400) \, \text{K}} (T - 400) \, \text{K} = 1.004 \, \frac{\text{m}^3}{\text{kg}}$$

$$T = 400 + 41 = 441 \, \text{K}$$

5.2-3 Graphical Presentation of Enthalpy Data

You have often heard the saying: A picture is worth a thousand words. Something similar might be said of two-dimensional charts, namely that you can get an excellent idea of the characteristics of the enthalpy and other properties of a substance in all regions of interest via a chart. Although the accuracy of the readings of values from a chart may be limited (depending on the scale of the chart), tracing out various processes on a chart enables you to rapidly visualize and analyze what is taking place. Charts are certainly a simple and quick method of getting data to compute enthalpy changes. Figure 5.8 is an example. A number of sources of charts are listed in the references at the end of the chapter. Appendix J contains charts for toluene and carbon dioxide.

Charts are drawn with various coordinates, such as p versus $\hat{H}$, p versus $\hat{V}$, or p versus T. Since a chart has only two dimensions, the coordinates can represent only two variables. The other variables of interest have to be plotted as lines of constant value across the face of the chart. Recall, for example, that in the p–$\hat{V}$ diagram for CO_2, Figure 4.7, lines of constant temperature were shown as parameters. Similarly, on a chart with pressure and enthalpy as the axes, lines of constant specific volume and/or temperature might be drawn as in Figure 5.8.

How many properties have to be specified for a single component gas to definitely fix the state of the gas? If you specify two intensive properties for a pure gas, you will ensure that all the other intensive properties will have definite values, and any two intensive properties can be chosen at will. Since a particular state for a gas can be defined by any two independent properties, a two-dimensional chart can be seen to be a handy way to present many combinations of physical properties.

EXAMPLE 5.11 Use of Pressure-Enthalpy Chart for Butane

Calculate $\Delta \hat{H}$, $\Delta \hat{V}$, and ΔT changes for 1 lb of saturated vapor of n-butane going from 2 atm to 20 atm (saturated).

Solution

Obtain the necessary data from Figure 5.8

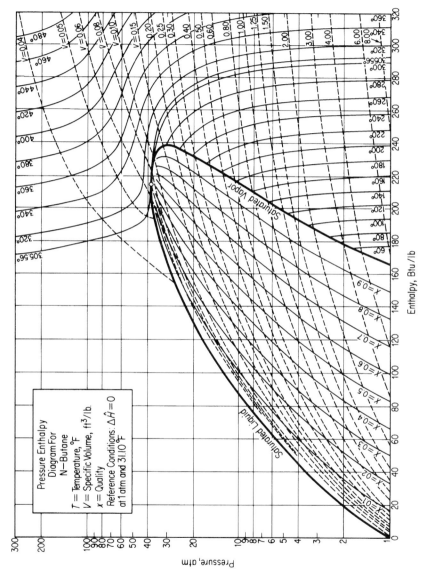

Figure 5.8 Pressure-enthalpy diagram for *n*-butane

	$\Delta\hat{H}$ (Btu/lb)	$\hat{V}$ (ft³/lb)	T (°F)
Saturated vapor at 2 atm:	179	3.00	72
Saturated vapor at 20 atm:	233	0.30	239

$$\Delta\hat{H} = 233 - 179 = 54 \text{ Btu/lb}$$
$$\Delta\hat{V} = 3.00 - 0.30 = 2.70 \text{ ft}^3\text{/lb}$$
$$\Delta T = 239 - 72 = 167°\text{F}$$

5.2-4 Retrieval of Data from Computer Databases

Values of the properties of thousands of substances are available in the form of computer programs that can provide the values at any given state, thus avoiding the need for interpolation and/or auxiliary computation, such as when only ΔH must be calculated from $\Delta U = \Delta H - \Delta pV$. One such program, that for the properties of water, is included on the disk in the back of this book. The program is abridged so that it does not have the accuracy or cover the range of commercially available programs, but it is quite suitable for solving problems in this text. Professor Yaws has been kind enough to provide data to calculate the heats of vaporization for your use for 700 compounds—check the disk in the back of the book.

Computer tapes can be purchased that provide information on the physical properties of large numbers of compounds, and immediate access to computer-based information systems can be obtained via the Internet. At the end of the chapter is a list of a number of programs and databases.

ADDITIONAL DETAILS

The Clapeyron equation is an exact thermodynamic relationship between the slope of the vapor-pressure curve and the molar heat of vaporization and the other variables listed below:

$$\frac{dp^*}{dT} = \frac{\Delta\hat{H}_v}{T(\hat{V}_g - \hat{V}_l)} \tag{5.18}$$

where p^* = vapor pressure
$\quad\quad\quad T$ = absolute temperature
$\quad\quad\quad \Delta\hat{H}_v$ = molar heat f vaporization at T
$\quad\quad\quad \hat{V}_i$ = molar volume of gas or liquid as indicated by the subscript g or l

Any consistent set of units may be used. Eq. (5.18) can be used to calculate $\Delta\hat{H}_v$ and to check the internal consistency of data if a function for the vapor pressure of a substance is known so you can evaluate dp^*/dT. If you assume that

(a) $\hat{V}_l$ is negligible in comparison with $\hat{V}_g$.

(b) The ideal gas law is applicable for the vapor: $\hat{V}_g = RT/p^*$

Then

$$\frac{dp^*}{p^*} = \frac{\Delta\hat{H}_v dT}{RT^2} \tag{5.19}$$

If you further assume that $\Delta\hat{H}_v$ is constant over the temperature range of interest, integration of Eq. (5.19) yields the Clausius-Clapeyron equation

$$\log_{10}\frac{p_1^*}{p_2^*} = \frac{\Delta\hat{H}_v}{2.303R}\left(\frac{1}{T_2} - \frac{1}{T_1}\right) \tag{5.20}$$

Two equations that can be used to estimate values of $\Delta\hat{H}_v$ are

Chen's equation.   An equation that yields values of $\Delta\hat{H}_v$ (in kJ/g mol) to within 2% is *Chen's equation:*

$$\Delta\hat{H}_v = \frac{T_b[0.0331(T_b/T_c) + 0.0297\log_{10}p_c - 0.0327]}{1.07 - T_b/T_c}$$

where T_b is the normal boiling point of the liquid in K, and p_c is the critical pressure in atmospheres.

Prediction using enthalpy of vaporization at the normal boiling point.   As an example, Watson[2] found empirically that

$$\frac{\Delta\hat{H}_{v_2}}{\Delta\hat{H}_{v_1}} = \left(\frac{1 - T_{r_2}}{1 - T_{r_2}}\right)^{0.38}$$

where $\Delta\hat{H}_{v_2}$ = heat of vaporization of a pure liquid at T_2
$\Delta\hat{H}_{v_1}$ = heat of vaporization of the same liquid at T_1

Yaws[3] lists various other values of the exponent for various substances.

LOOKING BACK

In this section we described four common ways in which values of the enthalpy for pure substances can be retrieved: heat capacity equations, tables, charts, and databases.

[2]K. M. Watson. *Ind. Eng. Chem. 23* (1931): 360, *35* (1943): 398.
[3]C. L. Yaws. *Physical Properties.* New York: McGraw-Hill, 1977.

Key Ideas

1. Enthalpy *changes* can be calculated or retrieved from various sources including heat capacity equations, tables, charts, and databases, among others.
2. Heat capacity equations can be developed given experimental data, and integrated to calculate enthalpy changes for a single phase.
3. Tables and charts give enthalpy changes for a single phase and include phase transitions.
4. Phase transitions such as the heat of vaporization or fusion should be as accurate as possible because they represent large quantities in the energy balance.

Key Terms

Chen's equation (p. 421)
Clapeyron equation (p. 420)
Clausius-Clapeyron equation (p. 421)
Enthalpy change (p. 412)
Heat capacity (p. 410)
Heat of condensation (p. 409)
Heat of fusion (p. 409)

Heat of sublimation (p. 409)
Heat of vaporization) (p. 409)
Latent heat (p. 409)
Phase transitions (p. 409)
Sensible heat (p. 409)
Steam tables (p. 416)
Watson's equation (p. 421)

Self-Assessment Test

1. A problem indicates that the enthalpy of a compound can be predicted by an empirical equation $H(\text{J/g}) = -30.2 + 4.25T + 0.001T^2$, where T is in kelvin. What is a relation for the heat capacity at constant pressure for the compound?
2. What is the heat capacity at constant pressure at room temperature of O_2 if the O_2 is assumed to be an ideal gas?
3. A heat capacity equation in cal/(g mol)(K) for ammonia gas is

$$C_p = 8.4017 + 0.70601 \times 10^{-2}T + 0.10567 \times 10^{-5}T^2 - 1.5981 \times 10^{-9}T^3$$

where T is in °C. What are the units of each of the coefficients in the equation?

4. Convert the following equation for the heat capacity of carbon monoxide gas, where C_p is in Btu/(lb mol)(°F) and T is in °F:

$$C_p = 6.865 + 0.08024 \times 10^{-2}T - 0.007367 \times 10^{-5}T^2$$

to yield C_p in J/(kg mol)(K) with T in kelvin.

5. Calculate the enthalpy change in 24 g of N_2 if heated from 300 K to 1500 K at constant pressure.
6. What is the enthalpy change when 2 lb of *n*-butane gas is cooled from 320°F and 2 atm to saturated vapor at 6 atm? Use the butane chart and the heat capacity equation.
7. You are told that 4.3 kg of water at 200 kPa occupies (a) 4.3, (b) 43, (c) 430, (d) 4300, and (e) 43,000 liters. State for each case whether the water is in the solid, liquid, liquid-vapor, or vapor regions.

8. Water at 400 kPa and 500 K is cooled to 200 kPa and 400 K. What is the enthalpy change? Use the steam tables.

9. Calculate the enthalpy change when 1 lb of benzene is changed from a vapor at 300°F and 1 atm to a solid at 0°F and 1 atm.

Thought Problems

1. Textbooks often indicate that for solids and liquids the difference $(C_p - C_v)$ is so small that you can say that $(C_p = C_v)$. Is this generally true?

2. Fire walkers with bare feet walk across beds of glowing coals without apparent harm. The rite is found in many parts of the world today and was practiced in classical Greece and ancient India and China, according to the *Encyclopaedia Britannica*. The temperature of glowing coals is about 1000°C. What are some of the possible reasons that fire walkers are not seriously burned?

3. A fire-induced BLEVE (boiling liquid expanding vapor explosion) in a storage tank can result in a catastrophe. The scenario is somewhat as follows: A pressure vessel (e.g., a pressurized storage tank), partially filled with liquid, is subjected to high heat flux from a fire. The temperature of the liquid starts to increase, causing an increase in pressure within the tank. When the vapor pressure reaches the safety relief valve pressure setting, the relief valve opens and starts to vent vapor (or liquid) to the outside. Concurrent with the previous step, the temperature of the portion of the tank shell not in contact with the liquid (i.e., the ullage space) increases dramatically.

   The heat weakens the tank shell around the ullage space. Thermally induced stresses are created in the tank shell near the vapor/liquid interface, and the heat-weakened tank plus the high internal pressure combine to cause a sudden, violent tank rupture. Fragments of the tank are propelled away from the tank at great force. Most of the remaining superheated liquid vaporizes rapidly due to the pressure release. The rest is mechanically atomized to small drops due to the force of the explosion. A fireball is created by the burning vapor and liquid.

   What steps would you recommend to prevent a BLEVE in the case of fire near a storage tank? [*Hint:* Two possible routes are: (1) prevent the fire from heating the tank; and (2) prevent the buildup of pressure in the tank.]

4. One way of increasing the temperature of a system is to transfer heat to it. Does this mean that heating a system always raises the temperature?

Discussion Questions

1. Dow Chemical sells Dowtherm Q, a heat transfer fluid that has an operating range of −30°F to 625°F, for $12/gal. Dowtherm Q competes with mineral oil, which costs $3/gal and operates up to 600°F. Why would a company pay so much more for Dowtherm Q?

2. An advertisement in the paper said you can buy a can of instant car cooler that you spray inside a car to reduce the temperature. The spray consists of a mixture of ethanol and water. The picture in the advertisement shows a thermometer "before" registering

41°C and the thermometer "after" registering 27°C. Explain to the owner of Auto Sport, an auto parts retailer, whether or not he should buy a case of the spray cans from the distributor.

3. A fermentation tank fitted with a 5-cm open vent was sterilized with steam at 140°C for one-half hour. Then the steam was shut off and the next batch of fluid was dumped into the tank. Shortly thereafter the tank imploded (burst inward). What is your explanation of the accident?

5.3 APPLICATIONS OF THE GENERAL ENERGY BALANCE WITHOUT REACTIONS OCCURRING

> ***Your objectives in studying this section are to be able to:***
>
> 1. Write down the general energy balance in words and symbols for open and closed systems.
> 2. Apply the general energy balance to solve particular problems.
> 3. Simplify the energy balance equation in conformity with the problem statement and other information.

LOOKING AHEAD

In this section we apply the energy balance, Eq. (5.10), written in symbols as Eq. (5.11) or Eq. (5.13), to closed and open systems, respectively. Material balances will not be important in the presentation.

MAIN CONCEPTS

In most problems you do not have to use all of the terms of the general energy balance equation because certain terms may be zero or may be so small that they can be neglected in comparison with the other terms. Several special cases of considerable industrial importance can be deduced from the general energy balance by introducing certain simplifying assumptions:

(1) No mass transfer (**closed or batch system**) ($m_1 = m_2 = 0$):

$$\Delta E = Q + W$$

(2) No accumulation ($\Delta E = 0$), no mass transfer ($m_1 = m_2 = 0$):

$$Q = -W$$

(3) No accumulation ($\Delta E = 0$), but with mass flow:

$$Q + W = \Delta[(\hat{H} + \hat{K} + \hat{P})m] \tag{5.21}$$

(4) No accumulation, $Q = 0$, $W = 0$, $\hat{K} = 0$, $\hat{P} = 0$:

$$\Delta H = 0 \tag{5.22}$$

[Equation (5.22) is sometimes called the **"enthalpy balance."**]

Take, for example, the flow system shown in Figure 5.9. Overall, between locations 1 and 5, we would find that $\Delta P = 0$. In fact, the only portion of the system where ΔP would be of concern would be between location 4 and some other location. Between 3 and 4, ΔP may be consequential, but between 2 and 4 it may be negligible in comparison with the work introduced by the pump. Between location 3 and any further downstream point, both Q and W are zero After reading the problem statements in the examples below but before continuing to read the solution, you should try to apply Eq. (5.13) yourself to test your ability to simplify the energy balance for particular cases.

Some special process names associated with energy balance problems are worth remembering:

(1) *Isothermal* ($dT = 0$): constant-temperature process

(2) *Isobaric* ($dp = 0$): constant pressure process

(3) *Isometric* or *isochoric* ($dV = 0$): constant-volume process

(4) *Adiabatic* ($Q = 0$): no heat interchange (i.e., an insulated system). If we inquire as to the circumstances under which a process can be called adiabatic, one of the following is most likely:

    (a) The system is insulated.

    (b) Q is very small in relation to the other terms in the energy equation and may be neglected.

    (c) The process takes place so fast that there is no time for heat to be transferred.

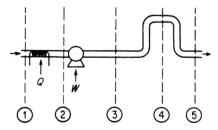

Figure 5.9   Flow system. The dashed lines delineate subsystems.

To assist you in solving problems involving energy balances, two additional steps should be added to your mental checklist for analyzing problems (see Table 3.1):

(1) *Step 7a*: Always write down the general energy balance, Eq. (5.11) or (5.13), below your sketch. By this step you will be certain not to neglect any of the terms in your analysis.

(2) *Step 7b*: Examine each term in the general energy balance and eliminate all terms that are zero or can be neglected. Write down why you do so.

We now examine some applications of the general energy balance, first to closed systems and then to open systems.

Closed Systems.   To further clarify the terms in the energy balance, consider a closed system for which the energy balance reduces $Q + W = \Delta E$. This equation applies with reference to the so called center of mass system. Suppose that the system is the fluid in a Dewar bottle (in order to make $Q = 0$). Since the fluid cannot move the bottle walls, $W = 0$. Furthermore, suppose that a ball affixed to the inside top of the Dewar is released somehow (with a negligible expenditure of energy), and drops to the bottom of the Dewar. What changes about the system? Recall that $\Delta E = \Delta U + \Delta P + \Delta K$. The dropping of the ball changes the center of mass of the fluid so that ΔP goes down, and hence ΔU and the temperature of the fluid go up, not by much, of course. The increase in temperature of the fluid was not caused by heat transfer because Q by definition must be transfer of energy across the system boundary. With the ball dropping, gravity does no work because the boundary must move for W to occur.

EXAMPLE 5.12 Application of the Energy Balance

Ten pounds of CO_2 at room temperature (80°F) are stored in a fire extinguisher having a volume of 4.0 ft³. How much heat must be removed from the extinguisher so that 40% of the CO_2 becomes liquid?

Solution

This problem involves a closed system (Figure E5.12) without reaction so that Eq. (5.11) applies. We can use the CO_2 chart in Appendix J to get the necessary property values.

    Steps 1, 2, 3, and 4   The specific volume of the CO_2 is 4.0/10 = 0.40 ft³/lb, hence CO_2 is a gas at the start of the process. The pressure is 300 psia and $\Delta\hat{H} = 160$ Btu/lb.

Step 5

Basis: 10 lb CO_2

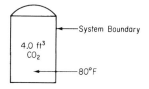

Figure E5.12

Steps 6 and 7   In the energy balance

$$\Delta E = Q + W$$

W is zero because the volume of the system is fixed, hence with $\Delta K = \Delta P = 0$,

$$Q = \Delta U = \Delta H - \Delta(pV)$$

We do not have values of $\Delta \hat{U}$, just values of $\Delta \hat{H}$, on the CO_2 chart. We can find $\Delta \hat{H}_{final}$ from the CO_2 chart by following the constant-volume line of 0.40 ft³/lb to the spot where the quality is 0.6. Hence the final state is fixed, and all the final properties can be identified, namely

$$\Delta \hat{H}_{final} = 81 \text{ Btu/lb}$$

$$p_{final} = 140 \text{ psia}$$

Steps 7, 8, and 9

$$Q = (81 - 160) - \left[\frac{(140)(144)(0.40)}{778.2} - \frac{(300)(144)(0.40)}{778.2} \right]$$

$$= -67.2 \text{ Btu (heat is removed)}$$

EXAMPLE 5.13 Application of the Energy Balance

Argon gas in an insulated plasma deposition chamber with a volume of 2 L is to be heated by an electric resistance heater. Initially the gas, which can be treated as an ideal gas, is at 1.5 Pa and 300 K. The 1000-ohm heater draws current at 40 V for 5 minutes (i.e., 480 J of work is done by the surroundings). What is the final gas temperature and pressure at equilibrium? The mass of the heater is 12 g and its heat capacity is 0.35 J/(g)(K). Assume that the heat transfer to the chamber from the gas at this low pressure and in the short time period is negligible.

Solution

No reaction occurs. The fact that the heater coil is "heated" inside the system does not mean that heat transfer to the system takes place. The system does not exchange mass with the surroundings, so Eq. (5.11) applies with $\Delta K = \Delta P = 0$:

$$\Delta E = Q + W = \Delta U$$

Steps 1, 2, 3, and 4   The system is the gas plus the heater as shown in Figure E5.13. Because of the assumption about the heat transfer, $Q = 0$. W is given as +480 J (work done on the system) in 5 minutes.
Step 5

$$\text{Basis} = 5 \text{ minutes}$$

Steps 6 and 7   For an ideal gas

$$pV = nRT$$

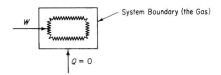

System Boundary (the Gas)

$Q = 0$

Figure E5.13

and initially we know p, V, and T, and thus can calculate the mass of the gas:

$$n = \frac{pV}{RT} = \frac{1.5 \text{ Pa}}{} \left| \frac{2 \text{ L}}{} \right| \frac{10^{-3} \text{ m}^3}{1 \text{ L}} \left| \frac{1 \ (\text{g mol})(\text{K})}{8.314 \ (\text{Pa})(\text{m}^3)} \right| \frac{}{300 \text{ K}}$$

$$= 1.203 \times 10^{-6} \text{ g mol}$$

The heater mass and heat capacity are given, and the C_v of the gas is (see Table 5.1) $C_v = C_p - R$. Since $C_p = \frac{5}{2}R$,

$$C_v = \tfrac{5}{2}R - R = \tfrac{3}{2}R$$

Assume that the given heat capacity of the heater is C_v also. We know that for the gas

$$\Delta U = n \int_{300}^{T} C_v \, dT' = nC_v(T - 300)$$

hence we can find T once we calculate ΔU from $\Delta U = W$.
Steps 7, 8, and 9

$$\Delta U = 480 \text{ J} = \overset{\text{heater}}{(12)(0.35)(T-300)} + \overset{\text{gas}}{(2.302 \times 10^{-6})(\tfrac{3}{2})(8.314)(T-300)}$$

$$T = 414.3$$

The final pressure is

$$\frac{p_2 V_2}{p_1 V_1} = \frac{n_2 R T_2}{n_1 R T_1}$$

or

$$p_2 = p_1 \left(\frac{T_2}{T_1}\right) = 1.5 \left(\frac{414.3}{300}\right) = 2.07 \text{ Pa}$$

EXAMPLE 5.14 Energy Balance

Ten pounds of water at 35°F, 4.00 lb of ice at 32°F, and 6.00 lb of steam at 250°F and 20 psia are mixed together in a container of fixed volume. What is the final temperature of the mixture? How much steam condenses? Assume that the volume of the vessel is constant with a value equal to the initial volume of the steam and that the vessel is insulated. Assume the process is a batch process rather than a flow process.

Solution

Steps 1, 2, 3, and 4   We can assume that the overall batch process takes place with $Q = 0$ and $W = 0$ if we define the system as in Figure E5.14. Let T_2 be the final temperature. The system consists of 20 lb of H_2O in one or two phases. ΔK and ΔP equal 0. The energy balance reduces to $\Delta U = 0$.

The initial properties can be obtained from the steam tables. Unfortunately, if we use American Engineering units we can only retrieve $\Delta \hat{H}$ values, not the $\Delta \hat{U}$ values that we want (and could get from the SI steam tables) for the energy balance, hence we need also to collect values of p and $\hat{V}$.

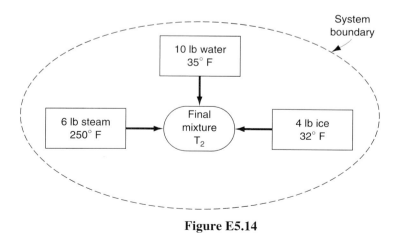

Figure E5.14

The properties needed are:

	$\Delta\hat{H}$ (Btu/lb)	$\hat{V}$ (ft³/lb)	p (psia)	T (°F)
Ice	−143.6*	—	—	32
Water	3.02	0.0162		35
Steam	1168	20.81	20	250

We know that $\Delta\hat{U} = \Delta\hat{H} - \Delta(p\hat{V})$ so that the energy balance becomes

$$\Delta H - \Delta(pV) = 0 \tag{a}$$

The volume of the container is fixed at the initial volume of the steam, namely $(6)(20.81) = 124.86$ ft³; we ignore the volumes of the ice and water in the calculations, as they are so small.

Step 5   The basis is 20 lb of water at the given conditions.

$$\text{Basis:} \begin{cases} 4 \text{ lb of ice at } 32°F \\ 10 \text{ lb of } H_2O \text{ at } 35°F \\ 6 \text{ lb of steam at } 250°F \text{ and } 20 \text{ psia} \end{cases}$$

Steps 6 and 7   If we assume that the final state of the water is liquid water in equilibrium with water vapor, an assumption to be checked out, we can calculate the final $\Delta\hat{H}$ of the 20 lb of water. We know that the final water is saturated. The second condition to fix the state (quality, temperature, and pressure) of the system can evolve from the energy balance, hence the problem has a unique solution. However, the final state must be calculated indirectly if tables are to be used.

Steps 7, 8, and 9

final ΔH   **initial ΔH**

$$\Delta H^f_{T_f} - (6\Delta\hat{H}^s_{T_s} + 10\Delta\hat{H}^w_{T_w} + 4\Delta\hat{H}^i_{T_i}) = (pV)^f_{T_f} - 6(p\hat{V})^s_{T_s} - 10(p\hat{V}^w_{T_w}) - 4(p\hat{V}^i_{T_i}) \tag{b}$$

The last two terms on the right-hand side of Eq. (b) cannot be more than 1 Btu at the very most and can safely be neglected. Be sure to check this assumption! Thus the equation to be used is

$$\Delta H^f_{T_f} - (pV)^f_{T_f} = (6\Delta\hat{H}^s_{T_s} + 10\Delta\hat{H}^w_{T_w} + 4\Delta\hat{H}^i_{T_i}) - 6(p\hat{V})^s_{T_s} \tag{c}$$

Because of the phase changes that take place as well as the nonlinearity of the heat capacities as a function of temperature, it is not possible to replace the enthalpies in Eq. (c) with functions of temperature and get a linear algebraic equation that is easy to solve. Consequently, the strategy we will use is first to assume a final temperature and pressure, and next check the assumption via Eq. (c). We want to bracket the temperature if possible, and then can interpolate for the desired answer.

The pressure in the vessel drops as more steams condenses. Data for the specific volume of the steam as a function of temperature and pressure can be taken from the steam tables for saturated conditions, and the mass of steam left in the vapor phase at any assumed temperature (or pressure) can be calculated by dividing 124.86 ft^3 by the specific volume of the steam.

The right-hand side of Eq. (c) is equal to

$$\frac{6\ lb}{}\ \left|\ \frac{1168.0\ Btu}{lb}\ +\ \frac{10\ lb}{}\ \right|\ \frac{3.02\ Btu}{lb}\ +\ \frac{4\ lb}{}\ \left|\ \frac{-143.6\ Btu}{lb}\right.$$

$$-\ \frac{6\ lb_m}{}\ \left|\ \frac{20\ lb_f}{in.^2}\ \right|\ \left(\frac{12\ in.}{1\ ft}\right)^2\ \left|\ \frac{20.81\ \ ft^3}{lb_m}\ \right|\ \frac{1\ Btu}{778\ (ft)(lb_f)}\ =\ 6001.6\ Btu$$

As an initial guess, suppose that one-half of the original steam does not condense. Then the specific volume of the final steam is (124.86 ft^3/3 lb) = 41.62 ft^3/lb. The closest integer line in the steam tables in the saturated steam column for pressure is 10 psia (T = 193.21°F) with a specific volume of 38.462 ft^3/lb and with $\Delta \hat{H}_{liquid}$ = 161.17 Btu/lb. Let us use these latter data to check the initial assumption. We calculate

$$S = \frac{124.86\ \ ft^3}{}\ \left|\ \frac{1\ lb\ steam}{38.462\ \ ft^3}\ \right.\ =\ 3.246\ lb\ steam\ not\ condensed$$

The final enthalpy of the system is:

Vapor: $$\frac{3.246\ lb\ steam}{}\ \left|\ \frac{1143.3\ Btu}{lb\ steam}\ \right.\ =\ 3711\ \frac{Btu}{}$$

Liquid: $$\frac{(20-3.246)\ lb\ liquid}{}\ \left|\ \frac{161.17\ Btu}{lb\ liquid}\ \right.\ =\ \frac{2700}{6411}$$

The final pV of the system is (ignoring the liquid phase):

Vapor: $$\frac{10\ lb_f}{in.^2}\ \left|\ \left(\frac{12\ in.}{1\ ft}\right)^2\ \right|\ \frac{124.86\ \ ft^3}{}\ \left|\ \frac{1\ Btu}{778\ (ft)(lb_f)}\ \right.\ =\ 231\ Btu$$

$$6411 - 231 = 6180 \neq 6001.6$$

The initial guess of p = 10 psia was a bit high. This time we will (for variety) guess the saturation temperature of T_f = 186°F (corresponding to p^* = 8.566 psia, $\hat{V}_{vapor}$ = 44.55 ft^3/lb, $\Delta \hat{H}_{liquid}$ = 153.93 Btu/lb, and $\Delta \hat{H}_{vapor}$ = 1140.5 Btu/lb).
For this second guess

$$S = \frac{124.86}{44.45} = 2.899\ lb\ not\ condensed$$

Repeat the use of Eq. (c)

$$(3306.3 + 2632.4) - 204) = 5734 \neq 6001.6$$

This time the guess was too low, however we have bracketed the solution. Linear interpolation for S gives

$$S = 2.899 + \left(\frac{6001.6 - 5734.29}{6180.28 - 5734.29} \right)(3.246 - 2.899) = 3.11$$

Consequently, the steam condensed is $6 - 3.11 = 2.89$ lb, and the assumption in steps 6 and 7 proved to be correct.

Open Systems.   We now present some examples in which mass flows in and out of the system.

EXAMPLE 5.15 Application of the Energy Balance

Air is being compressed from 100 kPa and 255 K (where it has an enthalpy of 489 kJ/kg) to 1000 kPa and 278 K (where it has an enthalpy of 509 kJ/kg). The exit velocity of the air from the compressor is 60 m/s. What is the power required (in kW) for the compressor if the load is 100 kg/hr of air?

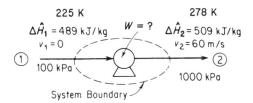

Figure E5.15

Solution

Steps 1, 2, 3, and 4   Figure E5.15 shows the known quantities. No reaction occurs. The process is clearly a flow process or open system. Let us assume that the entering velocity of the air is zero.
   Step 5

Basis: 100 kg of air = 1 hr

Steps 6 and 7   Let us simplify the energy balance (only one component exists):

$$\Delta E = Q + W - \Delta[(\hat{H} + \hat{K} + \hat{P})m]$$

(1) The process is in the steady state, hence $\Delta E = 0$.
(2) $m_1 = m_2 = m$.
(3) $\Delta(\hat{P}m) = 0$.
(4) $Q = 0$ by assumption (Q would be small even if the system were not insulated).
(5) $v_1 = 0$ (value is not known but would be small).

The result is

$$W = \Delta[(\hat{H} + \hat{K})m] = \Delta H + \Delta K$$

We have one equation and one unknown, W (ΔK, and ΔH can be calculated), hence the problem has a unique solution.

Steps 7, 8, and 9

$$\Delta H = \frac{(509 - 489) \text{ kJ}}{\text{kg}} \left| \frac{100 \text{ kg}}{} \right. = 2000 \text{ kJ}$$

$$\Delta K = 1/2 \; m(v_2^2 - v_1^2)$$

$$= \left(\frac{1}{2}\right) \frac{100 \text{ kg}}{} \left| \frac{(60 \text{ m})^2}{\text{s}^2} \right| \frac{1 \text{ kJ}}{\dfrac{1000(\text{kg})(\text{m}^2)}{(\text{s})^2}} = 180 \text{ kJ}$$

$$W = (2000 + 180) = 2180 \text{ kJ}$$

(*Note*: The positive sign indicates work is done on the air.)

To convert to power (work/time),

$$\text{kW} = \frac{2180 \text{ kJ}}{1 \text{ hr}} \left| \frac{1 \text{ kW}}{\dfrac{1 \text{ kJ}}{\text{s}}} \right| \frac{1 \text{ hr}}{3600 \text{ s}} = 0.61 \text{ kW}$$

EXAMPLE 5.16 Application of the Energy Balance

Water is being pumped from the bottom of a well 15 ft deep at the rate of 200 gal/hr into a vented storage tank to maintain a level of water in a tank 165 ft above the ground. To prevent freezing in the winter a small heater puts 30,000 Btu/hr into the water during its transfer from the well to the storage tank. Heat is lost from the whole system at the constant rate of 25,000 Btu/hr. What is the temperature of the water as it enters the storage tank, assuming that the well water is at 35°F? A 2-hp pump is being used to pump the water. About 55% of the rated horsepower goes into the work of pumping and the rest is dissipated as heat to the atmosphere.

Solution

Steps 1, 2, 3, and 4   Let the system consist of the well inlet, the piping, the pump, and the outlet at the storage tank. Assume the process is a steady-state flow process with material continually entering and leaving the system. See Figure E5.16.

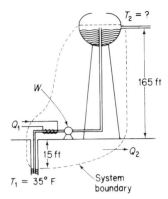

Figure E5.16

Step 5   Basis: 1 hr of operation

Steps 6 and 7   The material balance is 200 gal enter and 200 gal leave in an hour. The energy balance is

$$\Delta E = Q + W - \Delta[(\hat{H} + \hat{K} + \hat{P})m]$$

We will first simplify the energy balance:

(**1**)   The process is in the steady state, so that $\Delta E = 0$.

(**2**) $m_2 = m_2 = m$.

(**3**) $\Delta K \cong 0$ because we will assume that $v_1 = v_2 \cong 0$.

Then

$$0 = Q + W - \Delta[(\hat{H} + \hat{P})m]$$

Only the value of ΔH at the top of the tank is unknown, but it can be calculated from the energy balance. The temperature can be retrieved from

$$\Delta H = m\,\Delta\hat{H} = m \int_{T_1=35\text{°F}}^{T_2} C_p\,dT = mC_p(T_2 - 35)$$

if C_p is assumed to be constant. Hence the problem has a unique solution.

Steps 7, 8, and 9   The total mass of water pumped is

$$\frac{200\text{ gal}}{\text{hr}}\;\frac{8.33\text{ lb}}{1\text{ gal}} = 1666\text{ lb/hr}$$

The potential energy change is

$$\Delta P = m\Delta\hat{P} = mg\,\Delta h = \frac{1666\text{ lb}_\text{m}}{}\;\left|\;\frac{32.2\text{ ft}}{\text{s}^2}\;\right|\;180\text{ ft}\;\left|\;\frac{}{\dfrac{32.2\text{ (ft)(lb}_\text{m})}{(\text{s}^2)(\text{lb}_\text{f})}}\;\right|\;\frac{1\text{ Btu}}{778\text{ (ft)(lb}_\text{f})}$$

$$= 385.5\text{ Btu}$$

The heat lost by the system is 25,000 Btu while the heater puts 30,000 Btu into the system; hence the net heat exchange is

$$Q = 30,000 - 25,000 = 5000 \text{ Btu}$$

The rate of work being done on the water by the pump is

$$\dot{W} = \frac{2 \text{ hp}}{} \left| \frac{0.55}{} \right| \frac{33,000 \text{ (ft)(lb}_f)}{(\text{min})(\text{hp})} \left| \frac{60 \text{ min}}{\text{hr}} \right| \frac{\text{Btu}}{778 \text{ (ft)(lb}_f)} = 2800 \text{ Btu/hr}$$

ΔH can be calculated from: $Q + W = \Delta H + \Delta P$

$$5000 + 2800 = \Delta H + 386$$

$$\Delta H = 7414 \text{ Btu}$$

Because the temperature range considered is small, the heat capacity of liquid water may be assumed to be constant and equal to 1.0 Btu/(lb)(°F) for the problem. Thus,

$$7414 = \Delta H = mC_p \, \Delta T = 1666(1.0)(\Delta T)$$

$\Delta T \cong 4.5°F$ temperature rise, hence, $T = 39.5°F$.

EXAMPLE 5.17 Application of the Energy Balance

Steam (that is used to heat a biomass) enters the steam chest, which is segregated from the biomass, in a reactor, at 250°C saturated, and is completely condensed in the steam chest. The rate of the heat loss from the steam chest to the surroundings is 1.5 kJ/s. The reactants are placed in the vessel at 20°C and at the end of the heating the material is at 100°C . If the charge consists of 150 kg of material with an average heat capacity of C_p = 3.26 J/(g)(K), how many kilograms of steam are needed per kilogram of charge? The charge remains in the reaction vessel for 1 hr.

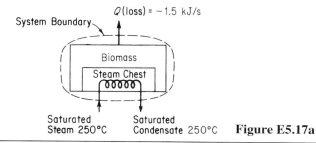

Saturated   Saturated
Steam 250°C   Condensate 250°C   **Figure E5.17a**

Solution

Steps 1, 2, 3, and 4   Figure E5.17a defines the system and lists the known conditions. If the system is the biomass plus the steam chest, the process is not a steady-state one because the temperature of the biomass increases.

Step 5   Basis: 1 hr of operation
Steps 6, 7, and 8   The energy balance is

$$\Delta E = Q + W - \Delta[(\hat{H} + \hat{K} + \hat{P})m] \tag{a}$$

Let us simplify the energy balance:

(1) The process is not in the steady state, so $\Delta E \neq 0$ (inside).
(2) We can assume that $\Delta K = 0$, and $\Delta P = 0$ inside the system.
(3) $W = 0$.
(4) The steam is the only material entering and leaving the system, and $m_1 = m_2$, and ΔK and ΔP of the entering and exit material are zero.

Consequently, Eq. (a) becomes

$$\Delta E = \Delta U = Q - \Delta[(\hat{H})m] \tag{b}$$

where ΔU refers only to the biomass and not the water because we will assume that there was no water in the steam chest at the start of the hour and none in the steam chest at the end of the hour. Let $m_1 = m_2 = m_{steam}$ that goes through the steam chest in one hour. Let the heat capacity of the biomass, $C_{p,biomass}$, be constant with respect to temperature. Then the terms in Eq. (b) are

(a) $\Delta U = \Delta H - \Delta(pV) = m_{biomass} C_{p,biomass} (373 - 293)$ °K [because we know that $\Delta(pV)$ for the liquid or solid charge will be negligible]. Thus

$$\Delta U = \Delta H_{biomass} = \frac{150 \text{ kg}}{} \frac{3.26 \text{ kJ}}{(\text{kg})(\text{K})} \frac{(373 - 293) \text{ K}}{} = 39,120 \text{ kJ}$$

(b) The heat loss is given as $\dot{Q} = -1.50$ kJ/s or

$$\frac{-1.50 \text{ kJ}}{\text{s}} \frac{3600 \text{ s}}{1 \text{ hr}} \frac{1 \text{ hr}}{} = -5400 \text{ kJ}$$

(c) The specific enthalpy change for the steam can be determined from the steam tables. The $\Delta\hat{H}_{vap}$ of saturated steam at 250°C is 1701 kJ/kg, so that

$$\Delta\hat{H}_{steam} = -1701 \text{ kJ/kg}$$

Introduction of all these values into Eq. (b) gives

$$39,120 \text{ kJ} = -\left(-1701 \frac{\text{kJ}}{\text{kg steam}}\right)(m_{steam} \text{kg}) - 5400 \text{ kJ} \tag{c}$$

from which the kilograms of steam per hour, m_{steam}, can be calculated as

$$m_{steam} = \frac{44,520 \text{ kJ}}{} \frac{1 \text{ kg steam}}{1701 \text{ kJ}} = 26.17 \text{ kg steam}$$

or

$$\frac{44{,}520 \text{ kJ}}{} \left| \frac{1 \text{ kg steam}}{1701 \text{ kJ}} \right| \frac{1}{150 \text{ kg charge}} = 0.17 \; \frac{\text{kg steam}}{\text{kg charge}}$$

Next, let us look at the problem from a different viewpoint. If the system had been chosen to be everything but the steam chest and lines, we would have a configuration as shown in Figure E5.17b. Under these circumstances heat would be transferred from the steam chest to the biomass. From a balance on the steam chest (no accumulation)

$$Q_{\text{system II}} = \Delta H_{\text{steam}} \tag{d}$$

As indicated in Figure E5.17b, the value of $Q_{\text{system II}}$ is negative.

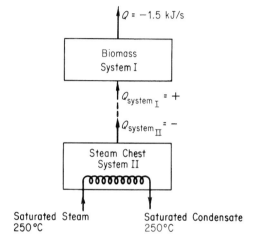

Saturated Steam
250°C

Saturated Condensate
250°C

Figure E5.17b

The energy balance for system I with no mass flow in or out of system I is

$$\Delta E_{\text{system I}} = \Delta U_{\text{biomass}} = \Delta H_{\text{biomass}} = Q_{\text{I}} - 5400 \tag{e}$$

Q_{II} and Q_{I} have opposite values because heat is removed from system II and added to system I. Because

$$Q_{\text{II}} = \Delta H_{\text{steam}} = (-1701)m_{\text{steam}}$$

we know that $Q_{\text{I}} = +(1701)m_{\text{steam}}$, and Eq. (e) becomes the same as Eq. (c).

LOOKING BACK

In this section we illustrated how to solve problems involving the energy balance via three examples for closed systems and three examples for open systems.

Key Ideas

1. The general energy balance equation can be simplified to represent a particular process so that each term in the equation needs to be examined carefully in view of the information given.
2. The energy balance provides one additional independent equation to add to the material balance equations.

Key Terms

Adiabatic (p. 425)
Batch system (p. 424)
Closed system (p. 424)
Enthalpy balance (p. 425)

Isobaric (p. 425)
Isometric (p. 425)
Isothermal (p. 425)

Self-Assessment Test

1. Liquid oxygen is stored in a 14,000-L storage tank. When charged, the tank contains 13,000 L of liquid in equilibrium with its vapor at 1 atm pressure. What is the **(a)** temperature, **(b)** mass, and **(c)** quality of the oxygen in the vessel? The pressure relief valve of the storage tank is set at 2.5 atm. If heat leaks into the oxygen tank at the rate of 5.0×10^6 J/hr, **(d)** when will the pressure relief valve open, and **(e)** what will be the temperature in the storage tank at that time?
   Data: at 1 atm, saturated, $\hat{V}_1 = 0.0281$ L/g mol, $\hat{V}_g = 7.15$ L/g mol, $\Delta\hat{H} = -133.5$ J/g; at 2.5 atm, saturated, $\Delta\hat{H} = -116.6$ J/g.

2. Suppose that you fill an insulated Thermos to 95% of the volume with ice and water at equilibrium and securely seal the opening.
   (a) Will the pressure in the Thermos go up, down, or remain the same after 2 hr?
   (b) After 2 weeks?
   (c) For the case in which after filling and sealing, the Thermos is shaken vigorously, what will happen to the pressure?

3. An 0.25-liter container initially filled with 0.225 kg of water at a pressure of 20 atm is cooled until the pressure inside the container is 100 kPa.
   (a) What are the initial and final temperatures of the water?
   (b) How much heat was transferred from the water to reach the final state?

4. In a shock tube experiment, the gas (air) is held at room temperature at 15 atm in a volume of 0.350 ft³ by a metal seal. When the seal is broken, the air rushes down the evacuated tube, which has a volume of 20 ft³. The tube is insulated. In the experiment:
   (a) What is the work done by the air?
   (b) What is the heat transferred to the air?
   (c) What is the internal energy change of the air?
   (d) What is the final temperature of the air after 3 min?
   (e) What is the final pressure of the air?

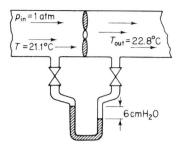

Figure SA5.3–7

5. In a refinery a condenser is set up to cool 1000 lb/hr of benzene that enters at 1 atm, 200°F, and leaves at 171°F. Assume negligible heat loss to the surroundings. How many pounds of cooling water are required per hour if the cooling water enters at 80°F and leaves at 100°F?

6. In a steady-state process, 10 g mol/s of O_2 at 100°C and 10 g mol/s of nitrogen at 150°C are mixed in a vessel which has a heat loss to the surroundings equal to 209 $(T - 25)$ J/s, where T is the temperature of the gas mixture in °C. Calculate the gas temperature of the exit stream in °C. Use the following heat capacity equations:

$$O_2: \ C_p = 6.15 + 3.1 \times 10^{-3}T$$
$$N_2: \ C_p = 6.5 + 1.25 \times 10^{-3}T$$

where T is in K and C_p is in cal/(g mol)(K).

7. An exhaust fan in a constant-area well-insulated duct delivers air at an exit velocity of 1.5 m/s at a pressure differential of 6 cm H_2O. Thermometers show the inlet and exit temperatures of the air to be 21.1°C and 22.8°C, respectively. The duct area is 0.60 m^2. Determine the actual power requirement for the fan. See Figure SA5.3–7.

8. A water system is fed from a very large tank, large enough so that the water level in the tank is essentially constant. A pump delivers 3000 gal/min in a 12-in.-ID pipe to users 40 ft below the tank level. The rate of work delivered to the water is 1.52 hp. If the exit velocity of the water is 1.5 ft/s and the water temperature in the reservoir is the same as in the exit water, estimate the heat loss per second from the pipeline by the water in transit.

Thought Problems

1. Do you save energy if you
   (a) Let ice build up inside your freezer?
   (b) Use extra laundry detergent?
   (c) Light a fire in your conventional fireplace?
   (d) Turn off your window air conditioner if you will be gone for a couple of hours?
   (e) Take baths rather than showers?
   (g) Use long-life incandescent light bulbs?

(h) Use fluorescent rather than incandescent lights?

(i) Install your refrigerator beside your range?

(j) Drive 55 instead of 70 miles per hour?

(k) Choose a car with air conditioning, power steering, and an automatic transmission over one without these features?

2. Liquid was transferred by gravity from one tank to another tank of about the same height several hundred meters away. The second tank overflowed. What might cause such overflow?

3. A proposal to the Department of Energy was as follows: The principal investigators planned to drive a turbine placed at the bottom of a tower high enough that the energy obtained from each pound of water, when converted to electricity by a generator run by the turbine, would be sufficient to electrolyze that pound of water. The resulting gas mixture, being lighter than air, would rise through an adjoining shaft to the top of the tower, where the gases would be burned to vaporize water. The water would be condensed, and returned down the tower. The fact that the system would not be 100 percent efficient would not prevent operation, as the tower could be made higher than the theoretical height, producing enough additional power to offset losses. Overall power would come from the lifting effect of the rising gases, from the heat generated from the burning gases, and by use of the superheated steam formed by the combustion to power a turbine. Explain whether or not this process would be successful, and the reasons for your answer.

4. When the use of heating equipment is temporarily discontinued because production is interrupted or because of the nature of the process, energy can usually be saved by allowing the equipment to cool down and then reheating. Some data have been collected as follows for an oil-fired furnace:

Temp. (°F)	Cooling Time (hr)	Reheating Time (hr)	Operating Power at T (MBtu/hr)	Energy Required to Reheat to T (MBtu)
2200	0	0	10.2	0
2100	0.7	0.6	9.7	12
2000	1.7	1.3	9.2	29
1900	2.4	1.6	8.8	40
1800	3.4	2.0	8.3	49
1700	4.0	2.3	7.3	60
1600	4.9	2.6	7.2	70
1500	5.8	2.9	6.8	79
1400	6.5	3.2	6.4	70

You are told that the furnace now at 2200°F is to be down for 8 hours but be prepared to have it back up to 2000°F then. What schedule would you recommend to the operator as being the most efficient?

Figure DQ5.3–1

Discussion Questions

1. A magazine [*Environ. Sci. Technol., 25* (1991): 1953] reported as follows:

   A novel scheme for generating electricity with compressed air produced by ocean wave action is being patented named MOTO (Motion of the Ocean). The scheme might also be engineered to produce hydrogen from water or potable water. MOTO consists of a network of 30-foot diameter toroids riding up and down on 8-foot diameter pilasters (see illustration). Each toroid can store up to 4500 cubic feet of compressed gas at 500 psi. According to the company, a system of three toroids could provide enough compressed air for to generate 3 MW of electricity. (See Figure DQ5.3–1)

   Can this be possible?

2. A headline in an advertisement says, "How a chemical company generates 2,500 kW from natural gas without burning one cubic foot." Is there a plausible explanation for this claim?

5.4 ENERGY BALANCES THAT ACCOUNT FOR CHEMICAL REACTION

Your objectives in studying this section are to be able to:

1. Compute heats of formation from experimental data for the enthalpy change (including phase changes) of a process with a reaction taking place.
2. Look up heats of formation in reference tables for a given compound.
3. List the standard conventions and reference states used for reactions associated with the standard heat of formation.

    4. Calculate the standard heat of reaction from tabulated standard heats of formation for a given reaction.

    5. Calculate the standard higher heating value from the lower heating value or the reverse.

    6. Determine the temperature of an incoming stream of material given the exit stream temperature (when a reaction occurs), or the reverse.

    7. Calculate how much material must be introduced into a system to provide a prespecified quantity of heat transfer for the system.

    8. Apply the general energy balance (and material balance) to processes involving reactions.

    9. Calculate the adiabatic reaction temperature.

LOOKING AHEAD

So far we have not discussed how to include the contribution of chemical reactions to the energy balance. In this section you will find how special enthalpies are added to each stream to account quite simply for such contributions.

MAIN CONCEPTS

As you know, the observed heat transfer that occurs in a closed system (with zero work) as a result of a reaction represents the energy associated with the rearrangement of the bonds holding together the atoms of the reacting molecules. For an **exothermic reaction**, the energy required to hold the products of the reaction together is less than that required to hold the reactants together. The reverse is true of an **endothermic reaction**. We will first describe the way you can accommodate the effects of chemical reaction in the energy balance, and then apply the concept to several examples.

5.4-1 Basic Information Needed to Accommodate Reactions in Enthalpy Calculations

To take account of energy changes caused by a reaction in the energy balance we incorporate in the enthalpy of each individual constituent an additional quantity termed the **standard heat** (really enthalpy) **of formation**,[4] $\Delta \hat{H}_f^{\,\circ}$, a quantity that is

---

[4]Historically, the name arose because the changes in enthalpy associated with chemical reactions were generally determined in a device called a calorimeter, to which heat is added or removed from the reacting system so as to keep the temperature constant.

discussed in detail below. (The superscript $^{\circ}$ denotes "standard state" and the subscript f denotes "formation.") Thus for the case of a single species A without any pressure effect on the enthalpy and omitting phase changes, the specific enthalpy change from the standard (reference) state is given by

$$\Delta \hat{H}_A = \Delta \hat{H}_{fA}^{\circ} + \int_{T_{ref}}^{T} C_{pA} \, dT' \tag{5.23}$$

For several species, we would have in a stream

$$\Delta H_{\text{mixture}} = \sum_{i=1}^{s} n_i \, \Delta \hat{H}_{fi}^{\circ} + \sum_{i=1}^{s} \int_{T_{\text{ref}}}^{T} n_i \, C_{pi} dT \tag{5.24}$$

where i designates each species, n_i is the number of moles of species i, and s is the total number of species. If phase changes take place, an additional term for the enthalpy of the phase change has to be added to the right-hand side of Eq. (5.23) to give the **total enthalpy** of substance A:

$$\Delta \hat{H}_A = \Delta \hat{H}^{\circ}_{fA} + (\hat{H}_{Tp} - \hat{H}^{\circ}_{ref})$$

where $(\hat{H}_{Tp} - \hat{H}^{\circ}_{ref})$ includes both "sensible heat" and phase changes.

If a mixture enters and leaves a system without a reaction taking place, we would find that the same species entered and left so that the enthalpy change would not be any different with the modification described above than what we have used before, because the terms that account for the heat of formation in the energy balance in the input and output would cancel. For example, for the case of two species in a flow system, the output enthalpy would be

$$\Delta H_{\text{output}} = \underbrace{n_1 \, \Delta \hat{H}_{f1}^{\circ} + n_1 \, \Delta \hat{H}_{f2}^{\circ}}_{\text{"heat of formation"}} + \underbrace{\int_{T_{\text{ref}}}^{T_{\text{out}}} (n_1 \, C_{p1} + n_2 \, C_{p2}) dT}_{\text{"sensible heat"}}$$

and the input enthalpy would be

$$\Delta H_{\text{input}} = \underbrace{n_1 \, \Delta \hat{H}_{f1}^{\circ} + n_1 \, \Delta \hat{H}_{f2}^{\circ}}_{\text{"heat of formation"}} + \underbrace{\int_{T_{\text{ref}}}^{T_{\text{in}}} (n_1 \, C_{p1} + n_2 \, C_{p2}) dT}_{\text{"sensible heat"}}$$

Without reaction, observe that $\Delta H_{\text{output}} - \Delta H_{\text{input}}$ would only involve the sensible heat terms that we have described before; the terms associated with the heat-of-formation would be exactly the same in the input and output in streams, and would cancel out.

However, if a reaction takes place, the quantities that enter and leave will differ, the inlet and exit temperatures may differ, and the terms involving the heat of formation will not cancel. For example, suppose that species 1 and 2 enter the system, react, and species 3 and 4 leave. Then

$$\Delta H_{\text{out}} - \Delta H_{\text{in}} = (n_3 \, \Delta \hat{H}_{\text{f3}}^\circ + n_4 \, \Delta \hat{H}_{\text{f4}}^\circ - (n_1 \, \Delta \hat{H}_{\text{f1}}^\circ + n_2 \, \Delta \hat{H}_{\text{f2}}^\circ)) \qquad (5.25)$$

$$+ \int_{T_{\text{ref}}}^{T_{\text{out}}} (n_3 \, C_{p3} + n_4 \, C_{p4}) \, \mathrm{d}T - \int_{T_{\text{ref}}}^{T_{\text{in}}} (n_1 \, C_{p1} + n_2 \, C_{p2}) \, \mathrm{d}T$$

$$\pm \text{ the enthalpies associated with phase changes}$$

In the special case of a steady-state flow process (the accumulation equals zero) in which the **reactants and products leave at the standard state** ($25°C$ and 1 atm), and **stochiometric quantities of reactants enter and react completely**, the sensible heat terms are zero, and the energy balance reduces to $Q = \Delta H$ where

$$\Delta H = \left(\sum_{\text{products}} n_\text{i} \, \Delta \hat{H}_{\text{fi}}^{\ \circ} - \sum_{\text{reactants}} n_\text{i} \, \Delta \hat{H}_{\text{fi}}^{\ \circ} \right) \equiv \Delta H_{\text{rxn}}^\text{o}$$

where $\Delta H^\circ_{\text{rxn}}$ is the symbol used for what is called the **heat of reaction at the standard state**, or sometimes the **standard heat of reaction**. Keep in mind that the "heat of reaction" is actually an *enthalpy change* and not necessarily equivalent to heat transfer.

Certain conventions exist with respect to calculating enthalpy changes accompanying chemical reactions:

(1) The reactants are shown on the left-hand side of the chemical equation, and the products are shown on the right: for example,

$$CH_4(g) + H_2O(l) \rightarrow CO(g) + 3H_2(g)$$

(2) The conditions of phase, temperature, and pressure must be specified unless the last two are the standard conditions, as presumed in the equation above, when only the phase is required. This is particularly important for compounds such as H_2O, which can exist as more than one phase under common conditions. If the reaction takes place at other than standard conditions, you might write

$$CH_4(g, \ 1.5 \ \text{atm}) + H_2O(l) \overset{50°C}{\rightarrow} \quad CO(g, \ 3 \ \text{atm}) + H_2 \ (g, \ 3 \ \text{atm})$$

(3) Unless the amounts of material reacting are stated, it is assumed that the quantities reacting are the stochiometric amounts shown in the chemical equation.

The **standard heat of formation** ("heat of formation"), $\Delta\hat{H}_f^o$, is the special enthalpy for the formation of 1 mole of a compound from its constituent elements, for example

$$C(s) + \tfrac{1}{2}O_2(g) \rightarrow CO(g)$$

in the standard state The initial reactants and final products must be at 25°C and 1 atm. The reaction does not necessarily represent a real reaction that would proceed at constant temperature, but can be a fictitious process for the formation of a compound from the elements. By defining **the heat of formation as zero in the standard state for each *element***, it is possible to design a system to express the heats of formation for all *compounds* at 25°C and 1 atm. If you use the conventions discussed above, then the thermochemical calculations will all be consistent, and you should not experience any confusion as to signs. Appendix F is a short table of the standard heats of formation. In the computer disk in the back of this book you will find heats of formation for 700 compounds provided through the courtesy of Professor Yaws.[5] Remember that the values for the standard heats of formation are negative for exothermic reactions.

EXAMPLE 5.18 Retrieval of Heats of Formation from Reference Data

What is the standard heat of formation of HCl(g)?

Solution

Look in Appendix F. The column heading is $\Delta\hat{H}_f^o$, and in the column opposite HCl(g) reads −92.311 kJ/g mol.

In the reaction to form HCl(g) at 25°C and 1 atm.

$$\tfrac{1}{2}H_2(g) + \tfrac{1}{2}Cl_2(g) \rightarrow HCl(g)$$

both $H_2(g)$ and $Cl_2(g)$ would be assigned $\Delta\hat{H}_f^o$ values of 0, and the value shown in Appendix F for HCl(g) of −92.311 J/g mol is the standard heat of formation for this compound as well as the standard heat of reaction for the reaction as written above:

$$\Delta H_{rxn}^o = 1(-92.311) - [\tfrac{1}{2}(O) + \tfrac{1}{2}(O)] = -92.311 \text{ kJ/g mol HCl(g)}$$

The value tabulated in Appendix F might actually be determined by carrying out the reaction shown for HCl(g) and measuring the heat transfer from a calorimeter, or by some other more convenient method.

---

[5]C. L. Yaws. "Correlation Constants for Chemical Compounds." *Chem. Engr.* (Aug. 15, 1976): 79.

EXAMPLE 5.19 Determination of a Heat of Formation from Experimental Data

Suppose that you want to find the standard heat of formation of CO from experimental data. Can you prepare pure CO from the reaction of C and O_2, and measure the heat transfer? This would be far too difficult. It would be easier experimentally to find the heat of reaction at standard conditions for the two reactions shown below and subtract them as follows:

Basis: 1 g mol of CO

$$\Delta \hat{H}^o_{rxn} (experimental)$$

A:	$C(\beta) + O_2(g)$	$\rightarrow$	$CO_2(g)$	-393.51 kJ/g mol
B:	$CO(g) + \frac{1}{2} O_2(g)$	$\rightarrow$	$CO_2(g)$	-282.99 kJ/g mol
$A - B$:	$C(\beta) + \frac{1}{2} O_2(g)$	$\rightarrow$	$CO(g)$	

$$\Delta \hat{H}^o_{rxn\ A-B} = (-393.51) - (-282.99) = \Delta \hat{H}^o_f = -110.52 \text{ kJ/g mol}$$

The energy change for the overall reaction scheme is the desired standard heat of formation per mole of CO(g). See Figure E5.19.

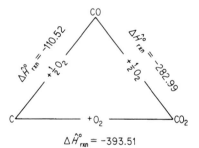

Figure E5.19

The calculations that we shall make in this book will all be for low pressures, and, although the effect of pressure upon the heats of reaction is relatively negligible under most conditions, if exceedingly high pressures are encountered, you should make the necessary corrections as explained in most texts on thermochemistry.

EXAMPLE 5.20 Calculation of the Standard Heat of Reaction from Standard Heats of Formation

Calculate $\Delta \hat{H}^o_{rxn}$ for the following reaction of 4 g mol of NH_3:

$$4NH_3(g) + 5O_2(g) \rightarrow 4NO(g) + 6H_2O(g)$$

Solution

Basis: 4 g mol of NH_3

Tabulated data	$NH_3(g)$	$O_2(g)$	$NO(g)$	$H_2O(g)$
$\Delta \hat{H}_f^o$ per mole at 25°C and 1 atm (kJ/g mol)	−46.191	0	+90.374	−241.826

We shall use Eq. (5.26) to calculate ΔH_{rxn}^o for 4 g mol of NH_3:

$$\Delta H_{rxn}^o = [4(90.374) + 6(-241.826)] - [5(0) + 4(-46.191)]$$

$$= -904.696 \text{ kJ/4 g mol } NH_3.$$

EXAMPLE 5.21 Heat of Formation for a Compound in Different Phases

If the standard heat of formation for $H_2O(l)$ is −285.838 kJ/g mol and the heat of evaporation is +44.012 kJ/g mol at 25°C and 1 atm, what is the standard heat of formation of $H_2O(g)$?

Solution

We shall proceed as in Example 5.19 to add the known chemical equations and the phase transitions to yield the desired chemical equation, and carry out the same operations on the enthalpy changes. For reaction A,

$$\Delta \hat{H}_{rxn}^o = \Sigma \Delta \hat{H}_{f \text{ products}}^o - \Sigma \Delta \hat{H}_{f \text{ reactants}}^o$$

A: $H_2(g) + \frac{1}{2}O_2(g)$ $\rightarrow$ $H_2O(l)$ $\Delta \hat{H}_{rxn}^o = -285.838$ kJ/g mol
B: $H_2O(l)$ $\rightarrow$ $H_2O(g)$ $\Delta \hat{H}_{vap}^o = +44.012$ kJ/g mol

$A + B$: $H_2(g) + \frac{1}{2}O_2 g)$ $\rightarrow$ $H_2O(g)$

$$\Delta \hat{H}_{rxnA}^o + \Delta \hat{H}_{vap}^o = \Delta \hat{H}_{rxn \ H_2O(g)}^o = \Delta \hat{H}_{fH_2O(g)}^o = -241.826 \text{ kJ/g mol}$$

You can see that any number of chemical equations can be treated by algebraic methods, and the corresponding standard heats of reaction can be added or subtracted in the same fashion as are the equations. By carefully following these results of procedure, you will avoid most of the common errors that arise in thermochemical calculations.

To simplify matters, the value used for $\Delta \hat{H}_{vap}^o$ of water in Example 5.21 was just the value at 25°C and 1 atm. To calculate the proper ΔH, if the final state for water is specified as $H_2O(g)$ at 25°C and 1 atm, the following enthalpy changes should be taken into account if you start with $H_2O(l)$ at 25°C and 1 atm:

$$
\Delta H_{\text{vap}} \text{ at } 25°C \text{ and 1 atm}
\left\{
\begin{array}{l}
\text{H}_2\text{O}(l) \ 25°C, \ 1 \text{ atm} \\
\qquad \downarrow \ \Delta H_1 \\
\text{H}_2\text{O}(l) \ 25°C, \text{ vapor pressure at } 25°C \\
\qquad \downarrow \ \Delta H_2 \ = \ \Delta H_{\text{vap}} \text{ at the vapor pressure of water} \\
\text{H}_2\text{O}(g) \ 25°C, \text{ vapor pressure at } 25°C \\
\qquad \downarrow \ \Delta H_3 \\
\text{H}_2\text{O}(g) \ 25°C, \ 1 \text{ atm}
\end{array}
\right.
$$

For practical purposes the value of ΔH_{vap} at 25°C and the vapor pressure of water, namely 43,911 J/g mol, will be adequate for engineering calculations at one atmosphere.

Another method of calculating enthalpy changes when chemical reactions occur is via **standard heats of combustion, $\Delta \hat{H}_c^{\circ}$,** which have a different set of reference conditions than do the standard heats of formation. The conventions used with the standard heats of combustion are:

(1) The compound is oxidized with oxygen or some other substance to the products $CO_2(g)$, $H_2O(l)$, HCl(aq), and so on.

(2) The reference conditions are still 25°C and 1 atm.

(3) Zero values of $\Delta \hat{H}_c^{\circ}$ are assigned to certain of the oxidation products as, for example, $CO_2(g)$, $H_2O(l)$, HCl(aq), and to $O_2(g)$ itself.

(4) If other oxidizing substances are present, such as S or N_2 or if Cl_2 is present, it is necessary *to make sure that states of the products are carefully specified and are identical to (or can be transformed into) the final conditions which determine the standard state as shown in Appendix F.*

Standard heats of reaction can be calculated from standard heats of combustion by an equation analogous to Eq. (5.26):

$$
\Delta \hat{H}_{\text{rxn}}^{o} \ = \ -\left(\sum \Delta H_{c \, \text{products}}^{o} - \sum H_{c \, \text{reactants}}^{o} \right) \tag{5.26}
$$

$$
= -\left(\sum_i n_{\text{prod } i} \, \Delta \hat{H}_{c \, \text{prod } i}^{o} - \sum_i n_{\text{react } i} \, \Delta \hat{H}_{c \, \text{react } i}^{o} \right) \tag{5.27}
$$

Note: The minus sign in front of the summation expression occurs because the choice of reference states is zero for the right-hand products of the standard reaction. Refer to Appendix F for values of $\Delta \hat{H}_c^{\circ}$.

For a fuel such as coal or oil, the negative of the standard heat of combustion is known as the **heating value** of the fuel. Both a **lower (net) heating value (LHV)** and a **higher (gross) heating value (HHV)** occur depending upon whether the water in the combustion products is in the form of a vapor (for the LHV) or a liquid (for the HHV).

You can estimate the heating value of a coal within about 3% from the Dulong formula:[6]

Higher heating value (HHV) in Btu per pound

$$= 14,544C + 62,028\left(H - \frac{O}{8}\right) + 4050S$$

where C is the weight fraction carbon, H is the weight fraction hydrogen, and S is the weight fraction sulfur. The values of C, H, S, and O can be taken from the fuel or flue-gas analysis. A general relation between the gross heating and net heating values is

net Btu/lb coal = gross Btu/lb coal − 91.23 (% total H by weight)

The HHV of fuel oils in Btu per pound can be approximated by

HHV = 17,887 + 57.5°API − 102.2 (%S).

EXAMPLE 5.22 Heating Value of Coal

Coal gasification consists of the chemical transformation of solid coal into gas. The heating values of coal differ, but the higher the heating value, the higher the value of the gas produced (which is essentially methane, carbon monoxide, hydrogen, etc.). The following coal has a reported heating value of 29,770 kJ/kg as received. Assuming that this is the gross heating value, calculate the net heating value.

Component	Percent
C	71.0
H_2	5.6
N_2	1.6
Net S	2.7
Ash	6.1
O_2	13.0
Total	100.0

---

[6]H. H. Lowry, ed. *Chemistry of Coal Utilization*, Chapter 4. New York: Wiley, 1945.

Solution

The corrected ultimate analysis listed above shows 5.6% hydrogen on the as-received basis.

Basis: 100 kg of coal as received.

The water formed on combustion is

$$\frac{5.6 \text{ kg H}_2}{} \left| \frac{1 \text{ kg mol H}_2}{2.02 \text{ kg H}_2} \right| \frac{1 \text{ kg mol H}_2\text{O}}{1 \text{ kg mol H}_2} \left| \frac{18 \text{ kg H}_2\text{O}}{1 \text{ kg mol H}_2\text{O}} \right. = 49.9 \text{ kg H}_2\text{O}$$

The energy required to evaporate the water is

$$\frac{49.9 \text{ kg H}_2\text{O}}{100 \text{ kg coal}} \left| \frac{2370 \text{ kJ}}{\text{kg H}_2\text{O}} \right. = \frac{1183 \text{ kJ}}{\text{kg coal}}$$

The net heating value is

$$29,770 - 1183 = 28,587 \text{ kJ/kg}$$

The value 2370 kJ/kg is not the latent heat of vaporization of water at 25°C (2440 kJ/kg) but includes the effect of a change from a heating value at constant volume to one at constant pressure (−70 kJ/kg).

Use the Dulong formula to calculate the HHV of this coal. Do you get 30,000 kJ/kg?

5.4-2 Energy Balances that Account for Chemical Reactions

Let us discuss specifically employing the heats of formation in an energy balance to answer questions such as

(1) What is the temperature of an incoming or exit stream?

(2) How much material must be introduced into an entering stream to provide for a specific amount of heat transfer?

Consider the process illustrated in Figure 5.10 for which the reaction is

$$aA + bB \rightarrow cC + dD$$

Nonstoichiometric amounts of reactants and products, respectively, enter and leave the system at different temperatures.

In employing Eq. (5.14) you should always **first choose a reference state** for the enthalpies at which the heats of formation are known, namely 25°C and 1 atm. (If no reaction takes place, the reference state can be the state of an inlet or outlet stream.)

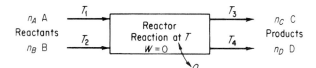

Figure 5.10   Process with reaction

The enthalpies of each stream entering and leaving are **calculated relative to the selected reference state**, and include (1) the standard heats of formation of the components, (2) the "sensible heats" of the components, and (3) the **phase changes of the components.**

Because enthalpy is a state function, you can choose any paths you want to execute the calculations for the overall enthalpy change in a process as long as you start and finish at the specified initial and final states, respectively. Figure 5.11 illustrates the idea. The reference state is chosen to be 25°C and 1 atm, the state for which the $\Delta \hat{H}_f^\circ$s are known. In the figure each component is treated separately via Eq. (5.23) taking into account the respective number of moles and adding in any phase changes. (For our purposes here we assume no enthalpy

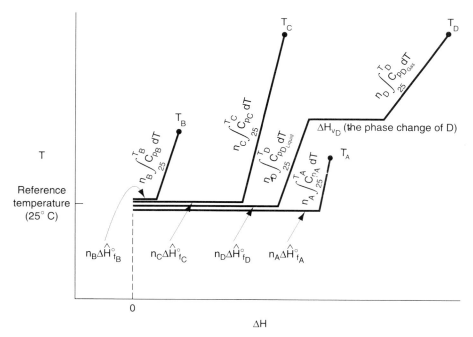

Figure 5.11   Calculation of the enthalpy changes for each component in a process using 25°C and 1 atm as the reference state.

change occurs on mixing. Refer to Sec. 5.6 if heats of mixing are involved in the process. In Figure 5.11, $T_C = T_D$. Pressure effects can be included along with temperature effects on the enthalpy, but we will omit in this book consideration of the effect of pressure except for problems in which the enthalpy data are retrieved from tables (such as the stream tables) or databases. The enthalpy change for each stream can be introduced into the general energy balance, Eq. (5.14).

Figure 5.12 illustrates the information flow for the calculations in the energy balance assuming a steady-state process ($\Delta E = 0$), no kinetic or potential energy changes, and $W = 0$. The general energy balance [Eq. (5.11)] reduces to

$$Q = \Delta H = \Delta H_{products} - \Delta H_{reactants}$$

Probably the easiest way to compute the necessary enthalpy changes is to use enthalpy data obtained directly from published tables or published formulas. Do not forget to take into account phase changes, if they take place, in the enthalpy calculations if the phase difference is not included in $\Delta \hat{H}_f^\circ$ or in the tables.

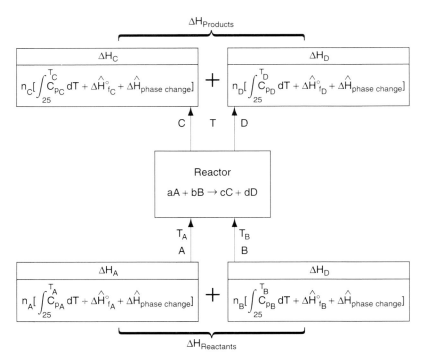

Figure 5.12   Information flow diagram showing how to calculate the enthalpies of the components entering and leaving a reactor.

EXAMPLE 5.23 Application of the Energy Balance with a Reaction Occurring

An iron pyrite ore containing 85.0% FeS_2 and 15.0% gangue (inert dirt, rock, etc.) is roasted with an amount of air equal to 200% excess air according to the reaction

$$4FeS_2 + 11O_2 \rightarrow 2Fe_2O_3 + 8SO_2$$

in order to produce SO_2. All the gangue plus the Fe_2O_3 end up in the solid waste product (cinder), which analyzes 4.0% FeS_2. Determine the heat transfer per kilogram of ore to keep the product stream at 25°C if the entering streams are at 25°C.

Solution

The material balance for the problem must be worked out prior to determining the heat transfer which is equal to the standard heat of reaction. This is a steady-state process with reaction; the system is the reactor.

    Steps 1, 2, 3, and 4   The process is a steady-state open system (see Figure E5.23). You calculate the excess air based on the stated reaction as if all the FeS_2 reacted to Fe_2O_3 even though some FeS_2 does not react. The molecular weights are Fe(55.85), Fe_2O_3(159.70), and FeS_2(120.0).

    Step 5

Basis: 100 kg of pyrite ore

    Steps 6 and 7   Six unknowns exist (including the N_2 in P) and five elemental balances can be written; the mass fraction of FeS_2 is given, hence the material balance problem has a unique solution. We will use five elemental balances, O, N, S, gangue, and Fe, plus the information about the FeS_2 to calculate the moles of SO_2, O_2, and N_2 in the gaseous product, and the kg of gangue, Fe_2O_3, and FeS_2 in the cinder.

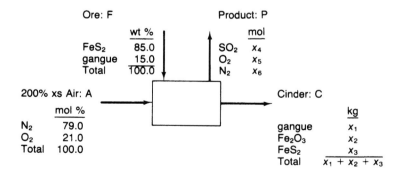

Figure E5.23

    Step 4 (continued)   The excess air is

$$\text{Mol FeS}_2 = \frac{85.0}{120.0} = 0.7083 \text{ kg mol}$$

$$\text{Required } O_2 = 0.7083(11/4) = 1.9479 \text{ kg mol}$$

$$\text{Excess } O_2 = 1.9479(2.00) = \underline{3.8958}$$

$$\text{Total } O_2 \text{ in} = \qquad\qquad 5.8437 \text{ kg mol}$$

$$\text{Total } N_2 \text{ in} = 5.8437(79/21) = 21.983 \text{ kg mol}$$

The element mass balances are

	In	*Out*
Gangue (kg):	15.0	$= x_1$
N2 (kg mol):	21.983	$= x_6$
S (kg mol):	2(85.0/120.0)	$= x_4 + (x_3/120.0)(2)$
Fe (kg mol):	1(85.0/120.0)	$= (x_2/159.70)2 + (x_3/120.0)(1)$
O2 (kg mol):	5.8437	$= x_4 + x_2 + (x_2/159.70)(1.5)$

Also,

$$\frac{x_3}{x_1 + x_2 + x_3} = 0.04$$

The solution of these equations is

In P	*In C*
$SO_2 = 1.368$ kg mol	Gangue $= 15.0$ kg
$O_2\ \ = 3.938$	$Fe_2O_3\ \ = 54.63 \rightarrow 0.342$ kg mol
N2 $= 21.983$	$FeS_2\qquad = 2.90 \rightarrow 0.0242$ kg mol

Next we use the energy balance to determine the heat transfer. The general energy balance reduces to ($\Delta E = 0$, $\Delta P = 0$, $\Delta K = 0$, $W = 0$) $Q = \Delta H$. Because all of the reactants and products are at the 25°C and 1 atm, we will choose the reference state to be 25°C and 1 atm with the result that all the "sensisble heat" terms in Eq. (5.24) become zero, and

$$Q = \underset{\text{products}}{\sum n_i \Delta \hat{H}_i^o} \ - \ \underset{\text{reactants}}{\sum n_i \Delta \hat{H}_i^o}$$

	Products			*Reactants*		
Comp.	$10^{-3} \times$ g mol	$\Delta\hat{H}_f^o$ (kJ/g mol)	$n_i\Delta\hat{H}_f^o$ (kJ)	$10^{-3} \times$ g mol	$\Delta\hat{H}_f^o$ (kJ/g mol)	$n_i\Delta\hat{H}_f^o$ (kJ)
FeS_2	0.0242	−177.9	−4.305	0.7083	−177.9	−126.007
Fe_2O_3	0.342	−822.156	−281.177	0	−822.156	0
N_2	21.9983	0	0	21.983	0	0
O_2	3.938	0	0	5.8437	0	0
SO_2	1.368	−296.90	−406.159	0	−296.90	0
		Total	−691.641			−126.007

$$Q = [-691.641 - (-126.007)](10^3) = -565.634 \times 10^3 \text{ kJ/100 kg ore}$$

or

$$Q = -5.656 \times 10^3 \text{ kJ/kg ore}$$

The negative sign indicates that heat is removed from the process.

EXAMPLE 5.24 Application of the Energy Balance to a Process in which the Temperatures of the Entering and Exit Streams Differ

Carbon monoxide at 50°F is completely burned at 2 atm pressure with 50% excess air that is at 1000°F. The products of combustion leave the combustion chamber at 800°F. Calculate the heat evolved from the combustion chamber expressed as Btu per pound of CO entering.

Solution

Steps 1, 2, 3, and 4   Refer to Figure E5.24. A material balance is needed before an energy balance can be made.

Step 5

Basis: 1 lb mol of CO (easier to use than 1 lb of CO)

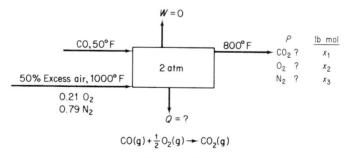

$$CO(g) + \tfrac{1}{2}O_2(g) \rightarrow CO_2(g)$$

Figure E5.24

Steps 6 and 7   We have three elements, hence can make three independent balances. Because we have three unknown compositions, a unique solution exists, a solution that can be obtained by direct addtion or subtraction.

Steps 3 and 4 (continued)   Amount of air entering:

$$\frac{1 \text{ lb mol CO}}{} \left| \frac{0.5 \text{ lb mol O}_2}{1 \text{ lb mol CO}} \right| \frac{1.5 \text{ lb mol O}_2 \text{ used}}{1.0 \text{ lb O}_2 \text{ mol needed}} \left| \frac{1 \text{ lb mol air}}{0.21 \text{ lb mol O}_2} \right. = 3.57 \text{ lb mol air}$$

composed of

$$3.57(0.79) = 2.82 \text{ lb mol } N_2$$

$$3.57(0.21) = 0.750 \text{ lb mol } O_2$$

Steps 7, 8, and 9   The element balances are

	In	Out	In P (lb mol)	
C	1	$= x_1$	CO_2	1
N2	2.82	$= x_3$	N_2	2.82
O2	$1(0.5) + 0.750$	$= x_1 + x_2 = 1 + x_2$	O_2	0.250

Next we make an energy balance, with $\Delta E = 0$, $W = 0$, $\Delta P = 0$, and $\Delta K = 0$, the balance reduces to $Q = \Delta H$.

$$\Delta \hat{H} \text{ (Btu/lb mol; reference is 32°F)}$$

Temp. (°F)	CO	O_2	N_2	CO_2
50	125.2	—	—	—
77	313.3	315.1	312.2	392.2
800	5484	5690	5443	8026
1000	6993	7288	6929	10,477

and the heats of formation from Appendix F multiplied by the conversion factor to convert from kJ/g mol to Btu/lb mol are as follows. The reference temperature is 77°F. We assume that the effect of pressure (2 atm) on the enthalpy changes is negligible.

	CO	O_2	N_2	CO_2
$\Delta \hat{H}_f^\circ$ Btu/lb mol:	$-47,587$	0	0	$-169,435$

We will use as a reference for the energy balance 77°F and 1 atm.

$$Q = [\Delta H_{CO_2} + \Delta H_{N_2} + \Delta H_{O_2}]_{\text{products}} - [\Delta H_{CO} + \Delta H_{N_2} + \Delta H_{O_2}]_{\text{reactants}}$$

Products

ΔH_{CO_2}:	$(1.00)[(8026 - 392) - 169,435]$	$=$	$-161,798$
ΔH_{N_2}:	$(2.82)[(5443 - 312) + 0]$	$=$	$14,469$
ΔH_{O_2}:	$(0.250)[(5690 - 315) + 0]$	$=$	$1,334$

Reactants

ΔH_{N_2}:	$(2.82)[(6929 - 312) + 0]$	$=$	$18,660$
ΔH_{O_2}:	$(0.750)[(7288 - 315) + 0]$	$=$	$5,230$
ΔH_{CO}:	$(1.00)[(125 - 313) - 47,587]$	$=$	$-47,775$
	Q	$=$	$-122,100$ Btu/lb mol CO

On the basis of 1 lb of CO:

$$Q = -\frac{122{,}100 \text{ Btu}}{1 \text{ lb mol CO}} \left| \frac{1 \text{ lb mol CO}}{28 \text{ lb CO}} = -4361 \text{ Btu/lb CO (removed)} \right.$$

Instead of using enthalpy data from tables for the "sensible heats," heat capacity equations could be used instead. The following equations give the enthalpy changes in Btu/lb mol of the compound.

Products

CO_2: $(1.00)\left[\int_{77}^{800} (8.448 + 0.5757 \times 10^{-2} T_{°F} - 0.2159 \times 10^{-5} T^2)\, dT - 169{,}435 \right]$

N_2: $(2.82)\left[\int_{77}^{800} (6.895 + 0.07624 \times 10^{-2} T_{°F} - 0.007009 \times 10^{-5} T_{°F}^2)\, dT \right]$

O_2: $(0.250)\left[\int_{77}^{800} (7.104 + 0.0785 T_{°F} \times 10^{-2} - 0.005528 \times 10^{-5} T_{°F}^2)\, dT \right]$

Reactants

CO: $(1.00)\left[\int_{77}^{50} (6.895 + 0.07624 \times 10^{-2} T_{°F} - 0.007367 \times 10^{-5} T_{°F}^2)\, dT - 47{,}587 \right]$

N_2: $(2.82)\left[\int_{77}^{1000} (6.895 + 0.07624 \times 10^{-2} T_{°F} - 0.007009 \times 10^{-5} T_{°F}^2)\, dT \right]$

O_2: $(0.250)\left[\int_{77}^{1000} (7.104 + 0.0785 T_{°F} \times 10^{-2} - 0.005528 \times 10^{-5} T_{°F}^2)\, dT \right]$

A special term called the **adiabatic reaction (flame) temperature** is defined as the temperature obtained inside the process when (1) the reacion is carried out under adiabatic conditions, that is, there is no heat interchange between the vessel in which the reaction is taking place and the surroundings; and (2) when there are no other effects present, such as electrical effects, work, ionization, free radical formation, and so on. We assume that the products leave at the temperature of the reaction, and thus if you know the temperature of the products, you automatically know the temperature of the reaction.

In calculations of flame temperatures for combustion reactions, the adiabatic reaction temperature *assumes complete combustion*. Equilibrium considerations may dictate less than complete combustion for an actual case. For example, the adiabatic flame temperature for the combustion of CH_4 with theoretical air has been calculated to be 2010°C; allowing for incomplete combustion, it would be 1920°C. The actual temperature when measured is 1885°C.

The adiabatic reaction temperature tells us the temperature ceiling of a process. We can do no better, but of course the actual temperature may be less. The adiabatic reaction temperature helps us select the types of materials that must be specified for the equipment in which the reaction is taking place. Chemical combustion with air produces gases at a maximum temperature of 2500 K, which can be increased to 3000 K with the use of oxygen and more exotic oxidants, and even this value can be exceeded although handling and safety problems are severe. Applications of such hot gases lie in the preparation of new materials, micromachining, welding using laser beams, and the direct generation of electricity using ionized gases as the driving fluid.

A computer code to calculate the adiabatic reaction temperature will be found on the disk in the back of this book. If tables rather than heat capacity equations are used to calculate the "sensible heats" of the various streams entering and leaving the reactor, the calculations will involve trial-and-error. The steady state energy balance with $Q = 0$ reduces to simply $\Delta H = 0$. To find the exit temperature for which $\Delta H = 0$, if tables are used as the source of the $\Delta \hat{H}$) values, the simplest procedure is to

(1) Assume a sequence of values of T selected to bracket 0 (+ and −) for the sum of the enthalpies of the products minus the reactants, and

(2) Once the bracket is obtained, interpolate within the bracket to get the desired value of T when $\Delta H = 0$.

If heat capacity equations are integrated, the energy balance reduces to solving a cubic or quadratic equation in the exit temperature.

EXAMPLE 5.25 Adiabatic Reaction (Flame) Temperature

Calculate the theoretical flame temperature for CO gas burned at constant pressure with 100% excess air, when the reactants enter at 100°C and one atm.

Solution

The solution presentation will be abbreviated to save space. The system is shown in Figure E5.25a. We will use data from Appendix F. The process is in the steady state.

$$CO(g) + \tfrac{1}{2}O_2(g) \to CO_2(g)$$

Basis: 1 g mol of CO(g); ref. temp. 25°C

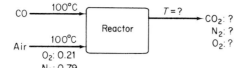

Figure E5.25a

A summary of the results of the material balance is:

Entering reactants		Exit products	
Component	g mol	Component	g mol
CO(g)	1.00	CO_2(g)	1.00
O_2(req.)	0.50	O_2(g)	0.50
O_2(xs)	0.50	N_2(g)	3.76
O_2(total)	1.00		
N_2	3.76		
Air	4.76		

The reference state will be 25°C and 1 atm. In the first approach to the solution of the problem, the "sensible heat" (enthalpy) values will be taken from the table of the enthalpy values for the combustion gases. The energy balance (with $Q = 0$) reduces to $\Delta H = 0$.

Reactants

Component	g mol	T(K)	$\Delta\hat{H}$ (J/g mol)	$\Delta\hat{H}_f^o$ (J/g mol)	ΔH (J)
CO	1.00	373	(2917 − 728)	−110,520	−108,331
O_2	1.00	373	(2953 − 732)	0	2,221
N_2	3.76	373	(2914 − 728)	0	8,219
		Total			−97,891

Products

Assume $T = 2000$K:

CO_2	1.00	2000	(92,466 − 912)	−393,510	−301,956
O_2	0.50	2000	(59,914 − 732)	0	29,591
N_2	3.76	2000	(56,902 − 728)	0	211,214
		Total			−61,151

$$\Delta H = \Delta H_{products} - \Delta H_{reactants} = (-61,151) - (-97,891) = 36,740 > 0$$

Assume $T = 1750$K:

CO_2	1.00	1750	$(77,455 - 912)$	$-393,510$	$-316,977$
O_2	0.50	1750	$(50,555 - 732)$	0	24,912
N_2	3.76	1750	$(47,940 - 728)$	0	177,517
		Total			$-114,548$

$$\Delta H = (-114,548) - (-97,891) = -16,657 < 0$$

Now that $\Delta H = 0$ is bracketed, we can carry out a linear interpolation to find the theoretical flame temperature (TFT) (see Figure E5.25b):

$$\text{TFT} = 1750 + \frac{0 - (-16,657)}{36,740 - (-16,657)} \; (250) = 1750 + 78 = 1828\text{K}(1555°\text{C})$$

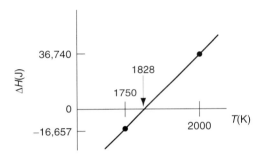

Figure E5.25b

If we use heat capacity equations to calculate the "sensible heats," we can obtain an analytic relation to be solved for the flame temperature as follows [C_p values are in J/(g mol)(K) and are good to ±1.5% up to 1500 K]; T is in K:

$$C_{pO_2} = 25.29 + 13.25 \times 10^{-3}T - 4.20 \times 10^{-6}T^2$$
$$C_{pN_2} = 27.02 + 5.81 \times 10^{-3}T - 0.29 \times 10^{-6}T^2$$
$$C_{pCO_2} = 26.75 + 42.26 \times 10^{-3}T - 14.25 \times 10^{-6}T^2$$
$$C_{pCO} = 27.11 + 6.552 \times 10^{-3}T - 0.999 \times 10^{-6}T^2$$

$$\Delta H_{\text{reactants}} = 1 \left[\overset{\text{CO}}{\int_{298}^{373} (27.11 + 6.552 \times 10^{-3}T - 0.999 \times 10^{-6}T^2)dT - 110,520} \right]$$
$$+ 1 \left[\overset{O_2}{\int_{298}^{373} (25.29 + 13.25 \times 10^{-3}T - 4.20 \times 10^{-6}T^2)dT + 0} \right]$$
$$+ 3.76 \left[\overset{N_2}{\int_{298}^{373} (27.02 + 5.81 \times 10^{-3}T - 0.29 \times 10^{-6}T^2)dT + 0} \right]$$

$$= -97,995 \text{ J}$$

$$\Delta H_{products} = 1\left[\overset{CO_2}{\int_{298}^{T}(26.75 + 42.26 \times 10^{-3}T - 14.25 \times 10^{-6}T^2)dT - 393.510}\right]$$

$$+ 0.5\left[\overset{O_2}{\int_{298}^{T}(25.29 + 13.25 \times 10^{-3}T - 4.20 \times 10^{-6}T^2)dT + 0}\right]$$

$$+ 3.76\left[\overset{N_2}{\int_{298}^{T}(27.02 + 5.81 \times 10^{-3}T - 0.29 \times 10^{-6}T^2)dT + 0}\right]$$

so that $-340,562 + 141.14T + 35.37 \times 10^{-3}T^2 - 5.81 \times 10^{-6}T^3 = 0$
the solution to which is (by Newton's method starting at $T = 2000$ K)

$$T = 1827 \text{ K}$$

Key Ideas

1. To account for the effect of exothermic or endothermic reactions in the energy balance, enthalpies called standard heats of formation, $\Delta \hat{H}_f^{\circ}$, accompany the "sensible heats" and the phase changes as part of the enthalpy of a component in a stream.

2. Tabulated standard heats of formation are enthalpy changes with reference to 25°C and 1 atm with the elements assigned zero vaues.

3. In the calculation of the standard "heat of reaction" you assume stoichiometric quantities of reactants react completely at 25°C and 1 atm.

4. The adiabatic reaction (flame) temperature is the temperature of the outlet stream of products assuming that $W = 0$ and $Q = 0$ for the process.

Key Terms

Adiabatic flame temperature (p. 457)   Higher heating value (HHV) (p. 449)
Adiabatic reaction temperature (p. 457)   Lower heating value (LHV) (p. 449)
Endothermic reaction (p. 442)   Standard heat of combustion (p. 448)
Exothermic reaction (p. 442)   Standard heat of formation (p. 442)
Heat of reaction (p. 444)   Standard heat of reaction (p. 444)
Heating value (p. 449)   Standard state (p. 444)

Self-Assessment Test

1. Calculate the standard heat of formation of CH_4 given the following experimental results at 25°C and 1 atm:

$$H_2(g) + \tfrac{1}{2}O_2(g) \rightarrow H_2O(l) \qquad \Delta H = -285.84 \text{ kJ/g mol } H_2$$
$$C(graphite) + O_2(g) \rightarrow CO_2(g) \qquad \Delta H = -393.51 \text{ kJ/g mol C}$$
$$CH_4(g) + 2O_2(g) \rightarrow CO_2(g) + 2H_2O(l) \qquad \Delta H = -890.36 \text{ kJ/g mol } CH_4$$

Compare your answer with that found in the table of the heats of formation listed in Appendix F.

2. Calculate the standard heat of reaction for the following reaction from the heats of formation:

$$C_6H_6(g) \rightarrow 3C_2H_2(g)$$

3. Calculate the standard heat of reaction for the Sachse process (in which acetylene is made by partial combustion of LPG) from heat of combustion data:

$$C_3H_8(l) + 2O_2(g) \rightarrow C_2H_2(g) + CO(g) + 3H_2O(l)$$

4. A synthetic gas analyzes 6.1% CO_2, 0.8% C_2H_4, 0.1% O_2, 26.4% CO, 30.2% H_2, 3.8% CH_4 and 32.6% N_2. What is the heating value of the gas measured at 60°F, saturated, when the barometer reads 30.0 in Hg?

5. A dry low-Btu gas with the analysis of CO: 20%, H_2: 20% and N_2: 60% is burned with 200% excess dry air which enters at 25°C. If the exit gases leave at 25°C, calculate the heat transfer from the process per unit volume of entering gas measured at standard conditions (25°C and 1 atm).

6. Methane is burned in a furnace with 100% excess dry air to generate steam in a boiler. Both the air and the methane enter the combustion chamber at 500°F and 1 atm, and the products leave the furnace at 2000°F and 1 atm. If the effluent gases contain only CO_2, H_2O, O_2, and N_2, calculate the amount of heat absorbed by the water to make steam per pound of methane burned.

7. A mixture of metallic aluminum and Fe_2O_3 can be used for high-temperature welding. Two pieces of steel are placed end to end and the powdered mixture is applied and ignited. If the temperature desired is 3000°F and the heat loss is 20% of ($\Delta H_{prod} - \Delta H_{react}$) by radiation, what weight in pounds of the mixture (used in the molecular proportions of $2Al + 1Fe_2O_3$) must be used to produce this temperature in 1 lb of steel to be welded? Assume that the starting temperature is 65°F.

$$2Al + Fe_2O_3 \rightarrow Al_2O_3 + 2Fe$$

Data	C_p, solid [Btu/(lb)(°F)]	Heat of fusion (Btu/lb)	Melting point (°F)	C_p, Liquid [Btu/(lb)(°F)]
Fe and steel	0.12	86.5	2800	0.25
Al_2O_3	0.20	—	—	—

8. Calculate the theoretical flame temperature when hydrogen burns with 400% excess dry air at 1 atm. The reactants enter at 100°C.

Thought Problems

1. A clipping from the *Wall Street Journal* read:

Technical Disputes

Furnace efficiency is a comparison of the energy that goes into a furnace with the usable heat that comes out. Because oil furnaces use blowers to burn the fuel more efficiently and because oil produces less water vapor than gas, less heat is vented up the chimney, the Oil Jobbers Council says. The American Gas Association complains that such calculations ignore variations in oil quality that hurt efficiency. And it contends that oil furnaces are apt to "lose efficiency over time," while it says gas furnaces don't.

Which organization is correct, or is neither correct?

2. A review of additives to gasoline to give blends that improve its octane rating shows that oxygenated compounds necessarily contain lower energy (Btu per gallon). Methanol was the lowest, having a heat of reaction approximately one-half that of gasoline. Methanol costs 40 cents/gal, whereas unleaded premium gas costs 80 cents/gal, so that the result may seem like a standoff (i.e., half the energy at half the cost). Are the two fuels really equivalent in practical use?

3. Degradation of performance and economic loss result from poor boiler performance. What are some of the steps you might take to improve boiler performance?

4. Does the vaporization of a given amount of liquid water at a given temperature require the same amount of heat transfer if the process is carried out in a closed system versus an open system?

5. Thermal destruction systems have become recognized over the past decade as an increasingly desirable alternative to the more traditional methods of disposing of hazardous wastes in landfills and injection wells. What are some of the problems in the combustion of substances such as methylene chloride, chloroform, trichloroethylene, waste oil, phenol, aniline, and hexachloroethane?

6. Would burning a fuel with oxygen or with air yield a higher adiabatic flame temperature?

7. A recent news article said:

> Two workers were killed and 45 others hurt when a blast at the _____ refinery shook a neighborhood and shot flames 500 feet into the air. Hydrogen from a "cracker unit" that separates crude oil into such products as gasoline and diesel fuel burned at temperatures from 4,000 to 5,000 degrees in the 9:50 A.M. accident at the refinery. The fire was put out about an hour later.

Is the temperature cited reasonable?

Discussion Questions

1. Many different opinions have been expressed as to whether gasohol is a feasible fuel for motor vehicles. An important economic question is: Does 10% grain-based-alcohol-in-gasoline gasohol produce positive net energy? Examine the details of the energy inputs and outputs, including agriculture (transport, fertilizer, etc.), ethanol processing (fermentation, distilling, drying, etc.), petroleum processing and distribution, and the use of byproducts (corncobs, stalks, mash, etc.). Ignore taxes and tax credits, and assume that economical processing takes place. Discuss octane ratings, heats of vaporization, flame temperatures, fuel-air ratios, volumetric fuel economy, effect of added water, and the effect on engine parts.

2. The Flameless Rotation Heater (FRH) was selected by *Research & Development* as one of the 100 most significant technical products of 1990. The FRH was designed for the MRE (Meal-Ready to Eat), and is considered a solution to the Armed Forces' "cold ration problem." The FRH is a 40/60 mixture of magnesium-iron powder and inert plastic powders. These materials are molded into a pad that is stable and weighs less than an ounce. The FRH comes in a bag that will also hold an MRE entrée.

The soldier in the field simply pops the 8-oz entrée into the FRH bag with a little water. The resulting exothermic reaction can raise the temperature of the entrée 100°F in about 12 minutes. The FRH is flameless, produces no noxious combustion products, and can be activated in shelters or even in the pocket of a soldier on the move.

How practical do you think the FRH is for the indicated purpose. Is the temperature cited correct?

3. A automobile owner sued his insurance company because his (lead-acid) battery exploded, damaging the hood and motor. The adjuster from the insurance company, after inspecting the remains of the battery, stated that his company was not liable to pay the claim because the battery grossly overheated.

Can a battery overheat? How? If overheating took place, why would the battery explode? What would be the most likely mechanism of the explosion?

4. As reported in the news, a tank car that normally carried ethylene oxide (EO) was turned over to a contractor for cleaning because off-color samples of EO were found during the off-loading of the tank car. At the start of the cleaning procedure the initial pressure in the car was 10 psig. A vent hose was connected to a caustic scrubber, and another hose was used to fill the tank with water that was stored for the fire-fighting system. After the operator filled the tank, he disconnected the vent line and noted an unexpected odor. Consequently, the cleaning procedure was stopped and the tank car valves closed. Then the car was moved to a storage track.

In the middle of the night the tank car exploded, sending pieces of the car as far as 2500 feet away, damaging 10 other tank cars, and several of the cleaning contractor's buildings. No one was injured.

On investigation, it was found that the Illinois Central Railroad had weighed the car in transit, and the weight indicated that the car had apparently contained 29,000 lb (1/6 of the volume) over the designated load weight that had been used in off-loading the EO in the car. Apparently this 29,000 lb was still in the tank car at the time cleaning was started (under the assumption that the tank was empty). Calculations showed that when filled with water, the EO, if mixed with the water, would lead to a 15% solution.

Investigators did not believe a 15% solution of EO in water could explode. The reaction is

$$EO \text{ (liquid)} + H_2O \rightarrow \text{ethylene glycol (mono- and digycols)}$$

Calculate the adiabatic reaction temperature rise for the tank car under the assumption that the EO was well mixed with the water.

Based on the results of these calculations, experiments were carried out on a similar tank car to see what happened in the cleaning process, and it was found that the EO did not mix much with the water. Instead the EO layer originally at the bottom of the tank rose to the top as the tank filled. A 50 cm layer of almost pure EO developed under which a 25 cm layer of a mixture of EO and water occurred; and underneath lay a 200 cm layer of water.

Reactions of pure EO occur as follows

$$EO \rightarrow CO + CH_4 + \text{a little } H_2$$

$$EO + 2.5\, O_2 \rightarrow 2CO_2 + 2H_2O \text{ (Flammability limits in air are 3-100\%)}$$

Estimate the adiabatic reaction temperature of pure EO for each of these two reactions. Prepare a figure showing the estimated temperature in the tank as a function of the EO concentration (use 0, 20, 40, 60, 80, and 100%) on the axis.

Ignition of EO liquid was shown to take place above 450°C. What might a possible mechanism have been for the explosion? What recommendations would you make in preparing a tank car for cleaning?

5. A plant operator was injured when a flexible hose line ruptured when he started to transfer thionyl chloride ($SOCl_2$) by vacuum from a 55 gal drum. He made sure the hose was clean before starting the process by washing it with water and draining the hose thoroughly.

You are asked to investigate this accident, report on the cause(s), and recommend steps to take to prevent reoccurrence. Prepare such a report.

5.5 REVERSIBLE PROCESSES AND THE MECHANICAL ENERGY BALANCE

> ### Your objectives in studying this section are to be able to:
>
> 1. Define a quasi-static and a reversible process.
> 2. Identify a process as reversible or irreversible given a description of the process.
> 3. Define efficiency and apply the concept to calculate the work for an irreversible process.
> 4. Write down the steady-state mechanical energy balance for an open system and apply it to a problem.

LOOKING AHEAD

In this section we briefly examine certain types of ideal processes called reversible processes.

MAIN CONCEPTS

The **reversible** or **quasi-static** process is a hypothetical ideal process that rarely occurs in practice. Why bother with it, then? The reason is that calculations for energy changes can be made for an ideal process, and then an **efficiency** can be used to convert the ideal work or energy change into the actual work or energy change. A process at equilibrium subjected to a differential external force (such as a higher temperature or pressure) so that a differential change occurs is called a reversible process. The system will pass through nonequilibrium states, but with only very slight deviations from equilibrium if the driving force is infinitesimal. Most industrial processes exhibit heat transfer over finite temperature differences, mixing of dissimilar substances, electrical resistance, sudden changes in phase, mass transport under finite concentration differences, free expansion, pipe friction, and other mechanical, chemical, and thermal nonidealities, and consequently are deemed **irreversible**. An irreversible process always involves a degradation of the potential of the process to do work, that is, will not produce the maximum amount of work that would be possible via a reversible process (if such a process could occur).

As an example of a reversible and irreversible process, examine the gas in the cylinder illustrated in Figure 5.13. During an expansion process the piston moves the distance x and the volume of gas confined in the piston increases from V_1 to V_2. Two forces act on the piston; one is the force exerted by the gas, equal to the pressure times the area of the piston, and the other is the force on the piston shaft and head from outside. If the force per unit area exerted by the gas equals the force (F) per unit area (A) exerted by the piston head, nothing happens. If F/A is greater than the pressure of the gas, the gas will be compressed, whereas if F/A is less than the force of the gas, the gas will expand.

In a reversible expansion, none of the available energy of the gas to do work is lost because of friction between the cylinder head and the cylinder wall, or because of turbulence in the gas caused by rapid movement of the gas, or by various viscous effects accompanying the expansion, or other reasons. If the process is ideal, the reversible (ideal) work done by the gas against the piston can be calculated from

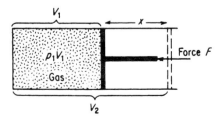

Figure 5.13   Gas expansion.

$$W_{rev} = -\int_{V_1}^{V_2} p\, dV$$

If the process was not ideal, as would most likely be the case, the actual work done would be less. The expansion process might slightly heat the cylinder wall, for example, if friction existed. Example 5.1 in Section 5.1 illustrates the calculation of the reversible work. Example 5.26 below illustrates one real process that is essentially reversible.

EXAMPLE 5.26 Evaporation

How much work is done by 1 liter of liquid water when it evaporates from an open vessel at 25°C and 100 kPa atmospheric pressure?

Solution

Steps 1, 2, 3, and 4   The system is the water. The process is a batch process. Does the water do work in evaporating? Certainly! It does work against the atmosphere. Furthermore, the process, diagrammed in Figure E5.26, is a reversible one because the evaporation takes place at constant temperature and pressure, and presumably the conditions in the atmosphere immediately above the open portion of the vessel are in equilibrium with the water surface. The atmospheric pressure is 100 kPa. The basis will be 1 L of liquid water at 25°C. The specific volume of water at 100 kPa is 1.694 m³/kg H_2O.

Step 5   Basis: 1 L of water (liquid)

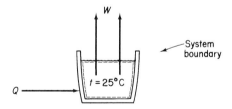

Figure E5.26

Step 6   The general energy balance

$$\Delta E = Q + W - \Delta[(\hat{H} + \hat{K} + \hat{P})m]$$

will not be useful in solving this problem because Q is unknown. However, imagine that an expansible bag is placed over the open face of the vessel so that the system stays a closed system. Because of the reversible conditions established for this problem, the work is

$$W = -\int_{V_1}^{V_2} p\, dV = -p\,\Delta V$$

which represents the reversible work done by the water in pushing back the atmosphere.

Steps 7, 8, and 9

$$W = \frac{-100 \times 10^3 \text{ Pa}}{} \left[\frac{1.694 \text{ m}^3}{\text{kg}} \middle| \frac{1 \text{ kg} - 1 \text{ L}}{} \middle| \frac{10^{-3} \text{ m}^3}{1 \text{ L}}\right] \frac{1 (\text{N})(\text{m}^{-2})}{1 \text{ Pa}} \middle| \frac{1 \text{ J}}{1(\text{N})(\text{m})}$$

$$= -1.693 \times 10^5 \text{ J}$$

Given the concept of an ideal (reversible) process and knowing the work in an actual process, two ways in which we can define mechanical efficiency are

$$\text{efficiency} = \eta_1 = \frac{\text{actual work output for the process}}{\text{work output for a reversible process}} \tag{5.28a}$$

and

$$\text{efficiency} = \eta_2 = \frac{\text{work input for a reversible process}}{\text{actual work input for the process}} \tag{5.28b}$$

depending on whether work is done by the system [Eq. (5.28a)] or work is being done on the system [Eq. (5.28b)].

Another type of efficiency is concerned just with the useful energy output divided by the total energy input:

$$\text{efficiency} = \eta_3 = \frac{\text{useful energy out}}{\text{energy in}} \tag{5.28c}$$

As an example, assume that the conversion of fuel in a power plant yields 88 kJ in the steam product per 100 kJ of available energy from the coal being burned. Also assume that the conversion of the energy in the steam to mechanical energy is 43% efficient, and the conversion of the mechanical to electrical energy is 97% efficient, all based on Eq. (5.28c). The overall efficiency is $(0.88)(0.43)(0.97) = 0.37$, meaning that two-thirds of the initial energy is dissipated as heat to the environment. These definitions provide a good way to compare process performance for energy conservation (but not the only way). Table 5.3 lists the efficiency of conversion for several common devices defined in terms of Eq. (5.28c).

EXAMPLE 5.27 Use of Efficiency

Calculate the reversible work required to compress 5 ft^3 of an ideal gas initially at 100°F from 1 to 10 atm in an adiabatic cylinder. Such a gas has an equation of state $pV^{1.40}$ = constant. Then calculate the actual work required if the efficiency of the process is 80%.

Solution

Steps 1, 2, 3, and 4   Figure 5.13 indicates the type of apparatus for the compression. The final volume is

$$V_2 = V_1 \left(\frac{p_1}{p_2} \right)^{1/1.40} = 5 \left(\frac{1}{10} \right)^{1/1.40} = 0.965 \text{ ft}^3$$

Step 5   The basis is 5 ft^3 at 100°F and 1 atm.

Steps 6–9

$$W_{\text{rev}} = -\int_{V_1=5}^{V_2=0.965} p \, dV = -\int_{V_1}^{V_2} p_1 \left(\frac{V_1}{V} \right)^{1.40} dV = -p_1 V_1^{1.4} \int_{V_1}^{V_2} V^{-1.40} \, dV$$

$$= -\frac{p_1 V_1^{1.40}}{1 - 1.40} (V_2^{-0.40} - V_1^{-0.40}) = -\frac{p_2 V_2 - p_1 V_1}{1 - 1.40}$$

$$= \frac{[(10)(0.965) - (1)(5)][\text{ft}^3)(\text{atm})]}{0.40} \left| \frac{1.987 \text{ Btu}}{0.7302 \ (\text{ft}^3)(\text{atm})} \right.$$

$$= 31.63 \text{ Btu}$$

The positive sign means that work is done on the system.
The actual work required is

$$\frac{31.63}{0.8} = 39.5 \text{ Btu}$$

TABLE 5.3   Energy Efficiencies

Device	Efficiency, %
Incandescent lamp	4–5
Gasoline automotive engine	25–26
Aircraft turbine engine	36–37
Diesel engine	37–38
Small electric motor	62–63
Storage battery	72–73
Home gas furnace	84–85

EXAMPLE 5.28 Calculation of Plant Efficiency

An analysis of energy usage takes place in all plants. This example uses data from M. Fehr, "An Auditor's View of Furnace Efficiency," in *Hydrocarbon Processing*,

Nov., 1988, p. 93. Further details about the process equipment and energy auditing can be found in the article.

Figure E5.28 illustrates a gas fired boiler. The data calculated from measurements on

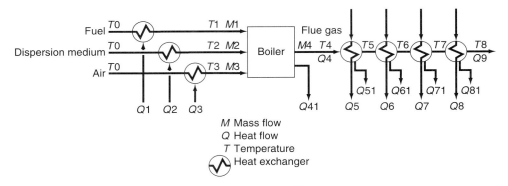

Figure E5.28

the heater were (all in kJ per m^3 at SC of fuel gas):

LHV of fuel:	36,654	$Q4$:	32,114
$Q1$:	16	$Q41$:	6,988
$Q2$:	0	$Q8$:	1,948
$Q3$:	2,432	$Q9$:	2,643

Calculations from measurements gave values for sums of the other Qs of

$$Q5 + Q6 + Q7 = 9{,}092$$
$$Q5 + Q6 + Q7 + Q51 + Q61 + Q71 = 27{,}119$$
$$Q8 + Q81 = 2{,}352$$

From these heat transfer calculations several efficiencies could be determined to evaluate plant performance:

(1) Gross efficiency

$$\frac{\text{LVH} + Q1 + Q2 + Q3 - Q9}{\text{LHV}} = \frac{36{,}654 + 16 + 0 + 2{,}432 - 2{,}643}{36{,}654} = 0.995$$

(2) Thermal efficiency

$$\frac{Q5 + Q6 + Q7 + Q8}{\text{LHV} + Q1 + Q2 + Q3} = \frac{9{,}092 + 1{,}948}{36{,}654 + 16 + 0 + 2{,}432} = 0.282$$

(3) Combustion efficiency

$$\frac{Q4}{Q4 + Q41} = \frac{32{,}114}{32{,}114 + 6{,}988} = 0.821$$

In some processes, such as distillation columns or reactors, heat transfer and enthalpy changes are *the* important energy components in the energy balance. Work, potential energy, and kinetic energy are zero or quite minor. However, in other processes, such as the compression of gases and pumping of liquids, work and the mechanical forms of energy are the important factors. For these processes, an energy balance treating only the mechanical forms of energy becomes a useful tool.

The energy balance of Sec. 5.1 is concerned with various classes of energy without inquiring into how "useful" each form of energy is to human beings. Our experience with machines and thermal processes indicates that some types of energy cannot be transformed completely into other types, and that energy in one state cannot be transformed to another state without the addition of extra work or heat. For example, internal energy cannot be completely converted into mechanical work. To account for these limitations on energy utilization, the second law of thermodynamics was eventually developed as a general principle.

One of the consequences of the second law of thermodynamics is that two categories of energy of different "quality" can be "envisioned"

(1) The so-called *mechanical* forms of energy, such as kinetic energy, potential energy, and work, which are *completely* convertible by an *ideal* (reversible) engine from one form to another within the class

(2) Other forms of energy, such as internal energy and heat, which are not so freely convertible

Of course, in any real process with friction, viscous effects, mixing of components, and other dissipative phenomena taking place that prevent the complete conversion of one form of mechanical energy to another, allowances will have to be made in making a balance on mechanical energy for these "losses" in quality.

A balance on mechanical energy can be written on a microscopic basis for an elemental volume by taking the scalar product of the local velocity and the equation of motion.[7] After integration over the entire volume of the system the **steady-state mechanical energy balance for a system with mass interchange with the surroundings** becomes, on a per unit mass basis,

$$\Delta(\hat{K} + \hat{P}) + \int_{p_1}^{p_2} \hat{V}\,dp - \hat{W} + \hat{E}_v = 0 \tag{5.29}$$

where $\hat{K}$ and $\hat{P}$ are associated with the mass in and out of the system, and E_v represents the loss of mechanical energy, that is, the irreversible conversion *by the*

---

[7]J. C. Slattery, *Momentum, Energy, and Mass Transfer*, 2d ed. New York: Krieger Publishing Co., 1981.

flowing fluid of mechanical energy to internal energy, a term that must in each individual process be evaluated by experiment (or, as occurs in practice, by use of already existing experimental results for a similar process). Equation (5.29) is sometimes called the **Bernoulli equation**, especially for the reversible process for which $\hat{E}_v$) = 0. The mechanical energy balance is best applied to fluid-flow calculations when the kinetic and potential energy terms and the work are of importance, and the friction losses can be evaluated from handbooks with the aid of *friction factors* or *orifice coefficients*.

Let us now look at two typical applications of the steady-state mechanical energy balance.

EXAMPLE 5.29 Calculation of Reversible Work for a Flow Process

We will repeat the solution of Example 5.27 except that in this problem the process will be an open system in the steady state.

Solution

Steps 1, 2, 3, and 4   Figure E5.29 designates the system and data. The moles of gas are

$$n_1 = \frac{p_1 V_1}{RT_1} = \frac{1 \text{ atm} \quad 5 \text{ ft}^3 \quad \quad 1 \text{ (lb mol)}(°R)}{560°R \quad 0.7302 \text{ (ft}^3)(\text{atm})} = 0.0122 \text{ lb mol}$$

$p_1 = 1$ atm
$T_1 = 560°R$

Adiabatic Reversible Compressor

$p_2 = 10$ atm

Figure E5.29

Step 5

$$\text{Basis} = 0.0122 \text{ lb mol}$$

Steps 3, 6, 7, 8, and 9   The mechanical energy balance (per unit mass, or mole, here for a single component)

$$\Delta(\hat{K} + \hat{P}) + \int_{p_1}^{p_2} \hat{V}\, dP - \hat{W} + \hat{E}_v = 0$$

can be simplified

$$\Delta\hat{K} = 0$$

$$\Delta\hat{P} = 0$$

$$\hat{E}_v = 0 \text{ (assumed reversible)}$$

so that

$$\hat{W}_{\text{rev}} = \int_{p_1}^{p_2} \hat{V}\, dp = \int_{p_1}^{p_2} \hat{V}_1 \left(\frac{p_1}{p}\right)^{1/1.40} dp$$

$$= \hat{V}_1\, p_1^{0.714}[(3.50(p_2^{0.286} - p_1^{0.286})]$$

$$W_{\text{rev}} = n_1\, \hat{W}_{\text{rev}} = n_1 \left(\frac{V_1}{n_1}\right) p_1^{0.714}\, [3.50(p_2^{0.286} - p_1^{0.286})]$$

$$= (5)\, (1)^{0.714}\, (3.50)[(10)^{0.286} - 1^{0.286}](\text{ft}^3)(\text{atm}) \left(\frac{1.987\ \text{Btu}}{0.7302\ (\text{ft}^3)(\text{atm})}\right)$$

$$= 44.3\ \text{Btu}$$

The actual work required to be done on the system is

$$\frac{44.3}{0.8} = 55.4\ \text{Btu}$$

EXAMPLE 5.30 Application of the Mechanical Energy Balance

Calculate the work per minute required to pump 1 lb of water per minute from 100 psia and 80°F to 1000 psia and 100°F. The exit stream is 10 ft above the entrance stream.

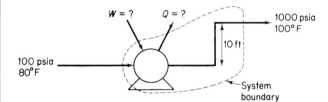

Figure E5.30

Solution

Steps 1, 2, 3, and 4   The system shown in Figure E5.30 is a steady-state process.
Steps 4 and 7   The general mechanical energy balance is

$$\Delta(\hat{K} + \hat{P}) + \int_{p_1}^{p_2} \hat{V}\, dp - \hat{W} + \hat{E}_v = 0 \qquad (a)$$

We will assume that $\Delta\hat{K}$ is insignificant, and also preliminarily assume that the process is reversible so that $E_v = 0$, and the pump is 100% efficient. (Subsequently, we will consider what to do if the process is not reversible.) Equation (a) reduces to

$$\hat{W} = \int_{p_1}^{p_2} \hat{V} \, dp + \Delta\hat{P} \tag{b}$$

Step 5

Basis: 1 min of operation = 1 lb H2O

Steps 6 and 8   From the stream tables, the specific volume of liquid water is 0.01607 ft^3/lb$_m$ at 80°F and 0.01613 ft^3/lb$_m$ at 100°F. For all practical purposes the water is incompressible, and the specific volume can be taken to be 0.0161 ft^3/lb$_m$. We have only one unknown in Eq. (b): ($\hat{W}$).

Step 9

$$\Delta\hat{P} = \frac{1 \text{ lb}_m}{} \left| \frac{10 \text{ ft}}{} \right| \frac{32.2 \text{ ft}}{\text{sec}^2} \left| \frac{}{\frac{32.2(\text{ft})(\text{lb}_m)}{(\text{lb}_f)(\text{sec}^2)}} \right| \frac{1 \text{ Btu}}{778(\text{ft})(\text{lb}_f)} = 0.0129 \text{ Btu}$$

$$1 \text{ lb}_m \int_{100}^{1000} 0.0161 \, dp = \frac{1 \text{ lb}_m}{} \left| \frac{0.0161 \text{ ft}^3}{\text{lb}_m} \right| \frac{(1000 - 100) \, (\text{lb}_f)}{\text{in.}^2} \left| \left(\frac{12 \text{ in.}}{1 \text{ ft}}\right)^2 \right| \frac{1 \text{ Btu}}{778(\text{ft})(\text{lb}_f)}$$

$$= 2.68 \text{ Btu}$$

$$\hat{W} = 2.68 + 0.0129 = 2.69 \, \frac{\text{Btu}}{\text{lb}_m}$$

About the same value can be calculated using Eq. (5.13) if $\hat{Q} = \hat{K} = 0$, because the enthalpy change for a reversible process for 1 lb of water going from 100 psia and 100°F to 1000 psia is 2.70 Btu. Make the computation yourself. However, usually the enthalpy data for liquids other than water are missing, or not of sufficient accuracy to be valid, which forces an engineer to turn to the mechanical energy balance.

We might now well inquire for the purpose of purchasing a pump-motor, say, as to what the work would be for a real process instead of the fictitious reversible process assumed above. First, you would need to know the efficiency of the combined pump and motor so that the actual input from the surroundings (the electric connection) to the system would be known. Second, the friction losses in the pipe, valves, and fittings must be estimated so that the term $\hat{E}_v$ could be reintroduced into Eq. (a). Suppose, for the purposes of illustration, that $\hat{E}_v$ was estimated to be, from an appropriate handbook, 320 (ft)(lb$_f$/lb$_m$ and the motor-pump efficiency was 60% (based on 100% efficiency for a reversible pump-motor). Then,

$$\hat{E}_v = \frac{320(\text{ft})(\text{lb}_f)}{1 \text{ lb}_m} \left| \frac{1 \text{ Btu}}{778(\text{ft})(\text{lb}_f)} \right. = 0.41 \text{ Btu/lb}_m$$

$$\hat{W} = 2.68 + 0.013 + 0.41 = 3.10 \text{ Btu/lb}_m$$

Remember that the positive sign indicates that work is done on the system. The pump motor must have the capacity

$$\frac{3.10 \text{ Btu}}{1 \text{ min}} \left| \frac{1}{0.60} \right| \frac{1 \text{ min}}{60 \text{ sec}} \left| \frac{1.415 \text{ hp}}{1 \text{ Btu/sec}} \right. = 0.122 \text{ hp}$$

LOOKING BACK

In this section we reviewed the concept of the reversible (ideal) process, and indicated some ways in which the calculation of reversible work or energy change is converted into the actual work or energy change.

Key Ideas

1. A reversible process is a process that occurs by means of differential displacements from equilibrium.
2. Most real processes are not reversible.
3. The actual work in a process can be calculated from the reversible work adjusted by a factor determined by experiment on the equipment being used.

Key Terms

Bernoulli equation (p. 472)
Efficiency (p. 468)
Friction factors (p. 472)
Ideal process (p. 465)
Irreversible (p. 466)

Mechanical energy balance (p. 471)
Mechanical work (p. 471)
Orifice coefficient (p. 472)
Quasi-static (p. 466)
Reversible (p. 466)

Self-Assessment Test

1. Which process will yield more work: (1) expansion of a gas confined by a piston against constant pressure, or (2) reversible expansion of a gas confined by a piston?

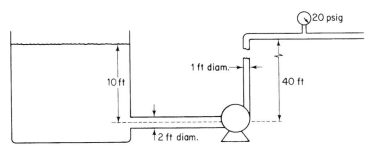

Figure SA5.5–5

2. Differentiate between thermal and mechanical energy.

3. Define a reversible process.

4. Find Q, W, ΔE, and ΔH for the reversible compression of 3 moles of an ideal gas from a volume of 100 dm^3 to 2.4 dm^3 at a constant T of 300 K.

5. Water is pumped from a very large storage reservoir as shown in Figure SA5.5–5 at the rate of 2000 gal/min. Determine the minimum power (i.e., that for a reversible process) required by the pump in horsepower.

Thought Problems

1. A news announcement in a professional journal described a 20-hp Stirling engine that was going to be connected to and drive a 68-kW generator. Would you buy one of these at $4500?

2. An author compared the effectiveness of gasoline-powered automobiles with electric-powered automobiles as in the following table, and concluded that electric-powered vehicles overall were more efficient. Is the comparison valid?

Efficiencies	Gasoline (bbl of oil)	Electric (bbl of oil)
kWh equivalent	1700	1700
Refinery efficiency	74% to gasoline	89% to heavy oil
Distribution efficiency	95%	
Power generation efficiency		40%
Power distribution efficiency		91%
Battery efficiency		70%
Motor, drive train efficiency	14.7%	80%
kWh available	175.7	308.8
Vehicle weight	2400 lb	2800 lb
Road load: 50 mph, level	8.4 kW	10 kW
Distance traveled	1045 miles	1545 miles

3. Examine Figure TP5.5–3 below. Is it possible by blowing in the stem of the funnel that the ping-pong ball will rise up toward the steam rather than fall down? Why?

Figure TP5.5–3

Discussion Questions

1. Which is more efficient in heating a given amount of water, an electric coffee pot, a microwave oven, or a pot on an electric stove?

2. Discuss the feasibility of the following novel proposal as a way to extract energy from river flow that would not require large dams. See Figure DQ5.5–2. The apparatus consists of a pair of tall chambers connected at the top by a tube containing an air turbine the diameter of a dinner plate. As the water flows through the system, one chamber fills, pushing air past one side of the turbine; the other side empties, sucking air from the tube.

   Whereas a standard turbine uses a conventional fan blade that spins according to the direction of the water, the counter-rotating turbine in the proposed system uses a blade shaped like an airplane wing that creates lift to spin the turbine when the airflow rushes over its front edge. The blades are placed around the turbine in paddlewheel fashion. As the first water chamber fills, air whooshes through the tube and hits the front of the paddlewheel, lifting each blade upward as it spins past the airflow. As the first chamber empties and the other fills, air races in the opposite direction through the tube and over the blades at the rear of the paddlewheel. The blades having spun halfway around the wheel, are now upside down, so the "lift" pulls them downward. Thus, even though the airflow reverses with each cycle, the turbine rotates in the same direction, generating a constant flow of electricity.

   Simple gates and counterbalances at the bottom coordinate the operation of the unit as the chambers alternately fill and empty. At the top of the cycle, water pressure in-

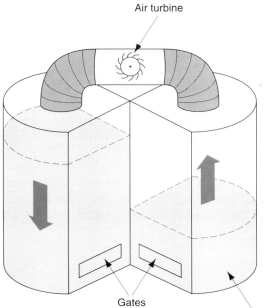

Air turbine

Gates

Water chamber   **Figure DQ5.5–2**

side the filled chamber exceeds the pressure exerted by a counterweight on the outflow door at the bottom of the chamber, popping it open. The same motion closes the outflow gate in the opposite chamber, thanks to a single shaft that controls the two gates.

5.6 HEATS OF SOLUTION AND MIXING

Your objectives in studying this section are to be able to:

1. Distinguish between ideal solutions and real solutions.
2. Calculate the heat of mixing, or the heat of dissolution, at standard conditions given the moles of the materials forming the mixture.
3. Calculate the standard integral heat of solution.
4. Define the standard integral heat of solution at infinite dilution.
5. Apply the energy balance to problems in which the heat of mixing is significant.
6. Use an enthalpy-concentration chart in solving material and energy balances.

LOOKING AHEAD

In this section we describe what happens to the enthalpy when pure components that do not form ideal solutions mix.

MAIN CONCEPTS

To this point we have assumed that when a stream consists of several components, the total properties of the stream are the appropriately weighted sum of the properties of the individual components. For such **ideal solutions**, we could write down, for the heat capacity of an ideal mixture, for example

$$C_{p \text{ mixture}} = x_A C_{pA} + x_B C_{pB} + x_C C_{pC} + \dots$$

or, for the enthalpy,

$$\Delta \hat{H}_{\text{mixture}} = x_A \, \Delta \hat{H}_A + x_B \, \Delta \hat{H}_B + x_C \, \Delta \hat{H}_C + \dots$$

Gas mixtures are good examples of ideal solutions.

However, for liquid mixtures we often find that heat is absorbed or evolved from the system upon mixing of the components. Such a solution would be called a **"real" solution**. Per mole of solute

$$\Delta H^{\circ}_{\text{final solution}} - \Delta H^{\circ}_{\text{initial components}} = \Delta H^{\circ}_{\text{mixing}} \qquad (5.30)$$

The specific **heat of mixing** ($\Delta \hat{H}^{\circ}_{\text{mixing}}$) (i.e., the enthalpy change on mixing per unit mass) has to be determined experimentally, but can be retrieved from tabulated experimental (smoothed) results, once such data are available. This type of energy change has been given the formal name **heat of solution** when one substance dissolves in another; and there is also the negative of the heat of solution, the **heat of dissolution**, for a substance that separates from a solution.

As an example, tabulated data for heats of solution for HCl appear in Table 5.4 in terms of energy *per mole of solute* for consecutively added quantities of solvent to the solute; the gram mole refers to the gram mole of solute. Heats of solution are somewhat similar to heats of reaction in that an energy change takes place because of differences in the forces of attraction of the solvent and solute molecules. Of course, these energy changes are much smaller than those we find accompanying the breaking and combining of chemical bonds. Heats of solution are conveniently handled the same way as are the heats of reaction in the energy balance.

The solution process can be represented by an equation such as the following:

$$HCl(g) + 5H_2O \rightarrow HCl \cdot 5H_2O$$

or

$$HCl(g) + 5H_2O \rightarrow HCl(5H_2O)$$

$$\Delta H^{\circ}_{\text{soln}} = -64,047 \text{ J/g mol HCl(g)}$$

The expression $HCl(5H_2O)$ means that 1 mole of HCl has been dissolved in 5 moles of water, and the enthalpy change for the process is $-64,047$ J/g mol of HCl. Table 5.4 shows the heat of solution for various cumulative numbers of moles of water added to 1 mole of HCl.

The **standard integral heat of solution** is the cumulative $\Delta H^{\circ}_{\text{soln}}$ as shown in the next to last column of Table 5.4 for the indicated number of molecules of water. As successive increments of water are added to the mole of HCl, the cumulative heat of solution (the integral heat of solution) increases, but the incremental enthalpy change decreases as shown in column 3. Note that both the reactants and products have to be at standard conditions. The heat of dissolution would be the negative of these values. The integral heat of the solution is plotted in Figure 5.14, and you can see that an asymptotic value is approached as the solution becomes more and more dilute. At infinite dilution this value is called the **standard**

TABLE 5.4   Heat of Solution of HCl (at 25°C and 1 atm)

Composition	Total moles H_2O added to 1 mole HCl	$-\Delta \hat{H}°$ for each incremental step (J/g mol HCl) $= -\Delta \hat{H}°_{dilution}$	Integral heat of solution: cumulative $-\Delta \hat{H}°$ (J/g mol HCl)	Heat of formation $-\Delta \hat{H}°_f$ of (J/g mol HCl)
HCl(g)	0			92,311
HCl · $1H_2O$(aq)	1	26,225	26,225	118,536
HCl · $2H_2O$(aq)	2	22,593	48,818	141,129
HCl · $3H_2O$(aq)	3	8,033	56,851	149,161
HCl · $4H_2O$(aq)	4	4,351	61,202	153,513
HCl · $5H_2O$(aq)	5	2,845	64,047	156,358
HCl · $8H_2O$(aq)	8	4,184	68,231	160,542
HCl · $10H_2O$(aq)	10	1,255	69,486	161,797
HCl · $15H_2O$(aq)	15	1,503	70,989	163,300
HCl · $25H_2O$(aq)	25	1,276	72,265	164,576
HCl · $50H_2O$(aq)	50	1,013	73,278	165,589
HCl · $100H_2O$(aq)	100	569	73,847	166,158
HCl · $200H_2O$(aq)	200	356	74,203	166,514
HCl · $500H_2O$(aq)	500	318	74,521	166,832
HCl · $1000H_2O$(aq)	1,000	163	74,684	166,995
HCl · $50,000H_2O$(aq)	50,000	146	75,077	167,388
HCl · ∞H_2O		67	75,144	167,455

SOURCE: *National Bureau of Standards Circular 500*. Washington, DC: U.S. Government Printing Office, 1952.

integral heat of solution at infinite dilution and is −75,144 J/g mol of HCl. What can you conclude about the reference state for the heat of solution of pure HCl from Figure 5.14? In Appendix H are other tables presenting standard integral heat of solution data and the heats of formation of solutions. Since the energy changes for heats of solution are point functions, you can easily look up any two concentrations of HCl and find the energy change caused by adding or subtracting water. For example, if you mix 1 mole of HCl · 15 H_2O and 1 mole of HCl · $5H_2O$ you obtain 2 moles of HCl · $10H_2O$, and the total enthalpy change at 25°C is

$$\Delta H° = [2(-69,486)] - [1(-70,989) + 1(-64,047)]$$

$$= -3936 \text{ J}$$

You would have to remove 3936 J to keep the temperature of the final mixture at 25°C.

To calculate the standard heat of formation of a solute in solution, you proceed as follows. What is the standard heat of formation of 1 g mol of HCl in 5 g

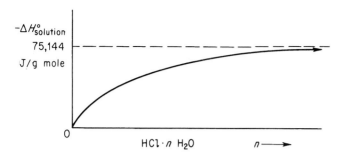

Figure 5.14 Integral heat of solution of HCl in water

mol of H_2O? We treat the solution process in an identical fashion to a chemical reaction:

				J/g mol		
$\frac{1}{2}H_2(g) + \frac{1}{2}Cl_2(g)$	$=$	$HCl(g)$	$A:$	$\Delta\hat{H}^{\circ}_{f}$	$=$	$-92{,}311$
$HCl(g) + 5H_2O$	$=$	$HCl(5H_2O)$	$B:$	$\Delta\hat{H}^{\circ}_{soln}$	$=$	$-64{,}047$
$\frac{1}{2}H_2(g) + \frac{1}{2}Cl_2(g) + 5H_2O$	$=$	$HCl(5H_2O)$	$A+B:$	$\Delta\hat{H}^{\circ}_{f}$	$=$	$-156{,}358$

It is important to remember that the heat of formation of H_2O itself does not enter into the calculation. The heat of formation of HCl in an infinitely dilute solution is

$$\Delta\hat{H}^{\circ}_{f} = -92{,}311 - 75{,}144 = -167{,}455 \text{ kJ/g mol}$$

Another type of heat of solution that is occasionally encountered is the partial molal heat of solution. Information about this thermodynamic property can be found in most standard thermodynamic texts or in books on thermochemistry, but we do not have the space to discuss it here.

One point of special importance concerns the formation of water in a chemical reaction. When water participates in a chemical reaction in solution as a reactant or product of the reaction, you must use the values for the heat of formation of the solvated water such as shown in the last column of Table 5.4.

EXAMPLE 5.31 Application of Heat of Solution Data

Hydrochloric acid is an important industrial chemical. To make aqueous solutions of it in a commercial grade (known as *muriatic acid*), purified HCl(g) is absorbed in water in a tantalum absorber in a continuous process. How much heat must be removed from the absorber per 100 kg of product if hot HCl(g) at 120°C is fed into water in the absorber as shown in Figure E5.31? The feed water can be assumed to be at 25°C, and the exit product HCl(aq) is 25% HCl (by weight) at 35°C.

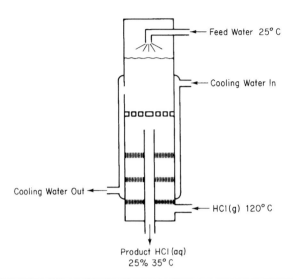

Figure E5.31

Solution

Steps 1, 2, 3, and 4   We need to obtain enthalpy data per mole of HCl. Consequently, we will first convert the product into moles of HCl and moles of H_2O.

Component	kg	Mol. wt.	kg mol	Mole fraction
HCl	25	36.37	0.685	0.141
H_2O	75	18.02	4.163	0.859
Total	100		4.848	1.000

The mole ratio of H_2O to HCl is $4.163/0.685 = 6.077$.

Step 5   The system will be the HCl and water (not including the cooling water).

Basis: 100 kg of product

Ref. temperature: 25°C

Steps 6 and 7   The energy balances reduces to $Q = \Delta H$, and both the initial and final enthalpies of all the streams are known or can be calculated directly, hence the problem has a unique solution. The kg and moles of HCl in and out, and the water in and out, have been calculated above.

Step 3 (continued)   The enthalpy values for the streams are [C_p for the HCl(g) is from Table E.l; C_p for the product is approximately 2.3 J/(g)(°C); $\Delta \hat{H}_f^\circ$ for HCl(6.077 H_2O) $\cong -157,753$ J/g mol HCl]:

Stream	g mol	$\Delta \hat{H}_f^o$ (J/g mol)	T (°C)	$\Delta \hat{H}$ (J/g mol)	ΔH (J)
Feed $H_2O(\ell)$	4.163	–285,840	25	0	–1,189,952
HCl(g)	0.685	– 92,311	120(0.685)$\int_{25}^{120} C_p \, dT = 2747$		–60,486
HCl(aq)	4.848	–267,741	35(100)$\int_{25}^{35} C_p \, dT = 2300$		–1,295,708

Note: The $\Delta \hat{H}°_f$ of the HCl(aq) in J/g mol of product is calculated as

$$[4.163(-285,840) + 0.685(-92,311) + 0.685(-65,442)](1/4.184) = -267,742 \text{ J/g mol product.}$$

$$Q = \Delta H_{out} - \Delta H_{in} = (-1,295,708) - (-1,189,952 - 60,480) = -45,275 \text{ J}$$

This value corresponds to the heat of mixing for 0.685 kg mol of HCl adjusted by the sensible heats.

The negative value of Q means heat is removed from the system.

A convenient way to represent enthalpy data for binary solutions is via an **enthalpy-concentration diagram**. Enthalpy-concentration diagrams ($H-x$) are plots of specific enthalpy versus concentration (usually weight or mole fraction) with temperature as a parameter. Figure 5.15 illustrates one such plot. If available,[8] such charts are useful in making combined material and energy balances calculations in distillation, crystallization, and all sorts of mixing and separation problems. You will find a few examples of enthalpy-concentration charts in Appendix I.

EXAMPLE 5.32 Application of the Enthalpy-Concentration Chart

Six hundred pounds of 10% NaOH per hour at 200°F are added to 400 lb/hr of 50% NaOH at the boiling point in an insulated vessel. Calculate the following:

(1) The final temperature of the exit solution

(2) The final concentration of the exit solution

(3) The pounds of water evaporated per hour during the process.

Solution

Use the steam tables and the NaOH-H_2O enthalpy-concentration chart in Appendix I as your source of data. (The reference conditions for the latter chart are $\Delta H = 0$ at 32°F

[8]For a literature survey as of 1957, See Robert Lemlich, Chad Gottschlich, and Ronald Hoke, *Chem. Eng. Data Ser.*, *2* (1957): 32. Additional references: for CCl_4, see M. M. Krishniah et al., *J. Chem. Eng. Data, 10* (1965): 117; for EtOH-EtAc, see Robert Lemlich, Chad Gottschlich, and Ronald Hoke, *Br. Chem. Eng., 10* (1965): 703; for methanol-toluene, see C. A. Plank and D. E. Burke, *Hydrocarbon Process, 45* (8), (1966): 167; for acetone-isopropanol, see S. N. Balasubramanian, *Br. Chem. Eng., 12* (1967): 1231; for acetonitrile-water-ethanol, see Reddy and Murti, *ibid.*, *13*, (1968): 1443; for alcohol-aliphatics, see Reddy and Murti, *ibid., 16* (1971): 1036; and for H_2SO_4, see D. D. Huxtable and D. R. Poole, *Proc. Int. Solar Energy Soc.*, Winnipeg, *8*, (August 15, 1976): 178.

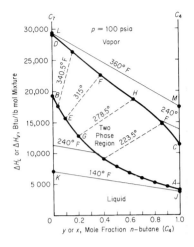

Figure 5.15 Enthalpy concentration chart of *n*-butane-*n*-heptane at 100 psia. Curve DFHC is the saturated vapor; curve BEGA is the saturated liquid. The dashed lines are equilibrium tie lines connecting *y* and *x* at the same temperature.

for liquid water and $\Delta H = 0$ for an infinitely dilute solution of NaOH, with pure caustic having an enthalpy at 68°F of 455 Btu/lb above this datum.) Treat the process as a flow process. The energy balance reduces to $\Delta H = 0$.

$$\text{Basis: 1000 lb of final solution} \equiv 1 \text{ hr}$$

We can write the following material balance:

Component	10% solution	+	50% solution	=	Final solution	wt %
NaOH	60		200		260	26
H_2O	540		200		740	74
Total	600		400		1000	100

Enthalpy Data:

	10% solution	*50% solution*
$\Delta \hat{H}$ (Btu/lb):	152	290

The energy balance is

10% solution		*50% solution*		*Final solution*
500(152)	+	400(290)	=	ΔH
91,200	+	116,000	=	207,200

Note that the enthalpy of the 50% solution at its boiling point is taken from the bubble-point at $\omega = 0.50$. The enthalpy per pound of the final solution is

$$\frac{207{,}200 \text{ Btu}}{1000 \text{ lb}} = 207 \text{ Btu/lb}$$

On the enthalpy-concentration chart for NaOH–H_2O, for a 26% NaOH solution with an enthalpy of 207 Btu/lb, you would find that only a two-phase mixture of (1) saturated H_2O vapor and (2) NaOH–H_2O solution at the boiling point could exist. To get the fraction H_2O vapor, we have to make an additional energy (enthalpy) balance. By interpolation, draw the tie line through the point $x = 0.26$, $H = 207$ (make it parallel to the 220° and 250°F tie lines). The final temperature appears from Figure E5.32 to be 232°F; the enthalpy of the liquid at the bubble point is about 175 Btu/lb. The enthalpy of the saturated water vapor (no NaOH is in the vapor phase) from the steam tables at 232°F is 1158 Btu/lb. Let x = lb of H_2O evaporated.

Basis: 1000 lb of final solution

$$x(1158) + (1000 - x)175 = 1000(207.2)$$

$$x = 32.6 \text{ of } H_2O \text{ evaporated/hr}$$

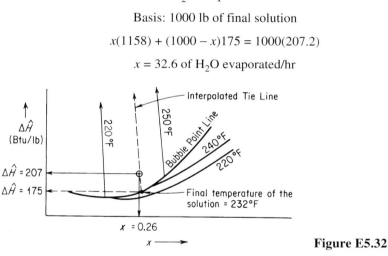

Figure E5.32

Enthalpy changes can also be calculated using graphical techniques that are described in texts treating the unit operations of chemical engineering.

LOOKING BACK

In this section we described how to carry out energy balances when enthalpy changes occur because of the heat of mixing. We also described the use of binary enthalpy-concentration charts.

Key Ideas

1. When pure components are mixed for nonideal solutions (usually liquids), enthalpy changes can occur so that the enthalpy of the solution is not the linearly weighted combination of the respective enthalpies of the pure components.

2. The energy balance calculations are the same for ideal and nonideal solutions; the standard heats of formation used for the mixtures are determined by experiment.

Key Words

Enthalpy-concentration chart (p. 483)   Heat of solution (p. 479)
Heat of dissolution (p. 479)   Ideal solution (p. 478)
Heat of mixing (p. 479)   Integral heat of solution (p. 479)

Self-Assessment Test

1. Is a gas mixture an ideal solution?

2. Give **(a)** two examples of exothermic mixing of two liquids and **(b)** two examples of endothermic mixing based on your experience.

3. **(a)** What is the reference state for H_2O in the table for the heat of solution of HCl?
   (b) What is the value of the enthalpy of H_2O in the reference state?

4. Use the heat of solution data in Appendix H to determine the heat transferred per mole of entering solution into or out of (state which) a process in which 2 g mol of a 50 *mole* % solution of sulfuric acid at 25°C is mixed with water at 25°C to produce a solution at 25°C containing a mole ratio of 10 H_2O to 1 H_2SO_4.

5. Calculate the heat that must be added or removed per ton of 50 wt % H_2SO_4 produced by the process shown in Figure SA5.6–5.

6. For the sulfuric acid-water system, what are the phase(s), composition(s), and enthalpy(ies) existing at $\Delta \hat{H}$) = 120 Btu/lb and $T = 260°F$?

7. Estimate the heat of vaporization of an ethanol-water mixture at 1 atm and an ethanol mass fraction of 0.50 from the enthalpy-concentration chart in Appendix I.

Thought Problems

1. A tanker truck of hydrochloric acid was inadvertently unloaded into a large storage tank used for sulfuric acid. After about one-half of the 3000-gal load had been discharged, a violent explosion occurred, breaking the inlet and outlet lines and buckling the tank. What might be the cause of the explosion?

2. A concentrated solution (73%) of sodium hydroxide was stored in a vessel. Under normal operations, solution was forced out by air pressure as needed. When application of air pressure did not work, apparently due to solidification of the caustic solution, water was poured through a manhole to dilute the caustic and free up the pressure line. An

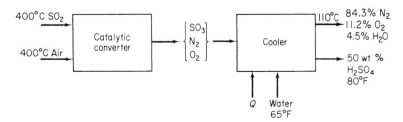

Figure SA5.6–5

explosion took place and splashed caustic out of the manhole 15 ft into the air. What caused the incident?

5.7 HUMIDITY CHARTS AND THEIR USE

Your objectives in studying this section are to be able to:

1. Define humidity, humid heat, humid volume, dry-bulb temperature, wet-bulb temperature, humidity chart, moist volume, and adiabatic cooling line.
2. Explain and show by use of equations why the slope of the wet-bulb lines are the same as the adiabatic cooling lines for water-air mixtures.
3. Use the humidity chart to determine the properties of moist air, and to calculate enthalpy changes and solve heating and cooling problems involving moist air.

LOOKING AHEAD

In this section we describe how to use humidity charts in making material and energy balances.

MAIN CONCEPTS

In Chapter 4 we discussed humidity, condensation, and vaporization. In this section we apply simultaneous material and energy balances to solve problems involving humidification, air conditioning, water cooling, and the like. Before proceeding, you should review briefly the sections in Chapter 4 dealing with vapor pressure and partial saturation. Table 5.5 summarizes the notation for the parameters involved in the energy balances.

Recall that the *humidity* $\mathcal{H}$ is the mass (lb or kg) of water vapor per mass (lb or kg) of bone-dry air (some texts use moles of water vapor per mole of dry air as the humidity)

$$\mathcal{H} = \frac{18 p_{H_2O}}{29(p_t - p_{H_2O})} = \frac{18 n_{H_2O}}{29(n_t - n_{H_2O})}$$

TABLE 5.5   Parameters Involved in Humidity Chart Calculations

Symbol	Meaning	SI value	American Engineering value
$C_{p\,air}$	Heat capacity of air	1.00 kJ/(kg)(K)	0.24 Btu/(lb)(°F)
$C_{p,H_2O\,vapor}$	Heat capacity of water vapor	1.88 kJ/(kg)(K)	0.45 Btu/(lb)(°F)
$\Delta \hat{H}_{vap}$	Specific heat of vaporization of water at 0°C (32°F)	4502 kJ/kg	1076 Btu/lb
$\Delta \hat{H}_{air}$	Specific enthalpy of air		
$\Delta \hat{H}_{H_2O\,vapor}$	Specific enthalpy of water vapor		

Other quantities involved in the preparation of humidity charts are:

(1) The **humid heat** is the heat capacity of an air-water vapor mixture expressed on the *basis of 1 lb or kg of bone-dry air*. Thus the humid heat C_S is

$$C_S = C_{p\,air} + (C_{p\,H_2O\,vapor})(\mathcal{H}) \tag{5.31}$$

where the heat capacities are all per mass and not per mole. Assuming that the heat capacities of air and water vapor are constant under the narrow range of conditions experienced for air-conditioning and humidification calculations, we can write in American Engineering units

$$C_S = 0.240 + 0.45(\mathcal{H}) \quad \text{Btu/(°F)(lb dry air)} \tag{5.32}$$

or in SI units,

$$C'_S = 1.00 + 1.88(\mathcal{H}) \quad \text{kJ/(K)(kg dry air)} \tag{5.32a}$$

(2) The **humid volume** is the volume of 1 lb or kg of dry air plus the water vapor in the air. In the American Engineering system,

$$\hat{V} = \frac{359 \text{ ft}^3}{1 \text{ lb mol}} \left| \frac{1 \text{ lb mol air}}{29 \text{ lb air}} \right| \frac{T_{°F} + 460}{32 + 460}$$

$$+ \frac{359 \text{ ft}^3}{1 \text{ lb mol}} \left| \frac{1 \text{ lb mol H}_2\text{O}}{18 \text{ lb H}_2\text{O}} \right| \frac{T_F + 460}{32 + 460} \left| \frac{\mathcal{H} \text{ lb H}_2\text{O}}{\text{lb air}} \right. \tag{5.33}$$

$$= (0.730 \, T_{°F} + 336) \left(\frac{1}{29} + \frac{\mathcal{H}}{18} \right)$$

where $\hat{V}$ is in ft³/lb dry air. In the SI system,

$$\hat{V} = \frac{22.4 \text{ m}^3}{1 \text{ kg mol}} \left| \frac{1 \text{ kg mol air}}{29 \text{ kg air}} \right| \frac{T_K}{273}$$

$$+ \frac{22.4 \ \text{m}^3}{1 \ \text{kg mol}} \left| \frac{1 \ \text{kg mol} \ H_2O}{18 \ \text{kg} \ H_2O} \right| \frac{T_K}{273} \left| \frac{\mathcal{H} \ \text{lb} \ H_2O}{\text{kg air}} \right. \tag{5.33a}$$

$$= 2.83 \times 10^{-3} T_K + 4.56 \times 10^{-3} \mathcal{H}$$

where $\hat{V}$ is in m^3/kg dry air.

(3) The **dry-bulb temperature** (T_{DB}) is the ordinary temperature you always have been using for a gas in °F or °C (or °R or K).

(4) The **wet-bulb temperature** (T_{WB}) you may guess, even though you may never have heard of this term before, has something to do with water (or other liquid, if we are concerned not with humidity but with saturation) evaporating from around an ordinary mercury thermometer bulb. Suppose that you put a wick, or porous cotton cloth, on the mercury bulb of a thermometer and wet the wick. Next you either (a) whirl the thermometer in the air as in Figure 5.16 (this apparatus is called a sling psychrometer when the wet-bulb and dry-bulb thermometers are mounted together), or (b) set up a fan to blow rapidly on the bulb at 1000 ft³/min or more. What happens to the temperature recorded by the wet-bulb thermometer?

As the water from the wick evaporates, the wick cools down and continues to cool until the rate of energy transferred to the wick by the air blowing on it equals the rate of loss of energy caused by the water evaporating from the wick. We say that the temperature of the bulb with the wet wick at equilibrium is the wet-bulb temperature. (Of course, if water continues to evaporate, it eventually will all disappear, and the wick temperature will rise.) The equilibrium temperature for the process described above will lie on the 100% relative humidity curve (saturated-air curve).

Suppose that we prepare a graph on which the vertical axis is the humidity and the horizontal axis is the dry-bulb temperature. We want to plot the temperature of the thermometer as it changes to reach T_{WB}. This line is the so called **wet-bulb line**.

The equation for the wet-bulb lines is based on a number of assumptions, a

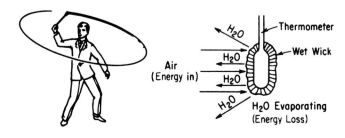

Figure 5.16   Wet-bulb temperature obtained with a sling psychrometer.

detailed discussion of which is beyond the scope of this book. Nevertheless, the idea of the wet-bulb temperature is based on the equilibrium between the *rates* of energy transfer to the bulb and evaporation of water. Rates of processes are a topic that we have not discussed. The fundamental idea is that a large amount of air is brought into contact with a little bit of water, and that presumably the evaporation of the water leaves the temperature and humidity of the air unchanged. Only the temperature of the water changes. The equation of the wet-bulb line is an energy balance

$$h_c(T - T_{WB}) = k'_g \Delta \hat{H}_{vap}(\mathcal{H}_{WB} - \mathcal{H}) \qquad (5.34)$$
$$\text{heat transfer to water} \quad \text{heat transfer from water}$$

where h_c = heat transfer coefficient for convection to the bulb
    T = temperature of moist air
    k'_g = mass transfer coefficient
    $\Delta\hat{H}_{vap}$ = latent heat of vaporization
    $\mathcal{H}$ = humidity of moist air

Next, we can form the ratio

$$\frac{\mathcal{H}_{WB} - \mathcal{H}}{T_{WB} - T} = - \frac{h_c}{(k'_g)\Delta\hat{H}_{vap}} \qquad (5.35)$$

to get the slope of the wet bulb line. For water only, it so happens that $h_c/k'_g \cong C_S$ (i.e., the numerical value is about 0.15), which gives the wet-bulb lines the slope of

$$\frac{\mathcal{H}_{WB} - \mathcal{H}}{T_{WB} - T} = - \frac{C_S}{\Delta\hat{H}_{vap}} \qquad (5.36)$$

For other substances, the value of h_c/k'_g can be as much as twice the value of C_S given for water.

Figure 5.17 shows the plot of the wet-bulb line as $T_{DB} \to T_{WB}$. The line is approximately straight and has a negative slope. Does this result agree with Eq. (5.36)?

Another type of process of some importance occurs when **adiabatic cooling or humidification** takes place between air and water that is recycled as illustrated in Figure 5.18. In this process the air is both cooled and humidified (its water content rises) while a little bit of the recirculated water is evaporated. At *equilibrium*, in the steady state, the temperature of the air is the same as the temperature of the water, and the exit air is saturated at this temperature. By making an overall energy balance around the process ($Q = 0$), we can obtain the equation for the adiabatic cooling of the air.

The equation, when plotted on the humidity chart, yields what is known as

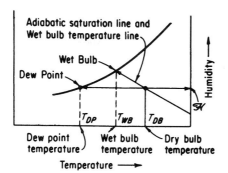

Figure 5.17 General layout of the humidity chart showing the location of the wet-bulb and dry-bulb temperatures, the dew point and dew point temperature, and the adiabatic saturation line and wet-bulb line.

an adiabatic cooling line We take the equilibrium temperature of the water, T_S, as a reference temperature rather than 0°C or 32°F. Do you see why? We ignore the small amount of makeup water or assume that it enters at T_S. The energy balance is

$$\underbrace{C_{p\,\text{air}}(T_{\text{air}} - T_S)}_{\substack{\text{enthalpy of air}\\\text{entering}}} + \underbrace{\mathcal{H}_{\text{air}}[\Delta\hat{H}_{\text{vap H}_2\text{O at}T_S} + C_{p\,\text{H}_2\text{O vapor}}(T_{\text{air}} - T_S)]}_{\substack{\text{enthalpy of water vapor}\\\text{in air entering}}}$$

$$= \underbrace{C_{p\,\text{air}}(T_S - T_S)}_{\substack{\text{enthalpy of air}\\\text{leaving}}} + \underbrace{\mathcal{H}_s\,[\Delta\hat{H}_{\text{vap H}_2\text{O at}T_S} + C_{p\,\text{H}_2\text{O vapor}}(T_S - T_S)]}_{\substack{\text{enthalpy of water vapor}\\\text{in air leaving}}} \tag{5.37}$$

Equation (5.37) can be reduced to

$$T_{\text{air}} = \frac{\Delta\hat{H}_{\text{vap H}_2\text{O at}T_S}\,(\mathcal{H}_S - \mathcal{H}_{\text{air}})}{C_{p\,\text{air}} + C_{p\,\text{H}_2\text{O vapor}}\,\mathcal{H}_{\text{air}}} + T_S \tag{5.38}$$

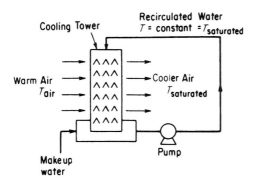

Figure 5.18 Adiabatic humidification with recycle of water.

which is the equation for adiabatic cooling.

Notice that this equation can be written as

$$\frac{\mathcal{H}_S - \mathcal{H}}{T_S - T_{air}} = -\frac{C_S}{\Delta \hat{H}_{vap\ at T_S}} \tag{5.39}$$

Compare Eq. (5.38) with Eq. (5.36). Can you conclude that the wet-bulb process equation, for water only, is essentially the same as the adiabatic cooling equation?

Of course! We have the nice feature that two processes can be represented by the same set of lines. For a detailed discussion of the uniqueness of this coincidence, consult any of the references at the end of the chapter. For most other substances besides water, the two equations will have different slopes.

Now that you have an idea of what the various features portrayed on the **humidity chart (psychrometric chart)** are, let us look at the chart itself (Figure 5.19). It is nothing more than a graphical means to assist in the execution of material and energy balances in water vapor-air mixtures. Its skeleton consists of a humidity ($\mathcal{H}$)-temperature (T_{DB}) set of coordinates together with the additional parameters (lines) of

(1) Constant relative humidity indicated in percent
(2) Constant moist volume (humid volume)
(3) Adiabatic cooling lines, which are the same (for water vapor only) as the wet-bulb or psychrometeric lines
(4) The 100% relative humidity (identical to the 100% absolute humidity) curve (i.e., saturated-air cure)

With any two values known, you can pinpoint the air-moisture condition on the chart and determine all the other associated values.

Off to the left of the 100% relative humidity line you will observe scales showing the enthalpy per mass of dry air of a saturated air-water vapor mixture. Enthalpy adjustments for air less than saturated (identified by minus signs) are shown on the chart itself by a series of curves. The enthalpy of the wet air, in energy/mass of dry air, is

$$\Delta \hat{H} = \Delta \hat{H}_{air} + \Delta \hat{H}_{H_2O\ vapor}(\mathcal{H})$$

We should mention at this point that the reference conditions for the humidity chart are liquid water at 32°F (0°C) and 1 atm (not the vapor pressure of H_2O) for water, and 0°F and 1 atm for air. The chart is suitable for use only at normal atmospheric conditions, and must be modified if the pressure is significantly different than 1 atm. If you wanted to, you could calculate the enthalpy values

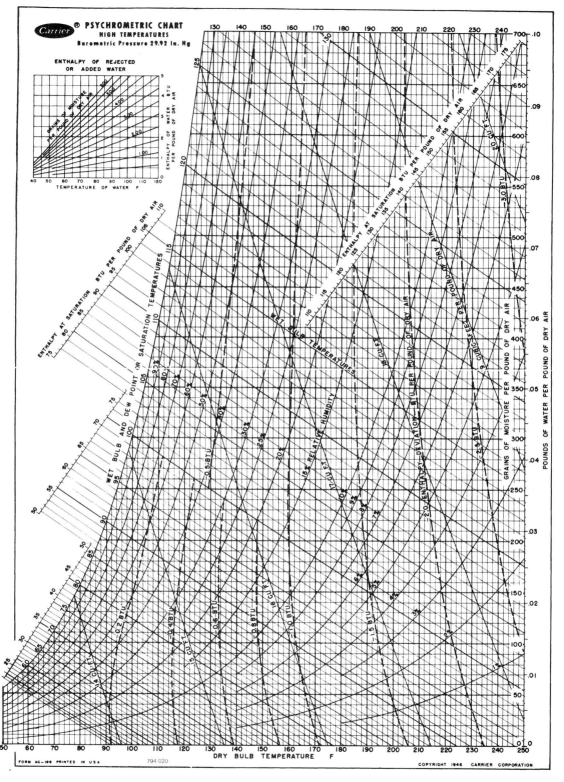

Figure 5.19a (Reprinted by permission of Carrier Corporation.)

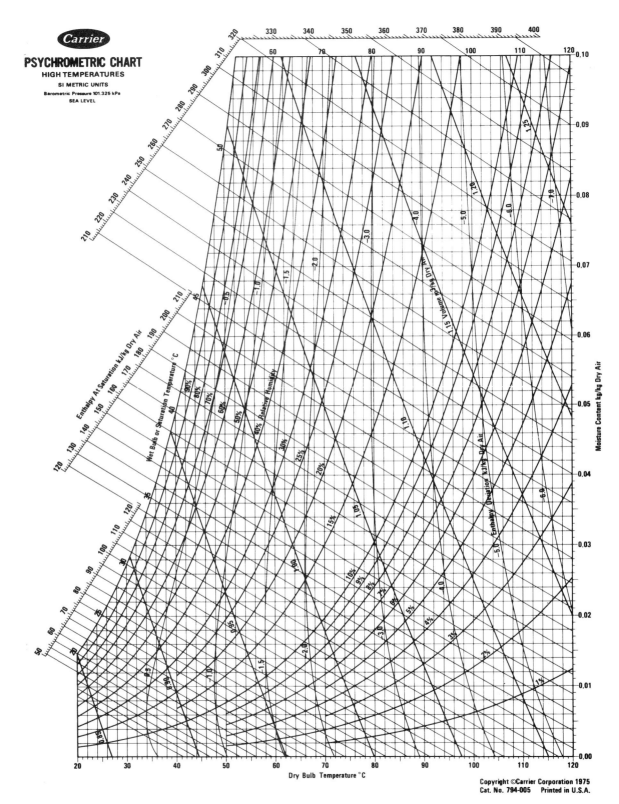

Figure 5.19b (Reprinted by permission of Carrier Corporation.)

shown on the chart directly from the tables listing the enthalpies of air and water vapor, or you could compute the enthalapies with reasonable accuracy from the following equation

$$\Delta\hat{H} = \underbrace{C_{pair}\,(T-T_{ref})_{air}}_{\text{enthalpy for air}} + \underbrace{\mathcal{H}\,[\Delta\hat{H}_{vap} + C_{pair}\,(T-T_{ref})_{water}]}_{\text{enthalpy for water vapor}} \qquad (5.40)$$

heat of vaporation of water at T_{ref}

In American Engineering units, Eq. (5.40) is

$$\Delta\hat{H} = 0.250T_{°F} + \mathcal{H}(1061 + 0.45T_{°F}) \qquad (5.41)$$

because $C_s = 0.240 + 0.45\mathcal{H}$ and $T_{WB} = T_S$.

The adiabatic cooling lines are lines of almost constant enthalpy for the entering air-water mixture, and you can use them as such without much error (1 or 2%). However, if you want to correct a saturated enthalpy value for the deviation that exists for a less-than-saturated air-water vapor mixture, you can employ the enthalpy deviation lines that appear on the chart and that can be used as illustrated in the examples below. Any process that is not a wet-bulb process or an adiabatic process with recirculated water can be treated by the usual material and energy balances, taking the basic data for the calculations from the humidity charts. If there is any increase or decrease in the moisture content of the air in a psychrometric process, the small enthalpy effect of the moisture added to the air or lost by the air may be included in the energy balance for the process to make it more exact as illustrated in the examples below.

Tables of the properties shown in the humidity charts exist if more accuracy is needed than can be obtained via the charts, and a computer program in the disk in the back of this book can be used to retrieve psychrometric data.

Although we shall be discussing humidity charts exclusively, charts can be prepared for mixtures of any two substances in the vapor phase, such as CCl_4 and air or acetone and nitrogen, if all the values of the physical constants for water and air are replaced by those of the desired gas and vapor.

EXAMPLE 5.33 Properties of Moist Air from the Humidity Chart

List all the properties you can find on the humidity chart in American Enginering units for moist air at a dry-bulb temperature of 90°F and a wet-bulb temperature of 70°F.

Solution

A diagram will help explain the various properties obtained from the humidity chart. See Figure E5.33. You can find the location of point A for 90°F DB (dry bulb) and

70°F WB (wet bulb) by following a vertical line at $T_{DB} = 90°F$ until it crosses the wet-bulb line for 70°F. This wet-bulb line can be located by searching along the 100% humidity line until the saturation temperature of 70°F is reached, or, alternatively, by proceeding up a vertical line at 70°F until it intersects the 100% humidity line. From the wet-bulb temperature of 70°F, follow the adiabatic cooling line (which is the same as the wet-bulb temperature line on the humidity chart) to the right until it intersects the 90°F DB line. Now that point A has been fixed, you can read the other properties of the moist air from the chart.

    (a) *Dew point.* When the air at A is cooled at constant pressure (and in effect at *constant humidity*), as described in Chapter 4, it eventually reaches a temperature at which the moisture begins to condense. This is represented by a horizontal line, a constant-humidity line, on the humidity chart, and the dew point is located at B, or about 60°F.

    (b) *Relative humidity.* By interpolating between the 40% $\mathcal{RH}$ and 30% $\mathcal{RH}$ lines you can find that point A is at about 37% $\mathcal{RH}$.

    (c) *Humidity* ($\mathcal{H}$). You can read the humidity from the right-hand ordinate as 0.0112 lb H_2O/lb dry air.

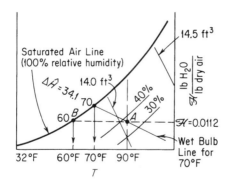

Figure E5.33

    (d) *Humid volume.* By interpolation again between the 14.0- and the 14.5-ft³ lines, you can find the humid volume to be 14.097 ft³/lb dry air.

    (e) *Enthalpy.* The enthalpy value for saturated air with a wet-bulb temperature of 70°F is $\Delta\hat{H} = 34.1$ Btu/lb dry air (a more accurate value can be obtained from psychrometric tables if needed). The enthalpy deviation (not shown in Figure E5.33—see Figure 5.19a instead) for less-than-saturated air is about −0.2 Btu/lb of dry air; consequently, the actual enthalpy of air at 37% $\mathcal{RH}$ is 34.0 − 0.2 = 33.9 Btu/lb of dry air.

EXAMPLE 5.34 Heating at Constant Humidity

Moist air at 38°C and 48% $\mathcal{RH}$ is heated in your furnace to 86°C. How much heat has to be added per cubic meter of initial moist air, and what is the final dew point of the air?

Solution

As shown in Figure E5.34, the process goes from point A to point B on a horizontal line of constant humidity. The initial conditions are fixed at $T_{DB} = 38°C$ and 48% $\mathcal{RH}$. Point B is fixed by the intersection of the horizontal line from A and the vertical line at 86°C. The dew point is unchanged in this process and is located at C at 24.8°C.

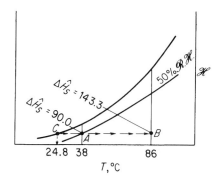

T, °C

Figure E5.34

The enthalpy values are as follows (all in kJ/kg of dry air):

Point	$\Delta\hat{H}_{satd}$	$\delta\hat{H}$	$\Delta\hat{H}_{actual}$
A	90.0	−0.5	89.5
B	143.3	−3.3	140.0

Also, at A the volume of the moist air is 0.91 m³/kg of dry air. Consequently, the heat added is ($Q = \Delta\hat{H}$) 140.0 − 89.5 = 50.5 kJ/kg of dry air.

$$\frac{50.5 \text{ kJ}}{\text{kg dry air}} \left| \frac{1 \text{ kg dry air}}{0.91 \text{ m}^3} \right. = 55.5 \text{ kJ/m}^3 \text{ initial moist air}$$

EXAMPLE 5.35 Cooling and Humidification

One way of adding moisture to air is by passing it through water sprays or air washers. See Figure E5.35a. Normally, the water used is recirculated rather than wasted. Then, in the steady state, the water is at the adiabatic saturation temperature, which is the same as the wet-bulb temperature. The air passing through the washer is cooled, and

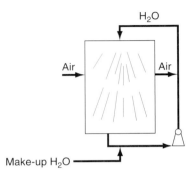

Figure E5.35a

if the contact time between the air and the water is long enough, the air will be at the wet-bulb temperature also. However, we shall assume that the washer is small enough so that the air does not reach the wet-bulb teperature; instead, the following conditions prevail:

	T_{DB}(°C)	T_{WB}(°C)
Entering air:	40	22
Exit air:	27	

Find the moisture added per kilogram of dry air.

Solution

The whole process is assumed to be *adiabatic*, and, as shown in Figure E5.35b, takes place between points A and B along the adiabatic cooling line. The wet-bulb temperature remains constant at 22°C. Humidity values are

$$\mathcal{H}\left(\frac{\text{kg } H_2O}{\text{kg dry air}}\right)$$

B	0.0145
A	0.0093
Difference:	0.0052 $\dfrac{\text{kg } H_2O}{\text{kg dry air}}$ added

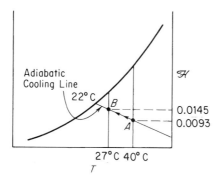

Figure E5.35b

EXAMPLE 5.36 Combined Material and Energy Balances for a Cooling Tower

You have been requested to redesign a water-cooling tower that has a blower with a capacity of 8.30×10^6 ft³/hr of moist air (at 80°F and a wet-bulb temperature of 65°F). The exit air leaves at 95°F and 90°F wet bulb. How much water can be cooled in pounds per hour if the water to be cooled is not recycled, enters the tower at 120°F, and leaves the tower at 90°F?

Solution

Enthalpy, humidity, and humid volume data for the air taken from the humidity chart are as follows (see Figure E5.36):

	A	B
$\mathcal{H}\left(\dfrac{\text{lb } H_2O}{\text{lb dry air}}\right)$	0.0098	0.0297
$\mathcal{H}\left(\dfrac{\text{grains } H_2O}{\text{lb dry air}}\right)$	69	208
$\Delta\hat{H}\left(\dfrac{\text{Btu}}{\text{lb dry air}}\right)$	$30.05 - 0.12 = 29.93$	$55.93 - 0.10 = 55.83$
$\hat{V}\left(\dfrac{\text{ft}^3}{\text{lb dry air}}\right)$	13.82	14.65

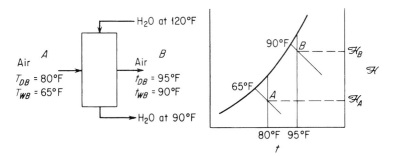

Figure E5.36

The cooling-water exit rate can be obtained from an energy balance around the process.

$$\text{Basis: } 8.30 \times 10^6 \text{ ft}^3/\text{hr of moist air}$$

$$\frac{8.30 \times 10^6 \text{ ft}^3}{\text{hr}} \left| \frac{1 \text{ lb dry air}}{13.82 \text{ ft}^3} \right. = 6.00 \times 10^5 \text{ lb dry air/hr}$$

The relative enthalpy of the entering water stream is (reference temperature is 32°F and 1 atm)

$$\Delta \hat{H} = C_{p\,H_2O}\,\Delta T = 1(120 - 32) = 88 \text{ Btu/lb } H_2O$$

and that of the exit stream is 58 Btu/lb H_2O. [The value from the steam tables at 120°F for liquid water of 87.92 Btu/lb H_2O is slightly different since it represents water at its vapor pressure (1.69 psia) based on reference conditions of 32°F and liquid water at its vapor pressure.] Any other datum could be used instead of 32°F for the liquid water. For example, if you chose 90°F, one water stream would not have to be taken into account because its relative enthalpy would be zero.

The loss of water to the air is

$$0.0297 - 0.0098 = 0.0199 \text{ lb } H_2O/\text{lb dry air}$$

(1) *Material balance for water stream*: Let W = lb H_2O enering the tower in the water stream per lb dry air. Then

$$W - 0.0199 = \text{lb } H_2O \text{ leaving tower in the water stream per lb dry air}$$

(2) *Energy balance (enthalpy balance) around the entire process:*

air and water in air entering   **water stream entering**

$$\frac{29.93 \text{ Btu}}{\text{lb dry air}} \left| 6.00 \times 10^5 \text{ lb dry air} \right. \quad + \quad \frac{88 \text{ Btu}}{\text{lb } H_2O} \left| \frac{W \text{ lb } H_2O}{\text{lb dry air}} \right| 6.00 \times 10^5 \text{ lb dry air}$$

air and water in air leaving

$$= \frac{55.83 \text{ Btu}}{\text{lb dry air}} \bigg| 6.00 \times 10^5 \text{ lb dry air}$$

water stream leaving

$$+ \frac{58 \text{ Btu}}{\text{lb H}_2\text{O}} \bigg| \frac{(W - 0.0199)\text{lb H}_2\text{O}}{\text{lb dry air}} \bigg| 6.00 \times 10^5 \text{ lb dry air}$$

$$29.93 + 88W = 55.83 + 58(W - 0.0199)$$

$$W = 0.825 \text{ lb H}_2\text{O/lb dry air}$$

$$W - 0.0199 = 0.805 \text{ lb H}_2\text{O/lb dry air}$$

The total water leaving the tower is

$$\frac{0.805 \text{ lb H}_2\text{O}}{\text{lb dry air}} \bigg| \frac{6.00 \times 10^5 \text{ lb dry air}}{\text{hr}} = 4.83 \times 10^5 \text{ lb/hr}$$

LOOKING BACK

In this section we described the structure of and information obtained from humidity charts, and how they can be used in material and energy balance calculations.

Key Ideas

1. Psychrometric (humidity) charts provide information to use in making material and energy balances for heating, cooling, humidification, dehumidification, condensation, and vaporization.
2. Given the values of two intensive variables, values of all of the others can be read from the charts (or tables or calculated via a computer code).

Key Terms

Adiabatic cooling (p. 490)   Humidity chart (p. 492)
Adiabatic humidification (p. 490)   Moist volume (p. 492)
Dry bulb temperature (p. 489)   Psychrometric chart (p. 492)
Humid heat (p. 488)   Wet bulb line (p. 489)
Humid volume (p. 488)   Wet bulb temperature (p. 489)
Humidity (p. 487)

Self-Assessment Test

1. What is the difference between the wet- and dry-bulb temperatures?
2. Can the wet-bulb temperature ever be higher than the dry-bulb temperature?

3. Explain why the slope of the wet-bulb lines are essentially the same as the slope of the adiabatic cooling lines for gaseous air and water mixtures.

4. Estimate for air at 70°C dry-bulb temperature, 1 atm, and 15% relative humidity the:
   (a) kg H_2O/kg of dry air
   (b) m^3/kg of dry air
   (c) Wet-bulb temperature (in °C)
   (d) Specific enthalpy
   (e) Dew point (in °C)

5. Calculate the following properties of moist air at 1 atm and compare with values read from the humidity chart.
   (a) The humidity of saturated air at 120°F
   (b) The enthalpy of air in part (a) per pound of dry air
   (c) The volume per pound of dry air of part (a)
   (d) The hmidity of air at 160°F with a wet-bulb temperature of 120°F

6. Humid air at 1 atm and 200°F, and containing 0.0645 lb of H_2O/lb of dry air, enters a cooler at the rate of 1000 lb of dry air per hour (plus accompanying water vapor). The air leaves the cooler at 100°F, saturated with water vapor (0.0434 lb of H_2O/lb of dry air). Thus 0.0211 lb H_2O is condensed per pound of dry air. How much heat is transferred to the cooler?

7. A cooling tower that uses a cold-water spray provides a method of cooling and dehumidifiying a school. During the day, the average number of students in the school is 100 and the average heat-generation rate per person is 800 Btu/hr. Suppose that the ambient conditions outside the school in the summer are expected to be 100°F and 95% $\mathcal{RH}$. You run this air through the cooler-dehumidifier and then mix the saturated exit air with recirculated air from the exhaust of the school building. You need to supply the mixed air to the building at 70°F and 60% $\mathcal{RH}$ and keep the recirculated air leaving the building at not more than 72°F. Leakage occurs from the building of the 72°F air also. Calculate:
   (a) The volumetric rate of air recirculation per hour in cubic feet at 70°F and 60% $\mathcal{RH}$
   (b) The volume of fresh air required at entering conditions
   (c) The heat transferred in the cooler-dehumidifier from the inlet air per hour.

Thought Problems

1. The use of home humidifers has recently been promoted in advertisements as a means of providing more comfort in houses with the thermostat turned down. "Humidification makes life more comfortable and prolongs the life of furniture." Many advertisers describe a humidifer as an energy-saving device because it allows lower temperatures (4 or 5°F lower) with comfort. Is this true?

2. Cooling systems known as "swamp coolers" are a low-cost environmental-friendly technology based on evaporative cooling—the natural evaporation of water, the same process that cools a sweating human body. In evaporative air-conditioning's simplest form, hot air is drawn through a water spray or through wet porous pads. The exit air

becomes cooler and more humid as it takes up water. The only moving parts requiring electrical power are a fan and a small water pump that moves the water to the top of the system. Will a swamp cooler be an effective cooler in a swamp?

Discussion Questions

1. In cold weather, water vapor exhausted from cooling towers condenses and fog is formed as a plume. What are one or two *economically practical* methods of preventing such cooling tower fog?

2. Many of the portable combustible-gas meters measure the concentration of combustibles in an atmospheric sample by drawing the sample across a heated catalytic filament. This filament forms one part of a balanced Wheatstone-bridge circuit. Burning the combustibles raises the temperature of the filament and increases its resistance—in proportion to the concentration of combustibles in the sample. This change in resistance unbalances the circuit, causing a meter deflection that indicates the concentration.

   An inaccurate combustible-gas meter can mislead its operator into believing that the atmosphere is safe when it is actualy combustible. How can you avoid this potential hazard? *Hint:* The meter is calibrated at 0 with no combustion taking place so no temperature rise occurs. What can interfere with the temperature rise on combustion that does not interfere with the calibration at 0?

SUPPLEMENTARY REFERENCES

Concepts and Units

BRADSHAW, J. A. "Historical Background and Foundations of Thermodynamics." in *Thermodynamics*, Vol. 5D, (p. 1). AIChE Modular Instruction Series. New York: American Institute of Chemical Engineers, 1984.

LEVI, B. G. (ed.). *Global Warming: Physics and Facts.* New York: American Institute of Physics, 1992.

LINDSAY, R. B. *Energy-Historical Development of the Concept.* New York: Academic Press, 1975.

National Research Council. *Energy in Transition, 1985–2010.* San Francisco: W. H. Freeman, 1980.

TAUBE, M. *Evolution of Matter and Energy.* New York: Springer-Verlag, 1985.

VERVALIN, C. H. "Energy Information Resources." *Hydrocarbon Processing* (November 1989): 119.

Heat Capacity

CEZAIRLIYAN, A. et al. *Specific Heat of Solids.* New York: Hemisphere Publishers, 1988.

DOMALSKI, E. S., W. H. EVANS, and E. D. HEARING. *Heat Capacities and Entropies of Or-*

ganic Compounds in the Condensed Phase. Washington, DC: American Chemical Society, 1984.

TOULOUKIAN, Y. S., and T. MAKITA. *Specific Heat*. New York: Plenum, 1977.

YAWS, C. L., H. M. NI, and P. Y. CHANG. "Heat Capacities for 700 Compounds." Chemical Engineering (May 9, 1988): 91.

YAWS, C. L. "Calculate Liquid Heat Capacity." *Hydrocarbon Processing*. (December, 1991): 73.

Enthalpy Properties

AMERICAN SOCIETY OF MECHANICAL ENGINEERS. *ASME Steam Tables*. New York, NY: ASME, 1993.

AMERICAN PETROLEUM INSTITUTE. *Technical Data Book—Petroleum Refining*, 2nd ed. Washington, DC: API, 1970.

BARNER, H. E., and R. V. SCHEUERMAN. *Handbook of Thermochemical Data for Compounds and Aqueous Species*. New York: Wiley-Interscience, 1978.

BENEDEK, P., and F. OGLI. *Computer Aided Chemical Thermodynamics of Gases and Liquids*. New York: Wiley-Interscience, 1985.

DAUBERT, T. E. and R. P. DAUBERT. *Data Compilation, Tables of Properties of Pure Compounds*. American Institute of Chemical Engineers, STN International (on-line access), Ohio.

EDMISTER, W. C. *Applied Hydrocarbon Thermodynamics*. 2nd ed. Houston: Gulf Publishing Company, 1984.

FREEMAN, R. D. *Chemical Thermodynamics*, Bulletin No. 54 in CODATA Directory of Data Sources for Science and Technology. Paris: CODATA Secretariat, 1983.

GLUSHKO, V. P. and L. V. GURVICH. *Thermodynamic Properties of Individual Substances*. New York: Hemisphere Publishers, 1989.

IHSAN, B. *Thermochemical Data of Pure Substances*. Weinheim, Germany: VCH, 1989.

IRVINE, R. T., and P. E. LILEY. *Steam and Gas Tables with Computer Equations*. Orlando, FL: Academic Press, 1984.

KEENAN, J. H., J. CHAO, and J. KAYE. *Gas Tables (English Units)*, 2nd ed. New York: Wiley, 1980.

LIDE, D. R., and H. V. KEHIAIAN. *ERC Handbook of Thermophysical and Thermochemical Data*. Boca Raton, FL: CRC Press, 1994. Includes disk.

MAJER, V., V. SVOBODA, and J. PICK. *Heats of Vaporation*. Amsterdam: Elsevier, 1989.

PERRY, J. H. *Chemical Engineers' Handbook*, 6th ed. New York: McGraw-Hill, 1986.

RIEMER, D. H., H. R. JACOBS, and R. F. BOEHM. *Computer Programs for Determining the Thermodynamic Properties*, IDO/1549-1, -2, -3, -4. Springfield, VA: National Technical Information Service, 1978.

ROSSINI, F. K., et al. "Tables of Selected Values of Chemical Thermodynamic Properties," *National Bureau of Standards, Circular 500*, 1952. (Revisions are being issued periodically under the Technical Note 270 series by other authors.)

SELOVER, T. B. *National Standard Reference Data Service of the USSR*. New York: Hemisphere, (continuing series).

STEPHENSON, F. M. and S. MALANOWSKI. *Handbook of the Thermodynamics of Organic Compounds*. New York: Elsevier, 1987.

TORQUATO, S., and P. SMITH. "The Latent Heat of Vaporization of a Widely Diverse Class of Fluids," *Trans. ASME, 106* (1984): 252.

TOULOUKIAN, Y. S., and E. H. BUYCO. *Thermophysical Properties of Matter—The TPRC Data Series*, New York: Plenum, 1983.

Energy Balance

BRUSH, S. G. *A History of Heat and Thermodynamics*. Amsterdam: North-Holland, 1976.

ERLICHSON, H. "Internal Energy in the First Law of Thermodynamics." *Am. J. Phys., 52* (1984): 623.

RAMAN, V. V. "Where Credit Is Due—The Energy Conservation Principle," *Phys. Teacher* (February 1975): 80.

SMITH, J. M., and H. C. VAN NESS. *Introduction to Chemical Engineering Thermodynamics*, 4th ed. New York: McGraw-Hill, 1987.

Energy Balances with Reaction

CARDOZO, R. L. "Enthalpies of Combustion, Formation, Vaporization, and Sublimation of Organics." *AIChE Journal, 37* (1991): 290.

HANBY, V. I. *Combustion and Pollution Control in Heating Systems*. London: Springer Verlag, 1994.

JANAF Thermocheical Tables, 2nd ed. and supplements 1-33 available from U.S. Government Printing Office, Cat. No. C13.48:37; supplements 34-52 available from American Chemical Society, Washington, DC (various dates). Also on-line on STN Intl. network.

WAGMAN, D. D. et al. *NBS Tables of Chemical Thermodynamic Properties*. On-line through STN Intl. and American Chemical Society, 1988.

YAWS, C. L., and P. Y. CHANG. "Enthalpy of Formation for 700 Major Organic Compounds," *Chem. Eng.* (September 26, 1988): 81.

Heat of Mixing and Enthalpy—Concentration Charts

BRANDANI, V., and F. EVANGELISTA. "Correlation and Predication of Enthalpies of Mixing for Systems Containing Alcohols with UNIQUAC Associated-Solution Theory." *Ind. Eng. Chem. Res., 26* (1987): 2423.

CHRISTENSEN, J. J., R. W. HANKS, and R. M. IZATT. *Handbook of Heats of Mixing*. New York: Wiley, 1982.

CHRISTENSEN, C., J. GMEHLING, and P. RASMUSSEN. *Heats of Mixing Data Collection*, Frankfurt, West Germany: Dechema, 1984.

DAN, D. and D. P. TASSIOS. "Prediction of Enthalpies of Mixing with a UNIFAC Model." *Ind. Eng. Chem. Process Des. Develop., 25* (1986): 22.

JOHNSON, J. E., and D. J. MORGAN. "Graphical Techniques for Process Engineering." *Chem. Eng.* (July 8, 1985): 72.

Psychrometric Charts

BULLOCK, C. E. "Psychrometric Tables." In *ASHRAE Handbook Product Directory*, paper No. 6. Atlanta: American Society of Heating, Refrigeration, and Air Conditioning Engineers, 1977.

MCMILLAN, H. K., and J. KIM. "A Computer Program to Determine Thermodynamic Properties of Moist Air." *Access* (January 1986): 36.

NELSON, R. "Material Properties in SI Units, Part 4." *Chem., Eng. Prog.* (May 1980): 83.

TREYBAL, R. E. *Mass Transfer Operations*. New York: McGraw-Hill, 1980.

WILWHELM, L. R. "Numerical Calculation of Psychrometric Properties in SI Units." *Trans. ASAE, 19* (1976): 318.

PROBLEMS

Section 5.1

5.1. Convert 45.0 Btu/lb_m to the following:
(a) cal/kg (b) J/kg
(c) kWh/kg (d) (ft)(lb_f)/lb_m

5.2. Convert the following physical properties of liquid water at 0°C and 1 atm from the given SI units to the equivalent values in the listed American Engineering units.
(a) Heat capacity of 4.184 J/(g)(K) Btu/(lb)(°F)
(b) Enthalpy of −41.6 J/kg Btu/lb
(c) Thermal conductivity of 0.59 (kg)(m)/(s^3)(K) Btu/(ft)(hr)(°F)

5.3. Convert the following quantities as specified.
(a) A rate of heat flow of 6000 Btu/(hr)(ft^2) to cal/(s)(cm^2).
(b) A heat capacity of 2.3 Btu/(lb)(°F) to cal/(g)(°C).
(c) A thermal conductivity of 200 Btu/(hr)(ft)(°F) to J/(s)(cm)(°C).
(d) The gas constant, 10.73 (psia)(ft^3)/(lb mol)(°R) to J/(g mol)(K).

5.4. A simplified equation for the heat transfer coefficient from a pipe to air is

$$h = \frac{0.026 G^{0.6}}{D^{0.4}}$$

where h = heat transfer coefficient, Btu/(hr)(ft^2)(°F)
G = mass rate flow, lb_m/(hr)(ft^2)
D = outside diameter of the pipe, (ft)

If h is to be expressed in J/(min)(cm^2)(°C), what should the new constant in the equation be in place of 0.026?

5.5. A problem for many people in the United States is excess body weight stored as fat. Many people have tried to capitalize on this problem with fruitless weight-loss schemes. However, since energy is conserved, an energy balance reveals only two real ways to lose weight (other than water loss): (1) reduce the caloric intake, and/or (2) increase the caloric expenditure. In answering the following questions, assume that fat contains approximately 7700 kcal/kg (1 kcal is called a "dietetic calorie" in nutrition, or commonly just a "calorie").

 (a) If a normal diet containing 2400 kcal/day is reduced by 500 kcal/day, how many days does it take to lose 1 lb of fat?

 (b) How many miles would you have to run to lose 1 lb of fat if running at a moderate pace of 12 km/hr expends 40 kJ/km?

 (c) Suppose that two joggers each run 10 km/day. One runs at a pace of 5 km/hr and the other at 10 km/hr. Which will lose more weight (ignoring water loss)?

5.6. The energy from the sun incident on the surface of the earth averages 32.0 cal/(min)(cm^2). It has been proposed to use space stations in synchronous orbits 36,000 km from earth to collect solar energy. How large a collection surface is needed (in m^2) to obtain 10^{11} watts of electricity? Assume that 10% of the collected energy is converted to electricity. Is your answer a reasonable size?

5.7. Solar energy has been suggested as a source of renewable energy. If in the desert the direct radiation from the sun (say for 320 days) is 975 W/m^2 between 10 AM and 3 PM, and the conversion efficiency to electricity is 21.0%, how many square meters are needed to collect an amount of energy equivalent to the annual U.S. energy consumption of 3×10^{20}J? Is the construction of such an area feasible?

 How many tons of coal (of heating value of 10,000 Btu/lb) would be needed to provide the 3×10^{20}J? What fraction of the total U.S. resources of coal (estimated at 1.7×10^{12} tons) is the calculated quantity?

5.8. Lasers are used in many technologies, but they are fairly large devices even in CD players. A materials scientist has been working on producing a laser from a microchip. He claims that his laser can produce a burst of light with up to 10,000 watts, qualifying it for use in eye surgery, satellite communications, and so on. Is it possible to have such a powerful laser in such a small package?

5.9. An examination question asked: "Is heat conserved?" Sixty percent of the students said "no", but 40 percent said "yes". The most common explanation was: (a) "Heat is a form of energy and therefore conserved." The next most common was: (b) "Heat is a form of energy and therefore is not conserved." Two other common explanations were: (c) "Heat is conserved. When something is cooled, it heats something else up. To get heat in the first place, though, you have to use energy. Heat is just one form of energy," and (d) "Yes, heat is transferred from a system to its surroundings and vice versa. The amount lost by one system equals the amount gained by the surroundings." Explain whether or not heat is conserved and criticize each of the four answers.

5.10. For the systems defined below, state whether Q, W, ΔE, and ΔU are 0, < 0, or > 0, and compare their relative values if not equal to 0:

(a) An egg (the system) is placed into boiling water.

(b) Gas (the system), initially at equilibrium with its surroundings, is compressed rapidly by a piston in an insulated nonconducting cylinder by an insulated nonconducting piston; give your answer for two cases: (1) before reaching a new equilibrium state, and (2) after reaching a new equilibrium state.

(c) A Dewar flask of coffee (the system) is shaken.

5.11. An article explaining the energy balance states that "When positive work is done on a system, its surroundings do an equal quantity of negative work, and vice versa." Is this statement true? Explain.

5.12. Often in books you read about "heat reservoirs" existing in two bodies at different temperatures. Can this concept be correct?

5.13. Draw a picture of the following processes, draw a boundary for the system, and state for each whether heat transfer, work, a change in internal energy, a change in enthalpy, a change in potential energy, and a change in kinetic energy occurs *for the system*. Also classify each system as open or closed, and as steady state or unsteady state.

(a) A pump, driven by a motor, pumps water from the first to the third floor of a building at a constant rate and temperature. The system is the pump.

(b) As in (a) except the system is the pump and the motor.

(c) A block of ice melts in the sun. The system is the block of ice.

(d) A mixer mixes a polymer into a solvent. The system is the mixer.

5.14. Explain specifically what the system is for each of the following processes; indicate if any energy transfer takes place by heat or work (use the symbols Q and W, respectively) or if these terms are zero.

(a) A liquid inside a metal can, well insulated on the outside of the can, is shaken very rapidly in a vibrating shaker.

(b) Hydrogen is exploded in a calorimetric bomb and the water layer outside the bomb rises in temperature by 1°C.

(c) A motor boat is driven by an outboard-motor propeller.

(d) Water flows through a pipe at 1.0 m/min, and the temperature of the water and the air surrounding the pipe are the same.

5.15. Draw a simple sketch of each of the following processes, and, in each, label the system boundary, the system, the surroundings, and the streams of material and energy that cross the system boundary.

(a) Water enters a boiler, is vaporized, and leaves as steam. The energy for vaporization is obtained by combustion of a fuel gas with air outside the boiler surface.

(b) Steam enters a rotary steam turbine and turns a shaft connected to an electric generator. The steam is exhausted at a low pressures from the turbine.

(c) A battery is charged by connecting it to a source of current.

5.16. Are the following variables intensive or extensive?
(a) Partial pressure   (e) Relative saturation
(b) Volume   (f) Specific volume
(c) Specific gravity   (g) Surface tension
(d) Potential energy   (h) Refractive index

5.17. Classify the following measurable physical characteristics of a gaseous mixture of two components as (1) an intensive property, (2) an extensive property, (3) both, or (4) neither:
(a) Temperature   (c) Composition
(b) Pressure   (d) Mass

5.18. Find the kinetic energy in $(ft)(lb_f)/(lb_m)$ of water moving at the rate of 10 ft/s through a pipe 2 in. ID.

5.19. A windmill converts the kinetic energy of the moving air into electrical energy at an efficiency of about 30%, depending on the windmill design and speed of the wind. Estimate the power in kW for a wind flowing perpendicular to a windmill with blades 15 m in diameter when the wind is blowing at 20 mi/hr at 27°C and 1 atm.

5.20. Before it lands, a vehicle returning from space must convert its enormous kinetic energy to heat. To get some idea of what is involved, a vehicle returning from the moon at 25,000 mi/hr can, in converting its kinetic energy, increase the internal energy of the vehicle sufficiently to vaporize it. Obviously, a large part of the total kinetic energy must be transferred from the vehicle. How much kinetic energy does the vehicle have (in Btu per lb)? How much energy must be transferred by heat if the vehicle is to heat up by only 20°F/lb?

5.21. One pound of liquid water is at its boiling point of 575°F. It is then heated at constant pressure to 650°F, and then compressed at constant temperature to one-half of its volume (at 650°F), and finally returns to its original state of the boiling point at 575°F. Calculate ΔH and ΔU for the overall process.

5.22. The world's largest plant that obtains energy from tidal changes is at Saint Malo, France. The plant uses both the rising and falling cycle (one period in and out is 6 hr 10 min in duration). The tidal range from low to high is 14 m, and the tidal estuary (the LaRance River) is 21 km long with an area of 23 km². Assume that the efficiency of the plant in converting potential to electrical energy is 85%, and estimate the average power produced by the plant. (*Note*: Also assume that after high tide, the plant does not release water until the sea level drops 7 m, and after a low tide does not permit water to enter the basin until the level outside the basin rises 7 m, and the level differential is maintained during discharge and charge.)

5.23. Steam is used to cool a polymer reaction. The steam in the steam chest of the apparatus is found to be at 250.5°C and 4000 kPa absolute during a routine measurement at the beginning of the day. At the end of the day the measurement showed that the temperature was 650°C and the pressure 10,000 kPa absolute. What was the internal energy change of 1 kg of steam in the chest during the day? Obtain your data from the steam tables.

5.24. **(a)** Ten pound moles of an ideal gas are originally in a tank at 100 atm and 40°F. The gas is heated to 440°F. The specific molar enthalpy, $\Delta \hat{H}$, of the ideal gas is given by the equation

$$\Delta \hat{H} = 300 + 8.00T$$

where $\Delta \hat{H}$ is in Btu/lb mol and T is the temperature in °F.

    (1) Compute the volume of the container (ft^3).
    (2) Compute the final pressure of the gas (atm).
    (3) Compute the enthalpy change of the gas.

    (b) Use the equation above to develop an equation giving the molar internal energy, ΔU, in cal/g mol as a function of temperature, T, in °C.

5.25. You have calculated that specific enthalpy of 1 kg mol of an ideal gas at 300 kN/m^2 and 100°C is 6.05×10^5 J/kg mol (with reference to 0°C and 100 kN/m^2). What is the specific internal energy of the gas?

5.26. Answer the following questions true or false:

    (a) In a process in which a pure substance starts at a specified temperature and pressure, goes through several temperature and pressure changes, and then returns to the initial state, $\Delta U = 0$.

    (b) The reference enthalpy for the steam tables ($\Delta H = 0$) is at 25°C and 1 atm.

    (c) Work can always be calculated as $\Delta(pV)$ for a process going from state 1 to state 2.

    (d) An isothermal process is one for which the temperature change is zero.

    (e) An adiabatic process is one for which the pressure change is zero.

    (f) A closed system is one for which no reaction occurs.

    (g) An intensive property is a property of material that increases in value as the amount of material increases.

    (h) Heat is the amount of energy liberated by the reaction within a process.

    (i) Potential energy is the energy a system has relative to a reference plane.

    (j) The units of the heat capacity can be (cal)/(g)(°C) or Btu/(lb)(°F), and the numerical value of the heat capacity is the same in each system of units.

5.27. C. D. Zinn and W. G. Lesso (*Interfaces*, $A(1)$ (1978): 68–71.) reported a study of the cost of reducing ethylene emissions from a petrochemical plant. The table shows the basic data for case studies as a function of decreasing amounts of C_2 emissions. Case A represents no emission control whatsoever.

	A	B	C	D	E	F
C_2 emissions, lb/yr $\times 10^{-3}$	182.50	91.25	36.50	18.224	9.125	4.56
Initial investment, $ $\times 10^{-3}$	0	110	166	396	821	1600
Operating cost, $/yr $\times 10^{-3}$	0	7.5	12	28	60	126
Purchased electricity, MWh/yr	0	166	267	617	1284	2734
Additional compressor power needed, hp	0	28	45	99	207	440

Electric power generation at the power station from which the purchased electricity comes required 10,300 Btu per kWh of electricity. The emissions of NO_x, SO_2, HC, CO, and PA for power generation by different fuels were

coal	10.24 lb/10^6 Btu
fuel oil	1.45 lb/10^6 Btu
natural gas	0.2246 lb/10^6 Btu

The emissions due to compressor operation were (in lb/Mhph): NO_x 24, SO_2 nil, HC 10, CO 3, and PA 0.12.

From the above data, prepare a table for the six cases that lists in order (a) the total emissions in lb/yr for each of the three fuels, (b) the annual cost assuming the plant operates for 10 years and has no salvage value (use a capital charge of 0.158 of the investment cost plus operating costs), and (c) the cost per pound of total reduction of emissions for the three fuels. How much emission control would you recommend be implemented?

5.28. Examine the simplified refinery power plant flow sheet in Figure P5.28 and calculate the energy loss (in Btu/hr) for the plant. Also calculate the energy loss as a percentage of the energy input.

5.29. Start with the general energy balance, and simplify it for each of the processes listed below to obtain an energy balance that represents the process. Label each term in the general energy balance by number, and list by their numbers the terms retained or deleted followed by your explanation. (You do not have to calculate any quantities in this problem.)

(a) One hundred kilograms per second of water is cooled from 100°C to 50°C in a heat exchanger. The heat is used to heat up 250 kg/s of benzene entering at 20°C. Calculate the exit temperature of benzene.

(b) A feedwater pump in a power generation cycle pumps water at the rate of 500 kg/min from the turbine condensers to the steam generation plant, raising the pressure of the water from 6.5 kPa to 2800 kPa. If the pump operates adiabatically with an overall mechanical efficiency of 50% (including both pump and its drive motor), calculate the electric power requirement of the pump motor (in kW). The inlet and outlet lines to the pump are of the same diameter. Neglect any rise in temperature across the pump due to friction (i.e., the pump may be considered to operate isothermally).

(c) A caustic soda solution is placed in a mixer together with water to dilute it from 20% NaOH in water to 10%. What is the final temperature of the mixture if the materials initially are at 25°C and the process is adiabatic?

Section 5.2

5.30. The enthalpy of Al_2O_3 can be represented by the equation

$$\Delta H_T - \Delta H_{273} = 92.38T + 1.877 \times 10^{-2} T^2 + \frac{2.186 \times 10^6}{T} - 34,635$$

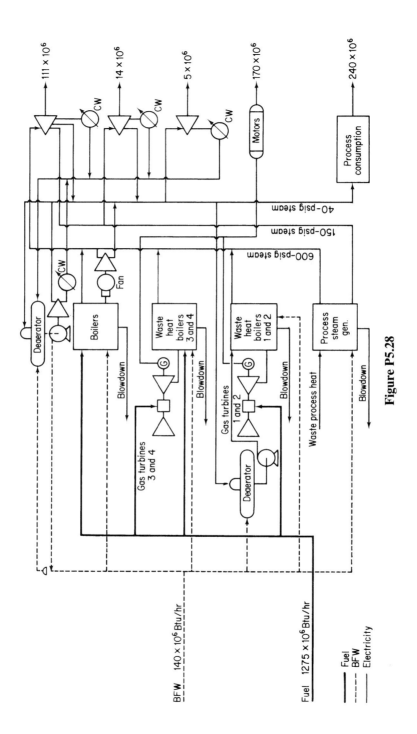

Figure P5.28

where ΔH is in J/g mol and T is in K. Determine an equation for the heat capacity of Al_2O_3 as a function of temperature.

5.31. Heat capacity data for gaseous NH_3 are listed below:

T(°C)	J/(g mol)(°C)
0	35.02
100	37.80
200	41.10
300	44.60
400	47.48
500	50.40
600	53.14
700	55.69
800	58.06
900	60.24
1000	62.63
1100	64.04
1200	65.66

Fit an equation of the form $C_p = a + bT + cT^2$ and an equation of the form $C_p = a + bT + cT^2 + dT^3$. Compare your results with the heat capacity equation for NH_3 gas in Appendix E.

5.32. Experimental values calculated for the heat capacity of ammonia from -40 to $1200°C$ are:

T (°C)	C_p^o [cal/(g mol)(°C)]	T (°C)	C_p^o [cal/(g mol)(°C)]
−40	8.180	500	12.045
−20	8.268	600	12.700
0	8.371	700	13.310
18	8.472	800	13.876
25	8.514	900	14.397
100	9.035	1000	14.874
200	9.824	1100	15.306
300	10.606	1200	15.694
400	11.347		

Fit the data by least squares for the following two functions:

$$\left.\begin{array}{l} C_p^o = a + bT + cT^2 \\ C_p^o = a + bT + cT^2 + dT^3 \end{array}\right\} \quad \text{where } T \text{ is in °C}$$

5.33. Experimental values of the heat capacity C_p have been determined in the laboratory as follows; fit a second-order polynomial in temperature to the data ($C_p = a + bT + cT^2$):

T (°C)	C_p [J/(g mol)(°C)]
100	40.54
200	43.81
300	46.99
400	49.33
500	51.25
600	52.84
700	54.14

(The data are for carbon dioxide.)

5.34. An equation for the heat capacity of acetone vapor is

$$C_p = 71.96 + 20.10 \times 10^{-2}\,T - 12.78 \times 10^{-5}\,T^2 + 34.76 \times 10^{-9}\,T^3$$

where C_p is in J/(g mol)(°C) and T is in °C. Convert the equation so that C_p is in Btu/(lb mol)(°F) and T is in °F.

5.35. Your assistant has developed the following equation to represent the heat capacity of air (with C_p in cal/(g mol) (K) and T in K):

$$C_p = 6.39 + 1.76 \times 10^{-3}\,T - 0.26 \times 10^{-6}\,T^2$$

(a) Derive an equation giving C_p but with the temperature expressed in °C.
(b) Derive an equation giving C_p in terms of Btu per pound per degree Fahrenheit with the temperature being expressed in degrees Fahrenheit.

5.36. An equation for the heat capacity of carbon (solid) was given in an article as

$$C_p = 1.2 + 0.0050\,\text{T} - 0.0000021\,T^2$$

where C_p is in Btu/(lb)(°F) and T is in °F. The calculated value of the enthalpy for C at 1000°F is 1510 Btu/lb. What is the reference temperature for the calculation of the enthalpy of carbon?

5.37. You have been asked to review the calculations of an assistant with respect to the calculation of the internal energy of CO_2. The assistant looked up the enthalpy of CO_2 at 600 kPa and 283K (16,660 kJ/gmol), and then calculated on the basis of 1 g mol

$$p\hat{V} = zRT \qquad \text{so } \hat{U} = \hat{H} - p\hat{V} = H - zRT$$

$$\hat{U} \;=\; 16{,}660 \;-\; \frac{0.96 \;\big|\; 8.314 \;\big|\; 283 \;\big|\; 1}{\big|\;\;\;\;\;\;\big|\;\;\;\;\;\;\big|\; 1000} \;=\; 16{,}658$$

Is this calculation correct?

5.38. The heat capacity of carbon monoxide is given by the following equation.

$$C_p = 6.395 + 6.77 \times 10^{-4}\,T + 1.3 \times 10^{-7}\,T^2$$

where C_p = cal/(g mol)(C)
  T = °C

What is the enthalpy change associated with heating carbon monoxide from 500°C to 1000°C?

5.39. Two gram moles of nitrogen are heated from 50°C to 250°C in a cylinder. What is ΔH for the process? The heat capacity equation is

$$C_p = 27.32 + 0.6226 \times 10^{-2}\,T - 0.0950 \times 10^{-5}\,T^2$$

where T is in kelvin and C_p is in J/(g mol)(°C).

5.40. Calculate the change in enthalpy for 5 kg mol of CO which is cooled from 927°C to 327°C.

5.41. Hydrogen sulfide is heated from 77°C to 227°C. What is the enthalpy change due to the heating?

5.42. What is the enthalpy change for acetylene when heated from 37.8°C to 93.3°C?

5.43. Use the steam tables to calculate the enthalpy change (in joules) of 2 kg mol of steam when heated from 400 K and 100 kPa to 900 K and 100 kPa. Repeat using the table in the text for the enthalpies of combustion gases. Repeat using the heat capacity for steam. Compare your answers. Which is most accurate?

5.44. A closed vessel contains steam at 1000.0 psia in a 4- to-1 vapor volume-to-liquid volume ratio. What is the steam quality?

5.45. What is the enthalpy change in British thermal units when 1 gal of water is heated from 60°F to 1150°F and 240 psig.

5.46. Use the CO_2 chart for the following calculations.
  (a) Four pounds of CO_2 are heated from saturated liquid at 20°F to 600 psia and 180°F.
    (1) What is the specific volume of the CO_2 at the final state?
    (2) Is the CO_2 in the final state gas, liquid, solid, or a mixture of two or three phases?
  (b) The 4 lb of CO_2 is then cooled at 600 psia until the specific volume is 0.07 ft³/lb.
    (1) What is the temperature of the final state?
    (2) Is the CO_2 in the final state gas, liquid, solid, or a mixture of two or three phases?

5.47. You are asked to calculate the electric power required (in kWh) to heat all of an aluminum wire (positioned in a vacuum similar to a light bulb filament) from 25°C to 660°C (liquid) to be used in a vapor deposition apparatus. The melting point of Al is 660°C. The wire is 2.5 mm in diameter and has a length of 5.5 cm. (The vapor deposition occurs at temperatures in the vicinity of 900°C).

5.48. Determine the enthalpy change when one gram mole of SO_2 gas is cooled from 538°C to −101°C at 1 atmosphere pressure. Data:

Boiling point:	−5°C
Melting point:	−75.5°C
Latent Heat of Vaporization:	24,940 J/g mol
Latent Heat of Fusion:	7,401 J/g mol
Use the average C_p of liquid SO_2 as:	1.28 J/(g mol)(°C)
Use the average C_p of solid SO_2 as:	0.958 J/(g mol)(°C)
Use the C_p of gaseous SO_2 from Appendix E.	

5.49. What is the enthalpy change that takes place when 3 kg of water at 101.3 kPa and 300 K are vaporized to 15,000 kPa and 800K?

5.50. Wet steam flows in a pipe at a pressure of 700 kPa. To check the quality, the wet steam is expanded to a pressure of 100 kPa in a separate pipe. A thermocouple inserted into the pipe indicates the expanded steam has a temperature of 125°C. What was the quality of the wet steam in the pipe prior to expansion?

5.51. If 30 m³ of combustion products of the following composition: 6.2% O_2, 1.0% H_2O, 12.3% CO_2, and the balance N_2 are cooled at atmospheric pressure from 1100K to 300K, what is the enthalpy change in kJ? Use the table for the enthalpies of the combustion gases.

5.52. Calculate the enthalpy change (in J/kg mol) that takes place in raising the temperature of 1 kg mol of the following gas mixture from 50°C to 550°C.

Component	Mol %
CH_4	80
C_2H_6	20

5.53. Use the tables for the heat capacities of the combustion gases to compute the enthalpy change (in Btu) that takes place when a mixture of 6.00 lb mol of gaseous H_2O and 4.00 lb mol of CO_2 is heated from 60°F to 600°F.

5.54. Chemical vapor deposition (CVD) is an important technique in producing solid-state materials because of its potential application to a variety of materials with controlled properties. Although this method has advantages, the film formation rates are so low, usually 0.1 to 1.0 nm/s, that the application of this method has been limited to date to very expensive materials.

In one trial to test film growth the temperature was controlled by use of steam on the underside of the film surface in the reactor assembly. Titanium tetraisopropoxide [Ti(OC$_3$H7)$_4$] was carried by a helium stream over the condensing surface, where it decomposed to TiO_2 plus gases.

The steam entered the reactor 523 K and 130 kPa and exited at 540 K and 100 kPa. What was the internal energy change per kg of steam going through the reactor? Use the steam tables.

5.55. In a proposed molten-iron coal gasification process (*Chemical Engineering*, July 1985: 17), pulverized coal of up to 3 mm size is blown into a molten iron bath, and oxygen and steam are blown in from the bottom of the vessel. Materials such as lime for settling the slag, or steam for batch cooling and hydrogen generation, can be injected at the same time. The sulfur in the coal reacts with lime to form calcium sulfide, which dissolves into the slag. The process operates at atmospheric pressure and 1400 to 1500°C. Under these conditions, coal volatiles escape immediately and are cracked. The carbon conversion rate is said to be above 98%, and the gas is typically 65 to 70% CO, 25 to 35% H_2, and less than 2% CO_2. Sulfur content of the gas is less than 20 ppm.

Assume that the product gas is 68% CO, 30% H_2, sand 2% CO_2, and calculate the enthalpy change that occurs on the cooling of 1000 m^3 of gaseous product from 1400°C to 25°C at 101 kPa. Use the table for the enthalpies of the combustion gases.

5.56. Calculate the enthalpy change (in joules) that occurs when 1 kg of benzene vapor at 150°C and 100 kPa condenses to a solid at −20.0°C and 100 kPa.

5.57. The vapor pressure of zinc in the range 600 to 985°C is given by the equation

$$\log_{10} p = -\frac{6160}{T} + 8.10$$

where p = vapor pressure, mm Hg
        T = temperature, K

Estimate the latent heat of vaporization of zinc at its normal boiling pint of 907°C.

5.58. A pure organic liquid with a molecular weight of 96 g/gmol is placed in a container and frozen. A vacuum is then established to remove all the air from the container. The material is then remelted and allowed to equilibrate at 10°C and 60°C. At 10°C the absolute pressure above the liquid is 28 mm Hg abs, and at 60°C, it is 100 mm Hg abs.

(a) Determine the pressure that would be present in the chamber at 30°C and the enthalpy change that would need to be input to vaporize 150 g of the material at 30°C.

(b) If dry air were bubbled through the chamber under isothermal conditions at 60°C and a total pressure of 2 atm at a rate of 100 cm^3/min, how long will it take to evaporate 25 g of the liquid?

(c) If the exiting saturated gas from the vessel were reduced in pressure to 1 atm, what would be the dew point temperature?

5.59. Use of the steam tables:

(a) What is the enthalpy change needed to change 3 lb of liquid water at 32°F to steam at 1 atm and 300°F?

(b) What is the enthalpy change needed to heat 3 lb of water from 60 psia and 32°F to steam at 1 atm and 300°F?

    (c) What is the enthalpy change needed to heat 1 lb of water at 60 psia and 40°F to steam at 300°F and 60 psia?

    (d) What is the enthalpy change needed to change 1 lb of a water-steam mixture of 60% quality to one of 80% quality if the mixture is at 300°F?

    (e) Calculate the ΔH value for an isobaric (constant pressure) change of steam from 120 psia and 500°F to saturated liquid.

    (f) Repeat part (e) for an isothermal change to saturated liquid.

    (g) Does an enthalpy change from saturated vapor at 450°F to 210°F and 7 psia represent an enthalpy increase or decrease? A volume increase or decrease?

    (h) In what state is water at 40 psia and 267.24°F? At 70 psia and 302°F? At 70 psia and 304°F?

    (i) A 2.5-ft³ tank of water at 160 psia and 363.5°F has how many cubic feet of liquid water in it? Assume that you start with 1 lb of H_2O. Could it contain 5 lb of H_2O under these conditions?

    (j) What is the volume change when 2 lb of H_2O at 1000 psia and 20°F expands to 245 psia and 460°F?

    (k) Ten pounds of wet steam at 100 psia has an enthalpy of 9000 Btu. Find the quality of the wet steam.

Section 5.3

5.60. Ten pounds of steam is placed in a tank at 300 psia and 480°F. After cooling the tank to 30 psia, some of the steam condenses. How much cooling was required and what was the final temperature in the tank? Assume the tank itself does not absorb energy.

5.61. One-tenth of a kg of steam at 650 K and 1000 kPa are enclosed in a metallic cylinder (mass 0.70 kg) by a piston (mass 0.46 kg) of area 118 cm². Both the piston and the cylinder have heat capacities of 0.34 J/(g of material) (K), and they are not insulated from the surroundings. The steam expands slowly pushing the piston up 80 cm where it is stopped and the pressure of the steam is 700 kPa. Find:

    (a) The work done by the system of steam-piston-cylinder on the surroundings.

    (b) The final temperature of the steam.

    (c) The heat transfer between the system and the surroundings (state which way).

    (d) The change in volume of the steam in its expansion.

5.62. One kg of gaseous CO_2 at 550 kPa and 25°C was compressed by a piston to 3500 kPa, and in so doing 4.016×10^3 J of work was done on the gas. To keep the container isothermal, the container was cooled by blowing air over fins on the outside of the container. How much heat (in J) was removed from the system?

5.63. A closed vessel having a volume of 100 ft³ is filled with saturated steam at 265 psia. At some later time, the pressure has fallen to 100 psia due to heat losses from the tank. Assuming that the contents of the tank at 100 psia are in an equilibrium state, how much heat was lost from the tank?

5.64. A large piston does 12,500 (ft)(lb$_f$) of work in compressing 3 ft^3 of air to 25 psia. Five pounds of water in a jacket surrounding the piston increased in temperature by 2.3°F during the process. What was the change in the internal energy of the air?

$$C_p \text{ water} = 8.0 \frac{\text{Btu}}{(\text{lb mol})(°F)}$$

5.65. An insulated cylinder contains two gases, A and B, each held separately in place by a single-fixed plug. Initially the pressure of gas A is higher than the pressure in gas B. The plug is then released, and the system allowed to equilibrate. For the system of the cylinder plus the two gases, what is
(a) the work done by the system?
(b) the heat transferred to or from the system?
(c) the change in internal energy of the system?

5.66. A small 3.1L gas cylinder of Ne at p_1 and T_1 is attached to a large 31.7L partially evacuated cylinder at p_2 and T_2. A valve between the cylinders is opened. Determine the work done, the heat transferred, and the change in the internal energy of the gas. Assume each cylinder is well insulated.

5.67. Carbon dioxide cylinders, initially evacuated, are being loaded with CO_2 from a pipeline in which the CO_2 is maintained at 200 psia and 40°F. As soon as the pressure in a cylinder reaches 200 psia and the cylinder is closed and disconnected from the pipeline. If the cylinder has a volume of 3 ft^3, and if the heat losses to the surroundings are small, compute:
(a) The final temperature of the CO_2 in a cylinder.
(b) The number of pounds of CO_2 in a cylinder.

5.68. One pound of steam at 130 psia and 600°F is expanded isothermally to 75 psia in a closed system. Thereafter it is cooled at constant volume to 50 psia. Finally, it is compressed adiabatically back to its original state. For each of the three steps of the process, compute ΔU and ΔH. For each of the three steps, where possible, also calculate Q and W.

5.69. Two 1-m^3 tanks submerged in a constant-temperature water bath of 77°C are connected by a globe valve. One tank contains steam at 40 kPa, while the second is completely evacuated. The valve is opened and the pressure in the two tanks equilibrates. Calculate:
(a) The work done in the expansion of the steam
(b) The heat transferred to the tanks from the water bath
(c) The change in internal energy of the combined steam system.

5.70. You have 0.37 lb of CO_2 at 1 atm abs and 100°F contained in a cylinder closed by a piston. The CO_2 is compressed to 70°F and 1000 psia. During the process, 40.0 Btu of heat is removed. Compute the work required for the compression.

5.71. A household freezer is placed inside an insulated sealed room. If the freezer door is left open with the freezer operating, will the temperature of the room increase or decrease? Explain your answer.

5.72. One pound of water at 70°F is put into a closed tank of constant volume. If the temperature is brought to 80°F by heating, find Q, W, ΔU, and ΔH for the process (in Btu). If the same temperature change is accomplished by a stirrer that agitates the water find Q, W, ΔE, and ΔH. What assumptions are needed?

5.73. A heating unit heats the air in a duct entering an auditorium by condensing saturated steam at 12 psia. The air (at 1 atm) is heated from 66°F to 76°F. Per 1000 ft³ of air entering the heater, how many pounds of saturated steam must be condensed? The heating unit loses to the surroundings 3 Btu per pound of steam condensed.

5.74. Find the power output of an insulated generator that uses 700 kg/hr of steam at 10 atm and 500K. The steam exits at 1 atm and is saturated.

5.75. Water at 180°F is pumped at a rate of 100 ft³/hr through a heat exchanger to reduce its temperature to 100°F. Find the rate at which heat is removed from the water in the heat exchanger.

5.76. For each process below, write down the general energy balance (include all terms). Based on the data given and any reasonable assumptions you make, simplify the balances as much as possible.
   (a) A fluid flows through a tiny orifice from a region where its pressure is 1300 kPa and 600K to a pipe where the pressure is 275 kPa.
   (b) A turbine directly connected to an electric generator operates adiabatically. The working fluid enters the turbine at 250 psia and 600°F, and leaves at 50 psia and 400°F. The entrance and exit velocities of the fluid are negligible.
   (c) A fluid leaves the nozzle of a hose at 200 kPa and 400°C, and is brought to rest by passing it through the blades of an adiabatic turbine rotor. The fluid leaves the blades at 50 psia and 250°F.

5.77. In one system, carbon dioxide is flashed across an *insulated* throttling valve. The inlet pressure and temperature of the carbon dioxide are 1800 psia and 250°F, and the outlet of the valve is set at 60 psia. Indicate on the CO_2 chart the position of the inlet and outlet points as points A & B, respectively. What are: (a) the temperature, (b) the quality and (c) the specific volume of the outlet stream at 60 psia?

5.78. A turbine is installed between the supply stream at 1800 psia and 250°F and the outlet stream. The exit stream from the turbine is at 800 psia and 80°F, and 25 Btu/lb of fluid are lost from the poorly insulated turbine. Indicate the outlet states from the turbine and the throttling valve as points on the CO_2 chart. See Figure P5.78.
   (a) How much useful work is extracted in the turbine?
   (b) The exhaust from the turbine enters a throttling valve and leaves at 140 psia and 30% liquid, but the valve is not perfectly insulated. What is the temperature of the exhaust stream at 140 psia?
   (c) How much heat is lost from the poorly insulated valve?

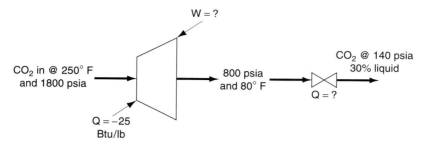

Figure P5.78

5.79. What is the heat that has to be transferred to a water heater of fixed volume to heat saturated water from 76°F to 220°F and 1 atm per pound of water?

5.80. In the vapor-recompression evaporator shown in Figure P5.80, the vapor produced on evaporation is compressed to a higher pressure and passed through the heating coil to provide the energy for evaporation. The steam entering the compressor is 98% vapor and 2% liquid, at 10 psia; the steam leaving the compressor is at 50 psia and 400°F; and 6 Btu of heat are lost from the compressor per pound of steam throughout. The condensate leaving the heating coil is at 50 psia, 200°F.

    (a) Compute: The Btu of heat supplied for evaporation in the heating coil per Btu of work needed for compression by the compressor.

    (b) If 1,000,000 Btu per hour of heat is to be transferred in the evaporator, what must be the intake capacity of the compressor in ft^3 of wet vapor per minute?

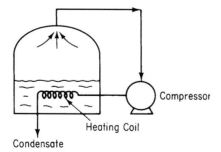

Figure P5.80

5.81. A 55-gal drum of fuel oil (15°API) is to be heated from 70°F to 180°F by means of an immersed steam coil. Steam is available at 220°F. How much steam is required if it leaves the coil as liquid water at 1 atm pressure?

5.82. Oil of average $C_p = 0.8$ Btu/(lb)(°F) flows at 2000 lb/min from an open reservoir standing on a hill 1000 ft high into another open reservoir at the bottom. To ensure rapid flow, heat is put into the pipe at the rate of 100,000 Btu/hr. What is the enthalpy change in the oil per pound? Suppose that a 1-hp pump (50% efficient) is added to the pipeline to assist in moving the oil. What is the enthalpy change per pound of the oil now?

5.83. Write the simplified energy balances for the following changes:

   (a) A fluid flows steadily through a poorly designed coil in which it is heated from 70°F to 250°F. The pressure at the coil inlet is 120 psia, and at the coil outlet is 70 psia. The coil is of uniform cross section, and the fluid enters with a velocity of 2 ft/sec.

   (b) A fluid is expanded through a well-designed adiabatic nozzle from a pressure of 200 psia and a temperature of 650°F to a pressure of 40 psia and a temperature of 350°F. The fluid enters the nozzle with a velocity of 25 ft/sec.

   (c) A turbine directly connected to an electric generator operates adiabatically. The working fluid enters the turbine at 1400 kPa absolute and 340°C. It leaves the turbine at 275 kPa absolute and at a temperature of 180°C. Entrance and exit velocities are negligible.

   (d) The fluid leaving the nozzle of part (b) is brought to rest by passing it through the blades of an adiabatic turbine rotor and it leaves the blades at 40 psia and at 400°F.

5.84. Your company produces small power plants that generate electricity by expanding waste process steam in a turbine. One way to ensure good efficiency in turbine operation is to operate adiabatically. For one turbine, measurements showed that for 1000 lb/hr steam at the inlet conditions of 500°F and 250 psia, the work output from the turbine was 86.5 hp and the exit steam leaving the turbine was at 14.7 psia with 15% wetness (i.e., with a quality of 85%).

   Check whether the turbine is operating adiabatically by calculating the value of Q.

5.85. Feedwater heaters are used to increase the efficiency of steam power plants. A particular heater is used to preheat 10 kg/s of boiler feed water from 20°C to 188°C at a pressure of 1200 kPa by mixing it with saturated steam bled from a turbine at 1200 kPa amd 188°C, as shown in Figure P5.85. Although insulated,

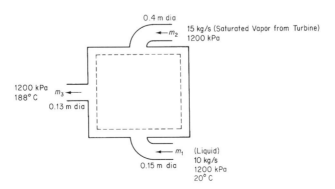

Figure P5.85

the heater loses heat at the rate of 50 J per gram of exiting mixture. What fraction of the exit stream is steam? See Figure P5.85.

5.86. Cell growth liberates energy which must be removed; otherwise, the temperature will rise so much that the cells may be killed. In a continuous fermenter for the production of *Penicillium chrysogenum*, the cells generate 27.6 kJ/L per hour, and the volume of the well-insulated fermenter is 2 L. The feed temperature is 25°C and the exit temperature is equal to the temperature in the fermenter. *Penicillium chrysosenum* cannot grow above 42°C. Will the cells survive? Assume for simplicity that the inlet and outlet streams have a heat capacity of 4 J/(g)°C and the mass flow rates of 1025 g/hr are constant.

5.87. Three hundred kilograms per hour of air flow through a countercurrent heat exchanger as shown in Figure P5.87. Two hundred thirty kilograms per hour of potassium carbonate solution are heated by the air. Assume that the heat exchanger has negligible heat losses. The terminal temperatures are given in Figure P5.87. Calculate the temperature, in kelvin, of the exit potassium carbonate stream.

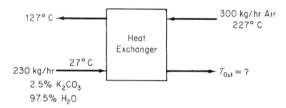

Figure P5.87

5.88. A chemical plant has just perfected a process for the manufacturing of a new revolutionary drug. A large plant must be designed even before complete data are available. The research laboratory has obtained the following data for the drug.

boiling point at 101.3 kPa	250°C
vapor pressure at 230°C	55 kPa
specific gravity, 28/4	1.316
melting point	122.4°C
solubility in water at 76°C	2.2/100 parts water by weight
molecular weight	122
heat capacity of liquid	at 150°C is 2.09 kJ/(kg)(°C)
heat capacity of liquid	at 240°C is 2.13 kJ/(kg)(°C)

Calculate the heat duty for a vaporizer that will be required to vaporize 10,000 kg/hr of this chemical at atmospheric pressure (assume that no superheating of the vapor occurs). The entering temperature of the drug will be 130°C.

5.89. A process involving catalytic dehydrogenation in the presence of hydrogen is known as *hydroforming*. Toluene, benzene, and other aromatic materials can be

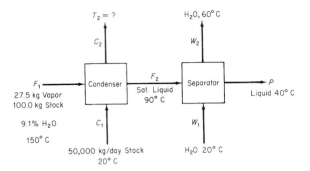

Figure P5.89

economically produced from naptha feed stocks in this way. After the toluene is separated from the other components, it is condensed and cooled in a process such as the one shown in Figure P5.89. For every 100 kg of stock charged into the system, 27.5 kg of a toluene and water mixture (9.1% water) are produced as overhead vapor and condensed by the charge stream. Calculate:

(a) The temperature of the charge stream after it leaves the condenser
(b) The kilograms of cooling water required per hour.
Additional data:

Stream	C_p [kJ/(kg)(°C)]	B.P. (°C)	ΔH_{vap} (kJ/kg)
$H_2O(l)$	4.2	100	2260
$H_2O(g)$	2.1	---	---
$C_7H_8(l)$	1.7	111	230
$C_7H_8(g)$	1.3	---	---
Stock (l)	2.1	---	---

5.90. Simplify the general energy balance so as to represent the process in each of the following cases. Number each term in the general balance, Eq. (5.13), and state why you retained or deleted it.

(a) A bomb calorimeter is used to measure the heating value of natural gas. A measured volume of gas is pumped into the bomb. Oxygen is then added to give a total pressure of 10 atm, and the gas mixture is exploded using a hot wire. The resulting heat transfer from the bomb to the surrounding water bath is measured. The final products in the bomb are CO_2 and water.

(b) Cogeneration (generation of steam both for power and heating) involves the use of gas turbines or engines as prime movers, with the exhausted steam going to the process to be used as a heat source. A typical installation is shown in Figure P5.90.

(c) In a mechanical refrigerator the Freon liquid is expanded through a small insulated orifice so that part of it flashes into vapor. Both the liquid and vapor exit at a lower temperature than the temperature of the liquid entering.

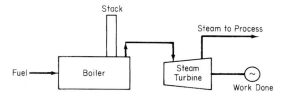

Figure P5.90

Section 5.4

5.91. Calculate the heat of reaction at the standard reference state for the following reactions:

(a) $CO_2(g) + H_2(g) \longrightarrow CO(g) + H_2O(l)$

(b) $2CaO(s) + 2MgO(s) + 4H_2O(l) \longrightarrow 2Ca(OH)_2(s) + 2Mg(OH)_2(s)$

(c) $Na_2SO_4(s) + C(s) \longrightarrow Na_2SO_3(s) + CO(g)$

(d) $NaCl(s) + H_2SO_4(l) \longrightarrow NaHSO_4(s) + HCl(g)$

(e) $NaCl(s) + 2SO_2(g) + 2H_2O(l) + O_2(g) \longrightarrow 2Na_2SO_4(s) + 4HCl(g)$

(f) $SO_2(g) + \frac{1}{2}O_2(g) + H_2O(l) \longrightarrow H_2SO_4(l)$

(g) $N_2(g) + O_2(g) \longrightarrow 2NO(g)$

(h) $Na_2CO_3(s) + 2Na_2S(s) + 4SO_2(g) \longrightarrow 3Na_2S_2O_3(s) + CO_2(g)$

(i) $CS_2(l) + Cl_2(g) \longrightarrow S_2Cl_2(l) + CCl_4(l)$

(j) $\underset{\text{ethylene}}{C_2H_4(g)} + HCl(g) \longrightarrow \underset{\text{ethylchloride}}{CH_3CH_2Cl(g)}$

(k) $\underset{\text{methyl alcohol}}{CH_3OH(g)} + \frac{1}{2}O_2(g) \longrightarrow \underset{\text{formaldehyde}}{H_2CO(g)} + H_2O(g)$

(l) $\underset{\text{acetylene}}{C_2H_2(g)} + H_2O(l) \longrightarrow \underset{\text{acetaldehyde}}{CH_3CHO(l)}$

(m) $\underset{\text{butane}}{n\text{-}C_4H_{10}(g)} \longrightarrow \underset{\text{ethylene}}{C_2H_4(g)} + \underset{\text{ethane}}{C_2H_6(g)}$

5.92. J. D. Park et al. [JACS 72, 331-3 (1950)] determined the heat of hydrobromination of propene and cycopropane. For hydrobromination (addition of HBr) of propene to 2-bromopropane, they found that $\Delta H = -84{,}441$ J/g mol. The heat of hydrogenation of propene to propane is $\Delta H = -126{,}000$ J/g mol. N.B.S. Circ. 500 gives the heat of formation of HBr(g) as $-36{,}233$ J/g mol when the bromine is liquid and the heat of vaporization of bromine as $30{,}710$ J/g mol. Calculate the heat of bromination of propane using gaseous bromine to form 2-bromopropane using the above data.

5.93. The following enthalpy changes are known for reactions a 25°C in the standard thermochemical state:

No.			$\Delta H°$(kJ)
1	$C_3H_6(g) + H_2(g)$	$\longrightarrow C_3H_8(g)$	−123.8
2	$C_3H_8(g) + 5O_2(g)$	$\longrightarrow 3CO_2(g) + 4H_2O(l)$	−2220.0
3	$H_2(g) + \frac{1}{2}O_2(g)$	$\longrightarrow H_2O(l)$	−285.8
4	$H_2O(l)$	$\longrightarrow H_2O(g)$	43.9
5	$C(diamond) + O_2(g)$	$\longrightarrow CO_2(g)$	−395.4
6	$C(graphite) + O_2(g)$	$\longrightarrow CO_2(g)$	−393.5

Calculate the heat of formation of propylene.

5.94. How would you determine the heat of formation of gaseous fluorine at 25°C and 1 atm?

5.95. Determine the standard heat of formation for FeO(s) given the following values for the heats of reaction at 25°C and 1 atm for the following reactions:

$$2Fe(s) + \tfrac{3}{2}O_2(g) \longrightarrow Fe_2O_3(s): -822,200 \text{ J}$$

$$2FeO(s) + \tfrac{1}{2}O_2(g) \longrightarrow Fe_2O_3(s): -284,100 \text{ J}$$

5.96. Use the table of the heats of formation in Appendix F of the text to calculate the standard heats of reaction per g mole of the compounds produced in the following reactions:
(a) $N_2(g) + 3H_2(g) \longrightarrow 2NH_3(g)$
(b) $Fe(s) + 1\tfrac{1}{2}Cl_2(g) \longrightarrow FeCl_3(s)$

5.97. Calculate the standard (25°C and 1 atm) heat of reaction per g mol of the first reactant on the left-hand side of the reaction equation for the following reactions:
(a) $NH_3(g) + HCl(g) \longrightarrow NH_4Cl(s)$
(b) $CH_4(g) + 2O_2(g) \longrightarrow CO_2(g) + 2H_2O(l)$
(c) $\underset{\text{cyclohexane}}{C_6H_{12}(g)} \longrightarrow \underset{\text{benzene}}{C_6H_6(l)} + 3H_2(g)$

5.98. Calculate the standard heat of reaction for the conversion of cyclohexane to benzene

$$C_6H_{12}(g) \longrightarrow C_6H_6(g) + 3H_2(g)$$

If the reactor for the conversion operates at 70% conversion of C_6H_{12}, what is the heat removed from or added to the reactor if
(a) the exit gases leave at 25°C?
(b) the exit gases leave at 300°C?
   and the entering materials consist of C_6H_{12} together with one-half mole of N_2 per mole of C_6H_{12} both at 25°C.

5.99. Estimate the higher heating value (HHV) and lower heating value of the following fuels in Btu/lb:
(a) Coal with the analysis C (80%), H (0.3%), O (0.5%), S (0.6%), and ash (18.60%).

(b) Fuel oil that is 30°API and contains 12.05% H and 0.5% S.

5.100. Is the higher heating value of a fuel ever equal to the lower heating value? Explain.

5.101. Find the higher (gross) heating value of $H_2(g)$ at 0°C.

5.102. The chemist for a gas company finds a gas analyzes 9.2% CO_2, 0.4% C_2H_4, 20.9% CO, 15.6% H_2, 1.9% CH_4, and 52.0% N_2. What should the chemist report as the gross heating value of the gas?

5.103. What is the higher heating value of 1 m^3 of *n*-propylbenzene measured at 25°C and 1 atm and with a relative humidity of 40%?

5.104. An off-gas from a crude oil topping plant has the following composition:

Component	Vol. %
Methane	88
Ethane	6
Propane	4
Butane	2

(a) Calculate the higher heating value on the following bases: (1) Btu per pound, (2) Btu per mole, and (3) Btu per cubic foot of off-gas measured at 60°F and 760 mm Hg.

(b) Calculate the lower heating value on the same three bases indicated in part (a).

5.105. Calculate the lower heating value of methane at 100°C.

5.106. The first administrator of the EPA was quoted as saying.

> . . . deficient building standards mean that more energy passes through the windows of buildings in the U.S. than flows through the Alaska pipeline.

What does this mean?

5.107. Can savings in energy be achieved by shutting down heating equipment when it is not in use in off-hours rather than leaving the equipment on all the time? For example, an oil burning furnace is used to reheat steel to 1230°C and requires 270L/hr on idle to maintain this temperature. If cooled to ambient temperature the furnace can be reheated in 6.5 hours while burning fuel oil at the average rate of 760L/hr. How much money can be saved by shutting the furnace down each weekend if fuel oil is $0.16/L? For a day? What problems might occur with the proposed procedure? *Data*: The heat of reaction at the furnace operating temperature is 40,000 MJ/L of fuel oil.

5.108. Physicians measure the metabolic rate of conversion of foodstuffs in the body by using tables that list the liters of O_2 consumed per gram of foodstuff. For a simple case, suppose that glucose reacts

$$C_6H_{12}O_6 \text{ (glucose)} + 6\,O_2\,(g) \longrightarrow 6\,H_2O\,(1) + 6\,CO_2\,(g)$$

How many liters of O_2 would be measured for the reaction of one gram of glucose (alone) if the conversion were 90% complete in your body? How many kJ/g of glucose would be produced in the body? *Data:* $\Delta \hat{H}_f^\circ$ of glucose is -1260 kJ/g mol of glucose.

5.109. The Neches Butane Products Co. makes butadiene, C_4H_6, by separating butenes, C_4H_8, from refinery gases piped to them, and then further dehydrogenating the butenes to butadiene. In order to increase capacity they would like to be able to dehydrogenate butane, C_4H_{10}, in one step without the intermediate separation of the butene. The C_4H_6 formed would be separated, and any unreacted C_4H_{10} or C_4H_8 would be recycled to the reactor. One of the large engineering companies claims to have such a process. You have just been employed by Neches B.P. Co.—fresh from school—and are requested to make some calculations on the proposed process:

$$C_4 H_{10} \, (1) \longrightarrow C_4 H_6 \, (g) + 2H_2 \, (g)$$

By heat exchange, the butane fed into the reactor is brought to the reaction temperature of 1000 K. It is desired to maintain an isothermal reactor. How much heat must be added or removed per g mole of butadiene formed? Assume the pressure is 100 kPa. What are some reasons why the heat load may be different from the value you calculate? *Data:* $\Delta \hat{H}_f^\circ$ of $C_4 H_6$ (g) is -165.5 kJ/g mole.

5.110. A molar mixture of 50% C_3H_8 and 50% C_2H_6 is fed to a compressor at a rate of 4700 kg mol/hr at a temperature of 35°C and a pressure of 120 kPa. The compressed gas leaves at a volumetric flowrate of 4500 m³/hr at 80°C.
   (a) Assuming the ideal gas law applies to the incoming gas, determine the lower heating value per m³ of feed gas (in kJ/m³).
   (b) Estimate the pressure at the exit of the compressor using Kay's rule for the pseudocritical constants to account for nonideal gas effects under these conditions.

5.111. Formaldehyde can be made by the oxidation of methanol (CH_3OH). If stochiometric amounts of $CH_3OH(g)$ and $O_2(g)$ enter the reactor at 100°C, the reaction is complete, and the products leave the reactor at 200°C, calculate the heat that is added or removed from the reactor per mole of $CH_3OH(g)$ fed to the reactor. The reaction is

$$CH_3OH(g) + \tfrac{1}{2}O_2(g) \longrightarrow H_2CO(g) + H_2O(g)$$

5.112. Liquid hydrazine (N_2H_4) is injected in a jet combustion chamber at 400K and burned with 100% excess air that enters at 700K. The combustion chamber is jacketed with water. The combustion products leave the jet exhaust at 900K. In the test, if 50 kg mol are burned per hour and water at 25°C enters the water jacket for cooling at the rate of 400 kg/min, will the exit line for the water be able accommodate that much vapor (at 1 atm)? The reaction is

$$N_2H_4(l) + O_2(g) \longrightarrow N_2(g) + 2H_2O(g)$$

Data:

$$\Delta \hat{H}_f^o \text{ of } N_2H_4(l) = 44.77 \text{ kJ/g mol} \qquad C_p \text{ of } N_2H_2(l) \approx 139 \text{ J/(g mol)(°C)}$$

5.113. A flue gas analysis was run on a natural gas (CH_4) fired boiler, and showed 8.2% CO_2 and 5.3% O_2 with the flue gas exiting the boiler at 515°F. By reducing the excess air until the flue gas showed 10.5% CO_2 and 2.0% O_2, the exit temperature of the flue gas was reduced to 490°F.

To maintain these latter operating conditions, it was proposed to install a control system that cost $31,000 installed. If the natural gas cost $1.55/$10^6$ Btu, and the process used 180,000 ft³/hr of gas (at SC), with 8000 hours of operation per year, how long would it take to save $31,000 in fuel costs if operations were carried out at the second set of conditions?

5.114. An engine that is used to drive a compressor discharges 4225 cfm of exhaust at 1100°F and 6 inches of water back pressure. The plant requires 2000 lb/hr of 15 psig steam, which was supplied by a fired boiler using #2 fuel oil (density = 7.2 lb/gal, LHV = 18,300 Btu/lb), and has available 220°F water for boiler feedwater. Calculate the exhaust heat recovered by a waste heat recovery muffler in 1b/hr of steam (thus allowing the output of steam from the fired boiler to be reduced resulting in the substantial fuel oil savings). Assume 80% efficiency in the boiler.

5.115. Although processes are designed to avoid releases of flammable and toxic substances into the surroundings, accidents do happen. Consequently, it is best to evaluate during the design phase the consequences of a hypothesized spill or leak so that emergency actions can be taken to mitigate the consequences of the accident. For liquid releases, the amount of vapor formed is critical, particularly for flammable vapors as the vapor can be blown by the wind beyond the site of the accident.

In a process under consideration, two possible liquid solvents are being considered: methanol and *n*-heptane. If an accident occurs, estimate the amount of vapor that would be generated for each solvent if it took 10 minutes before the flow of solvent could be shut off. Also estimate the total energy released if the vapor cloud burned completely. You want to chose the solvent that (a) generates the smallest vapor cloud, and (b) releases the least energy on combustion.

Process information: The normal solvent flow at some points in the process will be 500 gal/min, but the estimated flow through a broken pipe will be double the normal rate. The liquid temperature in the flowing stream will be as high as 400°F. The lower flammability limits are: CH_3OH (6.7%) and *n*-C_7H_{16} (1.05%).

Hints: To help in the choice of which solvent to use, first estimate the fraction of the liquid that will vaporize under three assumed conditions: (a) The liquid and vapor are at the normal boiling point following the release, (b) the liquid and vapor cool to an ambient temperature of 80°F following the release, and (c) the liquid and vapor cool to 25°F following the release. The last estimate is based on an analysis of heat transfer between the liquid-vapor mixture and the surrounding air. Get the enthalpy data for each solvent from a handbook or a computer database. Assume the vapor mixes with the air. Then determine the vapor cloud size based on having a mixture at its lower flammable limit, and assume the mixtures of vapor and air will be ideal because the mixture will be at a low pressure. Finally, calculate the energy released on combustion of the vapor release.

Adapted from Problem 48 in *Safety, Health, and Loss Prevention in Chemical Processes*. New York: American Institute of Chemical Engineers, 1990.

5.116. Complete the following sentences in the blank spaces provide with either the word increases or decreases:

(a) For a fixed outlet flue gas temperature from a heater, increasing the excess air _____ the outlet flue gas temperature.

(b) For a fixed outlet flue gas temperature, increasing the inlet air temperature _____ the thermal efficiency of the heater.

(c) For a fixed amount of excess air, increasing the return of flue gas to preheat the entering air _____ the fuel consumption for a fixed load on the heater.

5.117. Fluidized beds (see Figure P5.117) are used in the chemical industry as steam

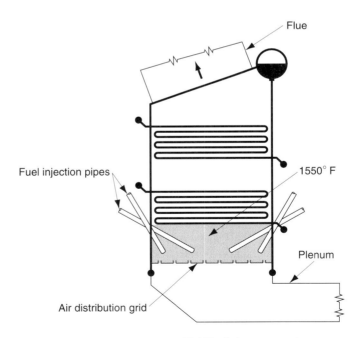

Fluid bed steam generator

Figure P5.117   Fluidized bed steam generator

generators Such equipment can burn almost any fuel, including low-quality coals such as lignite.

Coal burns rapidly in a fluidized bed, even at 1500°F. Based on bed volume, the heat release is 300,000–400,000 Btu/(hr)(ft^3). Counting the open furnace space above the bed the rate is 100,000–200,000 Btu/(hr)(ft^3).

Suppose that 200,000 Btu/(hr)(ft^3) are generated in a 40 ft^3 steam generator, that the water enters the two sets of coils at 70°F, and that in the upper set of coils the water flow rate is 3,000 lb/hr leaving as steam at 900°F and 380 psia. If the steam leaving the lower coils is at 1200°F and 400 psia, what is the flow rate of the water into the lower set of coils?

5.118.   M. Beck et al. (*Can J. Ch.E.*, *64* (1986): 553) described the use of immobilized enzymes (E) in a bioreactor to convert glucose (G) to fructose (F).

$$G + E \longleftrightarrow EG \longleftrightarrow E + F$$

At equilibrium the overall reaction can be considered to be G + E $\longleftrightarrow$ E + F.

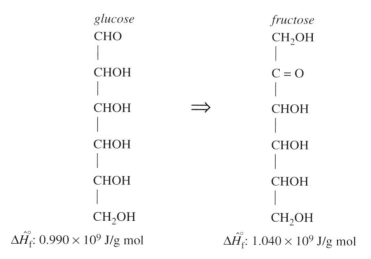

The fraction conversion is a function of the flow rate through the reactor and the size of the reactor, but for a flow rate of 3 × 10^{-3} m/s and a bed height of 0.44 m, the fraction conversion on a pass through the reactor was 0.48. Calculate the heat of reaction at 25°C per mole of G converted.

5.119.   A proposed method of producing ethanol (that might be used as an alternate fuel) is to react CO$_2$ with H$_2$

$$CO_2 + 3H_2 \longrightarrow CH_3OH + H_2O.$$

Assume that the gross feed enters the reactor in the stoichiometric quantities need for the reaction. Also 0.5% N$_2$ flows in with the fresh feed. On one pass

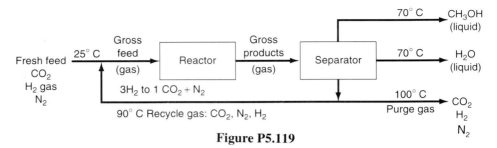

Figure P5.119

through the reactor 57% conversion is obtained. The concentration of N_2 in the gross feed into the reactor cannot exceed 2%.

Assume the process is in the steady state, that all gases are ideal, and the reaction system is as shown in the accompanying figure. How much heat must be added or removed from the entire system? The heat capacity of liquid CH_3OH from Perry in the range of 0 to 98°C is 0.68 cal/(g)(°C). See Figure P5.119.

5.120. Methanol can be made by catalytic oxidation of methane using the stoichiometric amount of O_2 as shown below with a yield of CH_3OH of 30% of the theoretical. Determine the quantity of heat required to be introduced into or is given off by the process per 100 kg mol of methanol formed. See Figure P5.120.

5.121. Sulfuric acid is a major bulk chemical used in a wide variety of industries. After sulfur is oxidized to SO_2, the SO_2 is further oxidized in the converters (reactors) to SO_3

$$SO_2(g) + 1/2\ O_2(g) \longrightarrow SO_3(g)$$

and the SO_3 is absorbed in dilute H_2SO_4 to form concentrated H_2SO_4.

In the first converter the entering gases at 400K and 1 atm are composed of 9.0% SO_2, 9.5% O_2, and 81.50% N_2. Only 75% of the entering SO_2 reacts on going through the first converter. If the maximum temperature of the gas before going to the next converter (where the reaction is completed) can be 700K, how much heat must be removed from the gas before it goes to the second converter per kg mol of S entering the process?

5.122. If for the process presented in problem 5.121 the entering gas is at 700K, what will the exit gas temperature be if the converter is well insulated?

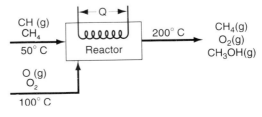

Figure P5.119

Figure P5.120

Data:

For air C_p is in Btu/(lb mol)(°F) and T is in °F

$C_p = 6.900 + 0.02884 \times 10^{-2}T + 0.02429 \times 10^{-5}T^2 - 0.08052 \times 10^{-9}T^3$

For these compounds C_p is in J/(g mol)(°C) and T is in °C

SO_2: $C_p = 38.91 + 3.904 \times 10^{-2}T - 3.104 \times 10^{-5}T^2 + 8.606 \times 10^{-9}T^3$

SO_3: $C_p = 48.50 + 9.188 \times 10^{-2}T - 8.540 \times 10^{-5}T^2 + 32.40 \times 10^{-9}T^3$

O_2: $C_p = 29.10 + 1.158 \times 10^{-2}T - 0.6076 \times 10^{-5}T^2 + 1.311 \times 10^{-9}T^3$

N_2: $C_p = 29.00 + 0.2199 \times 10^{-2}T - 0.5723 \times 10^{-5}T^2 - 2.871 \times 10^{-9}T^3$

5.123. A new process has been proposed to produce ethylene (C_2H_4) gas from propane (C_3H_8) gas at atmospheric pressure by the following reaction

$$C_3H_8(g) + 2O_2(g) \longrightarrow C_2H_4(g) + CO_2(g) + 2H_2O(g)$$

The products leave the system at 800 K and the C_3H_8 enters at 450 K while the O_2 enters at 300 K. In the process, $C_3H_8(g)$ is used in 25% excess (of the amount shown in the equation), but the overall fraction conversion of the C_3H_8 is only 40%. How much heat is added to or removed from the process per mole of C_3H_8 feed to the process?

5.124. A catalytic converter for the production of SO_3 from SO_2 is operating as illustrated in Fig. P5.124. The material balance for 1 hr of operation is also shown in Fig. P5.124.

$$SO_2(g) + 0.5O_2(g) \longrightarrow SO_3 \qquad \Delta H^{\,o}_{25} = -98{,}280 \text{ J/g mol}$$

The unit is insulated and heat losses are negligible. It has been found that corrosion can be reduced considerably by keeping the discharge temperature at approximately 400°C.

Determine the heat duty of a cooler for the converter that will accomplish this purpose.

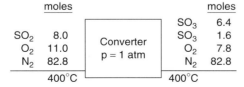

Figure P5.124

5.125. Propane, butane, or liquefied petroleum gas (LPG) has seen practical service in passenger automobiles for 30 years or more. Because it is used in the vapor phase, it pollutes less than gasoline but more than natural gas. A number of cars in the Clean Air Car Race ran on LPG. The table below lists their results and those for natural gas. It must be kept in mind that these vehicles were generally equipped with platinum catalyst reactors and with exhaust-gas recycle. Therefore, the gains in emission control did not come entirely from the fuels.

	Natural gas, avg. 6 cars	LPG avg. 13 cars	Fed std.
HC (g/mi)	1.3	0.49	0.22
CO (g/mi)	3.7	4.55	2.3
NO_x (g/mi)	0.55	1.26	0.6

Suppose that in a test butane gas at 100°F was burned completely with the stoichiometric amount of heated air which is at 400°F and a dew point of 77°F in an engine. To cool the engine, 12.5 lb of steam at 100 psia and 95% quality was generated from water at 77°F per pound of butane burned. It may be assumed that 7% of the gross heating value of the butane was lost as radiation from the engine. Did the exhaust gases leaving the engine exceed the temperature limit of the catalyst of 1500°F?

5.126. Compare prices of five fuels:
(a) Natural gas (all CH_4) at $2.20 per 10^3 ft^3
(b) No. 2 fuel oil (33°API) at $0.85 per gallon
(c) Dry yellow pine at $95 per cord
(d) High-volatile A bituminous coal at $55.00 per ton
(e) Electricity at $0.032 per kilowatt hour.

List the cost of each in descending order per 10^6 Btu. As far as energy consumption is concerned, would you be better off with a gas dryer or an electric dryer? Should you heat your house with wood, No. 2 fuel oil, or coal? (*Note*: A cord of wood is a pile 8 ft long, 4 ft high, and 4 ft wide.) You will have to look up enthalpy values.

5.127. Three fuels are being considered as a heat source for a metallurgical process:

	Components	Percent
Gas	CH_4	96.0
	CO_2	3.0
	N_2	1.0
Oil (liquid)	$C_{16}H_{34}$	99.0
	S	1.0
Coke	C	95.0
	ash	5.0

Calculate the maximum flame temperature (i.e., combustion with the theoretically required amount of air) for each of the three fuels assuming that the fuels and air enter at 18°C, and the ash (C_p = 1.15 J/(g)(°C)) leaves the combustion chamber containing no carbon and at 527°C. Which fuel gives the highest temperature?

5.128. Calculate the adiabatic flame temperature of C_3H_6(g) at 1 atm when burned with 20% excess air, and the reactants enter at 25°C.

5.129. Which substance will give the higher theoretical flame temperature if the inlet percentage of excess air and temperature conditions are identical: (a) CH_4, (b) C_2H_6, or (c) C_4H_8?

5.130. Calculate the adiabatic flame temperature of CH_4(g) at 1 atm when burned with 10% excess air. The air enters at 25°C and the CH_4 at 300K. The reaction is

$$CH_4(g) + 2O_2(g) \longrightarrow CO_2(g) + 2H_2O(g)$$

5.131. A gas is burned with 300% excess air with the gas and air entering the combustion chamber at 25°C. What is the theoretical adiabatic flame temperature achieved in °C? See Figure P5.131.

Figure P5.131

Section 5.5

5.132. A phase change (condensation, melting, etc.) of a pure component is an example of a reversible process because the temperature and pressure remain constant during the change. Use the definition of work to calculate the work done by butane when 1 kg of saturated liquid butane at 70 kPa vaporizes completely. Can you calculate the work done by the butane from the energy balance alone?

5.133. Calculate the work done when 1 lb mol of water in an open vessel evaporates completely at 212°F. Express your result in Btu.

5.134. One kilogram of steam goes through the following reversible process. In its initial state (state 1) it is at 2700 kPa and 540°C. It is then expanded isothermally to state 2, which is at 700 kPa. Then it is cooled at constant volume to 400 kPa (state 3). Next it is cooled at constant pressure to a volume of 0.4625 m³/kg (state 4). Then it is compressed adiabatically to 2700 kPa and 425°C (state 5), and finally it is heated at constant pressure back to the original state.
   (a) Sketch the path of each step in a p-V diagram.
   (b) Compute ΔU and ΔH for each step and for the entire process.
   (c) Compute Q and W whenever possible for each step of the process.

5.135. Calculate the work done when 1 lb mol of water evaporates completely at 212°F in the following cases. Express your results in Btu/lb mol.

(a) A steady-state flow process: water flowing in a pipeline at 1 lb mol/min neglecting the potential and kinetic energy changes.

(b) A nonflow process: water contained in a constant-pressure, variable-volume tank.

5.136. In a processing plant, milk flows from a storage tank maintained at 5°C through a valve to a pasteurizer via an insulated 10-cm-diameter pipe at the rate of 1000 L/min. The upstream pressure is 290 kPa and downstream pressure is 140 kPa. Determine the lost work (E_v) in J/min and the temperature change which occurs in the milk as a result of this throttling process. (Milk and water are sufficiently equivalent in properties for you to use those of water.)

5.137. A power plant is as shown in Figure P5.137. If the pump moves 100 gal/min into the boiler with an overall efficiency of 40%, find the horsepower required for the pump. List all additional assumptions required.

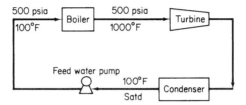

Figure P5.137

5.138. An office building requires water at two different floors. A large pipe brings the city water supply into the building in the basement level, where a booster pump is located. The water leaving the pump is transported by smaller insulated pipes to the second and fourth floors, where the water is needed. Calculate the *minimum* amount of work per unit time (in horsepower) that the pump must do in order to deliver the necessary water, as indicated in Figure P5.138. (*Minimum* refers to the fact that you should neglect the friction and pump energy losses in your calculations.) The water does not change temperature.

5.139. Water at 20°C is being pumped from a constant-head tank open to the atmosphere to an elevated tank kept at a constant pressure of 1150 kPa in an experi-

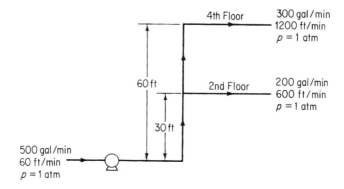

Figure P5.138

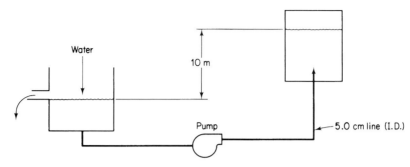

Figure 5.139

ment as shown in Figure P5.139. If water is flowing in the 5.0-cm line at a rate of 0.40 m³/min, find

(a) The rating of the pump in joules per kilogram being pumped

(b) The rating of the pump in joules per minute

The pump and motor have an overall efficiency of 70% and the energy loss in the line can be determined to be 60.0 J/kg flowing.

5.140. The following measurements have been made on a direct-fired heater using natural gas of the following composition

$$
\begin{array}{ll}
CH_4 & 96.4\% \\
C_2H_6 & 2.01 \\
C_3H_6 & 0.6 \\
N_2 & 0.99
\end{array}
$$

and 10.0% excess air.

Data: Flue gas temperature 450°C

Thermal heat loss 2% of LHV

Calculate the "efficiency" of the heater in percent

$$
\text{efficiency} = 100 - \left(\frac{\text{thermal heat loss}}{\text{LHV}} - \frac{\begin{array}{c}\text{enthalpy of flue gas}\\\text{relative to SC}\end{array}}{\text{LHV}} \right) 100
$$

Section 5.6

5.141. **(a)** From the data, plot the enthalpy of 1 *mole of solution* at 27°C as a function of the weight percent HNO_3. Use as reference states liquid water at 0°C and liquid HNO_3 at 0°C. You can assume that C_p for H_2O is 75 J/(g mol)(°C) and for HNO_3, 125 J/(g mol)(°C).

$-\Delta H_{soln}$ at 27°C (J/g mol HNO_3)	Moles H_2O added to 1 mole HNO
0	0.0
3,350	0.1
5,440	0.2
6,900	0.3
8,370	0.5
10,880	0.67
14,230	1.0
17,150	1.5
20,290	2.0
24,060	3.0
25,940	4.0
27,820	5.0
30,540	10.0
31,170	20.0

(b) Compute the energy absorbed or evolved at 27°C on making a solution of 4 moles of HNO_3, and 4 moles of water by mixing a solution of 33 1/3 mol % acid with one of 60 mol % acid.

5.142. *National Bureau of Standards Circular 500* gives the following data for calcium chloride (mol. wt. 111) and water:

Formula	State	$-\Delta H_f$ at 25°C (kcal/g mol)
H_2O	Liquid	68.317
	Gas	57.798
$CaCl_2$	Crystal	190.0
	in 25 moles of H_2O	208.51
	50	208.86
	100	209.06
	200	209.20
	500	209.30
	1000	209.41
	5000	209.60
	∞	209.82
$CaCl_2 \cdot H_2O$	Crystal	265.1
$CaCl_2 \cdot 2H_2O$	Crystal	335.5
$CaCl_2 \cdot 4H_2O$	Crystal	480.2
$CaCl_2 \cdot 6H_2O$	Crystal	623.15

Calculate the following:
(a) The energy evolved when 1 lb mol of $CaCl_2$ is made into a 20% solution at 77°F;
(b) The heat of hydration of the dehydrate to the hexahydrate;
(c) The energy evolved when a 20% solution containing 1 lb mol of $CaCl_2$ is diluted with water to 5% at 77°F.

5.143 A vessel contains 100 g of an NH_4OH–H_2O liquid mixture at 25°C and 1 atm that is 15.0% by weight NH_4OH. Just enough aqueous H_2SO_4 is added to the vessel from an H_2SO_4 liquid mixture at 25°C and 1 atm (25.0 mole % H_2SO_4) so that the reaction to $(NH_4)_2SO_4$ is complete. After the reaction, the products are again at 25°C and 1 atm. How much heat (in J) is absorbed or evolved by this process? It may be assumed that the final volume of the products is equal to the sum of the volumes of the two initial mixtures.

5.144. An ammonium hydroxide solution is to be prepared at 77°F by dissolving gaseous NH_3 in water. Prepare charts showing:
(a) The amount of cooling needed (in Btu) to prepare a solution containing 1 lb mol of NH_3 at any concentration desired;
(b) The amount of cooling needed (in Btu) to prepare 100 gal of a solution of any concentration up to 35% NH_3;
(c) If a 10.5% NH_3 solution is made up without cooling, at what temperature will the solution be after mixing?

5.145. An evaporator at atmospheric pressure is designed to concentrate 10,000 lb/hr of a 10% NaOH solution at 70°F into a 40% solution. The steam pressure inside the steam chest is 40 psig. Determine the pounds of steam needed per hour if the exit strong caustic preheats the entering weak caustic in a heat exchanger, leaving the heat exchanger at 100°F.

5.146. A 50% by weight sulfuric acid solution is to be made by mixing the following:
(a) Ice at 32°F
(b) 80% H_2SO_4 at 100°F
(c) 20% H_2SO_4 at 100°F

How much of each must be added to make 1000 lb of the 50% solution with a final temperature of 100°F if the mixing is adiabatic?

5.147. Saturated steam at 300°F is blown continuously into a tank of 30% H_2SO_4 at 70°F. What is the highest concentration of liquid H_2SO_4 that can result from this process?

5.148. One thousand pounds of 10% NaOH solution at 100°F is to be fortified to 30% NaOH by adding 73% NaOH at 200°F. How much 73% solution must be used? How much cooling must be provided so that the final temperature will be 70°F?

Section 5.7

5.149. Autumn air in the deserts of the southwestern United States during the day will typically be moderately hot and dry. If a dry-bulb temperature of 27°C and a wet-bulb temperature of 17°C is measured for the air at noon:

(a) What is the dew point?

(b) What is the percent relative humidity?

(c) What is the percent humidity?

5.150. The air supply for a dryer has a dry-bulb temperature of 32°C and a wet-bulb temperature of 25.5°C. It is heated to 90°C by coils and blown into the dryer. In the dryer, it cools along an adiabatic cooling line as it picks up moisture from the dehydrating material and leaves the dryer fully saturated.

(a) What is the dew point of the initial air?

(b) What is its humidity?

(c) What is its percent relative humidity?

(d) How much heat is needed to heat 100 m³ to 90°C?

(e) How much water will be evaporated per 100 m³ of air entering the dryer?

(f) At what temperature does the air leave the dryer?

5.151. Moist air at 100 kPa, a dry-bulb temperature of 90°C, and a wet-bulb temperature of 46°C is enclosed in a rigid container. The container and its contents are cooled to 43°C.

(a) What is the molar humidity of the cooled moist air?

(b) What is the final total pressure in atm in the container?

(c) What is the dew point in °C of the cooled moist air?

5.152. Calculate:

(a) The humidity of air saturated at 120°F

(b) The saturated volume at 120°F

(c) The adiabatic saturation temperature and wet-bulb temperature of air having a dry-bulb $T = 120°F$ and a dew point = 60°F

(d) The percent saturation when the air in (c) is cooled to 82°F

(e) The pounds of water condensed/100 lb of moist air in (c) when the air is cooled to 40°F

5.153. A rotary dryer operating at atmospheric pressure dries 10 tons/day of wet grain at 70°F, from a moisture content of 10% to 1% moisture. The air flow is counter current to the flow of grain, enters at 225°F dry-bulb and 110°F wet-bulb temperature, and leaves at 125°F dry-bulb. See Figure P5.153. Determine:

(a) The humidity of the entering and leaving air

(b) The water removal in pounds per hour

(c) The daily product output in pounds per day

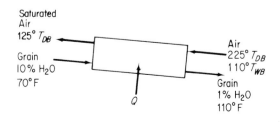

Figure 5.153

(d) The heat input to the dryer. Assume that there is no heat loss from the dryer, that the grain is discharged at 110°F, and that its specific heat is 0.18.

5.154. Temperatures (in °F) taken around a forced-draft cooling tower are as follows:

	In	Out
Air	85	90
Water	102	89

The wet-bulb temperature of the entering air is 77°F. Assuming that the air leaving the tower is saturated, calculate
(a) The humidity of the entering air
(b) The pounds of dry air through the tower per pound of water into the tower
(c) The percentage of water vaporized in passing through the tower.

5.155. A dryer produces 180 kg/hr of a product containing 8% water from a feed stream that contains 1.25 g of water per gram of dry material. The air enters the dryer at 100°C dry-bulb and a wet bulb temperature of 38°C; the exit air leaves at 53°C dry-bulb and 60% relative humidity. Part of the exit air is mixed with the fresh air supplied at 21°C, 52% relative humidity, as shown in Figure P5.155. Calculate the air and heat supplied to the heater, neglecting any heat lost by radiation, used in heating the conveyor trays, and so forth. The specific heat of the product is 0.18.

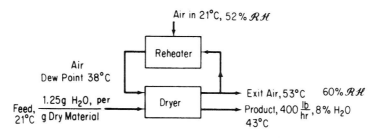

Figure 5.155

5.156. Air, dry-bulb 38°C, wet-bulb 27°C, is scrubbed with water to remove dust. The water is maintained at 24°C. Assume that the time of contact is sufficient to reach complete equilibrium between air and water. The air is then heated to 93°C by passing it over steam coils. It is then used in an adiabatic rotary drier from which it issues at 49°C. It may be assumed that the material to be dried enters and leaves at 46°C. The material loses 0.05 kg H_2O per kilogram of product. The total product is 1000 kg/hr.
(a) What is the humidity:

        (1) Of the initial air?

        (2) After the water sprays?

        (3) After reheating?

        (4) Leaving the drier?

(b) What is the percent humidity at each of the points in part (a)?

(c) What is the total weight of dry air used per hour?

(d) What is the total volume of air leaving the drier?

(e) What is the total amount of heat supplied to the cycle in joules per hour?

SOLVING SIMULTANEOUS MATERIAL AND ENERGY BALANCES

6

Now that you have accumulated some experience in making energy balances, it is time to apply this knowledge to more complex problems involving both material and energy balances. You have already encountered some simple examples of combined material and energy balances, as, for example, in the calculation of the adiabatic reaction temperature, where a material balance provides the groundwork for the implementation of an energy balance.

In Section 6.1 we first examine how to make sure that a problem is properly and completely specified by extending our earlier treatment in Chapter 3 of the analysis of the degrees of freedom. Then we review how computer codes and, in particular, flowsheeting codes can be employed to solve complex problems.

6.1 ANALYSIS OF THE DEGREES OF FREEDOM IN THE STEADY-STATE PROCESS

> ### *Your objectives in studying this section are to be able to:*
>
> 1. Identify the names and numbers of variables in the streams entering and leaving a processing unit, and the variables associated with the unit itself.
> 2. Determine the number of independent equations for each processing unit, the constraints.
> 3. Calculate the number of degrees of freedom (decision

variables) for single units and combinations of units both
without and with a reaction taking place.

4. Specify the values of variables equal to the number of degrees
of freedom for a unit.

LOOKING AHEAD

In this section we show how to calculate the degrees of freedom for a continuous
steady-state process.

MAIN CONCEPTS

An important aspect of combined material and energy balance problems is how to
ensure that the process equations or sets of modules are determinate, that is, have
at least one solution, and hopefully no more than one solution. The question is:
How many variables are unknown, and consequently must have their values specified in any problem? The **number of degrees of freedom** is the number of variables in a set of *independent* equations to which values must be assigned so that
the equations can be solved.

Let N_d = number of degrees of freedom, N_v = number of variables, and N_r =
number of equations (restrictions, constraints). Then for N_r independent equations
in general

$$N_d = N_v - N_r \tag{6.1}$$

and we conclude that $N_v - N_r$ variables must be specified **as long as the N_r equations are still independent**. You do not have to write down the equations during
the analysis but just identify them. Whether the equations are linear or nonlinear
makes no difference.

In this chapter the analysis of the degrees of freedom for a process assumes
that the process is a steady-state flow process as commonly assumed in design. If
operations or control is of interest, you would base the analysis on an unsteady
process in which the accumulation term would be taken into consideration. (Also,
not all of the variables in a process can be manipulated so the selection of which
variables to specify is limited.)

Both extensive and intensive variables are included in the analysis in contrast with the degrees of freedom obtained from application of the phase rule
(Sec. 4.5), which treats only intensive variables. What kinds of variables do you
have to consider? Typical ones are

(1) Temperature

(2) Pressure

(3) Either mass (mole) flow rate of each component in a stream, or the concentration of each component plus the total flow rate

(4) Specific enthalpies

(5) Heat flow rate, work (in the energy balance)

(6) Recycle ratio

Some variables can be substituted for others, such as temperature for specific enthalpies, and stream flows for the recycle ratio.

Examine the flow stream in Figure 6.1. Two modes of specifying the number of variables associated with a process stream (**stream variables**) exist (we assume the stream is a single phase in which no reactions occur; with more than one phase, each phase would be treated as a separate stream):

Using moles (or mass) flow rate	No.	*Using compositions and total flow rate*	No.
Temperature (T)	1	Temperature (T)	1
Pressure (p)	1	Pressure (p)	1
Component flow rates (n_i) or m_i)	N_{sp}	Compositions (x_i or ω_i)	$N_{sp} - 1$
		Total flow rate (F)	1
Total	$N_{sp} + 2$		$N_{sp} + 2$

where N_{sp} is the number of **components (species)** in the stream. The count for the number of compositions is $N_{sp} - 1$ and not N_{sp} because of the implicit constraint that the sum of the mole (or mass) fractions is equal to 1.

Thus, we can conclude that the number of variables N_v needed to specify the condition of a stream completely is given by

$$N_v = N_{sp} + 2 \tag{6.2}$$

You should keep in mind that in a binary system, for example, in which one stream component is zero, for consistency you would count $N_{sp} = 2$ with the one component having a zero value treated as a constraint.

What kinds of constraints (equations) are involved in the analysis of the degrees of freedom? Usually

Figure 6.1   Stream variables

(1) independent material balances for each species (a total balance could be substituted for one species balance)

(2) energy balance

(3) phase equilibria relations, that is, equations that give the compositions between one species that exists in two (or more) phases

(4) chemical equilibrium relations. The number of such equations is equal to the minimum number of independent stoichiometric relations (see Appendix L) that can represent the species present in a single phase

(5) implicit relations, such that the concentration of a species is zero in a stream

(6) explicit relations, such that a given fraction of a stream condenses.

To illustrate analysis for the degrees of freedom, examine Figures 6.2A and B, which show a simple steady-state isothermal, isobaric process involving three streams plus heat transfer. The count of variables and constraints is

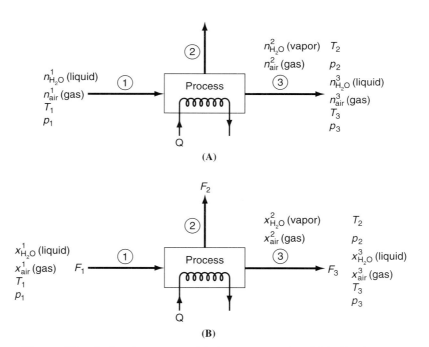

Figure 6.2   A simple process with three streams in which (A) the stream flows are expressed as mole flow rates; and (B) the stream flow is the total flow and the compositions of the species are given in mole fractions.

Variables:

$(N_{sp} + 2) \times 3 = (2 + 2) \times 3 =$ 12

Q 1

        Total   13

Constraints:

   Material balances   2

   Energy balance   1

   Phase equilibrium for H_2O 1

   T is the same in all 3 streams

       $(T_1 = T_2 = T_3)$ 2 independent equations   2

   p is the same in all 3 streams

       $(T_1 = T_2 = T_3)$ 2 independent equations   2

                         Total   8

Degrees of freedom:

   $13 - 8 =$ 5

The way in which the compositions are specified makes no difference in the analysis.

We now take up five simple processes in Example 6.1, and evaluate the number of degrees of freedom for each.

EXAMPLE 6.1 Determining the Degrees of Freedom in a Process

We shall consider five typical processes, as depicted by the respective figures below, and for each ask the question: How many variables have to be specified [i.e., what are the degrees of freedom (N_d)] to make the problem of solving the combined material and energy balances determinate? All the processes will be steady-state, and the entering and exit streams single-phase streams.

(a) Stream splitter (Figure E6.1a): We assume that $Q = W = 0$, and that the energy balance is not involved in the process. By implication of a splitter, the temperatures, pressures, and the compositions of the inlet and outlet streams are identical. The count of the total number of variables, total number of constraints, and degrees of freedom is as follows:

   Total number of variables

       $N_v = 3(N_{sp} + 2) =$ $3(N_{sp} + 2)$

   Number of independent equality constraints

       Material balances   1

       Compositions of Z, P_1, and P_2

           are the same $2(N_{sp} - 1)$

$$T_{P1} = T_{P2} = T_Z \qquad\qquad 2$$
$$p_{P1} = p_{P2} = p_Z \qquad\qquad 2 \qquad\qquad \dfrac{2N_{sp} + 3}{N_{sp} + 3}$$

Total number of degrees of freedom

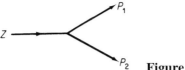

Figure E6.1a

Did you note that we only counted N_{sp} material balances once? Why? Let us look at some of the balances:

$$\text{Component 1: } Zx_{1Z} = P_1 x_{1P_1} + P_2 x_{1P_2}$$
$$\text{Component 2: } Zx_{2Z} = P_1 x_{2P_1} + P_2 x_{2P_2}$$
$$\text{etc.}$$

If $x_{1Z} = x_{1P_1} = x_{1P_2}$ and the same is true for x_2, and so on, only one independent material exists.

Do you understand the counts resulting from making the compositions equal in Z, P_1, and P_2? Write down the expressions for each component: $x_{iZ} = x_{iP_1} = x_{iP_2}$. Each set represents $2N_{sp}$ constraints. But you cannot specify every x_i in a stream, only $N_{sp} - 1$ of them. Do you remember why?

To make the problem determinate we might specify the values of the following decision variables:

Flow rate Z	1
Composition of Z	$N_{sp} - 1$
T_Z	1
p_Z	1
Ratio of split $\alpha = P_1/P_2$	1
Total number of degrees of freedom	$N_{sp} + 3$

(b) Mixer (Figure E6.1b): For this process we assume that $W = 0$, but Q is not.

Total number of variables (3 streams + Q)	$3(N_{sp} + 2) + 1$
Number of independent equality constraints	
Material balances	N_{sp}
Energy balance	1
Total number of degrees of freedom $= 3(N_{sp} + 2) + 1 - (N_{sp} + 1) = 2N_{sp} + 6$	

(c) Heat exchanger (Figure E6.1c): For this process we assume that $W = 0$ (but not Q).

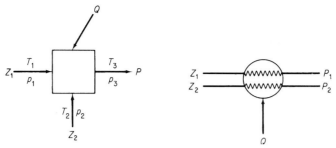

Figure E6.1b   **Figure E6.1c**

Total number of variables
  $N_v = 4(N_{sp} + 2) + 1 =$ $4N_{sp} + 9$
Number of independent equality constraints
  Material balances (streams 1 and 2)   2
  Energy balance   1
Composition of inlet and outlet
  streams the same $2(N_{sp} - 1)$ $2N_{sp} + 1$

Total number of degrees of freedom = $2N_{sp} + 8$

One might specify four temperatures, four pressures, $2(N_{sp} - 1)$ compositions in Z_1 and Z_2, and Z_1 and Z_2 themselves to use up $2N_{sp} + 8$ degrees of freedom.

(d) Pump (Figure E6.1d): Here $Q = 0$ but W is not; $N_p = 0$.

Total number of variables
  $N_v = 2(N_{sp} + 2) + 1$ $2N_{sp} + 5$
Number of independent equality constraints
  Material balances   1
  Composition of inlet and outlet
    streams the same $N_{sp} - 1$
  Energy balance   1 $N_{sp} + 1$

Total number of degrees of freedom $N_{sp} + 4$

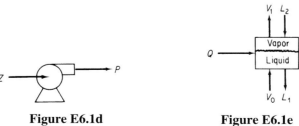

Figure E6.1d   **Figure E6.1e**

(e) Two-phase well-mixed tank (stage) at equilibrium (Figure E6.1e) (L, liquid phase; V, vapor phase): Here $W = 0$. Although two phases exist (V_1 and L_1) at equilibrium inside the system, the streams entering and leaving are single-phase streams. By equilibrium we mean that both phases are at the same temperature and pressure and that an equation is known that relates the composition in one phase to that in the other for each component.

Total number of variables (4 streams + Q)	$4(N_{sp} + 2) + 1$
Number of independent equality constraints	
Material balances	N_{sp}
Energy balance	1
Composition relations at equilibrium	N_{sp}
Temperatures of the streams V_1 and L_1 in the two phases equal	1
Pressures of the streams V_1 and L_1 in the two phases equal	1

Total number of degrees of freedom $= 4(N_{sp} + 2) + 1 - 2N_{sp} - 3 = 2N_{sp} + 6$

In general you might specify the following variables to make the problem determinate:

Input stream L_2	$N_{sp} + 2$
Input stream V_0	$N_{sp} + 2$
Pressure	1
Q	1
Total	$2N_{sp} + 6$

Other choices are of course possible, but such choices must leave the equality constraints independent.

EXAMPLE 6.2 Proper Process Specification

Figure E6.2 shows an isothermal separations column. At present the column specifications call for the feed mass fractions to be $\omega_{C_4} = 0.15$, $\omega_{C_5} = 0.20$, $\omega_{iC_5} = 0.30$, and $\omega_{C_6} = 0.35$; the overhead mass fractions to be $\omega_{C_5} = 0.40$ and $\omega_{C_6} = 0$; and the residual product mass fraction to be $\omega_{C_4} = 0$. Unless specified as 0, the component exists in a stream.

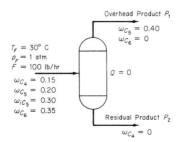

Figure E6.2

Given that $F = 100$ lb/hr, is the separator completely specified, that is, are the degrees of freedom $N_d = 0$? The streams P_1 and P_2 are not in equilibrium.

Solution

First we calculate N_v and then N_r. We will assume that all the stream temperatures and pressures are identical. Number of variables N_v:

$$N_v = (N_{sp} + 2)(3) = (4 + 2)(3) = \underline{18}$$

Number of restrictions and equations for N_r:

Component mass balances (no reaction) $= N_{sp} =$	4
$T_F = T_{P1} = T_{P2}$	2
$p_F = p_{P1} = p_{P2}$	2

Initial column specifications:

(4 in F, 2 in P_1, and 1 in P_2 plus $T_F = 30°C$ and $p_F = 1$ atm)	9
Total for N_r	$\underline{17}$

Number of degrees of freedom N_d: $18 - 17 = \quad\quad 1$

Note that only four factors can be specified in stream F, namely the rate of F itself and three ω's; one of the ω's is redundant. One more variable must be specified for the process, but one that will not reduce the number of independent equations and restrictions already enumerated.

So far we have examined single units without a reaction occurring in the unit. How is the count for N_d affected by the presence of a reaction in the unit? The way N_v is calculated does not change. As to N_r, all restrictions and constraints are deducted from N_v that represent *independent* restrictions on the unit. Thus the number of material balances is not necessarily equal to the number of species (H_2O, O_2, CO_2, etc.) but instead is the number of independent material balances that exist determined in the same way as we did in Secs. 3.2 to 3.4, usually (but not always) equal to the number of elemental balances (H, O, C, etc.). Fixed ratios of materials such as the O_2/N_2 ratio in air or the CO/CO_2 ratio in a product gas would be a restriction, as would be a specified conversion fraction or a known molar flow rate of a material. If some degrees of freedom exist still to be specified, improper specification of a variable may disrupt the independence of equations and/or specifications previously enumerated in the enumeration of N_r, so be careful.

EXAMPLE 6.3 Degrees of Freedom with a Reaction Taking Place in the System

A classic reaction for producing H_2 is the so-called "water gas shift" reaction:

$$CO + H_2O \rightleftharpoons CO_2 + H_2$$

Figure E6.3 shows the process data and the given information. How many degrees of freedom remain to be satisfied? For simplicity assume that the temperature and pressures of all entering and exit streams are the same and that all streams are gases. The amount of water in excess of that needed to convert all the CO completely to CO_2 is prespecified.

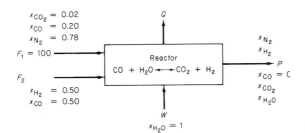

Figure E6.3

Solution

$$N_v = 4(N_{sp} + 2) + 1 = 4(5 + 2) + 1 = \qquad \underline{29}$$
(+1 is for Q)

N_r: Independent material balances
(C, O, N, H)	4	
Energy balance	1	
$T_{F_1} = T_{F_2} = T_W = T_P$	3	
$p_{F_1} = p_{F_2} = p_W = p_P$	$\underline{3}$	11

Compositions and flows specified:
In F_1 ($x_{H_2O} = x_{H_2} = 0$)	5	
In F_2 ($x_{N_2} = x_{CO_2} = x_{H_2O} = 0$)	4	
In W (all but x_{H_2O} are 0)	4	
In P ($x_{CO} = 0$)	$\underline{1}$	14
Excess W given		$\underline{1}$
		$\underline{26}$

$$N_d = 29 - 26 = \qquad 3$$

In the streams F_2 and W, only four compositions can be specified; a fifth specification is redundant. The total flows are not known. The given value of the excess water provides the information about the reaction products. Certainly, the temperature and pressure need to be specified, absorbing two degrees of freedom. The remaining degree of freedom might be the N_2/H_2 ratio in P, or the value of F_2, or the ratio of F_1/F_2, and so on.

EXAMPLE 6.4 Degrees of Freedom for the Case of Multiple Reactions

Methane burns in a furnace with 10% excess air, but not completely, so some CO exits the furnace, but no CH_4 exits. The reactions are:

$$CH_4 + 1.5O_2 \rightarrow CO + 2H_2O$$

$$CH_4 + 2O_2 \rightarrow CO_2 + 2H_2O$$

$$CO + 0.5O_2 \rightarrow CO_2$$

Carry out a degree-of-freedom analysis for this combustion problem.

Solution

Figure E6.4 shows the process; all streams are assumed to be gases. Only two of the reactions are independent. Q is a variable here. For simplicity assume that the entering and exit streams are at the same temperatures and pressures.

$$N_v = 3(6 + 2) + 1 = \underline{\qquad 25}$$
$$(+1 \text{ is for } Q)$$

N_r:

Material balances (C, H, O, N)	4	
Energy balance	1	
$T_A = T_F = T_P$	2	
$p_A = p_F = p_P$	2	
Compositions specified:		
In A ($N_{sp} - 1$)	5	
In F ($N_{sp} - 1$)	5	
In P	1	11
Percent excess air:		$\underline{1}$
		$\underline{21}$

$$N_d = 25 - 21 = 4$$

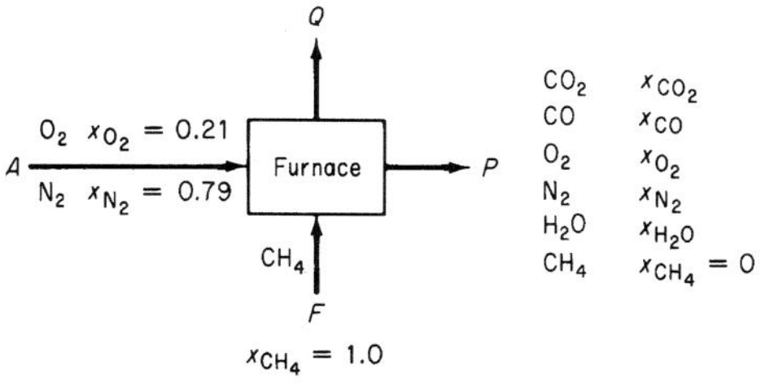

Figure E6.4

To have a well-defined problem you should specify (a) the temperature, (b) the pressure, (c) either the feed rate, or the air rate, or the product rate, and (d) either the CO/CO_2 ratio or the fraction of CH_4 converted to CO or alternatively to CO_2.

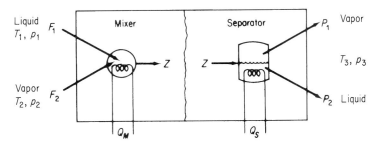

Figure 6.3   Degrees of freedom in combined units.

You can compute the degrees of freedom for combinations of like or different simple processes by proper combination of their individual degrees of freedom. In adding the degrees of freedom for units, you must eliminate any double counting either for variables or constraints, and take proper account of interconnecting streams whose characteristics are often fixed only by implication.

Examine the mixer-separator in Figure 6.3. For the mixer consider as a separate unit, from Example 6.1b, $N_d = 2N_{sp} + 6$. For the separator, an equilibrium unit:

$$N_v = 3(N_{sp} + 2) + 1 = \qquad\qquad 3N_{sp} + 7$$
$$N_r:$$

Material balances	N_{sp}	
Equilibrium relations	N_{sp}	
Energy balance	1	
$T_Z = T_{P_1} = T_{P_2}$	2	
$p_Z = p_{P_1} = p_{P_2}$	2	$2N_{sp} + 5$

$$N_d = (3N_{sp} + 7) - (2N_{sp} + 5) = \qquad N_{sp} + 2$$

The sum of the mixer and separator is $3N_{sp} + 8$.

We must deduct redundant variables and add redundant restrictions as follows:

Redundant variables:	
Remove 1 Q	1
Remove Z	$N_{sp} + 2$
Redundant constraints:	
1 energy balance	1

Then $N_d = (3N_{sp} + 8) - (N_{sp} + 3) + 1 = 2N_{sp} + 6$, the same as in Example 6.1e.

EXAMPLE 6.5 Degrees of Freedom in a System Composed of Several Units

Ammonia is produced by reacting N_2 and H_2:

$$N_2 + 3H_2 \rightarrow 2NH_3$$

Figure E6.5a shows a simplified flowsheet. All the units except the separator and lines are adiabatic. The liquid ammonia product is essentially free of N_2, H_2, and A. Assume that the purge gas is free of NH_3. Treat the process as four separate units for a degree-of-freedom analysis, and then remove redundant variables and add redundant constraints to obtain the degrees of freedom for the overall process. The fraction conversion in the reactor is 25%.

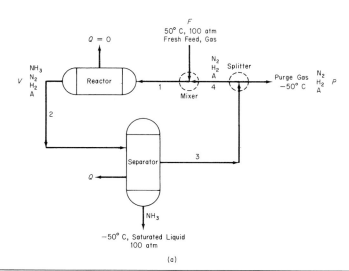

(a)   **Figure E6.5a**

Solution

    Mixer:

$$N_v = 3(N_{sp} + 2) + 1 = 3(6) + 1= \qquad\qquad 19$$

N_r:

Material balances (H_2, N_2, A only)	3
Energy balance	1
Specifications:	
NH_3 concentration is zero	3
$T_P = -50°C$	1
$T_F = 50°C$	1
Assume that $p_F = p_{\text{mix out}} = p_{\text{split}} = 100$	3
$Q = 0$	1

N_d: $19 - 13 =$ $\underline{13}$ $\underline{\underline{6}}$

Reactor:

$$N_v = 2(N_{sp} + 2) + 1 = 2(6) + 1 = \qquad\qquad 13$$

N_r:

Material balances (H, N, A)	3	
Energy balances	1	
Specifications:		
NH_3 entering = 0	1	
$Q = 0$	1	
Fraction conversion	1	
$p_{in} = p_{out} = 100$ atm	2	
Energy balance	1	$\underline{10}$

N_d: $13 - 10 = \qquad\qquad\qquad \underline{3}$

Separator:

$$N_v = 3(N_{sp} + 2) + 1 = 3(6) + 1 = \qquad\qquad 19$$

N_r:

Material balances	4	
Energy balance	1	
Specifications:		
$T_{out} = -50°C$	1	
$p_r = p_{in} = p_{NH_3} = 100$	3	
NH_3 concentration is 0 in		
recycle gas	1	
N_2, H_2, A are 0 in liquid NH_3	3	$\underline{13}$

N_d: $19 - 13 = \qquad\qquad\qquad \underline{6}$

Splitter:

$$N_v = 3(N_{sp} + 2) = 3(6) = \qquad\qquad 18$$

N_r:

Material balances	1	
Specifications:		
NH_3 concentration = 0	1	
Compositions same $2(N_{sp} - 1)$	6	
Stream temperatures same = $-50°C$	3	
Stream pressures same = 100 atm	3	$\underline{14}$

$N_d = 18 - 14 \qquad\qquad\qquad \underline{4}$

The total number of degrees of freedom is 19 less the redundant information, which is as follows:

Redundant variables in interconnecting streams being eliminated:

$$\text{Stream 1: } (4 + 2) = \ 6$$

$$\text{Stream 2: } (4 + 2) = \ 6$$

$$\text{Stream 3: } (4 + 2) = \ 6$$

$$\text{Stream 4: } (4 + 2) = \ \underline{6}$$

$$24$$

Redundant constraints being eliminated:

Stream 1:
NH_3 concentration = 0   1
$p = 100$ atm   1
Stream 2:
$p = 100$ atm   1
Stream 3:
NH_3 concentration = 0   1
$p = 100$ atm   1
$T = -50°C$ 1
Stream 4:
NH_3 concentration = 0   1
$T = -50°C$ 1
$p = 100$ atm $\underline{1}$
9

Overall the number of degrees of freedom should be

$$N_d = 19 - 24 + 9 = 4$$

We can check the count for N_d by making a degrees-of-freedom analysis about the entire process as follows:
Examine Figure E6.5b.

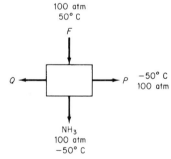

Figure E6.5b

$N_v = 3(4 + 2) + 1 =$ 19
N_r:
Material balances
(H, N, A)   3
Energy balance   1
Specifications:
Stream F ($T = 50°C$, $p =$
100 atm, $NH_3 = 0$)   3

NH$_3$ stream ($T = -50°C$,
    $p = 100$ atm, three components
    have 0 concentration)   5
Purge stream ($T = -50°C$,
    $p = 100$ atm, NH$_3 = 0$)   <u>3</u>   <u>15</u>
$N_d = 19 - 15 =$ <u>4</u>

We do not have the space to illustrate additional combinations of simple units to form more complex units, but Kwauk[1] prepared several excellent tables summarizing the variables and degrees of freedom for distillation columns, absorbers, heat exchangers, and the like. Also read the references at the end of this chapter.

LOOKING BACK

In this section we described how to determine the number of degrees of freedom involved in a process, i.e., the number of additional values of variables that have to be specified to get a solution to a problem.

Key Ideas

1. The number of degrees of freedom is the total number of variables in a process less the number of independent equations involved in the process.
2. Specification of values of variables should not confound the original count of variables and *independent* constraints.
3. Subsystems can be combined to get the overall degrees of freedom if redundant variables and associated constraints are eliminated.

Key Terms

Components (p. 545)   Species (p. 545)
Degrees of freedom (p. 544)   Stream variables (p. 545)

Self-Assessment Test

1. Is there any difference between the number of species present in a process and the number of components in the process?
2. Why are there $N_{sp} + 2$ variables associated with each stream?
3. Determine the number of degrees of freedom for a still (see Figure SA6.1–3).

[1]M. Kwauk. *AIChE J.*, 2, (1956): 2.

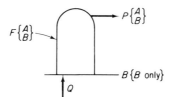

Figure SA6.1–3

4. Determine the number of degrees of freedom in the following process (Figure SA6.1–4):

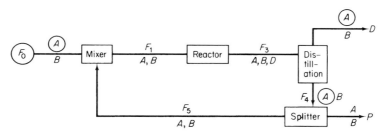

Figure SA6.1–4

The encircled variables have known values. The reaction parameters in the reactor are known as is the fraction split at the splitter between F_4 and F_5. Each stream is a single phase.

5. Figure SA6.1–5 represents the schematic flowsheet of a distillation tower used to recover gasoline from the products of a catalytic cracker. Is the problem completely specified, that is, is the number of degrees of freedom equal to zero for the purpose of calculating the heat transfer to the cooling water in the condenser?

Thought Problems

1. If one or more of the variables in a process can take on only integer values (such as number of stages in a column or number of reactors in a series of reactors), will the analysis of degrees of freedom have to be changed?
2. What other variables in some processes might have to be included in the count of variables for a stream to add to $(N_{sp} + 2)$.

Discussion Question

1. How should a computer code handle the calculation of the degrees of freedom so that a naive user does not overspecify or underspecify the problem?

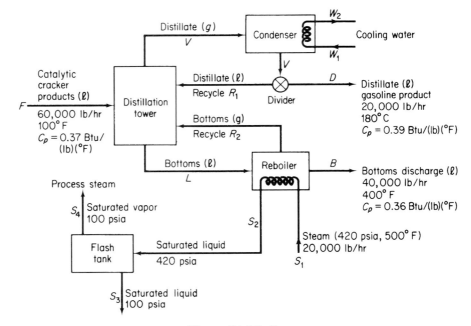

Figure SA6.1–5

6.2 SOLVING MATERIAL AND ENERGY BALANCES USING FLOWSHEETING CODES

> *Your objectives in studying this section are to be able to:*
>
> 1. Understand the difference between equation-based and modular-based flowsheeting.
> 2. Describe how material and energy balances are carried out both in equation- and modular-based flowsheeting codes.

LOOKING AHEAD

In this section the structure of and strategy in using flowsheeting codes to solve material and energy balances is explained.

MAIN CONCEPTS

In Sec. 3.5 we discussed combining units from the viewpoint of making material balances. As more and more units are connected together in a plant, you can understand that the degree of complexity requires that the solution of material and energy balances be carried out via a computer code. Such a program can also, at the same time, determine the size of equipment and piping, evaluate costs, and optimize performance. A plant **flowsheet** mirrors the stream network and equipment performance subject to all sort of constraints.

Once the process flowsheet is specified, the solution of the appropriate steady-state process material and energy balances is referred to as **flowsheeting**, and the computer code used in the solution is known as a **flowsheeting package or code**. The essential problem in flowsheeting without associated optimization is to solve (satisfy) a large set of linear and nonlinear equations to an acceptable degree of precision, normally by an iterative procedure. In flowsheeting without optimization, you must make sufficient specifications to take up *all* the degrees of freedom. Table 6.1 lists some commercial codes used to execute flowsheeting. Individual process units that make up the process flowsheet are represented in the form of modules (building blocks) or as equation sets. A greater degree of detail is needed in the flowsheeting code when it is used to solve plant operating problem than in carrying out the initial design of the plant.

Figure 6.4 illustrates the main features of a flowsheeting package. The code must facilitate the transfer of information between equipment and streams, have access to a reliable database, and be flexible enough to accommodate equipment specifications provided by the user to supplement the library of programs that come with the code. Fundamental to all flowsheeting codes is the calculation of mass and energy balances for the entire process. Valid inputs to the material and

TABLE 6.1   Flowsheeting Codes

Name	Source
ASPENPLUS	Aspen Technology Corp., Cambridge, MA
CHEMCAD	Chemstations, Houston, TX
DESIGN/2000	Chem Share, Houston, TX
HYSIM HYSYS	Hyprotech, Calgary, Alberta
PROCESS PRO/II	Simulation Sciences, Fullerton, CA
SPEEDUP	Aspen Technology Corp., Cambridge, MA

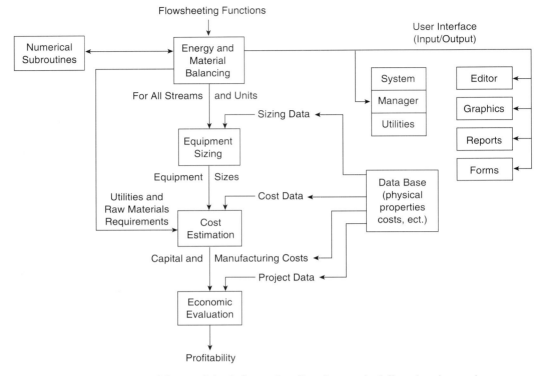

Figure 6.4   Information flow in a typical flowsheeting code.

energy balance phase of the calculations for the flowsheet must be defined in sufficient detail to determine all the intermediate and product streams and the unit performance variables for all units.

Frequently, process plants contain recycle streams and control loops, and the solution for the stream properties requires iterative calculations. Thus, efficient numerical methods for convergence must be used. In addition, appropriate physical properties and thermodynamic data have to be retrieved from a database. Finally, a master program must exist that links all the building blocks, physical property data, thermodynamic calculations, subroutines, and numerical subroutines, and that also supervises the information flow. You will find that optimization and economic analysis are often the ultimate goal in the use of flowsheeting codes.

Two extremes are encountered in flowsheeting software. At one extreme the entire set of equations (and inequalities) representing the process is written down, including the material and energy balances, the stream connections, and the relations representing the equipment functions. This representation is known as the **equation-oriented method** of flowsheeting. The equations can be solved in a sequential fashion analogous to the modular representation described below or si-

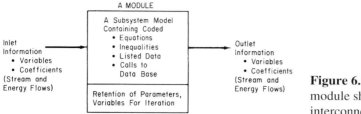

Figure 6.5   A typical process module showing the necessary interconnections of information.

multaneously by Newton's method (or the equivalent), or by employing sparse matrix techniques to reduce the extent of matrix manipulations; references can be found at the end of this chapter.

At the other extreme, the process can be represented by a collection of modules (the **modular method**) in which the equations (and other information) representing each subsystem or piece of equipment are collected together and coded so that the module may be used in isolation from the rest of the flowsheet and hence is portable from one flowsheet to another or can be used repeatedly in a given flowsheet. A module is a model of an individual element in a flowsheet (such as a reactor) that can be coded, analyzed, debugged, and interpreted by itself. Examine Figures 6.5 and 6.6. Each module contains the equipment sizes, material and energy balance relations, the component flow rates, and the temperatures, pressures, and phase conditions of each stream that enters and leaves the physical equipment represented by the module. Values of certain of these parameters and variables determine the capital and operating costs for the units. Of course, the interconnections set up for the modules must be such that information can be transferred from module to module concerning the streams, compositions, flow rates, coefficients, and so on. In other words, the modules comprise a set of building blocks that can be arranged in general ways to represent any process.

In addition to the two extremes, combinations of equations and modules can be used. Equations can be lumped into modules, and modules can be represented

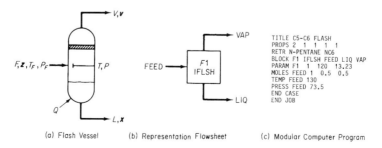

Figure 6.6   A module that represents a Flash Unit. (From J. D. Seader, W. D. Seider, and A. C. Pauls. *Flowtran Simulation-An Introduction.* Austin, TX: CACHE, 1987.)

by their basic equations or by polynomials that fit the input–output information. In current practice, modular based codes prevail because historically they were the first to be developed and improved as time went on.

Another classification of flowsheeting codes focuses on how the equations or modules are solved. One treatment is to solve sequentially and the other to solve simultaneously. Either the program and/or the user must select the decision variables for recycle and provide estimates of certain stream values to make sure that convergence of the calculations occurs, especially in a process with many recycle streams.

A third classification of flowsheeting codes is whether they solve steady-state or dynamic problems. We consider only the former here.

We will review equation-based flowsheeting first because it is much closer to the techniques used up to this point in this book, and then turn to consideration of modular-based flowsheeting.

6.2-1 Equation-Based Flowsheeting

Sets of linear and/or nonlinear equations can be solved simultaneously using an appropriate computer code by one of the methods described in Appendix L. But equation-based flowsheeting codes have some advantages in that the physical property data needed for the coefficients in the equations are transparently transmitted from a database at the proper time in the sequence of calculations.

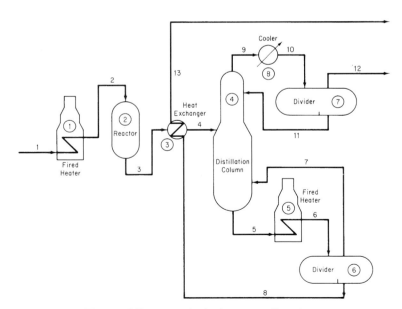

Figure 6.7   Hypothetical process flowsheet.

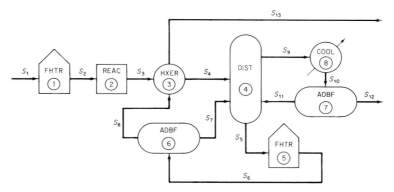

Figure 6.8 Information flow sheet for the hypothetical process (S stands for stream; module or computer code number is encircled).

Whatever the code used to solve material and energy balance problems, you must provide certain input information to the code in an acceptable format. All flowsheeting codes require that you convert the information in the flowsheet (see Figure 6.7) to an information flowsheet as illustrated in Figure 6.8, or something equivalent. In the information flowsheet, you use the name of the mathematical model (subroutine for modular-based flowsheeting) that will be used for the calculations instead of the name of the process unit.

Once the information flowsheet is set up, the determination of the process topology is easy, that is, you can immediately write down the stream interconnections between the modules (or subroutines) that have to be included in the input data set. For Figure 6.8 the matrix of stream connections (the **process matrix**) is (a negative sign designates an exit stream):

Unit	Associated streams			
1	1	−2		
2	2	−3		
3	3	8	−4	−13
4	4	7	11	−9	−5
5	5	−6		
6	6	−8	−7	
7	10	−11	−12	
8	9	−10		

System diagram

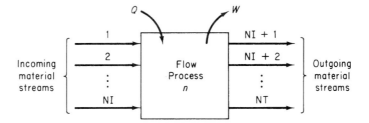

Total mass balance (or mole balance without reaction)

$$\sum_{i=1}^{NI} F_i = \sum_{i=NI+1}^{NT} F_i$$

Energy balance

$$\sum_{i=1}^{NI} F_i H_i + Q_n - W_{s,n} = \sum_{i=NI+1}^{NI} F_i H_i$$

Vapor-liquid equilibrium distribution

$$y_j = K_j x_j \quad \text{for} \quad j = 1, 2, \ldots, NC$$

Equilibrium vaporization coefficients

$$K_j = K(T_i, P_i, \overline{W}_i) \quad j = 1, 2, \text{for} \ldots, NC$$

Total mole balance (with reaction)

$$\sum_{i=1}^{NI} F_i + \sum_{l=1}^{NR} R_l \left[\sum_{j=1}^{NC} V_{j,l} \right] = \sum_{i=NI+1}^{NI} F_i$$

Component mole balances (with reaction)

$$\sum_{i=1}^{NI} F_i w_{i,j} + \sum_{l=1}^{NR} V_{j,l} R_l = \sum_{i=NI+1}^{NT} F_i w_{i,j} \quad \text{for } j = 1, 2, \ldots, NC$$

Molar atom balances

$$\sum_{i=1}^{NI} F_i \left[\sum_{j=1}^{NC} w_{i,j} a_{j,k} \right] = \sum_{i=NI+1}^{NT} F_i \left[\sum_{j=1}^{NC} w_{i,j} a_{j,k} \right] \quad \text{for } k = 1, 2, \ldots, NE$$

Mechanical energy balance

$$\sum_{i=1}^{NI} (K_i + P_i) + \sum_{i=1}^{NI} \int_{P_{1,i}}^{P_{2,i}} V_i \, dp_i = \sum_{i=NI+1}^{N} (K_i + P_i) + \sum_{i=NI+1}^{NT} \int_{P_{1,i}}^{P_{2,i}} V_i \, dP_i + W_{s,n} + E_{v,n}$$

Component mass or mole balances (without reaction)

$$\sum_{i=1}^{NI} F_i w_{i,j} = \sum_{i=NI+1}^{NT} F_i w_{i,j}$$

$$\text{for } j = 1, 2, \ldots, NC$$

Summation of mole or mass fractions

$$\sum_{j=1}^{NC} w_{i,j} = 1.0 \quad \text{for} \quad i = 1, 2, \ldots, NI$$

Physical property functions

$$\begin{aligned} H_i &= H_{VL}(T_i, P_i, \overline{W}_i) \\ S_i &= S_{VL}(T_i, P_i, \overline{W}_i) \end{aligned} \quad i = 1, 2, \ldots, NI$$

Figure 6.9   Generic equations for a steady-state open system

The interconnections between the unit modules may represent information flow as well as material and energy flow. In the mathematical representation of the plant, the interconnection equations are the material and energy balance flows between model subsystems. Equations for models such as mixing, reaction, heat exchange, and so on, must also be listed so that they can be entered into the computer code used to solve the equation. Figure 6.9 (and Table 6.2) lists the common type of equations that might be used for a single subsystem. In general, similar process units repeatedly occur in a plant and can be represented by the same set of equations, which differ only in the names of variables, the number of terms in the summations, and the values of any coefficients in the equations.

Equation-based codes can be formulated to include inequality constraints along with the equations. Such constraints might be of the form $a_1 x_1 + a_2 x_2 + \ldots \leq b$, and might arise from such factors as

1. Conditions imposed in linearizing nonlinear equations
2. Process limits for temperature, pressure, concentration
3. Requirements that variables be in a certain order
4. Requirements that variables be positive or integer

TABLE 6.2   Notation for Figure 6.8

$a_{j,k}$	number of atoms of the kth chemical element in the jth component
F_i	total flow rate of the ith stream
H_i	relative enthalpy of the ith stream
K_j	vaporation coefficient of the jth component
NC	number of chemical components (compounds)
NE	number of chemical elements
NI	number of incoming material streams
NR	number of chemical reactions
NT	total number of material streams
p_i	pressure of the ith stream
Q_n	heat transfer for the nth process unit
R_l	reaction expression for the lth chemical reaction
T_i	temperature of the ith stream
$V_{j,l}$	stoichiometric coefficient of the jth component in the lth chemical reaction
$w_{i,j}$	fractional composition (mass of mole) of the jth component in the ith stream
$\overline{W}_i$	average composition in the ith stream
$W_{s,n}$	work for the nth process unit
x_j	mole fraction of component j in the liquid
y_j	mole fraction of component j in the vapor

Two important aspects of solving the sets of nonlinear equations in flow-sheeting codes, both equation-based and modular-based, are (1) the procedure for establishing the precedence order in solving the equations, and (2) the treatment of recycle (feedback) of information, material, and/or energy. One method for solving sets of equations is to use Newton's or secant methods combined with sparse matrix methods to convert the nonlinear algebraic equations in the model to linearized approximates. Then the linearized equations can be solved iteratively taking advantage of their structure. The alternative is to use **tearing**. Tearing is discussed in detail in Sec. 6.2-2, which treats modular-based flowsheeting, but briefly, by "tearing" we mean selecting certain output variables from a set of equations as known values so that the remaining variables can be solved by serial substitution. A residual set of equations equal to the number of tear variables will remain, and if these are not satisfied, new guesses are made for the values of the tear variables, and the sequence repeated. Examine Figure 6.10c. To avoid having to solve an entire set of equations simultaneously, a method for **precedence or-**

$$h_1: x_1^2 x_2 - 2x_3^{1.5} + 4 = 0$$
$$h_2: x_2 + 2x_5 - 8 = 0$$
$$h_3: x_1 x_4 x_5^2 - 2x_3 - 7 = 0$$
$$h_4: -2x_2 + x_5 + 5 = 0$$
$$h_5: x_2 x_4^2 x_5 + x_2 x_4 - 6 = 0$$

(a) The n independent equations involving n variables ($n = 5$).

(b) The occurrence matrix (the I's represent the occurrence of a variable in an equation).

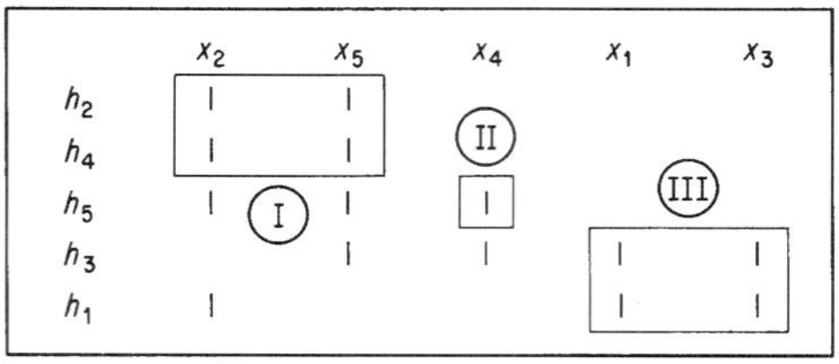

(c) The rearranged (partitioned) occurrence matrix with groups of equations (sets I, II, and III) that have to be solved simultaneously collected together in a precedence order for solution.

Figure 6.10   Partitioning and tearing. The equations are partitioned into blocks containing common variables, as in (c). Equations h_2 and h_4 (set I) are solved simultaneously for x_2 and x_5 first, then h_5 (set II) is solved for x_4, and lastly h_1 and h_3 are solved simultaneously. For example, assume a value for x_3; solve h_1 for x_1; then check to see if Equation h_3 is satisfied. If not, adjust x_3, solve h_1 for x_1, recheck h_3, and so on until both h_1 and h_3 are satisfied.

dering is used to partition the set of equations into a sequence of smaller sets of irreducible equations (equations that have to be solved simultaneously) as illustrated in Figure 6.10.

In addition to determining the precedence order and tears, you have to be able to select **initial guesses** for the iterative solution procedure for the algebraic equations. Poor choices may lead to unsatisfactory results. You want the initial guesses to be as close to the correct answer as possible so that the procedure will converge. You can perhaps solve approximate models, and then pass to more complex ones. Also you have to **scale** the variables and equations. By scaling of variables we mean introducing transformations that make all the variables have ranges in the same order of magnitude. By scaling of equations we mean multiplying each equation by a factor that causes the value of the deviation of each equation from zero to be of the same order of magnitude. Both of these steps speed convergence of the solution.

EXAMPLE 6.6 Sequential Solution of Material Balances

Figure E6.6a shows 10 units that are part of a larger plant. The 10 independent linear equations relating the flows of materials in and out of the units are listed in the figure.

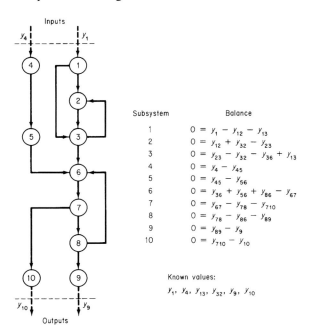

Subsystem	Balance
1	$0 = y_1 - y_{12} - y_{13}$
2	$0 = y_{12} + y_{32} - y_{23}$
3	$0 = y_{23} - y_{32} - y_{36} + y_{13}$
4	$0 = y_4 - y_{45}$
5	$0 = y_{45} - y_{56}$
6	$0 = y_{36} + y_{56} + y_{86} - y_{67}$
7	$0 = y_{67} - y_{78} - y_{710}$
8	$0 = y_{78} - y_{86} - y_{89}$
9	$0 = y_{89} - y_9$
10	$0 = y_{710} - y_{10}$

Known values:

$y_1,\ y_4,\ y_{13},\ y_{32},\ y_9,\ y_{10}$

Figure E6.6a   An example system containing recycle and bypass streams.

Figure E6.6b shows the incident matrix.

							Variables								Interface variables		
		y_{12}	y_{13}	y_{23}	y_{32}	y_{36}	y_{45}	y_{56}	y_{67}	y_{86}	y_{78}	y_{710}	y_{89}	y_1	y_4	y_9	y_{10}
	1	−1	−1	0	0	0	0	0	0	0	0	0	0	+1	0	0	0
	2	+1	0	−1	+1	0	0	0	0	0	0	0	0	0	0	0	0
	3	0	+1	+1	−1	−1	0	0	0	0	0	0	0	0	0	0	0
Equations	4	0	0	0	0	0	−1	0	0	0	0	0	0	0	+1	0	0
	5	0	0	0	0	0	+1	−1	0	0	0	0	0	0	0	0	0
	6	0	0	0	0	+1	0	+1	−1	+1	0	0	0	0	0	0	0
	7	0	0	0	0	0	0	0	+1	0	−1	−1	0	0	0	0	0
	8	0	0	0	0	0	0	0	0	−1	+1	0	−1	0	0	0	0
	9	0	0	0	0	0	0	0	0	0	0	0	+1	0	0	−1	0
	10	0	0	0	0	0	0	0	0	0	0	+1	0	0	0	0	−1

Incidence matrix

Figure E6.6b   The incidence matrix for the process in Figure E6.6a.

Suppose that instead of a simultaneous solution of the equations, a sequential solution is wanted. In what order should the equations be solved? Partition the equations so that a sequential solution can be executed. Lump together blocks of equations that still have to be solved simultaneously.

Solution

From Figure E6.6a you can see that it is not necessary to solve all 10 equations simultaneously. The system of equations can be broken up into lower-order subsystems, some of which can be comprised of individual equations or small groups of equations. These groups are associated with the so-called *irreducible* sets of equations. By inspection you can establish the following precedence order:

Step	Subsystem equation(s)
1	1
2	4
3	5
4	9
5	10
6	2
7	3
8	6, 7, and 8 simultaneously

In the case of more complicated sets of equations that cannot easily be decomposed by inspection, refer to some of the supplementary references at the end of this chapter for suitable algorithms.

6.2-2   Modular-Based Flowsheeting

The **sequential modular** method of flowsheeting is the one most commonly encountered in computer packages. A module exists for each process unit in the information flowsheet. Given the values of each input stream composition, flow rate, temperature, pressure, enthalpy, and the equipment parameters, the module calculates the properties of its outlet streams. The output stream for a module can become the input stream for another module for which the calculations proceed until the material and energy balances are resolved for the entire process.

The underlying concept of modularity in flowsheeting packages for design and analysis is to represent equipment or unit operations by portable computer subroutines. By portable we mean that such a subroutine can be assembled as an element of a large group of subroutines and successfully represent a certain type of equipment in any process. Figure 6.11 shows typical standardized unit operations modules together with their flowsheet symbols. Other modules take care of equipment sizing and cost estimation, perform numerical calculations, handle recycle calculations (described in more detail below), optimize, and serve as controllers

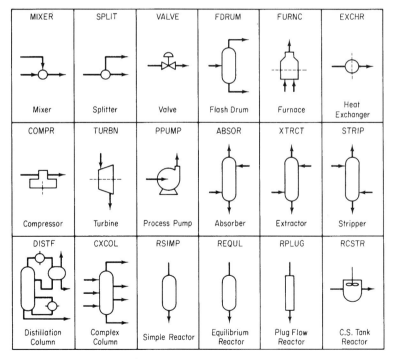

MIXER	SPLIT	VALVE	FDRUM	FURNC	EXCHR
Mixer	Splitter	Valve	Flash Drum	Furnace	Heat Exchanger
COMPR	TURBN	PPUMP	ABSOR	XTRCT	STRIP
Compressor	Turbine	Process Pump	Absorber	Extractor	Stripper
DISTF	CXCOL	RSIMP	REQUL	RPLUG	RCSTR
Distillation Column	Complex Column	Simple Reactor	Equilibrium Reactor	Plug Flow Reactor	C.S. Tank Reactor

Figure 6.11   Typical process modules used in sequential modular-based flow-sheeting codes with their subroutine names.

(executives) for the whole set of modules so that they function in the proper sequence. Internally, a very simple module might just be a table look-up program. However, most modules consist of Fortran or C subroutines that execute a sequence of calculations. Subroutines may consist of 100 to thousands of lines of code.

Information flows between modules via the material streams. Associated with each stream is an ordered list of numbers that characterize the stream. Table 6.3 lists a typical set of parameters associated with a stream. As a user of a modular-based code, you have to provide

1. The process topology
2. Input stream information including physical properties, connections
3. Design parameters needed in the modules and equipment specifications
4. Convergence criteria

In addition, you may have to insert a preferred calculation order for the modules. When economic evaluation and optimization are being carried out, you must also provide cost data and optimization criteria.

Modular-based flowsheeting exhibits several advantages in design. The flowsheet architecture is easily understood because it closely follows the process flowsheet. Individual modules can easily be added and removed from the computer package. Furthermore, new modules may be added to or removed from the flowsheet without any difficulty or affecting other modules. Modules at two different levels of accuracy can be substituted for one another as mentioned above.

TABLE 6.3   Stream Parameters

1. Stream number*
2. Stream flag (designates type of stream)
3. Total flow, lb mol/hr
4. Temperature, °F
5. Pressure, psia
6. Flow of component 1, lb mol/hr
7. Flow of component 2, lb mol/hr
8. Flow of component 3, lb mol/hr
9. Molecular weight
10. Vapor fraction
11. Enthalpy
12. Sensitivity

*Corresponds to an arbitrary numbering scheme used on the information flowsheet.

Modular-based flowsheeting also has certain drawbacks:

1. The output of one module is the input to another. The input and output variables in a computer module are fixed so that you cannot arbitrarily introduce an output and generate an input as can be done with an equation-based code.
2. The modules require extra computer time to generate reasonably accurate derivatives or their substitutes, especially if a module contains tables, functions with discrete variables, discontinuities, and so on. Perturbation of the input to a module is the primary way in which a finite-difference substitute for a derivative can be generated.
3. The modules may require a fixed precedence order of solution, that is, the output of one module must become the input of another; hence convergence may be slower than in an equation-solving method, and the computational costs may be high.
4. To specify a parameter in a module as a design variable, you have to place a control block around the module and adjust the parameter such that design specifications are met. This arrangement creates a loop. If the values of many design variables are to be determined, you might end up with several nested loops of calculation (which do, however, enhance stability). A similar arrangement must be used if you want to impose constraints.
5. Conditions imposed on a process (or a set of equations for that matter) may cause the unit physical states to move from two-phase to single-phase operation, or the reverse. As the code shifts from one module to another to represent the process properly, a severe discontinuity occurs and physical property values may jump about.

To obtain a solution for the material and energy balances in a flowsheet by the sequential modular method, you or the code must partition the flowsheet, select tear streams, nest the computations, and thus determine the computation sequence.

Partitioning of the modules in a block information diagram into minimum-sized subsets of modules that have to be solved simultaneously can be executed by many methods. As with solving sets of equations, you want to obtain the smallest block of modules in which the individual modules are tied together by the information flow of outputs and inputs. Between blocks, the information flow occurs serially. A simple algorithm is to trace a path of the flow of information (material usually but possibly energy or a signal) from one module to the next through the module output streams. The tracing continues until (1) a module in the path is encountered again, in that case all the modules in the path up to the repeated module form a group together that is collapsed and treated as a single

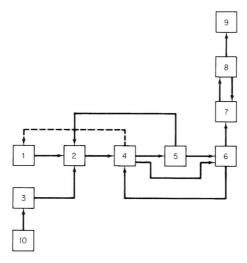

Figure 6.12 Block diagram to be partitioned.

module in subsequent tracing, or (2) a module or group with no output is encountered, in which case the module or group of modules is deleted from the block diagram. As an example, the block diagram in Figure 6.12 can be partitioned by the following steps.

Start with an arbitrary unit, say 4, and any sequence.

Path 1:   Start: $4 \rightarrow 5 \rightarrow 6 \rightarrow 4$ collapse as (456)
         Continue: $(456) \rightarrow 2 \rightarrow 4$ collapse as (4562)
         Continue: $(4562) \rightarrow 1 \rightarrow 2$ collapse as (45621)
         Continue: $(45621) \rightarrow 7 \rightarrow 8 \rightarrow 7$ collapse as (78)
         Continue: $(45621) \rightarrow (78) \rightarrow 9$ terminate (no output)

Put number 9 on the precedence order list and delete module 9 from the information block diagram.

    Continue: $(45621) \rightarrow (78) \rightarrow 9$ Terminate (put 78 on the list before 9)

Delete (78) from the block diagram.

    Continue: (45621)   Terminate (put 45621 before 78 in the list)

Delete (45621) from the block diagram.

This completes the search in path 1, as no more modules exist.

Path 2:   Start: $10 \rightarrow 3$ Terminate (put 3 in list before 45621)

         Continue: 10   Terminate (put 10 in list before 3)

All of the modules have been deleted from the block diagram, and no more paths have to be searched. Computer techniques to partition complex sets of modules besides the one described above can be found in the supplementary references at the end of this chapter. Simple sets can be partitioned by inspection.

From a computational viewpoint, the presence of recycle streams is one of the impediments in the sequential solution of a flowsheeting problem. Without recycle streams, the flow of information would proceed in a forward direction, and the calculational sequence for the modules could easily be determined from the precedence order analysis outlined above. With recycle streams present, large groups of modules have to be solved simultaneously, defeating the concept of a sequential solution module by module. For example, in Figure 6.13, you cannot make a material balance on the reactor without knowing the information in stream S6, but you have to carry out the computations for the cooler module first to evaluate S6, which in turn depends on the separator module, which in turn depends on the reactor module. Partitioning will identify those collections of modules that have to be solved simultaneously (termed **maximal cyclical subsystems** or **irreducible nets**).

To execute a sequential solution for a set of modules, you have to tear certain streams. **Tearing** in connection with modular flowsheeting involves decoupling the interconnections between the modules so that sequential information flow can take place. Tearing is required because of the loops of information created by recycle streams. What you do in tearing is to provide initial guesses for values of some of the unknowns (the tear variables), usually but not necessarily the recycle streams, and then calculate the values of the tear variables from the modules. These calculated values form new guesses, and so on, until the differences between the estimated and calculated values are sufficiently small. **Nesting** of the computations determines which tear streams are to be converged simultaneously and in which order collections of tear streams are to be converged.

Physical insight and experience in numerical analysis are important in selecting which variables to tear. For example, Figure 6.14 illustrates an equilibrium vapor–liquid separator for which the combined material and equilibrium equations give the relation

$$\sum_{j=1}^{C} \frac{z_j(1-K_j)}{1-(V/F)+VK_j/F} = 0 \tag{6.3}$$

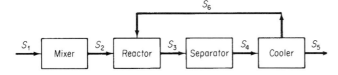

Figure 6.13   Modules in which recycle occurs; information (material) from the cooler module is fed back to the reactor.

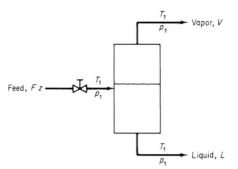

Figure 6.14   Vapor–liquid separator.

where z_j is the mole fraction of species j out of C components in the feed stream, $K_j = y_j/x_j$ is the vapor–liquid equilibrium coefficient, a function of temperature, and the stream flow rates are noted in the figure. For narrow-boiling systems, you can guess V/F, y_j, and x_j, and use Eq. (6.3) to calculate K_j and hence the temperature. This scheme works well because T lies within a narrow range. For wide-boiling materials, the scheme does not converge well. It is better to solve Eq. (6.3) for V/F by guessing T, y, and x, because V/F lies within a narrow range even for large changes in T. Usually, the convergence routines for the code comprise a separator module whose variables are connected to the other modules via the tear variables. Examine Figure 6.15.

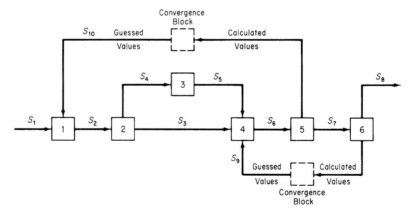

Figure 6.15   A computational sequence for modular flowsheeting. Initial values of both recycles are guessed, then the modules are solved in the order 1, 2, 3, 4, 5, and 6. Calculated values for recycle streams S9 and S10 are compared to guessed values in a convergence block, and unless the difference is less than some prescribed tolerance, another iteration takes place with the calculated values, or estimates based on them, forming the new initial guessed values of the recycle streams.

An engineer can usually carry out the partitioning, tearing, nesting, and determine the computational sequence for a flowsheet by inspection if the flowsheet is not too complicated. In some codes, the user supplies the computational sequence as input. Other codes determine the sequence automatically. In ASPEN, for example, the code is capable of determining the entire computational sequence, but the user can supply as many specifications as desired, up to and including the complete sequence. Consult one of the supplementary references at the end of this chapter for detailed information on optimal techniques of partitioning and tearing, techniques beyond our scope in this text.

Once the tear streams are identified and the sequence of calculations specified, everything is in order for the solution of material and energy balances. All that has to be done is to calculate the correct values for the stream flow rates and their properties. To execute the calculations, some computer codes use the method of successive substitution, which is described in Appendix L. The output(s) of each module on interation k is expressed as an explicit function of the input(s) calculated from the previous iteration, $k - 1$. For example, in Figure 6.15 for module 1,

$$S2^{(k)} = f(S1^{(k-1)}, S10^{(k-1)}, \text{coefficients})$$

To accelerate recycle stream convergence, the Wegstein method (refer to Appendix L) has been the mainstay of sequential modular flowsheeting for almost 25 years. Although this method neglects possible interaction between variables in tear streams, for most systems it works very well. A number of heuristics have been developed to improve convergence by delaying application of the Wegstein acceleration step until a specified number of direct substitution steps have been made and by setting bounds on the maximum acceleration allowed. With the use of Wegstein's method or direct substitution, it is necessary to control the convergence of recycle streams separately from the convergence of module specifications. Newton or quasi-Newton methods can also be used to solve for recycle streams.

LOOKING BACK

In this section we described the two main ways of solving the material and energy balances in a flowsheet: (a) equation based, and (b) modular based.

Key Ideas

1. A flowsheet is a graphical representation of the material and energy flows (and information flows) in a process.

2. Engineers solve the material and energy balances in a process via a computer code termed a flowsheeting code.

3. In the computer code, the material and energy balances can be represented by (a) equations and/or (b) modules. Each representation requires different strategies for solution, some of which are proprietary.

Key Terms

Equation oriented code (p. 563)	Partitioning (p. 573)
Flowsheet (p. 561)	Precedence order (p. 568)
Flowsheeting code (p. 561)	Process matrix (p. 565)
Irreducible nets (p. 575)	Scale (p. 569)
Maximal cyclical subsystems (p. 575)	Sequential modular (p. 571)
Modular method (p. 563)	Tearing (p. 568)
Nesting (p. 575)	

Discussion Question

1. A number of articles have been written of the subject of "paper vs. polystyrene" as materials for paper cups. Set up the flowsheets for the production of each, and include all of the quantitative and qualitative factors, both positive and negative, for the production from basic raw materials to the final product. Indicate what material and energy balances are needed, and, if possible, collect data so that they can be solved. Summarize the material and energy usage in the manufacturing of a cup.

SUPPLEMENTARY REFERENCES

Degrees of Freedom

PHAM, Q. T. "Degrees of Freedom of Equipment and Processes." *Chem. Eng. Sci.*, *49* (1994): 2507.

PONTON, J. W. "Degrees of Freedom Analysis in Process Control." *Chem. Eng. Sci.*, *49* (1994): 2089.

SMITH, B. D. *Design of Equilibrium Stage Processes*, Chapter 3. New York: McGraw-Hill, 1963.

SOMMERFELD, J. T. "Degrees of Freedom and Precedence Orders in Engineering Calculations." *Chem. Eng. Educ.* (Summer, 1986): 138.

Flowsheeting Codes

American Institute of Chemical Engineers. *CEP Software Directory*. New York: AIChE, issued annually.

CANFIELD, F. B. and P. K. NAIR. "The Key of Computed Integrated Processing." In *Proceed. ESCAPE-1*. Elsinore, Denmark (May 1992).

CHEN, H. S., and M. A. STADTHERR. "A Simultaneous-Modular Approach to Process Flowsheeting and Optimization: I. Theory and Implementation." *AIChE J.*, *30* (1984).

DIMIAN, A. "Use Process Simulation to Improve Plant Operations." *Chem. Eng. Progress*, (Sept. 1994): 58.

EDGAR, T. F. "CACHE Survey of PC-Based Flowsheet Software." *CACHE NEWS, 30,* (1990): 32.

GALLUN, S. E., R. H. LUECKE, D. E. SCOTT, and A. M. MORSHEDI. "Use Open Equations for Better Models." *Hydrocarbon Processing* (July 1992): 78.

HILLESTAD, M., and T. HERTZBERG. "Dynamic Simulation of Chemical Engineering Systems by the Sequential Modular Approach." *Comput. Chem. Eng., 10* (1986): 377.

MAH, R. S. H. *Chemical Process Structures and Information Flows.* Seven Oaks, UK: Butterworths, 1990.

MONTAGNA, J. M., and O. A. IRIBARREN. "Optimal Computation Sequence in the Simulation of Chemical Plants." *Comput. Chem. Eng., 12* (1988): 12.

SARGENT, R. W. H., J. D. PERKINS, and S. THOMAS. "Speedup: Simulation Program for Economic Evaluation and Design of Unified Processes." In *Computer-Aided Process Plant Design*, M. E. Leesley, ed. Houston: Gulf Publishing Company, 1982.

THOMÉ, B. (ed.). *Systems Engineering—Principles and Practice of Computer-Based Systems Engineering.* New York: Wiley, 1993.

WESTERBERG, A. W., and H. H. CHIEN, eds. "Thoughts on a Future Equation-Oriented Flowsheeting System." *Comput. Chem. Eng., 9* (1985): 517.

WESTERBERG, A. W., H. P. HUTCHINSON, R. L. MOTARD, and P. WINTER. *Process Flowsheeting.* Cambridge: Cambridge University Press, 1979.

PROBLEMS

Section 6.1

6.1. Determine the number of degrees of freedom for the condenser shown in Figure P6.1.

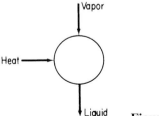

Figure P6.1

6.2. Determine the number of degrees of freedom for the reboiler shown in Figure P6.2. What variables should be specified to make the solution of the material and energy balances determinate?

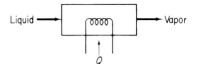

Figure P6.2

6.3. If to the equilibrium stage shown in Example 6.1 you add a feed stream, determine the number of degrees of freedom. See Figure P6.3.

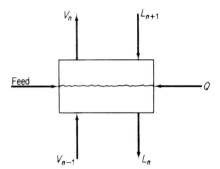

Figure P6.3

6.4. How many variables must be specified for the furnace shown in Figure P6.4 to absorb all the degrees of freedom?

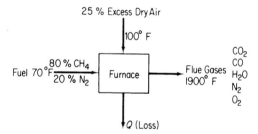

Figure P6.4

6.5. Figure P6.5 shows a simple absorber or extraction unit. S is the absorber oil (or fresh solvent), and F is the feed from which material is to be recovered. Each stage has a Q (not shown); the total number of equilibrium stages is N. What is the number of degrees of freedom for the column? What variables should be specified?

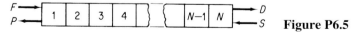

Figure P6.5

6.6. In a reactor model, rather than assume that the components exit from the reactor at equilibrium, an engineer will specify the r independent reactions that occur in the reactor, and the extent of each reaction, ξ_i. The reactor model must also provide for heating or cooling. How many degrees of freedom are associated with such a reactor model? See Figure P6.6.

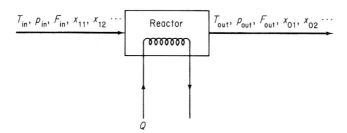

$T_{in}, p_{in}, F_{in}, x_{11}, x_{12} \cdots$ Reactor $T_{out}, p_{out}, F_{out}, x_{01}, x_{02} \cdots$

Q

Figure P6.6

6.7. Set up decomposition schemes for the processes shown in Figure P6.7. What additional variables must be specified to make the system determinate if (a) the feed conditions are known; or (b) the product conditions are specified?

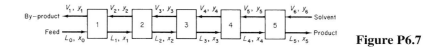

Figure P6.7

6.8. Determine whether or not the following problems are determinate in the sense that all the values of the material flows can be calculated.

(a) A vapor mixture containing 45 weight percent ammonia, the balance being water, and having an enthalpy of 1125 Btu per pound, is to be fractioned in a bubble-cap column operating at a pressure of 250 psia. The column is to be equipped with a total condenser. The distillate product is to contain 99.0 weight percent ammonia and the bottom product is to contain 10.0 weight percent ammonia. The distillate and the reflex leaving the condenser will have an enthalpy of 18 Btu/lb (Figure P6.8a).

(b) An engineer designed an extraction unit (Figure P6.8b) to recover oil from a pulp using alcohol as a solvent. The inerts refer to oil-free and solvent-free pulp. Several of the streams are shown as F_0 and F_1. Notice that the extracts from the first two stages were not clear but contained some inerts. (Both V_1

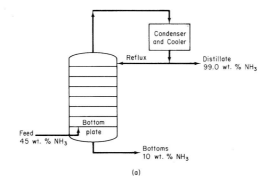

(a)

Figure P6.8a

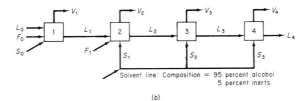

Solvent line: Composition = 95 percent alcohol
                              5 percent inerts

(b)   **Figure P6.8b**

and V_2 contain all three components: oil, solvent, and inerts.) Equal amounts of S_1 and L_1 were added to stage 32. There are 2 lb of L_2 for each lb of V_2 leaving the second stage.

The raffinate from stage 1, L_1, contains 32.5 percent alcohol and also in this same stream the weight ratio of inerts to solution is 60 lb inert/100 lb solution. The remaining raffinate streams, L_2, L_3, and L_4, contain 60 lb inert/100 lb alcohol. The L_2 stream contains 15 percent oil.

6.9. Examine Figure P6.9. Values of F_1, x_{11}, x_{12}, x_{13}, x_{14}, and x_{15} are known. Streams F_2 and F_3 are in equilibrium, and the three streams all have the same (known) temperatures and pressures. Is the problem completely specified, underspecified, or overspecified? Assume that the values of K in the equilibrium relations can be calculated from the given temperatures and pressures.

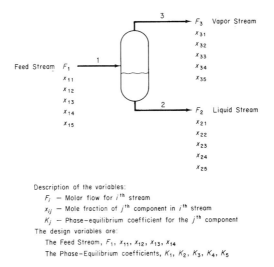

Description of the variables:
 F_i — Molar flow for i^{th} stream
 x_{ij} — Mole fraction of jth component in i^{th} stream
 K_j — Phase-equilibrium coefficient for the jth component
The design variables are:
 The Feed Stream, F_1, x_{11}, x_{12}, x_{13}, x_{14}
 The Phase-Equilibrium coefficients, K_1, K_2, K_3, K_4, K_5

Figure P6.9

6.10. Book[2] describes a mixer-heat exchanger section of a monoethylamine plant that is illustrated in Figure P6.10 along with the notation. Trimethylamine recycle enters

[2]N. L. Book. "Structural Analysis and Solution of Systems of Algebraic Design Equations." Ph.D. dissertation, University of Colorado, 1976.

in stream 4, is cooled in the heat exchanger, and is mixed with water from stream 1 in mixer 1. The trimethylamine-water mixture is used as the cold-side fluid in the heat exchanger and is then mixed with the ammonia-methanol stream from the gas absorber in mixer 3. The mixture leaving mixer 3 is the reaction mixture which feeds into the preheater of the existing plant.

Table P6.10 lists the 31 equations that represent the process in Figure P6.10. C_i is a heat capacity (a constant), F_i a flow rate, A area (a constant), ΔT_{lm} a log mean temperature difference

$$\frac{(T_4 - T_5) - (T_2 - T_3)}{\ln\left[(T_4 - T_5)/(T_2 - T_3)\right]}$$

U is a heat transfer coefficient (a constant), V_i a volume, y_i the molar flow rate of component i, x_i the mole fraction of component i, Q the heat transfer rate, and ρ_i the molar density (a constant). The question is: How many degrees of freedom exist for the process?

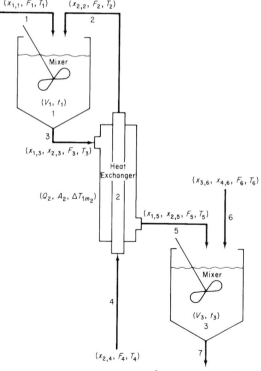

Figure P6.10

Table P6.10 List of Equations to Model the Process

Material Balance Equations

Mole fraction equations

1. $x_{1,1} - 1 = 0$
2. $x_{2,2} - 1 = 0$
3. $x_{1,3} + x_{2,3} - 1 = 0$
4. $x_{2,4} - 1 = 0$
5. $x_{1,5} + x_{2,5} - 1 = 0$
6. $x_{3,6} + x_{4,6} - 1 = 0$
7. $x_{1,7} + x_{2,7} + x_{3,7} + x_{4,7} - 1 = 0$

Flow balance equations

8. $F_1 + F_2 - F_3 = 0$
9. $F_2 - F_4 = 0$
10. $F_3 - F_5 = 0$
11. $F_5 + F_6 - F_7 = 0$

Component balance equations

12. $x_{1,1}F_1 - y_1 = 0$
13. $x_{1,3}F_3 - y_1 = 0$
14. $x_{1,5}F_5 - y_1 = 0$
15. $x_{1,7}F_7 - y_1 = 0$
16. $x_{2,2}F_2 - y_2 = 0$
17. $x_{2,3}F_3 - y_2 = 0$
18. $x_{2,4}F_4 - y_2 = 0$
19. $x_{2,5}F_5 - y_2 = 0$
20. $x_{2,7}F_7 - y_2 = 0$
21. $x_{3,6}F_6 - y_3 = 0$
22. $x_{3,7}F_7 - y_3 = 0$
23. $x_{4,6}F_6 - y_4 = 0$
24. $x_{4,7}F_7 - y_4 = 0$

Energy Balance Equations

25. $C_1F_1T_1 + C_2F_2T_2 - C_3F_3T_3 = 0$
26. $C_4F_4T_4 - Q_2 - C_2F_2T_2 = 0$
27. $C_3F_3T_3 + Q_2 - C_5F_5T_5 = 0$
28. $C_5F_5T_5 + C_6F_6T_6 - C_7F_7T_7 = 0$

Equipment Specification Equations

29. $V_1 - (F_3)\left(\dfrac{V_1}{F_1}\right)/\rho_3 = 0$

30. $Q_2 - U_2A_2\,\Delta T_{lm_2} = 0$

31. $V_3 - (F_7)\left(\dfrac{V_3}{F_3}\right)/\rho_7 = 0$

6.11. Cavett proposed the following problems as a test problem for computer-aided design. Four flash drums are connected as shown in Figure P6.11. The temperature in each flash drum is specified, and equilibrium is assumed to be independent of composition so that the vapor-liquid equilibrium constants are truly constant. Is the problem properly specified, or do additional variables have to be given? If the latter, what should they be? The feed is as follows

Component	Feed
1. Nitrogen and helium	358.2
2. Carbon dioxide	4,965.6
3. Hydrogen sulfide	339.4
4. Methane	2,995.5
5. Ethane	2,395.5
6. Propane	2,291.0
7. Isobutane	604.1
8. n-Butane	1,539.9
9. Isopentane	790.4
10. n-Pentane	1,192.9
11. Hexane	1,764.7
12. Heptane	2,606.7
13. Octane	1,844.5
14. Nonane	1,669.0
15. Decane	831.7
16. Undecane plus	1,214.5
Total	27,340.6

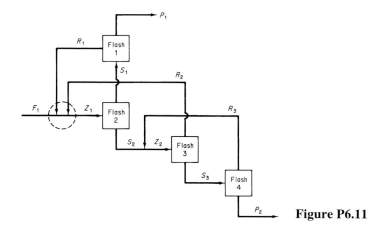

Figure P6.11

6.12. The flowsheet (Figure P6.12) has a high-pressure feed stream of gaseous component A contaminated with a small amount of B. It mixes first with a recycle stream consisting mostly of A and passes into a reactor, where an exothermic reaction to form C from A takes place. The stream is cooled to condense out component C and passed through a valve into a flash unit. Here most of the unreacted A and the contaminant B flash off, leaving a fairly pure C to be withdrawn as the liquid stream. Part of the recycle is bled off to keep the concentration of B from building up in the system. The rest is repressurized in a compressor and mixed, as stated earlier, with the feed stream. The number of parameters/variables for each unit are designated by the number within the symbol for the unit. How many degrees of freedom exist for this process?

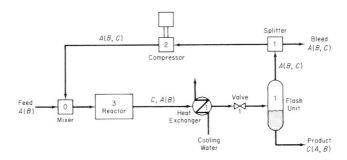

Figure P6.12

6.13. In petroleum refining, lubricating oil is treated with sulfuric acid to remove unsaturated compounds, and after settling, the oil and acid layers are separated. The acid layer is added to water and heated to separate the sulfuric acid from the sludge contained in it. The dilute sulfuric acid, now 20% H_2SO_4 at 82°C, is fed to a Simonson-Mantius evaporator, which is supplied with saturated steam at 400 kPa gauge to lead coils submerged in the acid, and the condensate leaves at the saturation temperature. A vacuum is maintained at 4.0 kPa by means of a barometric leg. The acid is concentrated to 80% H_2SO_4; the boiling point at 4.0 kPa is 121°C. How many kilograms of acid can be concentrated per 1000 kg of steam condensed?

6.14. You are asked to perform a feasibility study on a continuous stirred tank reactor shown in Figure P6.14 (which is presently idle) to determine if it can be used for the second-order reaction

$$2A \longrightarrow B + C$$

Since the reaction is exothermic, a cooling jacket will be used to control the reactor temperature. The total amount of heat transfer may be calculated from an overall heat transfer coefficient (U) by the equation

$$Q = UA \, \Delta T$$

where Q = total rate of heat transfer from the reactants to the water jacket in the steady state

U = empirical coefficient
A = area of transfer
ΔT = temperature difference (here $T_4 - T_2$)

Some of the energy released by the reaction will appear as sensible heat in stream F_2, and some concern exists as to whether the fixed flow rates will be sufficient to keep the fluids from boiling while still obtaining good conversion. Feed data is as follows.

Component	Feed rate (lb mol/hr)	C_p [Btu/(lb mol)(°F)]	MW
A	214.58	41.4	46
B	23.0	68.4	76
C	0.0	4.4	6

The consumption rate of A may be expressed as

$$-2k(C_A)^2 V_R$$

where

$$C_A = \frac{(F_{1,A})(\rho)}{\Sigma(F_{1,i})(MW_i)} = \text{concentration of } A, \text{ lb mol/ft}^3$$

$$k = k_0 \exp\left(\frac{-E}{RT}\right)$$

k_0, E, R are constants and T is the absolute temperature.

Solve for the temperatures of the exit streams and the product composition of the steady-state reactor using the following data:

Fixed parameters

$$\text{Reactor volume} = V_R = 13.3 \text{ ft}^3$$
$$\text{Heat transfer area} = A = 29.9 \text{ ft}^2$$
$$\text{Heat transfer coefficient} = U = 74.5 \text{ Btu/(hr)(ft}^2)(°F)$$

Variable input

$$\text{Reactant feed rate} = F_i \text{ (see table above)}$$
$$\text{Reactant feed temperature} = T_1 = 80°F$$
$$\text{Water feed rate} = F_3 = 247.7 \text{ lb mol/hr water}$$
$$\text{Water feed temperature} = T_3 = 75°F$$

Physical and thermodynamic data

$$\text{Reaction rate constant} = k_0 = 34 \text{ lb mol/(hr)(ft}^2)$$
$$\text{Activation energy/gas constant} = E/R = 1000°F$$

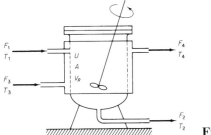

Figure P6.14

$$\text{Heat of reaction} = \Delta H = -5000 \text{ Btu/lb mol } A$$
$$\text{Heat capacity of water} = C_{p_w} = 18 \text{ Btu/(lb mol)(°F)}$$
$$\text{Product component density} = \rho = 55 \text{ lb/ft}^3$$

The densities of each of the product components are essentially the same. Assume that the reactor contents are perfectly mixed as well as the water in the jacket, and that the respective exit stream temperatures are the same as the reactor contents or jacket contents.

6.15. The steam flows for a plant are shown in Figure P6.15. Write the material and energy balances for the system and calculate the unknown quantities in the diagram (A to F). There are two main levels of steam flow: 600 psig and 50 psig. Use the steam tables for the enthalpies.

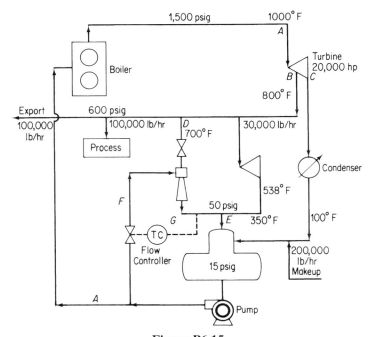

Figure P6.15

6.16. Figure P6.16 shows a calciner and the process data. The fuel is natural gas. How can the energy efficiency of this process be improved by process modification? Suggest at least two ways based on the assumption that the supply conditions of the air and fuel remain fixed (but these streams can be possibly passed through heat exchangers). Show all calculations.

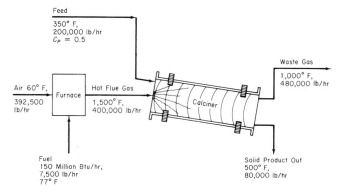

Feed
350° F,
200,000 lb/hr
C_p = 0.5

Waste Gas
1,000° F,
480,000 lb/hr

Air 60° F,
392,500
lb/hr

Furnace

Hot Flue Gas
1,500° F,
400,000 lb/hr

Calciner

Fuel
150 Million Btu/hr,
7,500 lb/hr
77° F

Solid Product Out
500° F,
80,000 lb/hr

Figure P6.16

6.17. Limestone ($CaCO_3$) is converted into CaO in a continuous vertical kiln (see Figure P6.17). Heat is supplied by combustion of natural gas (CH_4) in direct contact with the limestone using 50% excess air. Determine the kilograms of $CaCO_3$ that can be processed per kilogram of natural gas. Assume that the following average heat capacities apply:

$$C_p \text{ of } CaCO_3 = 234 \text{ J/(g mol)(°C)}$$

$$C_p \text{ of } CaO = 111 \text{ J/(g mol)(°C)}$$

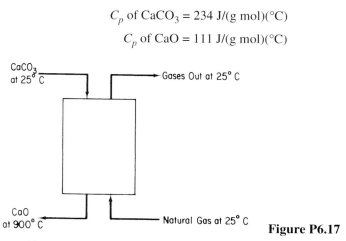

$CaCO_3$
at 25° C

Gases Out at 25° C

CaO
at 900° C

Natural Gas at 25° C

Figure P6.17

6.18. A feed stream of 16,000 lb/hr of 7% by weight NaCl solution is concentrated to a 40% by weight solution in an evaporator. The feed enters the evaporator, where it is heated to 180°F. The water vapor from the solution and the concentrated solution leave at 180°F. Steam at the rate of 15,000 lb/hr enters at 230°F and leaves as condensate at 230°F. See Figure P6.18.

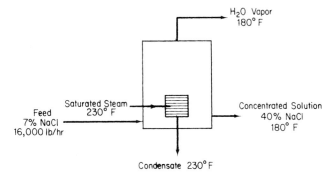

Figure P6.18

(a) What is the temperature of the feed as it enters the evaporator?
(b) What weight of 40% NaCl is produced per hour?
Assume that the following data apply:

Average C_p 7% NaCl soln: 0.92 Btu/(lb)(°F)
Average C_p 40% NaCl soln: 0.85 Btu/(lb)(°F)
$\Delta \hat{H}_{vap}$ at H_2O at 180°F = 990 Btu/lb
$\Delta \hat{H}_{vap}$ at H_2O at 230°F = 959 Btu/lb

6.19. The Blue Ribbon Sour Mash Company plans to make commercial alcohol by a process shown in Figure P6.19. Grain mash is fed through a heat exchanger where it is heated to 170°F. The alcohol is removed as 60% by weight alcohol from the first fractionating column; the bottoms contain no alcohol. The 60% alcohol is further fractionated to 95% alcohol and essentially pure water in the second column. Both stills operate at a 3:1 reflux ratio and heat is supplied to the bottom of the columns by steam. Condenser water is obtainable at 80°F. The operating data and physical properties of the streams have been accumulated and are listed for convenience:

Stream	State	Boiling point (°F)	C_p[Btu/(lb)(°F)] Liquid	C_p[Btu/(lb)(°F)] Vapor	Heat of vaporization (Btu/lb)
Feed	Liquid	170	0.96	—	950
60% alcohol	Liquid or vapor	176	0.85	0.56	675
Bottoms I	Liquid	212	1.00	0.50	970
95% alcohol	Liquid or vapor	172	0.72	0.48	650
Bottoms II	Liquid	212	1.0	0.50	970

Prepare the material balances for the process, calculate the precedence order for solution, and
(a) Determine the weight of the following streams per hour:
    (1) Overhead product, column I

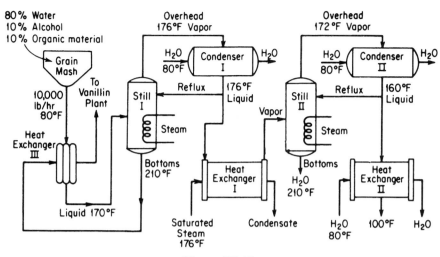

Figure P6.19

    (2) Reflux, column I
    (3) Bottoms, column I
    (4) Overhead product, column II
    (5) Reflux, column II
    (6) Bottoms, column II
  (b) Calculate the temperature of the bottoms leaving heat exchanger III.
  (c) Determine the total heat input to the system in Btu/hr.
  (d) Calculate the water requirements for each condenser and heat exchanger II in gal/hr if the maximum exit temperature of water from this equipment is 130°F.

6.20. Toluene, manufactured by the conversion of *n*-heptane with a Cr_2O_3-on-Al_2O_3 catalyst

$$CH_3CH_2CH_2CH_2CH_2CH_2CH_3 \rightarrow C_6H_5CH_3 + 4H_2$$

by the method of hydroforming is recovered by use of a solvent. See Figure P6.20 for the process and conditions.

    The yield of toluene is 15 mole % based on the *n*-heptane charged to the reactor. Assume that 10 kg of solvent are used per kilogram of toluene in the extractors.

  (a) Calculate how much heat has to be added or removed from the catalytic reactor to make it isothermal at 425°C.
  (b) Find the temperature of the *n*-heptane and solvent stream leaving the mixer-settlers if both streams are at the same temperature.
  (c) Find the temperature of the solvent stream after it leaves the heat exchanger.
  (d) Calculate the heat duty of the fractionating column in kJ/kg of *n*-heptane feed to the process.

	$-\Delta H_f^{oa}$ (kJ/g mol)	$C_p[\text{J}/(\text{g})(°\text{C})]$		$\Delta H_{\text{vaporization}}$ (kJ/kg)	Boiling point (K)
		Liquid	Vapor		
Touleneᵇ	12.00	2.22	2.30	364	383.8
n-Heptane	−224.4	2.13	1.88	318	371.6
Solvent	—	1.67	2.51	—	434

ᵃAs liquids.
ᵇThe heat of solution of toluene in the solvent is −23 J/g toluene.

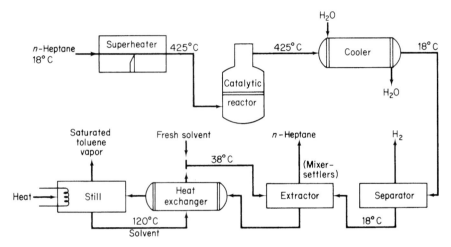

Figure P6.20

6.21. One hundred thousand pounds of a mixture of 50% benzene, 40% toluene, and 10% *o*-xylene is separated every day in a distillation-fractionation plant as shown on the flowsheet for Figure P6.21.

	Boiling point (°C)	C_p liquid [cal/(g)(°C)]	Latent heat of vap. (cal/g)	C_p vapor [cal/(g)(°C)]
Benzene	80	0.44	94.2	0.28
Toluene	109	0.48	86.5	0.30
o-Xylene	143	0.48	81.0	0.32
Charge	90	0.46	88.0	0.29
Overhead T_I	80	0.45	93.2	0.285
Residue T_I	120	0.48	83.0	0.31
Residue T_{II}	413	0.48	81.5	0.32

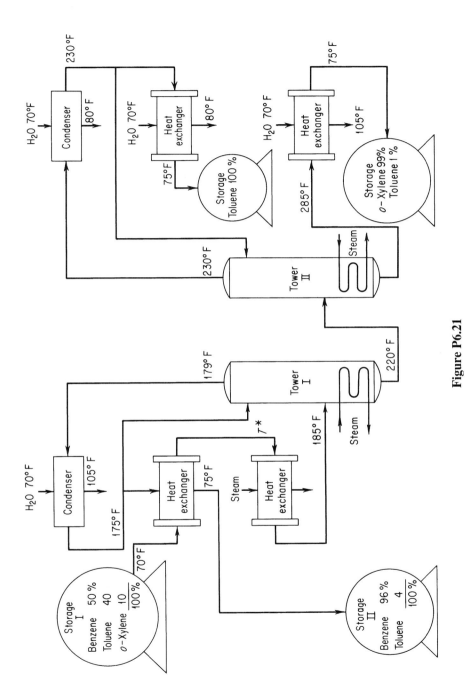

Figure P6.21

The reflux ratio for tower I is 6:1; the reflux ratio for tower II is 4:1; the charge to tower I is liquid; the chart to tower II is liquid. Compute:

(a) The temperature of the mixture at the outlet of the heat exchanger (marked as T^*)

(b) The Btu supplied by the steam reboiler in each column

(c) The quantity of cooling water required in gallons per day for the whole plant

(d) The energy balance around tower I

6.22. Sulfur dioxide emissions from coal-burning power plants causes serious atmospheric pollution in the eastern and midwestern portions of the United States. Unfortunately, the supply of low-sulfur coal is insufficient to meet the demand. Processes presently under consideration to alleviate the situation include coal gasification followed by desulfurization and stack-gas cleaning. One of the more promising stack-gas-cleaning processes involves reacting SO_2 and O_2 in the stack gas with a solid metal oxide sorbent to give the metal sulfate, and then thermally regenerating the sorbent and absorbing the result SO_3 to produce sulfuric acid. Recent laboratory experiments indicate that sorption and regeneration can be carried out with several metal oxides, but no pilot or full-scale processes have yet been put into operation.

You are asked to provide a preliminary design for a process that will remove 95% of the SO_2 from the stack gas of a 1000-MW power plant. Some data are given below and in the flow diagram of the process (Figure P6.22). The sorbent consists of fine particles of a dispersion of 30% by weight CuO in a matrix of inert porous Al_2O_3. This solid reacts in the fluidized-bed absorber at 315°C. Exit solid is sent to the regenerator, where SO_3 is evolved at 700°C, converting all the $CuSO_4$ present back to CuO. The fractional conversion of CuO to $CuSO_4$ that occurs in the sorber is called α and is an important design variable. You are asked to carry out your calculations for $\alpha = 0.2$, 0.5, and 0.8. The SO_3 produced in the regenerator is swept out by recirculating air. The SO_3-laden air is sent to the acid tower, where the SO_3 is absorbed in recirculating sulfuric acid and oleum, part of which is withdrawn as salable byproducts. You will notice that the sorber, regenerator, and perhaps the acid tower are adiabatic; their temperatures are adjusted by heat exchange with incoming streams. Some of the heat exchangers (nos. 1 and 3) recover heat by countercurrent exchange between the feed and exit streams. Additional heat is provided by withdrawing flue gas from the power plant at any desired high temperature up to 1100°C and then returning it as a lower temperature. Cooling is provided by water at 25°C. As a general rule, the temperature difference across heat-exchanger walls separating the two streams should average about 28°C. The nominal operating pressure of the whole process is 10 kPa. The three blowers provide 6 kPa additional head for the pressure losses in the equipment, and the acid pumps have a discharge pressure of 90 kPa gauge. You are asked to write the material and energy balances and some equipment specifications as follows:

(a) Sorber, regenerator, and acid tower. Determine the flow rate, composition, and temperature of all streams entering and leaving.

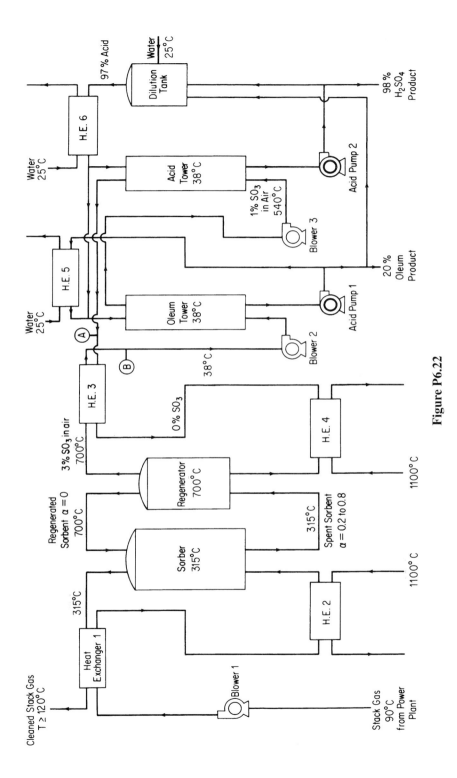

Figure P6.22

(b) Heat exchangers. Determine the heat load, and flow rate, temperature, and enthalpy of all streams.

(c) Blowers. Determine the flow rate and theoretical horsepower.

(d) Acid pump. Determine the flow rate and theoretical horsepower.

Use SI units. Also, use a basis of 100 kg of coal burned for all your calculations; then convert to the operating basis at the end of the calculations.

Power plant operation. The power plant burns 340 metric tons/hr of coal having the analysis given below. The coal is burned with 18% excess air, based on complete combustion to CO_2, H_2O, and SO_2. In the combustion only the ash and nitrogen is left unburned; all the ash has been removed from the stack gas.

Element	Wt.%
C	76.6
H	5.2
O	6.2
S	2.3
N	1.6
Ash	8.1

Data on Solids (Units of C_p are J/(g mol)(K); units of H are k J/g mol.)

	Al_2O_3		CuO		$CuSO_4$	
T (K)	C_p	$H_T - H_{298}$	C_p	$H_T - H_{298}$	C_p	$H_T - H_{298}$
298	79.04	0.00	42.13	0.00	98.9	0.00
400	96.19	9.00	47.03	4.56	114.9	10.92
500	106.10	19.16	50.04	9.41	127.2	23.05
600	112.5	30.08	52.30	14.56	136.3	36.23
700	117.0	41.59	54.31	19.87	142.9	50.25
800	120.3	53.47	56.19	25.40	147.7	64.77
900	122.8	65.65	58.03	31.13	151.0	79.71
1000	124.7	77.99	59.87	37.03	153.8	94.98

6.23. When coal is distilled by heating without contact with air, a wide variety of solid, liquid, and gaseous products of commercial importance is produced, as well as some significant air pollutants. The nature and amounts of the products produced depend on the temperature used in the decomposition and the type of coal. At low temperatures (400 to 750°C) the yield of synthetic gas is small relative to the yield of liquid products, whereas at high temperatures (above 900°C) the reverse is true. For the typical process flowsheet, shown in Figure P6.23:

(a) How many tons of the various products are being produced?

(b) Make an energy balance around the primary distillation tower and benzol tower.

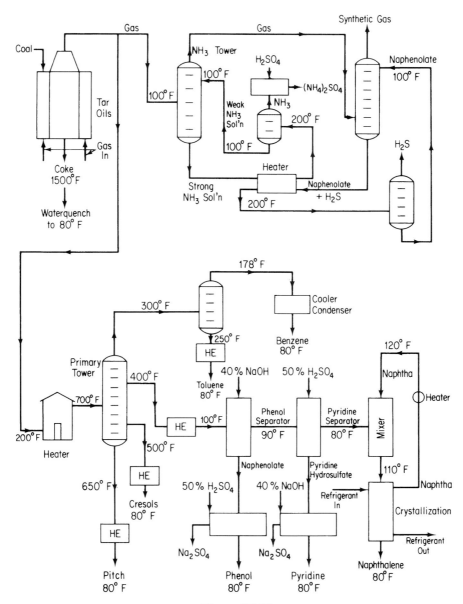

Figure P6.23

(c) How much (in pounds) of 40% NaOH solution is used per day for the purification of the phenol?

(d) How much 50% H_2SO_4 is used per day in the pyridine purification?

(e) What weight of Na_2SO_4 is produced per day by the plant?

(f) How many cubic feet of gas per day are produced? What percent of the gas (volume) is needed for the ovens?

Products Produced Per Ton of Coal Charged	Mean C_p Liquid (cal/g)	Mean C_p Vapor (cal/g)	Mean C_p Solid (cal/g)	Melting Point (°C)	Boiling Point (°C)
Synthetic gas–10,000 ft³ (555 Btu/ft³)					
$(NH_4)_2SO_4$, 22 lb					
Benzol, 15 lb	0.50	0.30	—	—	60
Toluol, 5 lb	0.53	0.35	—	—	109.6
Pyridine, 3 lb	0.41	0.28	—	—	114.1
Phenol, 5 lb	0.56	0.45	—	—	182.2
Naphthalene, 7 lb	0.40	0.35	0.281 $+ 0.00111\ T_{°F}$	80.2	218
Cresols, 20 lb	0.55	0.50	—	—	202
Pitch, 40 lb	0.65	0.60	—	—	400
Coke, 1500 lb	—	—	0.35		—

	ΔH_{vap} (cal/g)	ΔH_{fusion} (cal/g)
Benzol	97.5	—
Toluol	86.53	—
Pyridine	107.36	—
Phenol	90.0	—
Naphthalene	75.5	35.6
Cresols	100.6	—
Pitch	120	—

6.24. A gas consisting of 95 mol % hydrogen and 5 mol % methane at 100°F and 30 psia is to be compressed to 569 psia at a rate of 440 lb mol/hr. A two-stage compressor system has been proposed with intermediate cooling of the gas to 100°F via a heat exchanger. See Figure P6.24. The pressure drop in the heat exchanger from the inlet stream (S1) to the exit stream (S2) is 2.0 psia. Using a process simulator program analyze all of the steam parameters subject to the following constraints: The exit stream from the first stage is 100 psia; both compressors are positive-displacement type and have a mechanical efficiency of 0.8, a polytropic efficiency of 1.2, and a clearance fraction of 0.05.

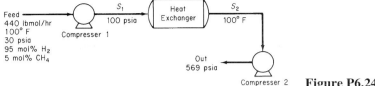

Figure P6.24

6.25. A gas feed mixture at 85°C and 100 psia having the composition shown in Figure P6.25 is flashed to separate the majority of the light components from the heavy. The flash chamber operates at 5°C and 25 psia. To improve the separation process, it has been suggested to introduce a recycle as shown in Figure P6.25. Will a significant improvement be made by adding a 25% recycle of the bottoms? 50%? With the aid of a computer process simulator, determine the molar flow rates of the streams for each of the three cases.

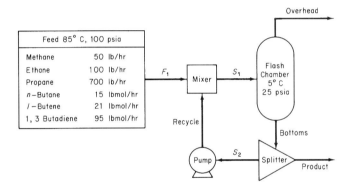

Figure P6.25

6.26. A mixture of three petroleum fractions containing lightweight hydrocarbons is to be purified and recycled back to a process. Each of the fractions is denoted by its normal boiling point: BP135, BP260, and BP500. The gases separated from this feed are to be compressed as shown in Figure P6.26. The inlet feed stream (1) is at 45°C and 450 kPa, and has the composition shown. The exit gas (10) is compressed to 6200 kPa by a three-stage compressor process with intercooling of the vapor streams to 60°C by passing through a heat exchanger. The exit pressure for compressor 1 is 1100 kPa and 2600 kPa for compressor 2. The efficiencies for compressors 1, 2, and 3, with reference to an adiabatic compression are, 78, 75, and 72%, respectively. Any liquid fraction drawn off from a separator is recycled to the previous stage. Estimate the heat duty (in kJ/hr) of the heat exchangers and the various stream compositions (in kg mol/hr) for the system. Note that the separators may be considered as adiabatic flash tanks in which the pressure decrease is zero. This problem has been formulated from *Application Briefs of Process*, the user manual for the computer simulation software package of Simulation Science, Inc.

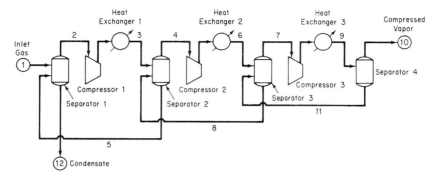

Figure P6.26

Component	kg mol/hr	M.W.	sp gr	Normal boiling point (°C)
Nitrogen	181			
Carbon dioxide	1,920			
Methane	14,515			
Ethane	9,072			
Propane	7,260			
Isobutane	770			
n-Butane	2,810			
Isopentane	953			
n-Pentane	1,633			
Hexane	1,542			
BP135	11,975	120	0.757	135
BP260	9,072	200	0.836	260
BP500	9,072	500	0.950	500

6.27. A demethanizer tower is used in a refinery to separate natural gas from a light hydrocarbon gas mixture stream (1) having the composition listed below. However,
initial calculations show that there is considerable energy wastage in the process.
A proposed improved system is outlined in Figure P6.27. Calculate the temperature (°F), pressure (psig), and composition (lb mol/hr) of all the process streams in
the proposed system.

   Inlet gas at 120°F and 588 psig stream (1) is cooled in the tube side of a gas-
gas heat exchanger by passing the tower overhead stream (8) through the shell
side. The temperature difference between the exit streams (2) and (10) of the heat
exchanger is to be 10°F. Note that the pressure drop through the tube side is 10
psia and 5 psia on the shell side. The feed stream (2) is then passed through a
chiller in which the temperature drops to −84°F and a pressure loss of 5 psi results.
An adiabatic flash separator is used to separate the partially condensed vapor from
the remaining gas. The vapor then passes through an expander turbine and is fed

to the first tray of the tower at 125 psig. The liquid stream (5) is passed through a valve, reducing the pressure to that of the third tray on the lower side. The expander transfers 90% of its energy output to the compressor. The efficiency with respect to an adiabatic compression is 80% for the expander and 75% for the compressor. The process requirements are such that the methane-to-ethane ratio in the demethanizer liquids in stream (9) is to be 0.015 by volume; the heat duty on the reboiler is variable to achieve this ratio. A process rate of 23.06×10^6 standard cubic feet per day of feed stream 1 is required.

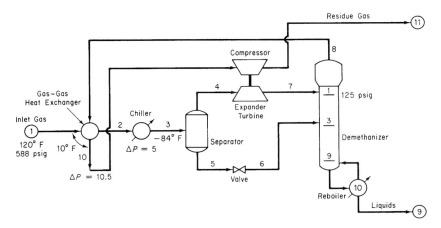

Figure P6.27

Component	Mol %
Nitrogen	7.91
Methane	73.05
Ethane	7.68
Propane	5.69
Isopropane	0.99
n-butane	2.44
Isopentane	0.69
n-pentane	0.82
C_6	0.42
C_7	0.31
Total	100.00

The tower has 10 trays, including the reboiler. *Note:* To reduce the number of trials, the composition of stream 3 may be referenced to stream 1, and if the exit stream of the chiller is given a dummy symbol, the calculations sequence can begin at the separator, thus eliminating the recycle loop.

Carry out the solution of the material and energy balances for the flowsheet in Figure P6.27, determine the component and total mole flows, and determine the enthalpy flows for each stream. Also find the heat duty of each heat exchanger.

This problem has been formulated from *Application Briefs of Process*, the user manual for the computer simulation software package of Simulation Sciences, Inc.

6.28. Determine the values of the unknown quantities in Figure P6.28 by solving the following set of linear material and energy balances that represent the steam balance:

(a) $181.60 - x_3 - 132.57 - x_4 - x_5 = -y_1 - y_2 + y_5 + y_4 = 5.1$

(b) $1.17x_3 - x_6 = 0$

(c) $132.57 - 0.745x_7 = 61.2$

(d) $x_5 + x_7 - x_8 - x_9 - x_{10} + x_{15} = y_7 + y_8 = y_3 = 99.1$

(e) $x_8 + x_9 + x_{10} + x_{11} - x_{12} - x_{13} = -y_7 = -8.4$

(f) $x_6 - x_{15} = y_{12} = y_5 = 24.2$

(g) $-1.15(181.60) + x_3 - x_6 + x_{12} + x_{16} = 1.15y_1 - y_9 + 0.4 = -19.7$

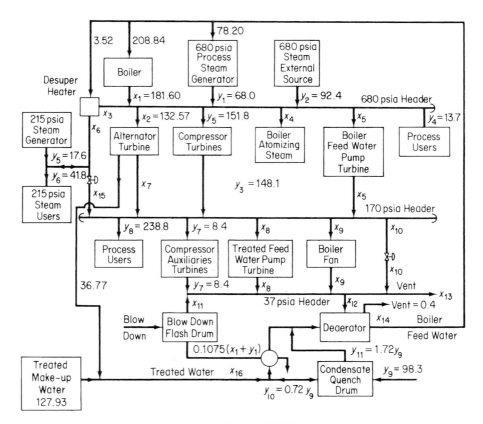

Figure P6.28

(h) $181.60 - 4.594x_{12} - 0.11x_{16} = -y_1 + 1.0235y_9 + 2.45 = 35.05$

(i) $-0.0423(181.60) + x_{11} = 0.0423y_1 = 2.88$

(j) $-0.016(181.60) + x_4 = 0$

(k) $x_8 - 0.0147x_{16} = 0$

(l) $x_5 - 0.07x_{14} = 0$

(m) $-0.0805(181.60) + x_9 = 0$

(n) $x_{12} - x_{14} + x_{16} = 0.4 - y_9 = -97.9$

There are four levels of steam: 680, 215, 170, and 37 psia. The 14 x_i, $i = 3, \ldots, 16$, are the unknowns and the y_i are given parameters for the system. Both x_i and y_i have the units of 10^3 lb/hr.

UNSTEADY-STATE MATERIAL AND ENERGY BALANCES

7

Unsteady-state problems in previous chapters have used the overall or integrated accumulation term in the material and energy balances. Now we focus our attention briefly on **unsteady-state** processes in which the value of the state (dependent variable) as a function of time is of interest. Recall that the term unsteady state refers to processes in which quantities or operating conditions within the system *change with time*. Sometimes you will hear the word *transient state* applied to such processes. The unsteady state is somewhat more complicated than the steady state, and in general problems involving unsteady-state processes are somewhat more difficult to formulate and solve than those involving steady-state processes. However, a wide variety of important industrial problems fall into this category, such as the startup of equipment, batch heating or reactions, the change from one set of operating conditions to another, and the perturbations that develop as process conditions fluctuate.

7.1 UNSTEADY-STATE MATERIAL AND ENERGY BALANCES

> ### *Your objectives in studying this section are to be able to:*
>
> 1. Write down the macroscopic unsteady-state material and energy balances in words and in symbols.
> 2. Solve simple ordinary differential material or energy balance equations given the initial conditions.
> 3. Take a word problem and translate it into a differential equation(s).

LOOKING AHEAD

In this section we describe how lumped macroscopic material and energy balances are developed for unsteady-state processes in which time is the independent variable.

MAIN CONCEPTS

The macroscopic balance ignores all the detail within a system and consequently results in a balance about the entire system. Time is the independent variable in the balance. The dependent variables, such as concentration and temperature, are not functions of position but represent overall averages throughout the entire volume of the system. In effect, the system is assumed to be sufficiently well mixed so that the output concentrations and temperatures are equivalent to the concentrations and temperatures inside the system.

To assist in the translation of Eq. (7.1)

$$\left\{ \begin{array}{c} \text{accumulation} \\ \text{or depletion} \\ \text{within} \\ \text{the} \\ \text{system} \end{array} \right\} = \left\{ \begin{array}{c} \text{transport into} \\ \text{system through} \\ \text{system} \\ \text{boundary} \end{array} \right\} - \left\{ \begin{array}{c} \text{transport out} \\ \text{of sytem} \\ \text{through system} \\ \text{boundary} \end{array} \right\} \quad (7.1)$$

$$+ \left\{ \begin{array}{c} \text{generation} \\ \text{within} \\ \text{system} \end{array} \right\} - \left\{ \begin{array}{c} \text{consumption} \\ \text{within} \\ \text{system} \end{array} \right\}$$

into mathematical symbols, refer to Figure 7.1. Equation (7.1) can be applied to the mass of a single species or to the total amount of material or energy in the system. Let us write each of the terms in Eq. (7.1) in mathematical symbols for a very small time interval Δt. Let the accumulation be positive in the direction in

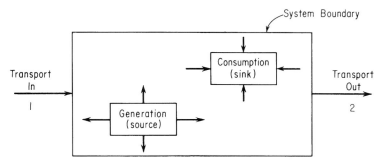

Figure 7.1   General unsteady-state process with transport in and out and internal generation and consumption.

which time is positive, that is, as time increases from t to $t + \Delta t$. Then, using the component mass balance as an example, the accumulation will be the mass of A in the system at time $t + \Delta t$ minus the mass of A in the system at time t,

$$\text{accumulation} = \rho_A V|_{t+\Delta t} - \rho_A V|_t$$

where ρ_A = mass of component A per unit volume
           V = volume of the system

and the symbol $|_t$ means that the quantities preceding the vertical line are evaluated at time t, or time $t + \Delta t$ or at surface S_1, or at surface S_2, as the case may be as denoted by the subscript. Note that the net dimensions of the accumulation term are the mass of A.

We shall split the mass transport across the system boundary into two parts, transport through defined surfaces S_1 and S_2 whose areas are known, and transport across the system boundary through other (undefined) surfaces. The net transport of A into (through S_1) and out of (through S_2) the system through defined surfaces can be written as:

$$\text{net flow across boundary via } S_1 \text{ and } S_2 = \rho_A \upsilon S \Delta t|_{S_1} - \rho_A \upsilon S \Delta t|_{S_2}$$

where υ = fluid velocity in a duct of cross section S
           S = defined cross-sectional area perpendicular to material flow

Again note that the net dimensions of the transport term are the mass of A. Other types of transport across the system boundary can be represented by

$$\text{net residual flow across boundary} = \dot{w}_A \Delta t$$

where $\dot{w}_A$ is the rate of mass flow of component A through the system boundaries other than the defined surfaces S_1 and S_2.

Finally, the net generation-consumption term will be assumed to be due to a chemical reaction r_A:

$$\text{net generation-consumption} = \dot{r}_A \Delta t$$

where $\dot{r}_A$ is the net rate of generation-consumption of component A by chemical reaction.

Introduction of all these terms into Eq. (7.1) gives Eq. (7.2) Equations (7.3) and (7.4) can be developed from exactly the same type of analysis. Try to formulate Eqs. (7.3) and (7.4) yourself.

Species material balance for species A:

$$\rho_A V|_{t+\Delta t} - \rho_A V|_t = \rho_A \upsilon S \Delta t|_{S_1} - \rho_A \upsilon S \Delta t|_{S_2} + \dot{w}_A \Delta t + \dot{r}_A \Delta t \qquad (7.2)$$

accumulation   transport through defined   transport   genera-
                          boundaries   through   tion or
                                                          other   consump-
                                                          boundaries   tion

Total material balance:

$$\rho V|_{t+\Delta t} - \rho V|_t = \rho \upsilon S \, \Delta t|_{S_1} - \rho \upsilon S \, \Delta t|_{S_2} + \dot{w} \, \Delta t \tag{7.3}$$

accumulation   transport through defined   transport
                              boundaries   through
                                                        other
                                                        boundaries

By an analogous procedure we can formulate the energy balance:

$$E|_{t+\Delta t} - E|_t = \left|\left(\hat{H} + \frac{\upsilon^2}{2} + gh\right)\dot{m}\,\Delta t\right|_{S_1} - \left|\left(\hat{H} + \frac{\upsilon^2}{2} + gh\right)\dot{m}\,\Delta t\right|_{S_2}$$

accumulation   transport through defined boundaries

$$+ \dot{Q}\,\Delta t + \dot{W}\,\Delta t + \dot{B}\,\Delta t \tag{7.4}$$

heat   work   transport
                    through
                    other
                    boundaries

where $\dot{B}$ = rate of energy transfer accompanying $\dot{w}$
       $\dot{m}$ = rate of mass transfer through defined surfaces S_1 in and S_2 out,
            respectively
       $\dot{Q}$ = rate of heat transfer to system
       $\dot{W}$ = rate of work done on the system
       $\dot{w}$ = rate of total mass flow through the system boundaries other than
            through the defined surfaces S_1 and S_2

The other notation for the material and energy balances is identical to that of Chapters 3 and 5; note that the work, heat, generation, and mass transport are now all expressed as rate terms (mass or energy per unit time).

If each side of Eq. (7.2) is divided by Δt we obtain

$$\frac{\rho_A V|_{t+\Delta t} - \rho_A V|_t}{\Delta t} = \rho_A \upsilon S|_{S_1} - \rho_A \upsilon S|_{S_2} + \dot{w}_A + \dot{r}_A \tag{7.5}$$

Similar relations can be obtained from Eqs. (7.3) and (7.4). If we take the limit of each side of Eq. (7.5) as $\Delta t \to 0$, we get a differential equation

$$\frac{d(\rho_A V)}{dt} = -\Delta(\rho_A \upsilon S) + \dot{w}_A + \dot{r}_A \tag{7.6}$$

and analogously the total mass balance and the energy balance, respectively, are

$$\frac{d(\rho V)}{dt} = -\Delta(\rho \upsilon S) + \dot{w} \tag{7.7}$$

$$\frac{\mathrm{d}(E)}{\mathrm{d}t} = \dot{Q} + \dot{W} + \dot{B} - \Delta\left[\left(\hat{H} + \frac{v^2}{2} + gh\right)\dot{m}\right] \tag{7.8}$$

Can you get Eqs. (7.7) and (7.8) from (7.3) and (7.4), respectively? Try it.

The relation between the energy balance given by Eq. (7.8), which has the units of energy per unit time, and the energy balance given by Eq. (5.13),which has the units of energy, should now be fairly clear. Equation (5.13) represents the integration of Eq. (7.8) with respect to time, expressed formally as follows:

$$E_{t_2} - E_{t_1} = \int_{t_1}^{t_2} \left\{\dot{Q} + \dot{B} + \dot{W} - \Delta\left[\left(\hat{H} + \hat{K} + \hat{P}\right)\dot{m}\right]\right\} \mathrm{d}t \tag{7.9}$$

The quantities designated in Eqs. (5.13) without the superscript dot are the respective integrated values from Eq. (7.9).

To solve one or any combination of the very general equations (7.6), (7.7), or (7.8) analytically may be quite difficult, and in the following examples we shall have to restrict our analyses to simple cases. If we make enough (reasonable) assumptions and work with simple problems, we can consolidate or eliminate enough terms of the equations to be able to integrate them and develop some analytical answers. If analytical solutions are not possible, then a computer code can be used to get a numerical solution for a specific case.

In the formulation of unsteady-state equations you apply the usual procedures of problem solving discussed in previous chapters. You can set up the equation as in Eqs. (7.2)–(7.4) or use the differential equations (7.5)–(7.7) directly. To complete the problem formulation you must include some known value of the dependent variable (or its derivative) at a specified time, usually the **initial condition**.

We are now going to examine some very simple unsteady-state problems. You can (and will) find more complicated examples in texts dealing with all phases of mass transfer, heat transfer, and fluid dynamics.

EXAMPLE 7.1 Unsteady-State Material Balance without Generation

A tank holds 100 gal of a water-salt solution in which 4.0 lb of salt is dissolved. Water runs into the tank at the rate of 5 gal/min and salt solution overflows at the same rate. If the mixing in the tank is adequate to keep the concentration of salt in the tank uniform at all times, how much salt is in the tank at the end of 50 min? Assume that the density of the salt solution is essentially the same as that of water.

Solution

We shall set up from scratch the differential equations that describe the process.
    Step 1   Draw a picture, and put down the known data. See Figure E7.1.

Figure E7.1

Step 2   Choose the independent and dependent variables. Time, of course, is the independent variable, and either the salt quantity or concentration of salt in the tank can be the dependent variable. Suppose that we make the mass (quantity) of salt the dependent variable. Let x = lb of salt in the tank at time t.

Step 3   Write the known value of x at a given value of t. This is the initial condition:

$$\text{at } t = 0 \quad x = 4.0 \text{ lb}$$

Step 4   It is easiest to make a total material balance and a component material balance on the salt. (No energy is needed because the system can be assumed to be isothermal.)

Total balance:

accumulation   **in**

$$[m_{\text{tot}} \text{ lb}]_{t+\Delta t} - [m_{\text{tot}} \text{ lb}]_t \quad = \quad \frac{5 \text{ gal}}{\text{min}} \left| \frac{1 \text{ ft}^3}{7.48 \text{ gal}} \right| \frac{\rho_{\text{H}_2\text{O}} \text{ lb}}{\text{ft}^3} \right| \Delta t \text{ min}$$

out

$$- \frac{5 \text{ gal}}{\text{min}} \left| \frac{1 \text{ ft}^3}{7.48 \text{ gal}} \right| \frac{\rho_{\text{soln}} \text{ lb}}{\text{ft}^3} \right| \Delta t \text{ min}$$

This equation tells us that the flow of water into the tank equals the flow of solution out of the tank if $\rho_{\text{H}_2\text{O}} = \rho_{\text{soln}}$ as assumed.

Salt balance:

accumulation   **in**   **out**

$$[x \text{ lb}]_{t+\Delta t} - [x \text{ lb}]_t \quad = \quad 0 \quad - \quad \frac{5 \text{ gal}}{\text{min}} \left| \frac{x \text{ lb}}{100 \text{ gal}} \right| \Delta t \text{ min}$$

Dividing by Δt and taking the limit as Δt approaches zero,

$$\lim_{\Delta t \to 0} \frac{[x]_{t+\Delta t} - [x]_t}{\Delta t} = -0.05x$$

or

$$\frac{dx}{dt} = -0.05x \tag{a}$$

Notice how we have kept track of the units in the normal fashion in setting up these equations. Because of our assumption of uniform concentration of salt in the tank, the concentration of salt leaving the tank is the same as that in the tank, or x lb/100 gal of solution.

Step 5   Solve the unsteady-state material balance on the salt. By separating the independent and dependent variables we get

$$\frac{dx}{x} = -0.05 \, dt$$

This equation is easily integrated between the definite limits of

$$
\begin{array}{ll}
t = 0 & x = 4.0 \\
t = 50 & X = \text{the unknown value of } x \text{ lb}
\end{array}
$$

$$\int_{4.0}^{X} \frac{dx}{x} = -0.05 \int_{0}^{50} dt$$

$$\ln \frac{X}{4.0} = -2.5 \qquad\qquad X = \frac{4.0}{12.2} = 0.328 \text{ lb salt}$$

An equivalent differential equation to Eq. (a) can be obtained directly from the component mass balance in the form of Eq. (7.6) if we let ρ_A = concentration of salt in the tank at any time t in terms of lb/gal:

$$\frac{d(\rho_A V)}{dt} = -\left(\frac{5 \text{ gal}}{\text{min}} \, \bigg| \, \frac{\rho_A \text{ lb}}{\text{gal}} \right) - 0$$

If the tank holds 100 gal of solution at all times, V is a constant and equal to 100, so that

$$\frac{d\rho_A}{dt} = -\frac{5\rho_A}{100} \tag{b}$$

The initial conditions are

$$\text{at } t = 0 \quad \rho_A = 0.04$$

The solution of Eq. (b) would be carried out exactly as the solution of Eq. (a).

EXAMPLE 7.2 Unsteady-State Material Balance without Generation

A square tank 4 m on a side and 10 m high is filled to the brim with water. Find the time required for it to empty through a hole in the bottom 5 cm² in area.

Solution

Step 1   Draw a diagram of the process, and out down the data. See Figure E7.2a.

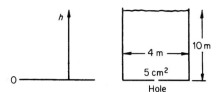

Figure E7.2a

Step 2   Select the independent and dependent variables. Again, time will be the independent variable. We could select the quantity of water in the tank as the dependent variable, but since the cross section of the tank is constant, let us choose h, the height of the water in the tank, as the dependent variable.

Step 3   Write the known value of h at the given time t:

$$\text{at } t = 0 \;\; h = 10 \text{ m}$$

Step 4   Develop the unsteady-state balance(s) for the process. In an elemental of time Δt, the height of the water in the tank drops Δh. The mass of water leaving the tank is in the form of a cylinder 5 cm² in area, and we can calculate the quantity as

$$\frac{5 \text{ cm}^2}{} \left| \left(\frac{1 \text{ m}}{100 \text{ cm}}\right)^2 \right| \frac{v * \text{ m}}{\text{s}} \left| \frac{\rho \text{ kg}}{\text{m}^3} \right| \frac{\Delta t \text{ s}}{} = 5 \times 10^{-4} v * \rho \, \Delta t \text{ kg}$$

where ρ = density of water

$v*$ = average velocity of the water leaving the tank

The depletion of water inside the tank in terms of the variable h, expressed in kg,

is

$$\frac{16 \text{ m}^2}{} \left| \frac{\rho \text{ kg}}{\text{m}^3} \right| h \text{ m} \Big|_{t+\Delta t} - \frac{16 \text{ m}^2}{} \left| \frac{\rho \text{ kg}}{\text{m}^3} \right| h \text{ m} \Big|_{t} = 16\rho\Delta h \;\; \text{kg}$$

An overall material balance is

accumulation   in   out

$$16\rho\Delta h \;\; = \;\; 0 \;\; - \;\; 5 \times 10^{-4} \rho v * \Delta t \tag{a}$$

Note Δh has a negative value, but our equation takes account of this automatically; the accumulation is really a depletion. You can see that the term ρ, the density of water, cancels out, and we could just have well made our material balance on a volume of water.

Equation (a) becomes

$$\frac{\Delta h}{\Delta t} = -\frac{5 \times 10^{-4} v *}{16}$$

Taking the limit as Δh and Δt approach zero, we get the differential equation

$$\frac{dh}{dt} = -\frac{5 \times 10^{-4} v*}{16} \qquad \text{(b)}$$

Step 5 Unfortunately, this is an equation with one independent variable, t, and two dependent variables, h and $v*$. We must find another equation to eliminate either h or $v*$ if we want to obtain a solution. Since we want our final equation to be expressed in terms of h, the next step is to find some function to relate $v*$ to h and t, and then we can substitute for $v*$ in the unsteady-state equation, Eq. (b).

We shall employ the steady-state mechanical energy balance equation for an incompressible fluid, discussed in Sect. 5.5, to relate $v*$ and h. See Figure E7.2b. With $W = 0$ and $E_v = 0$ the steady state mechanical energy balances reduce to

$$\Delta\left(\frac{v^2}{2} + gh\right) = 0 \qquad \text{(c)}$$

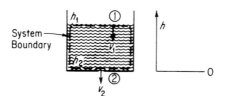

Figure E7.2b

We assume that the pressures are the same at section 1—the water surface—and section 2—the hole—for the system consisting of the water in the tank. Equation (c) can be rearranged to

$$\frac{v_2^2 - v_1^2}{2} + g(h_2 - h_1) = 0 \qquad \text{(d)}$$

where v_2 = exit velocity through the 5 cm^2 hole at boundary 2
        v_1 = velocity of water in the tank at boundary 1

If $v_1 \cong 0$, a reasonable assumption for the water in the large tank at any time, at least compared to v_2, we have

$$v_2^2 = -2g(0 - h_1) = 2gh$$

$$v_2 = c\sqrt{2gh} \qquad \text{(e)}$$

Because the exit-stream flow is not frictionless and because of turbulence and orifice effects in the exit hole, we must correct the value of v given by Eq. (e) for frictionless flow by an empirical adjustment factor as follows:

$$v_2^* = c\sqrt{2gh} \qquad \text{(f)}$$

where c is an orifice correction that we could find (from a text discussing fluid dynamics) with a value of 0.62 for this case. Thus

$$v_2^* = 0.62\sqrt{2(9.80)h} = 2.74\sqrt{h} \ \text{m/s}$$

Let us substitute this relation into Eq. (b) in place of v^*. Then we obtain

$$\frac{dh}{dt} = -\frac{\left(5.0\times10^{-4}\right)(2.74)(h)^{1/2}}{16} \tag{g}$$

Equation (g) can be integrated between

$$h = 10 \ \text{m at} \ t = 0$$

and

$$h = 0 \ \text{m at} \ t = \theta, \quad \text{the unknown time}$$

$$-1.17\times10^4 \int_{10}^{0} \frac{dh}{h^{1/2}} = \int_{0}^{\theta} dt$$

to yield θ

$$\theta = 1.17\times10^4 \int_{0}^{10} \frac{dh}{h^{1/2}} = 1.17\times10^4 \left[2\sqrt{h}\right]_{0}^{10} = 7.38\times10^4 \ \text{s}$$

Now suppose that in addition to the loss of fluid through the hole in the bottom of the tank additional fluid is poured continuously into the top of the tank at varying rates. Numerical integration of the differential equations yields a varying height of fluid as illustrated in Figure E7.2c.

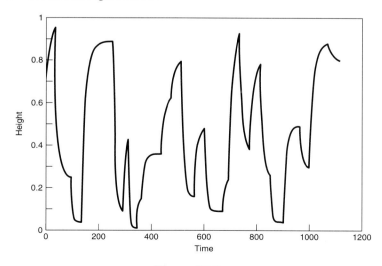

Figure E7.2c

EXAMPLE 7.3 Material Balance in Batch Distillation

A small still is separating propane and butane at 135°C, and initially contains 10 kg moles of a mixture whose composition is $x = 0.30$ (x = mole fraction butane). Additional mixture ($x_F = 0.30$) is fed at the rate of 5 kg mol/hr. If the total volume of the liquid in the still is constant, and the concentration of the vapor from the still (x_D) is related to x_S as follows:

$$x_D = \frac{x_S}{1 + x_S} \tag{a}$$

how long will it take for the value of x_S to change from 0.3 to 0.4? What is the steady-state ("equilibrium") value of x_S in the still (i.e., when x_S becomes constant?) See Figure E7.3

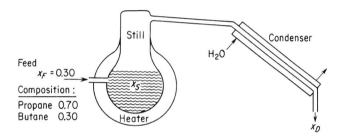

Figure E7.3

Solution

Since butane and propane form ideal solutions, we do not have to worry about volume changes on mixing or separation. Only the material balance is needed to answer the questions posed. If t is the independent variable and x_S is the dependent variable, we can say that

Butane Balance (C_4): The input to the still is:

$$\frac{5 \text{ mol feed}}{\text{hr}} \left| \frac{0.30 \text{ mol } C_4}{\text{mol feed}} \right.$$

The out put from the still is equal to the amount condensed

$$\frac{5 \text{ mol cond.}}{\text{hr}} \left| \frac{x_D \text{ mol } C_4}{\text{mol condensed}} \right.$$

The accumulation is $\dfrac{10 \, dx_S}{dt}$

Our unsteady state equation is then

accumulation   in   out

$$\frac{dx_S}{dt} = 0.15 - 0.5x_D \qquad (b)$$

As in Example 7.2, it is necessary to reduce the equation to one dependent variable by substituting for x_D from Eq. (a) in Eq. (b)

$$\frac{dx_S}{dt} = 0.15 - \frac{x_S}{1 + x_S}(0.5) \qquad (c)$$

Integration of Eq. (c) between the following limits

$$at\ t = 0 \qquad x_S = 0.30$$
$$t = \theta \qquad x_S = 0.40$$

$$\int_{0.30}^{0.40} \frac{dx_S}{0.15 - [0.5x_S/(1 + x_S)]} = \int_0^\theta dt = \theta$$

$$\int_{0.30}^{0.40} \frac{(1 + x_S)\, dx_S}{0.15 - 0.35x_S} = \theta = \left[-\frac{x_S}{0.35} - \frac{1}{(0.35)^2} \ln(0.15 - 0.35x_S) \right]_{0.30}^{0.40}$$

gives $\theta = 5.85$ hr

If you only had experimental data for x_D as a function of x_S instead of the given equation, you could always integrate the equation numerically.

The steady-state value of x_S is established at infinite time or, alternatively, when the accumulation is zero. At that time,

$$0.15 = \frac{0.5x_S}{1 + x_S} \qquad \text{or} \qquad x_S = 0.428$$

The value of x_S could never be any greater than 0.428 for the given conditions.

EXAMPLE 7.4 Oscillating Reactions

In an isothermal reactor the following reactions take place

$$A + X \xrightarrow{k_1} 2X$$

$$X + Y \xrightarrow{k_2} 2Y$$

$$Y \xrightarrow{k_3} B$$

where A is the initial reactant, X and Y are intermediate species, and B is the final product of the reaction. A is maintained at a constant value by starting with so much A that only X, Y, and B vary with time.

Develop the unsteady state material balance equations that predict the change of X and Y as a function of time for the initial conditions $c_X(0) = 30$ and $c_Y(0) = 90$ (c designates concentration).

Solution

The macroscopic balance for both species X and Y, is

accumulation = in − out + generation − consumption

For a batch (closed) system, the reactor, the in and out terms are zero. By following the steps outlined in the previous problems, the generation and consumption terms are

$$\begin{array}{ccc} & \underline{\text{X}} & \underline{\text{Y}} \\ \text{generation:} & k_1 c_A c_X & k_2 c_X c_Y \\ \text{consumption:} & k_2 c_X c_Y & k_3 c_Y \end{array}$$

and the derivatives represent the accumulation. We can merge $k_1 c_A$ into a constant k_1^* since c_A is constant. Then the differential equations for X and Y are

$$\frac{dc_X}{dt} = k_1^* c_X - k_2 c_X c_Y \tag{a}$$

$$\frac{dc_Y}{dt} = k_2 c_X c_Y - k_3 c_Y \tag{b}$$

Figure E7.4a shows the concentrations of X and Y for the values $k_1^* = 70$, $k_2 = 1$, and $k_3 = 70$ when (a) and (b) are solved using a computer code.

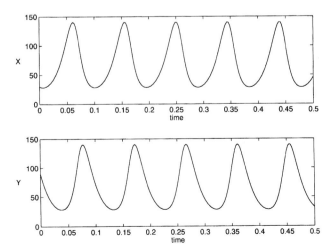

Figure E7.4a

In the steady state, the intermediates exhibit an interesting phenomena. Divide Eq. (b) by Eq. (a) to get

$$\frac{dc_Y}{dc_X} = \frac{c_Y\left[k_2 c_X - k_3\right]}{c_X\left[k_1^* - k_2 c_Y\right]}$$

which can be arranged to

$$\frac{\left(k_1^* - k_2 c_Y\right)}{c_Y} dc_Y = \frac{\left(k_2 c_X - k_3\right)}{c_X} dc_X \tag{c}$$

Integration of Eq. (c) yields

$$k_1^* \ln c_Y - k_2 c_Y + k_3 \ln c_X - k_2 c_X = \text{constant} \tag{d}$$

Eq. (d) represents a series of closed loops when c_Y is plotted against c_X; the constant can be evaluated from the initial conditions for c_X and c_Y. Examine Figure E7.4b.

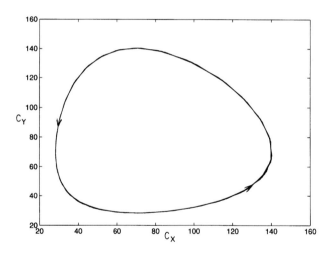

Figure E7.4b

EXAMPLE 7.5 Unsteady-State Energy Balance

Oil initially at 60°F is being heated in a stirred (perfectly mixed) tank by saturated steam which is condensing in the steam coils at 40 psia. If the rate of heat transfer is given by Newton's heating law,

$$\dot{Q} = \frac{dQ}{dt} = h\left(T_{\text{steam}} - T_{\text{oil}}\right)$$

where Q is the heat transferred in Btu and h is the heat transfer coefficient in the proper units, how long does it take for the discharge from the tank to rise from 60°F to 90°F? What is the maximum temperature that can be achieved in the tank?

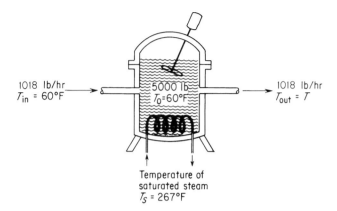

1018 lb/hr
$T_{in} = 60°F$

5000 lb
$T_0 = 60°F$

1018 lb/hr
$T_{out} = T$

Temperature of
saturated steam
$T_S = 267°F$

Figure E7.5

Additional Data:

Motor horsepower	=	1 hp; efficiency is 0.75
Initial amount of oil in tank	=	5000 lb
Entering oil flow rate	=	1018 lb/hr at a temperature of 60°F
Discharge oil flow rate	=	1018 lb/hr at a temperature of T
h	=	291 Btu/ (hr)(°F)
$C_{p\,oil}$	=	0.5 Btu/ (lb)(°F)

Solution

The process is shown in Figure E7.5. The system is the oil in the tank. The independent variable will be t, the time; the dependent variable will be the temperature of the oil in the tank, which is the same as the temperature of the oil discharged. The material balance is not needed because the process, insofar as the quantity of oil is concerned, is assumed to be in the steady-state.

The first step is to set up the unsteady-state energy balance. Let T_S = the steam temperature and T = the oil temperature. The balance per unit time is

accumulation = input − output

$$\frac{\mathrm{d}E}{\mathrm{d}t} = \frac{\mathrm{d}U}{\mathrm{d}t} = Q - \Delta H$$

The rate of change of energy inside the tank is nothing more than the rate of change of internal energy, which, in turn, is essentially the same as the rate of change of enthalpy (i.e., $\mathrm{d}E/\mathrm{d}t = \mathrm{d}U/\mathrm{d}t = \mathrm{d}H/\mathrm{d}t$) because $\mathrm{d}(pV)/\mathrm{d}t \cong 0$.

A good choice for a reference temperature for the enthalpies is 60°F because the choice makes the input enthalpy zero.

$$
\left.
\begin{array}{l}
\begin{array}{ll}
\text{rate of enthalpy of} & \dfrac{1018\ \text{lb}}{\text{hr}}\left|\dfrac{0.5\ \text{Btu}}{(\text{lb})(°\text{F})}\right|\dfrac{\overset{60°\text{F}-60°\text{F}}{(T_{in}-T_{ref})}}{} = 0 \\
\text{the input stream} &
\end{array} \\[2em]
\begin{array}{ll}
\text{rate of} & h\,(T_s - T) = \dfrac{291\ \text{Btu}}{(\text{hr})(°\text{F})}\left|(267 - T)°\text{F}\right. \\
\text{heat transfer} &
\end{array}
\end{array}
\right\}\quad \textbf{input (Btu)}
$$

$$
\begin{array}{ll}
\text{rate of enthalpy of} & \dfrac{1018\ \text{lb}}{\text{hr}}\left|\dfrac{0.5\ \text{Btu}}{(\text{lb})(°\text{F})}\right|(T - 60)°\text{F} \\
\text{the output stream} &
\end{array}\qquad \textbf{output (Btu)}
$$

$$
\begin{array}{ll}
\text{rate of enthalpy change} & \dfrac{5000\ \text{lb}}{}\left|\dfrac{0.5\ \text{Btu}}{(\text{lb})(°\text{F})}\right|\dfrac{dT\ (°\text{F})}{dt\ (\text{hr})} \\
\text{inside the tank} &
\end{array}\qquad \textbf{accumulation (Btu)}
$$

The energy introduced by the motor enters the tank as rate of work, $\dot{W}$.

$$
\text{rate of work}\qquad \dot{W} = \dfrac{3\ \text{hp}}{4}\left|\dfrac{0.7608\ \text{Btu}}{(\text{sec})(\text{hp})}\right|\dfrac{3600\ \text{sec}}{1\ \text{hr}} = \dfrac{1910\ \text{Btu}}{\text{hr}}
$$

Then

$$
2500 = 291(267 - T) - 509(T - 60) + 1910
$$

$$
\frac{dT}{dt} = 44.1 - 0.32T
$$

$$
\int_{60}^{90} \frac{dT}{44.1 - 0.32T} = \int_0^{\theta} dt = \theta
$$

$$
\theta = 1.52\ \text{hr.}
$$

EXAMPLE 7.6 Modeling a Calcination Process

The process flowsheet illustrated in Figure E7.6a is a fluidized bed in which granular solids are calcinated. Water vapor and gaseous decomposition products leave the bed with the fluidizing gas while metallic oxide products deposit on the particles in the fluidized bed. In Figure E7.6a, x is the mass concentration of the component of interest; F is the mass flow rate of the feed; x_0 is the concentration of x in F; Q is the mass flow rate of stream 1 or 2, respectively; P is the mass flow rate of product; and V is the mass of inventory at time t in vessel 1 or 2, respectively.

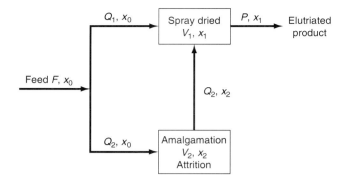

Figure E7.6a

The mass balances for each vessel after introduction of x_0 into stream F become: accumulation = in – out if we ignore any reactions

Vessel 1

Total mass

$$\frac{dV_1}{dt} = Q_1 + Q_2 - P = 0 \tag{a}$$

Component mass

$$V_1 \frac{dx_1}{dt} = Q_1 x_0 + Q_2 x_2 - P x_1 \tag{b}$$

Initial condition

$$x_1(0) = 0$$

Vessel 2

Total mass

$$\frac{dV_2}{dt} = Q_2 - Q_2 = 0 \tag{c}$$

Component mass

$$V_2 \frac{dx_2}{dt} = Q_2 x_0 - Q_2 x_2 \tag{d}$$

Initial condition

$$x_2(0) = 0$$

After x_0 is removed from stream F, the concentration of x_0 becomes 0 in the model.

Figure E7.6b shows the value of x_1 as a function of time t for the values $P = 16.5$ g/min, $Q_2 = 6.5$ g/min, $V_1 = 600$ g, and $V_2 = 2.165 \times 10^4$ g when the component of interest is added and then removed from the feed F, and the stated values of the parameters were obtained from experiments.

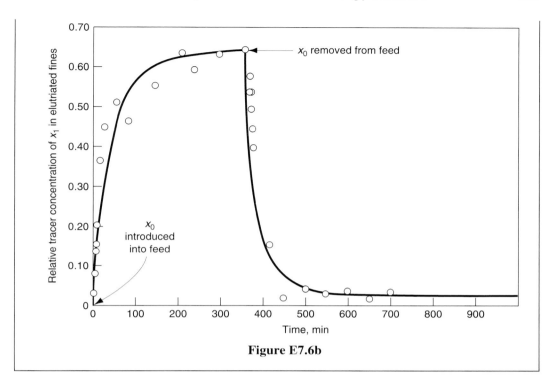

Figure E7.6b

LOOKING BACK

In the section we explained how to formulate unsteady-state material and energy balances in which the variation with time is of interest.

Key Ideas

1. Unsteady-state material and energy balances in this chapter are formulated in terms of differential quantities in contrast with the integrated quantities in previous chapters.

2. When so formulated, you have to solve differential equations rather than algebraic equations.

3. The concept underlying the equations is still the same: accumulation = in – out + generation – consumption.

Key Words

Initial condition (p. 608)   Transient state (p. 604)
Lumped balances (p. 605)   Unsteady-state (p. 604)
Macroscopic balances (p. 605)   Well mixed (p. 605)

Self-Assessment Test

1. In a batch type of chemical reactor, do we have to have the starting conditions to predict the yield?

2. Is time the independent or dependent variable in macroscopic unsteady-state equations?

3. How can you obtain the steady-state balances from Eqs. (7.6)–(7.8)?

4. Which of the following plots in Figure SA7.1–4 of response vs. time are not transient-state processes?

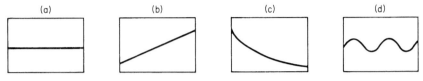

Figure SA7.1–4

5. Group the following words in sets of synonyms: (1) response, (2) input variable, (3) parameter, (4) state variable, (5) system parameter, (6) initial condition, (7) output, (8) independent variable, (9) dependent variable, (10) coefficient, (11) output variable, (12) constant.

6. A chemical inhibitor must be added to the water in a boiler to avoid corrosion and scale. The inhibitor concentration must be maintained between 4 and 30 ppm. The boiler system always contains 100,000 kg of water and the blowdown (purge) rate is 15,000 kg/hr. The makeup water contains no inhibitor. An initial 2.8 kg of inhibitor is added to the water and thereafter 2.1 kg is added periodically. What is the maximum time interval (after the 2.8 kg) until the first addition of the inhibitor?

7. In a reactor a small amount of undesirable byproduct is continuously formed at the rate of 0.5 lb/hr when the reactor contains a steady inventory of 10,000 lb of material. The reactor inlet stream contains traces of the byproduct (40 ppm), and the effluent stream is 1400 lb/hr. If on startup, the reactor contains 5000 ppm of the byproduct, what will be the concentration in ppm at 10 hr?

Thought Problems

1. Refer to the tank in Example 7.2. If the square tank is replaced by (a) a vertical cylindrical tank, (b) a conical tank with vertex at the bottom, (c) a horizontal cylindrical tank, or (d) a spherical tank with the parameters shown in the following figure, if h and D are the same in each tank at $t = 0$, which tank will be the first to drain completely? See Figure TP7.1–1.

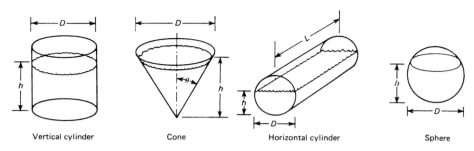

Vertical cylinder   Cone   Horizontal cylinder   Sphere

Figure TP7.1–1

2. Many chemical plant operations require that vessels and piping be inerted or purged. If a vessel is to be opened for maintenance or repair, for example, and if the vessel has contained either toxic or flammable materials, purging is required before workers can enter the vessel. For vessel entry, piping leading to or from the vessel will have to be blanked off and at least the portions of the vessels that are open to the pipe will have to be purged as well. Purging must be continued until the atmosphere in the vessel is safe for entry.

     If the chemical in the vessel is flammable, purging must be accomplished in two steps: first, the flammable material is purged from the vessel with an inert gas such as nitrogen, and the nitrogen is then purged with air. When a vessel that is to be used for storing or processing flammable chemicals is initially put into service, it must be purged with an inert gas before flammable chemical are put in the vessel. This step is required to assure that a flammable mixture of the chemical and the air in the tank does not form.

     Regardless of the method used to purge the equipment, the final step should be to check the atmosphere in the equipment to make certain that the concentration of the flammable or toxic material has been reduced to safe levels. For tank entry, the oxygen concentration must also be checked before the tank is entered. The measurement of residual concentrations is sometimes required by law; it is always needed for good practice.

     Suppose that a 150-ft^3 tank containing air is to be inerted to a 1% oxygen concentration. Pure nitrogen is available for the job. The tank has a maximum allowable working pressure of 150 psia, so either of two methods is possible. In the first method, air is purged by a continuous sweep of nitrogen. The nitrogen is simply allowed to flow into the tank at essentially atmospheric pressure. It is assumed that the nitrogen mixes rapidly and completely with the air in the tank, so the gas leaving the tank has the same concentration of oxygen as the gas in the tank.

     In the second technique, the tank is pressurized, the pure nitrogen inlet stream is turned off, and the gas mixture in the tank is then exhausted, lowering the pressure in the tank to atmospheric pressure. If the pressurization technique is used, multiple pressurization cycles may be required, with the tank returned to atmospheric pressure at the end of each cycle. Complete mixing is assumed for each cycle.

In this problem, you may assume that both nitrogen and air behave as ideal gases and that the temperature remains constant at 80°F throughout the process. Determine the volume of nitrogen (measured at standard conditions of 1.0 atm and 0°C) required to purge the tank using each purging technique. For the pressurization technique, assume the pressure in the tank is raised to 140 psig (a little below its maximum working pressure) with nitrogen and then vented to 0 psig. This problem was adapted from the publication *Safety, Health, and Loss Prevention in Chemical Processes*, American Institute of Chemical Engineers, New York, 1990.

SUPPLEMENTARY REFERENCES

BUNGAY, H. R. *Computer Games and Simulation for Biochemical Engineers*. New York: Wiley, 1985.

CLEMENTS, W. C. *Unsteady-State Balances*. AIChE Chemi Series No. F5.6. New York: American Institute of Chemical Engineers, 1986.

HIMMELBLAU, D. M., and K. B. BISCHOFF. *Process Analysis and Simulation*. Autsin TX: Swift Publishing Co., 1980.

JENSON, V. G., and G. V. JEFFREYS. *Mathematical Methods in Chemical Engineering*. New York: Academic Press, 1963.

PORTER, R. L. *Unsteady-State Balances—Solution Techniques for Ordinary Differential Equations*. AIChE Chemi Series No. F5.5. New York: American Institute of Chemical Engineers, 1986.

RIGGS, J. B. *An Introduction to Numerical Methods for Chemical Engineers*, 2nd ed. Lubbock, TX: Texas Tech University Press, 1994.

RUSSEL, T. W. F., and M. M. DENN. *Introduction to Chemical Engineering Analysis*. New York: American Institute of Chemical Engineers, 1968.

PROBLEMS

7.1. A tank containing 100 kg of a 60% brine (60% salt) is filled with a 10% salt solution at the rate of 10 kg/min. Solution is removed from the tank at the rate of 15 kg/min. Assuming complete mixing, find the kilograms of salt in the tank after 10 min.

7.2. A defective tank of 1500 ft³ volume containing 100% propane gas (at 1 atm) is to be flushed with air (at 1 atm) until the propane concentration reduced to less than 1%. At that concentration of propane the defect can be repaired by welding. If the flow rate of air into the tank is 30 ft³/min, for how many minutes must the tank be flushed out? Assume that the flushing operation is conducted so that the gas in the tank is well mixed.

7.3. A 2% uranium oxide slurry (2 lb UO_2/100 lb H_2O) flows into a 100-gal tank at the rate of 2 gal/min. The tank initially contains 500 lb of H_2O and no UO_2. The slurry is well mixed and flows out at the same rate at which it enters. Find the concentration of slurry in the tank at the end of 1 hr.

7.4. The catalyst in a fluidized-bed reactor of 200-m^3 volume is to be regenerated by contact with a hydrogen stream. Before the hydrogen can be introduced in the reactor, the O_2 content of the air in the reactor must be reduced to 0.1%. If pure N_2 can be fed into the reactor at the rate of 20 m^3/min, for how long should the reactor be purged with N_2? Assure that the catalyst solids occupy 6% of the reactor volume and that the gases are well mixed.

7.5. An advertising firm wants to get a special inflated sign out of a warehouse. The sign is 20 ft in diameter and is filled with H_2 at 15 psig. Unfortunately, the door to the warehouse is only 19 ft high by 20 ft wide. The maximum rate of H_2 that can be safely vented from the balloon is 5 ft^3/min (measured at room conditions). How long will it take to get the sign small enough to just pass through the door?
   (a) First assume that the pressure inside the balloon is constant so that the flow rate is constant.
   (b) Then assume the amount of amount of H_2 escaping is proportional to the volume of the balloon, and initially is 5 ft^3 min.
   (c) Could a solution to this problem be obtained if the amount of escaping H_2 were pro-portional to the pressure difference inside and outside the balloon?

7.6. A plant at Canso, Nova Scotia, makes fish-protein concentrate (FPC). It takes 6.6 kg of whole fish to make l kg of FPC, and therein is the problem—to make money, the plant must operate most of the year. One of the operating problems is the drying of the FOC. It dries in the fluidized dryer rate a rate approximately proportional to its moisture content. If a given batch of FPV loses one-half of its initial moisture in the first 15 min, how long will it take to remove 90% of the water in the batch of FPC?

7.7. Water flows from a conical tank at the rate of 0.020(2 +h^2) m^3/min, as shown in Figure P7.7. If the tank is initially full, how long will it take for 75% of the water to flow out of the tank? What is the flow rate at that time?

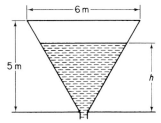

                                                        Figure P7.7

7.8. A sewage disposal plant has a big concrete holding tank of l00,000 gal capacity. It is three-fourths full of liquid to start with and contains 60,000 lb of organic material in suspension. Water runs into the holding tank at the rate of 20,000 gal/hr and the solution leaves at the rate of 15,000 gal/hr. How much organic material is in the tank at the end of 3 hr?

7.9. Suppose that in problem 7.8 the bottom of the tank is covered with sludge (precipitated organic material) and that the stirring of the tank causes the sludge to go into suspension at a rate proportional to the difference between the concentration of sludge in the tank at any time and 10 lb of sludge/gal. If no organic material were present, the sludge would go into suspension at the rate of 0.05 lb/(min) (gal solution) when 75,000 gal of solution are in the tank. How much organic material is in the tank at the end of 3 hr?

7.10. In a chemical reaction the products X and Y are formed according to the equation

$$C \rightarrow X + Y$$

The rate at which each of these products is being formed is proportional to the amount of C present. Initially: $C = 1$, $X = 0$, $Y = 0$. Find the time for the amount of X to equal the amount of C.

7.11. A tank is filled with water. At a given instant two orifices in the side of the tank are opened to discharge the water. The water at the start is 3 m deep and one orifice is 2 m below the top while the other one is 2.5 m below the top. The coefficient of discharge of each orifice is known to be 0.61. The tank is a vertical right circular cylinder 2 m in diameter. The upper and lower orifices are 5 and 10 cm in diameter, respectively. How long will be required for the tank to be drained so that the water level is at a depth of 1.5 m?

7.12. Suppose that you have two tanks in series, as diagrammed in Figure P7.12. The volume of liquid in each tank remains constant because of the design of the overflow lines. Assume that each tank is filled with a solution containing 10 lb of A, and that the tanks each contain 100 gal of aqueous solution of A. If fresh water enters at the rate of 10 gal/hr, what is the concentration of A in each tank at the end of 3 hr? Assume complete mixing in each tank and ignore any change of volume with concentration.

7.13. A well-mixed tank has a maximum capacity of 100 gal and it is initially half full. The discharge pipe at the bottom is very long and thus it offers resistance to the flow of water through it. The force that causes the water to flow is the height of the water in the tank, and in fact that flow is just proportional to the height. Since the height is proportional to the total volume of water in the tank, the volumetric flow rate of water out, q_o, is

$$q_o = kV$$

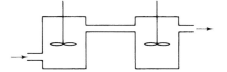

Figure P7.12

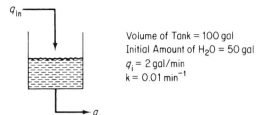

Volume of Tank = 100 gal
Initial Amount of H_2O = 50 gal
q_i = 2 gal/min
k = 0.01 min^{-1}

Figure P7.13

The flow rate of water into the tank, q_i, is constant. Use the information given in Figure P7.13 to decide whether the amount of water in the tank increases, decreases, or remains the same. If it changes, how much time is required to completely empty or fill the tank, as the case may be?

7.14. A stream containing a radioactive fission product with a decay constant of 0.01 hr^{-1} (i.e., $dn/dt = 0.01\ n$), is run into a holding tank at the rate of 100 gal/hr for 24 hr. Then the stream is shut off for 24 hr. If the initial concentration of the fission product was 10 mg/liter and the tank volume is constant at 10,000 gal of solution (owing to an overflow line), what is the concentration of fission product:
(a) At the end of the first 24-hr period?
(b) At the end of the second 24-hr period?
What is the maximum concentration of fission product? Assume complete mixing in the tank.

7.15. A radioactive waste that contains 1500 ppm of ^{92}Sr is pumped at the rate of 1.5×10^{-3} m^3/min into a holding tank that contains 0.40 m^3. ^{92}Sr decays as follows:

$$^{92}\text{Sr} \longrightarrow\ ^{92}\text{Y} \longrightarrow\ ^{92}\text{Zr}$$
$$\text{half-life: 2.7 hr} \qquad \text{3.5 hr}$$

If the tank contains clear water initially and the solution runs out at the rate of 1.5×10^{-3} m^3/min, assuming perfect mixing:
(a) What is the concentration of Sr, Y, and Zr after 1 day?
(b) What is the equilibrium concentration of Sr and Y in the tank?
The rate of decay of such isotopes is $dN/dt = -\lambda N$, where $\lambda = 0.693/t_{1/2}$ and the half-life is $t_{1/2}$. N = moles.

7.16. A cylinder contains 3 m^3 of pure oxygen at atmospheric pressure. Air is slowly pumped into the tank and mixes uniformly with the contents, an equal volume of which is forced out of the tank. What is the concentration of oxygen in the tank after 9 m^3 of air has been admitted?

7.17. Suppose that an organic compound decomposes as follows:

$$C_6H_{12} \rightarrow C_4H_8 + C_2H_4$$

If 1 mol of C_6H_{12} exists at $t = 0$, but no C_4H_8 and C_2H_4, set up equations showing the moles of C_4H_8 and C_2H_4 as a function of time. The rates of formation of C_4H_8 and C_2H_4 are each proportional to the number of moles of C_6H_{12} present.

7.18. A large tank is connected to a smaller tank by means of a valve. The large tank contains N_2 at 690 kPa while the small tank is evacuated. If the valve leaks between the two tanks and the rate of leakage of gas is proportional to the pressure difference between the two tanks $(p_1 - p_2)$, how long does it take for the pressure in the small tank to be one-half its final value? The instantaneous initial flow rate with the small tank evacuated is 0.091 kg mol/hr.

	Tank 1	Tank 2
Initial pressure (kPa)	700	0
Volume (m^3)	30	15

Assume that the temperature in both tanks is held constant and is 20°C.

7.19. The following chain reactions take place in a constant-volume batch tank:

$$A \overset{k_1}{\rightarrow} B \overset{k_2}{\rightarrow} C$$

Each reaction is first order and irreversible. If the initial concentration of A is C_{A_0} and if only A is present initially, find an expression for C_B as a function of time. Under what conditions will the concentration of B be dependent primarily on the rate of reaction of A?

7.20. Consider the following chemical reaction in a constant-volume batch tank:

$$A \underset{k_2}{\overset{k_1}{\rightleftharpoons}} B$$
$$\downarrow k_3$$
$$C$$

All the indicated reactions are first order. The initial concentration of A is C_{A_0}, and nothing else is present at that time. Determine the concentrations of A, B, and C as functions of time.

7.21. Tanks A, B, and C are each filled with 1000 gal of water. See Figure P7.21. Workers have instructions to dissolve 2000 lb of salt in each tank. By mistake, 3000 lb is dissolved in each of tanks A and C and none in B. You wish to bring all the compositions to within 5% of the specified 2 lb/gal. If the units are connected A–B–C–A by three 50-gpm pumps,
   (a) Express the concentrations C_A, C_B, and C_C in terms of t (time).
   (b) Find the shortest time at which all concentrations are within the specified range. Assume the tanks are all well mixed.

7.22. Determine the time required to heat a 10,000-lb batch of liquid from 60°F to 120°F using an external, counterflow heat exchanger having an area of 300 ft^2. Water at 180°F is used as the heating medium and flows at a rate of 6000 lb/hr. An overall heat transfer coefficient of 40 Btu/(hr)(ft^2)(°F) may be assumed; use New-

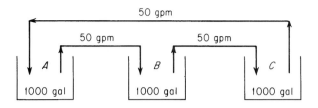

Figure P7.21

ton's law of heating. The liquid is circulated at a rate of 6000 lb/hr, and the specific heat of the liquid is the same as that of water (1.0). Assume that the residence time of the liquid in the external heat exchanger is very small and that there is essentially no holdup of liquid in this circuit.

7.23. A ground material is to be suspended in water and heated in preparation for a chemical reaction. It is desired to carry out the mixing and heating simultaneously in a tank equipped with an agitator and a steam coil. The cold liquid and solid are to be added continuously, and the heated suspension will be withdrawn at the same rate. One method of operation for starting up is to (1) fill the tank initially with water and solid in the proper proportions, (2) start the agitator, (3) introduce fresh water and solid in the proper proportions and simultaneously begin to withdraw the suspension for reaction, and (4) turn on the steam. An estimate is needed of the time required, after the steam is turned on, for the temperature of the effluent suspension to reach a certain elevated temperature.

  (a) Using the nomenclature given below, formulate a differential equation for this process. Integrate the equation to obtain n as a function of B and ϕ (see nomenclature).

  (b) Calculate the time required for the effluent temperature to reach $180°F$ if the initial contents of the tank and the inflow are both at $120°F$ and the steam temperature is $220°F$. The surface area for heat transfer is 23.9 ft^2, and the heat transfer coefficient is $100 \text{ Btu/(hr)(ft}^2)(°F)$. The tank contains 6000 lb, and the rate of flow of both streams is 1200 lb/hr. In the proportions used, the specific heat of the suspension may be assumed to be 1.00.

If the area available heat transfer is doubled, how will the time required be affected? Why is the time with the larger area less than half that obtained previously? The heat transferred is $Q = UA(T_{\text{tank}} - T_{\text{steam}})$.

Nomenclature

W = weight of tank contents, lb
G = rate of flow of suspension, lb/hr
T_S = temperature of steam, °F
T = temperature in tank at any instant, perfect mixing assumed, °F
T_0 = temperature of suspension introduced into tank; also initial temperature of tank contents, °F
U = heat-transfer coefficient, Btu/(hr)(ft^2)(°F)

A = area of heat-transfer surface, ft^2
C_p = specific heat of suspension, Btu/(lb)(°F)
t = time elapsed from the instant the steam is turned on, hr
n = dimensionless time, Gt/W
B = dimensionless ratio, UA/GC_p
ϕ = dimensionless temperature (relative approach to the steam temperature) $(T - T_0)/(T_S - T_0)$

7.24. Consider a well-agitated cylindrical tank in which the heat transfer surface is in the form of a coil that is distributed uniformly from the bottom of the tank to the top of the tank. The tank itself is completely insulated. Liquid is introduced into the tank at a uniform rate, starting with no liquid in the tank, and the steam is turned on at the instant that liquid flows into the tank.

(a) Using the nomenclature of Problem 7.23, formulate a differential equation for this process. Integrate the differential equation to obtain an equation for ϕ as a function of B and f, where f = fraction filled = W/W_{filled}.

(b) If the heat transfer surface consists of a coil of 10 turns of 1-in.-OD tubing 4 ft in diameter, the feed rate is 1200 lb/hr, the heat capacity of the liquid is 1.0 Btu/(lb)(°F), the heat transfer coefficient is 100 Btu/(hr)(°F)(ft²) of covered area, the steam temperature is 200°F, and the temperature of the liquid introduced into the tank is 70°F, what is the temperature in the tank when it is completely full? What is the temperature when the tank is half full? The heat transfer is given by $Q = UA(T_{\text{tank}} - T_{\text{steam}})$.

7.25. A cylindrical tank 5 ft in diameter and 5 ft high is full of water at 70°F. The water is to be heated by means of a steam jacket around the sides only. The steam temperature is 230°F, and the overall coefficient of heat transfer is constant at 40 Btu/(hr)(ft²)(°F). Use Newton's law of cooling (heating) to estimate the heat transfer. Neglecting the heat losses from the top and the bottom, calculate the time necessary to raise the temperature of the tank contents to 170°F. Repeat, taking the heat losses from the top and the bottom into account. The air temperature around the tank is 70°F, and the overall coefficient of heat transfer for both the top and the bottom is constant at 10 Btu/(hr)(ft²)(°F).

Appendix A

ANSWERS TO SELF-ASSESSMENT TESTS

Section 1.1

1. 2.10×10^{-5} m^3/s
2. **(a)** 252 lb$_m$; **(b)** 29.6 lb$_f$
3. Examine the conversion factors inside the cover.
4. c is dimensionless.
5. 32.174 (ft)(lb$_m$)/(s^2)(lb$_f$)
6. (a)
7. A has the same units as k; B has the units of T.

Section 1.2

1. 60.05
2. A kilogram mole contains 6.02×10^{26} molecules, whereas a pound mole contains $(6.02 \times 10^{26})(0.454)$ molecules.
3. 0.123 kg mol NaCl/kg mol H$_2$O
4. 1.177 lb mol
5. 0.121 kg/s

Section 1.3

1. **(a)** F; **(b)** T; **(c)** F; **(d)** T
2. 13.6 g/cm^3
3. 62.4 lb$_m$/ft^3 (10^3 kg/m^3)

4. This means that the density at 10°C of liquid HCN is 1.2675 times the density of water at 4°C.

5. 0.79314 g/cm^3 (Note that you need the density of water at 60°F.)

6. 9

7. **(a)** 63%; **(b)** 54.3; **(c)** 13.8

8. 8.11 ft^3

9. **(a)** 0.33; **(b)** 18.7

10. **(a)** C_4 – 0.50, C_5 – 0.30, C_6 – 0.20; **(b)** C_4 – 0.57, C_5 – 0.28, C_6 – 0.15; **(c)** C_4 – 57, C_5 – 28, C_6 – 15; **(d)** 66.4 kg/kg mol

Section 1.4

1. See text.

Section 1.5

1. **(a)** 0°C and 100°C; (b) 32°F and 212°F

2. $\Delta°F(1.8) = \Delta°C$.

3. Yes. Yes.

4. $92.76 + 0.198T_{°F}$.

5.

°C	°F	K	°R
−40.0	−40.0	233	420
25.0	77.0	298	537
425	796	698	1256
−234	−390	38.8	69.8

6. Immerse in ice-water bath and mark 0°C. Immerse in boiling water at 1 atm pressure and mark 100°C. Interpolate between 0°C and 100°C in desired intervals.

7. **(a)** 1°C; **(b)** 1°C; **(c)** 1 $\Delta°C$

Section 1.6

1. Gauge pressure + barometric pressure = absolute pressure.

2. See p. 56.

3. Barometric pressure − vacuum pressure = absolute pressure.

4. **(a)** 15.5; **(b)** 106.6 kPa; **(c)** 1.052; **(d)** 35.6

5. **(a)** Gauge pressure; **(b)** barometric pressure, absolute pressure; **(c)** 50 in. Hg.

6. In the absence of the barometric pressure, assume 101.3 kPa; then absolute pressure is 61.3 kPa.

7. **(a)** T; **(b)** T

Section 1.7

1. **(a)** $C_9H_{18} + \frac{27}{2} O_2 \rightarrow 9CO_2 + 9H_2O$
   (b) $4FeS_2 + 11O_2 \rightarrow 2Fe_2O_3 + 8SO_2$

2. 3.08

3. 323

4. **(a)** $CaCO_3 - 43.4\%$; $CaO - 56.6\%$; **(b)** 0.308

5. **(a)** H_2O; **(b)** NaCl; **(c)** NaCl, H_2O, NaOH (assuming that the gas escapes)

Section 2.1

1. A serial sequence is adequate; some steps can be in parallel. Feedback may be inserted at various stages.

3. **(a)** Find a reason to remember; talk to others who are interested to ascertain what motivated them. Write down the reason for periodic pep talks with yourself.

   (b) Get a "bird's-eye view" of the whole section. Use the figure numbered 0 to see the relation to other sections. Look at a similar textbook. Ask for help; do not wait.

   (c) After scanning the serious reading count.

   (d) Take a few minutes as you read to mentally test yourself on what you have just read to reinforce learning. Continue to test at periodic intervals.

   (e) Practice using the information for an assumed quiz; predict quiz questions; use text examples or problems as samples.

4. Some possible options are: (1) obtain a rope or a net or a piece of clothing, and tow the cantaloupes across; (2) juggle the cantaloupes keeping one in the air—violates the laws of physics; (3) throw the cantaloupes across—even if they get to the other side, they will be smashed; (4) get a better raft; and many other possibilities.

Section 2.3

1. See Table 2.5.

2. **(a)** *Chemical Engineers' Handbook* and/or *Properties of Gases and Liquids*

   (b) *Technical Data Book—Petroleum Refining*

(c) *Handbook of Physics and Chemistry* and others

(d) *Chemical Engineers' Handbook*

Section 3.1

1, 2. The systems are somewhat arbitrary, as are the time intervals selected, but (a) and (c) can be closed systems (ignoring evaporation) and (b) open.

3. See text.

4. (a) batch; (b) flow; (c) flow; (d) flow.

5. Accumulation is zero.

6. Yes

7. Equation (3.1) or simplifications thereof.

Section 3.2

1. A set of unique values for the variables in the equations representing the problem.

2. (a) one; (b) three; (c) three

3. (a) two; (b) two of these three: acetic acid, water, total; (c) two; (d) feed of the 10% solution and mass fraction ω of the acetic acid in P; (e) 14% acetic acid; 86% water

4. Not for a unique solution. Only two are independent.

5. Substitution; Gauss-Jordan elimination; use of determinants.

6. Select the most accurate equations that will provide a unique solution. (Or use least squares if all must be used.)

7. The sum of the mass or mole fractions in one stream is unity.

8. Collect additional data so as to specify or evaluate the values of the excess unknown variables.

9. F, D, P, ω_{D2}, ω_{P1}.

10. Only two independent material balances can be written. The sum of the mass fractions for streams D and P yields two additional independent equations. One value of F, D, or P must be specified to obtain a solution. Note that specifying W_{D2} or W_{P1} will not help.

Section 3.3

1. 33.3 kg

2. 178 kg/hr

3. (a) 28% Na_2SO_4; (b) 33.3

4. 0.994

Section 3.4

1. Orsat (dry basis) does not include the water vapor in the analysis.

2. SO_2 is not included in the gas analysis.

3. See Eqs. (3.4)–(3.6).

4. Yes. See Eq. (3.4).

5. More

6. 4.5%

7. 9.1% CO_2, 8.9% O_2, 82.0% N_2

8. **(a)** T; **(b)** T if C is regarded as an element; F if the exit compound is CO_2 and the entering reactant is C; **(c)** F

9. No chemical reaction takes place, or with reaction, the moles of reactants may equal the moles of product by accident.

10. 1

11. **(a)** 252; **(b)** 1.063; **(c)** 2.31; **(d)** 33.8%

12. 186 kg

13. 0.81

Section 3.5

1.

	lb	fr.
tol	396	0.644
bz	19.68	0.032
xyl	200.00	0.325

2. 863 lb air/lb S

	Converter	Burner
SO_2	0.5%	9.5%
SO_3	9.4	—
O_2	7.4	11.5
N_2	82.7	79.0

Section 3.6

1. Recycle is feedback from downstream to upstream in the materials flow bypassing one or more units. Bypassing is feedforward bypassing one or more units.

2. Purge is a side stream usually small in quantity relative to main stream that is bled from the main stream to remove impurities from the system.

3. The stoichiometric ratio

4. **(a)** $R = 3000$ kg/hr; **(b)** air = 85,100 kg/hr

5. **(a)** 890 recycled and 3.2 purged; **(b)** 9.2% conversion (errors can be caused by loss of significant figures)

6. **(a)** 1200.6 kg/hr; **(b)** 194.9 kg/hr

Section 4.1

1. $pV = nRT$

2. T: absolute temperature in degrees; p: absolute pressure in mass/(length)

$(time)^2$; V: $(length)^3/mole$; n: mole; R: $(mass)(length)^2/(time)^2(mole)$ $(degree)$

3. See Table 4.1.
4. 1883 ft^3
5. 2.98 kg
6. 1.32
7. 28.3 m^3/hr
8. 0.0493 kg/m^3
9. 1.02 (lb CH_4/ft^3 CH_4 at 70°F and 2 atm)/(lb air/ft^3 at S.C.)
10. N_2, 0.28 psia; CH_4, 10.9 psia; C_2H_6, 2.62 psia
11. **(a)** 11.12 psia at 2 ft^3 and 120°F; **(b)** 0.28 psia at 2 ft^3 and 120°F
12. **(a)** 2,735 ft^3/hr; **(b)** 5,034 ft^3/hr, **(c)** 22,429 ft^3/hr, **(d)** 30,975 ft^3/hr
13. 118,400 ft^3/hr

Section 4.2-1

1. $\hat{V}_{c_i} = RT_c/P_c$. It can be used to calculate $\hat{V}_{r_i}$, which is a parameter on the Nelson and Obert charts.
2. **(a)** No; **(b)** 5.08 ft^3; **(c)** 1.35 ft^3
3. 1.65 kg
4. 14.9 atm

Section 4.2-2

1. See the last paragraph in Sec. 4.2-2. Also, they can be used for interpolation and extrapolation.
2. b is m^3; a is $(K)^{1/2}(m)^6(kPa)$.
3. $V = 0.60$ ft^3
4. 314 K
5. **(a)** 50.7 atm; **(b)** 34.0 atm
6. 0.316 m^3/kg mol

Section 4.2-3

1. 262 atm

Section 4.3

1. See Figure 4.1.
2. Ice at its vapor pressure changes to liquid water at 32°F (at 0.0886 psia) and

the pressure increases as shown in Figure 4.11 with vapor and liquid in equilibrium. At 250°F, the pressure is 29.82 psia.
3. **(a)** 75 psia (5.112 atm); **(b)** it sublimes.
4. Experimental $p^* = 219.9$ mm Hg; predicted 220.9 mm Hg
5. 80.1°C

Section 4.4

1. The partial pressure of the vapor equals the vapor pressure of the gas. Liquid and vapor are in equilibrium.
2. Yes; yes
3. 21°C; benzene
4. 0.0373
5. 4.00 lb
6. **(a)** Both gas; **(b)** some liquid water, residual is gas; **(c)** both gas; **(d)** some liquid water, residual is gas.

Section 4.5

1. 190 psia; $C_2H_6 = 0.0677$, $C_3H_8 = 0.660$, $i\text{-}C_4H_{10} = 0.2415$, $n\text{-}C_4H_{10} = 0.0308$
2. No, gas and liquid in equilibrium. The triple point in the p–T projection is actually a line on the p-V-T surface. The pressure and temperature are fixed but the volume is not fixed.
3. **(a)** $C = 3$, $\mathcal{P} = 1$, $F = 4$; **(b)** $C = 2$, $\mathcal{P} = 2$, $F = 2$

Section 4.6

1. 0.063   2. 57.3%   3. 86% $\mathcal{RH}$

Section 4.7

1. 53°C (126°F)   2. 10.5% H_2O 3. 94.0%; 100%

Section 5.1

1. Intensive—independent of quantity of material; extensive—dependent of quantity of material. Measurable—temperature, pressure; unmeasurable—internal energy, enthalpy. State variable—difference in value between two states depends only on the states; path variable—difference in two states depends on trajectory reaching in the final state.
2. Heat: energy transfer across a system boundary due to a temperature differ-

ence. Work: energy transfer across a system boundary by means of a vector force acting through a vector displacement on the boundary.

3. **(a)** Open; **(b)** open; **(c)** open; **(d)** closed; **(e)** closed

4. **(a)**

6. Decrease

7. **(a)** $K_0 = 0$, $P_0 = 4900$ J; **(b)** $K_f = 0$; $P_f = 0$; **(c)** $\Delta K = 0$, $\Delta P = -4900$ J (a decrease); **(d)** 1171 cal, 4.64 Btu

8. Assume $P_{atm} = 1$ atm is constant; -1.26×10^5 J

Section 5.2

1. $C_p = 4.25 + 0.002T$

2. $\frac{7}{2}R$

3. **(a)** cal/(g mol)(ΔK); **(b)** cal/(g mol)(ΔK)(K); **(c)** cal/(g mol)(ΔK)(K^2); **(d)** cal/(g mol)(ΔK)(K^3)

4. $27{,}115 + 6.55T - 9.98 \times 10^{-4}T^2$

5. 32,970 J (7880 cal)

6. 192 Btu for 2 lb. Use the butane chart and the heat capacity equation.

7. **(a)** Liquid; **(b)** two phase; **(c)** two phase; **(d)** vapor; **(e)** vapor

8. 19,013 J/kg

9. −361 Btu to −334 Btu (depending on data used)

Section 5.3

1. **(a)** $T = 90.19$ K; **(b)** $m = 1.48 \times 10^7$ g; **(c)** $x = 3.02 \times 10^{-4}$; **(d)** 49.6 hr; **(e)** from a handbook, $T = 100$ K approx (for process at constant volume)

2. **(a)** Down; **(b)** probably up, depending on room temperature; **(c)** down

3. **(a)** $T_1 = 166.2°C$, $T_2 = 99.63°C$; **(b)** $Q = -6.40 \times 10^4$ J

4. **(a)** 0; **(b)** 0; **(c)** 0; **(d)** if ideal gas $\Delta T = 0$, $T_2 =$ room temperature; **(e)** 0.26 atm

5. 12,200 lb/hr

6. 99°C

7. 1847 watts (2.48 hp)

8. −19.7 Btu/s

Section 5.4

1. −74.83 kJ/g mol CH_4 (−17.88 kcal)

2. 597.32 kJ/g mol C_6H_6 (142.76 kcal)

3. −148.53 k cal/g mol

4. 230 Btu/ft^3 at 60°F and 30.0 in. Hg

5. 125 Btu/ft^3 at S.C.

6. 6020 Btu/lb methane

7. 0.52 lb

8. 975 K

Section 5.5

1. (2)
2. Thermal—presumably internal energy or heat—not freely convertible to mechanical energy. The latter types of energy are freely convertible one to the other.
3. Process in which total of freely convertible forms of energy are not reduced.
4. $W = 27.6$ kJ; $Q = -27.6$ kJ; $\Delta E = \Delta H = 0$
5. 38.8 hp

Section 5.6

1. Almost always
2. (a) HNO_3, HCl, or H_2SO_4 in water; (b) NaCl, KCl, NH_4NO_3 in water
3. (a) 25°C and 1 atm; (b) 0
4. −4,655 cal/g mol soln (heat transfer is from solution to surroundings)
5. $Q = -1.61 \times 10^6$ Btu
6. Two phase; for the liquid $\omega_{H_2O} = 0.50$ and $\Delta\hat{H} = 8$ Btu/lb; for the vapor $\omega_{H_2O} = 1.00$ and $\Delta\hat{H} = 1174$ Btu/lb
7. 500 Btu/lb

Section 5.7

1. $t_{DB} =$ gas temperature; $t_{WB} =$ temperature registered at equilibrium between evaporating water around the temperature sensor and surroundings at t_{DB}.
2. No
3. See Eqs. (5.36)–(5.39).
4. (a) 0.03 kg/kg dry air; (b) 1.02 m³/kg dry air; (c) 38°C$_{in}$; (d) 151 kJ/kg dry air; (e) 31.5°C
5. (a) $\mathcal{H} = 0.0808$ lb H_2O /lb dry air; (b) $H = 118.9$ Btu/lb dry air; (c) $V = 16.7$ ft³/lb dry air; (d) $\mathcal{H} = 0.0710$ lb H_2O/lb dry air
6. 49,700 Btu
7. (a) 1.94×10^6 ft³/hr; (b) 3.00×10^4 ft³/hr; (c) 2.44×10^5 Btu/hr

Section 6.1

1. Yes. See Appendix L.
2. For one phase, $N_{sp} - 1 + 2$ intensive variable (from the phase rule) plus one extensive variable are needed to completely specify a stream.
3. 9

4. Any number from 7 to 11 may be acceptable, depending on assumptions concerning the relations among the pressures in the reactor and mixer streams.

5. Yes

Chapter 7

1. The solution of the differential equation requires the initial conditions to be able to predict the response at other times.

2. Independent

3. Let the derivatives vanish (delete them from the equations).

4. (a)

5. (1), (4), (7), (9), (11); (2), (8); (3) (5), (10), (12); (6)

6. 13 hr

7. 0.212 lb

Appendix B

ATOMIC WEIGHTS AND NUMBERS

TABLE B.1   Relative Atomic Weights, 1965 (Based on the Atomic Mass of $^{12}C = 12$)
The values for atomic weights given in the table apply to elements as they exist in nature, without artificial alteration of their isotopic composition, and, further, to natural mixtures that do not include isotopes of radiogenic origin.

Name	Symbol	Atomic Number	Atomic Weight	Name	Symbol	Atomic Number	Atomic Weight
Actinium	Ac	89	—	Mercury	Hg	80	200.59
Aluminum	Al	13	26.9815	Molybdenum	Mo	42	95.94
Americium	Am	95	—	Neodymium	Nd	60	144.24
Antimony	Sb	51	121.75	Neon	Ne	10	20.183
Argon	Ar	18	39.948	Neptunium	Np	93	—
Arsenic	As	33	74.9216	Nickel	Ni	28	58.71
Astatine	At	85	—	Niobium	Nb	41	92.906
Barium	Ba	56	137.34	Nitrogen	N	7	14.0067
Berkelium	Bk	97	—	Nobelium	No	102	
Beryllium	Be	4	9.0122	Osmium	Os	75	190.2
Bismuth	Bi	83	208.980	Oxygen	O	8	15.9994
Boron	B	5	10.811	Palladium	Pd	46	106.4
Bromine	Br	35	79.904	Phosphorus	P	15	30.9738
Cadmium	Cd	48	112.40	Platinum	Pt	78	195.09
Caesium	Cs	55	132.905	Plutonium	Pu	94	
Calcium	Ca	20	40.08	Polonium	Po	84	
Californium	Cf	98	—	Potassium	K	19	39.102
Carbon	C	6	12.01115	Praseodym	Pr	59	140.907
Cerium	Ce	58	140.12	Promethium	Pm	61	—
Chlorine	Cl	17	35.453^b	Protactinium	Pa	91	—
Chromium	Cr	24	51.996^b	Radium	Ra	88	—
Cobalt	Co	27	58.9332	Radon	Rn	86	
Copper	Cu	29	63.546^b	Rhenium	Re	75	186.2
Curium	Cm	96	—	Rhodium	Rh	45	102.905
Dysprosium	Dy	66	162.50	Rubidium	Rb	37	84.57
Einsteinium	Es	99	—	Ruthenium	Ru	44	101.07
Erbium	Er	68	167.26	Samarium	Sm	62	150.35
Europium	Eu	63	151.96	Scandium	Sc	21	44.956
Fermium	Fm	100	—	Selenium	Se	34	78.96
Flourine	F	9	18.9984	Silicon	Si	14	28.086
Francium	Fr	87	—	Silver	Ag	47	107.868
Gadolinium	Gd	64	157.25	Sodium	Na	11	22.9898
Gallium	Ga	31	69.72	Strontium	Sr	38	87.62
Germanium	Ge	32	72.59	Sulfur	S	16	32.064
Gold	Au	79	196.967	Tantalum	Ta	73	180.948
Hafnium	Hf	72	178.49	Technetium	Tc	43	—
Helium	He	2	4.0026	Tellurium	Te	52	127.60
Holmium	Ho	67	164.930	Terbium	Tb	65	158.924
Hydrogen	H	1	1.00797	Thallium	Tl	81	204.37
Indium	In	49	114.82	Thorium	Th	90	232.038
Iodine	I	53	126.9044	Thulium	Tm	59	168.934
Iridium	Ir	77	192.2	Tin	Sn	50	118.69
Iron	Fe	26	55.847	Titanium	Ti	22	47.90
Krypton	Kr	36	83.80	Tungsten	W	74	183.85
Lanthanum	La	57	138.91	Uranium	U	92	238.03
Lawrencium	Lr	103	—	Vanadium	V	23	50.942
Lead	Pb	82	207.19	Xenon	Xe	54	131.30
Lithium	Li	3	6.939	Ytterbium	Yb	70	173.04
Lutetium	Lu	71	174.97	Yttrium	Y	39	88.905
Magnesium	Mg	12	24.312	Zinc	Zn	30	65.37
Manganese	Mn	25	54.9380	Zirconium	Zr	40	91.22
Mendelevium	Md	101	—				

SOURCE: *Comptes Rendus*, 23rd IUPAC Conference, 1965, Butterworth's, London, 1965, pp. 177–178.

Appendix C

STEAM TABLES

Note: Tables in SI and American Engineering units for superheated steam are on the chart in the pocket in the back of the book. A computer code for steam properties is on the disk in the back of the book.

TABLE C.1   Saturated Steam: Temperature Table

Temp. Fahr.	Absolute pressure		Specific volume			Enthalpy		
	Lb/in.²	In. Hg 32°F	Sat. Liquid	Evap.	Sat. Vapor	Sat. Liquid	Evap.	Sat. Vapor
t	p		v_f	v_{fg}	v_g	h_f	h_{fg}	h_g
32	0.0886	0.1806	0.01602	3305.7	3305.7	0	1075.1	1075.1
34	0.0961	0.1957	0.01602	3060.4	3060.4	2.01	1074.9	1076.0
36	0.1041	0.2120	0.01602	2836.6	2836.6	4.03	1072.9	1076.9
38	0.1126	0.2292	0.01602	2632.2	2632.2	6.04	1071.7	1077.7
40	0.1217	0.2478	0.01602	2445.1	2445.1	8.05	1070.5	1078.6
42	0.1315	0.2677	0.01602	2271.8	2271.8	10.06	1069.3	1079.4
44	0.1420	0.2891	0.01602	2112.2	2112.2	12.06	1068.2	1080.3
46	0.1532	0.3119	0.01602	1965.5	1965.5	14.07	1067.1	1081.2
48	0.1652	0.3364	0.01602	1829.9	1829.9	16.07	1065.9	1082.0
50	0.1780	0.3624	0.01602	1704.9	1704.9	18.07	1064.8	1082.9
52	0.1918	0.3905	0.01603	1588.4	1588.4	20.07	1063.6	1083.7
54	0.2063	0.4200	0.01603	1482.4	1482.4	22.07	1062.5	1084.6
56	0.2219	0.4518	0.01603	1383.5	1383.5	24.07	1061.4	1085.5
58	0.2384	0.4854	0.01603	1292.7	1292.7	26.07	1060.2	1086.3
60	0.2561	0.5214	0.01603	1208.1	1208.1	28.07	1059.1	1087.2
62	0.2749	0.5597	0.01604	1129.7	1129.7	30.06	1057.9	1088.0
64	0.2949	0.6004	0.01604	1057.1	1057.1	32.06	1056.8	1088.9
66	0.3162	0.6438	0.01604	989.6	989.6	34.06	1055.7	1089.8
68	0.3388	0.6898	0.01605	927.0	927.0	36.05	1054.5	1090.6

TABLE C.1   (*cont.*)

Temp. Fahr. *t*	Absolute pressure Lb/in.^{2} *p*	Absolute pressure In. Hg 32°F	Specific volume Sat. Liquid v_f	Specific volume Evap. v_{fg}	Specific volume Sat. Vapor v_g	Enthalpy Sat. Liquid h_f	Enthalpy Evap. h_{fg}	Enthalpy Sat. Vapor h_g
70	0.3628	0.7387	0.01605	868.9	868.9	38.05	1053.4	1091.5
72	0.3883	0.7906	0.01606	814.9	814.9	40.04	1052.3	1092.3
74	0.4153	0.8456	0.01606	764.7	764.7	42.04	1051.2	1093.2
76	0.4440	0.9040	0.01607	718.0	718.0	44.03	1050.1	1094.1
78	0.4744	0.9659	0.01607	674.4	674.4	46.03	1048.9	1094.9
80	0.5067	1.032	0.01607	633.7	633.7	48.02	1047.8	1095.8
82	0.5409	1.101	0.01608	595.8	595.8	50.02	1046.6	1096.6
84	0.5772	1.175	0.01608	560.4	560.4	52.01	1045.5	1097.5
86	0.6153	1.253	0.01609	527.6	527.6	54.01	1044.4	1098.4
88	0.6555	1.335	0.01609	497.0	497.0	56.00	1043.2	1099.2
90	0.6980	1.421	0.01610	468.4	468.4	58.00	1042.1	1100.1
92	0.7429	1.513	0.01611	441.7	441.7	59.99	1040.9	1100.9
94	0.7902	1.609	0.01611	416.7	416.7	61.98	1039.8	1101.8
96	0.8403	1.711	0.01612	393.2	393.2	63.98	1038.7	1102.7
98	0.8930	1.818	0.01613	371.3	371.3	65.98	1037.5	1103.5
100	0.9487	1.932	0.01613	350.8	350.8	67.97	1036.4	1104.4
102	1.0072	2.051	0.01614	331.5	331.5	69.96	1035.2	1105.2
104	1.0689	2.176	0.01614	313.5	313.5	71.96	1034.1	1106.1
106	1.1338	2.308	0.01615	296.5	296.5	73.95	1033.0	1107.0
108	1.2020	2.447	0.01616	280.7	280.7	75.94	1032.0	1107.9
110	1.274	2.594	0.01617	265.7	265.7	77.94	1030.9	1108.8
112	1.350	2.749	0.01617	251.6	251.6	79.93	1029.7	1109.6
114	1.429	2.909	0.01618	238.5	238.5	81.93	1028.6	1110.5
116	1.512	3.078	0.01619	226.2	226.2	83.92	1027.5	1111.4
118	1.600	3.258	0.01620	214.5	214.5	85.92	1026.4	1112.3
120	1.692	3.445	0.01620	203.45	203.47	87.91	1025.3	1113.2
122	1.788	3.640	0.01621	193.16	193.18	89.91	1024.1	1114.0
124	1.889	3.846	0.01622	183.44	183.46	91.90	1023.0	1114.9
126	1.995	4.062	0.01623	174.26	174.28	93.90	1021.8	1115.7
128	2.105	4.286	0.01624	165.70	165.72	95.90	1020.7	1116.6
130	2.221	4.522	0.01625	157.55	157.57	97.89	1019.5	1117.4
132	2.343	4.770	0.01626	149.83	149.85	99.89	1018.3	1118.2
134	2.470	5.029	0.01626	142.59	142.61	101.89	1017.2	1119.1
136	2.603	5.300	0.01627	135.73	135.75	103.88	1016.0	1119.9
138	2.742	5.583	0.01628	129.26	129.28	105.88	1014.9	1120.8
140	2.887	5.878	0.01629	123.16	123.18	107.88	1013.7	1121.6
142	3.039	6.187	0.01630	117.37	117.39	109.88	1012.5	1122.4

v = specific volume, ft$_3$/lb. h = enthalpy, Btu/lb.

TABLE C.1   (*cont.*)

Temp. Fahr. t	Absolute pressure Lb/in.² p	In. Hg 32°F	Sat. Liquid v_f	Evap. v_{fg}	Sat. Vapor v_g	Sat. Liquid h_f	Evap. h_{fg}	Sat. Vapor h_g
	Absolute pressure		*Specific volume*			*Enthalpy*		
144	3.198	6.511	0.01631	111.88	111.90	111.88	1011.3	1123.2
146	3.363	6.847	0.01632	106.72	106.74	113.88	1010.2	1124.1
148	3.536	7.199	0.01633	101.82	101.84	115.87	1009.0	1124.9
150	3.716	7.566	0.01634	97.18	97.20	117.87	1007.8	1125.7
152	3.904	7.948	0.01635	92.79	92.81	119.87	1006.7	1126.6
154	4.100	8.348	0.01636	88.62	88.64	121.87	1005.5	1127.4
156	4.305	8.765	0.01637	84.66	84.68	123.87	1004.4	1128.3
158	4.518	9.199	0.01638	80.90	80.92	125.87	1003.2	1129.1
160	4.739	9.649	0.01639	77.37	77.39	127.87	1002.0	1129.9
162	4.970	10.12	0.01640	74.00	74.02	129.88	1000.8	1130.7
164	5.210	10.61	0.01642	70.79	70.81	131.88	999.7	1131.6
166	5.460	11.12	0.01643	67.76	67.78	133.88	998.5	1132.4
168	5.720	11.65	0.01644	64.87	64.89	135.88	997.3	1133.2
170	5.990	12.20	0.01645	62.12	62.14	137.89	996.1	1134.0
172	6.272	12.77	0.01646	59.50	59.52	139.89	995.0	1134.9
174	6.565	13.37	0.01647	57.01	57.03	141.89	993.8	1135.7
176	6.869	13.99	0.01648	56.64	54.66	143.90	992.6	1136.5
178	7.184	14.63	0.01650	52.39	52.41	145.90	991.4	1137.3
180	7.510	15.29	0.01651	50.26	50.28	147.91	990.2	1138.1
182	7.849	15.98	0.01652	48.22	48.24	149.92	989.0	1138.9
184	8.201	16.70	0.01653	46.28	46.30	151.92	987.8	1139.7
186	8.566	17.44	0.01654	44.43	44.45	153.93	986.6	1140.5
188	8.944	18.21	0.01656	42.67	42.69	155.94	985.3	1141.3
190	9.336	19.01	0.01657	40.99	41.01	157.95	984.1	1142.1
192	9.744	19.84	0.01658	39.38	39.40	159.95	982.8	1142.8
194	10.168	20.70	0.01659	37.84	37.86	161.96	981.5	1143.5
196	10.605	21.59	0.01661	36.38	36.40	163.97	980.3	1144.3
198	11.057	22.51	0.01662	34.98	35.00	165.98	979.0	1145.0
200	11.525	23.46	0.01663	33.65	33.67	167.99	977.8	1145.8
202	12.010	24.45	0.01665	32.37	32.39	170.01	976.6	1146.6
204	12.512	25.47	0.01666	31.15	31.17	172.02	975.3	1147.3
206	13.031	26.53	0.01667	29.99	30.01	174.03	974.1	1148.1
208	13.568	27.62	0.01669	28.88	28.90	176.04	972.8	1148.8
210	14.123	28.75	0.01670	27.81	27.83	178.06	971.5	1149.6
212	14.696	29.92	0.01672	26.81	26.83	180.07	970.3	1150.4
215	15.591		0.01674	25.35	25.37	186.10	968.3	1151.4
220	17.188		0.01677	23.14	23.16	188.14	965.2	1153.3
225	18.915		0.01681	21.15	21.17	193.18	961.9	1155.1

TABLE C.1   (*cont.*)

| Temp. Fahr. | Absolute pressure | | | Specific volume | | | Enthalpy | | |
| | Lb/in.^{2} | In. Hg 32°F | Sat. Liquid | Evap. | Sat. Vapor | Sat. Liquid | Evap. | Sat. Vapor |
t	p		v_f	v_{fg}	v_g	h_f	h_{fg}	h_g
230	20.78		0.01684	19.371	19.388	198.22	958.7	1156.9
235	22.80		0.01688	17.761	17.778	203.28	955.3	1158.6
240	24.97		0.01692	16.307	16.324	208.34	952.1	1160.4
245	27.31		0.01696	15.010	15.027	213.41	948.7	1162.1
250	29.82		0.01700	13.824	13.841	218.48	945.3	1163.8
255	32.53		0.01704	12.735	12.752	223.56	942.0	1165.6
260	35.43		0.01708	11.754	11.771	228.65	938.6	1167.3
265	38.54		0.01713	10.861	10.878	233.74	935.3	1169.0
270	41.85		0.01717	10.053	10.070	238.84	931.8	1170.6
275	45.40		0.01721	9.313	9.330	243.94	928.2	1172.1
280	49.20		0.01726	8.634	8.651	249.06	924.6	1173.7
285	53.25		0.01731	8.015	8.032	254.18	921.0	1175.2
290	57.55		0.01735	7.448	7.465	259.31	917.4	1176.7
295	62.13		0.01740	6.931	6.948	264.45	913.7	1178.2
300	67.01		0.01745	6.454	6.471	269.60	910.1	1179.7
305	72.18		0.01750	6.014	6.032	274.76	906.3	1181.1
310	77.68		0.01755	5.610	5.628	279.92	902.6	1182.5
315	83.50		0.01760	5.239	5.257	285.10	898.8	1183.9
320	89.65		0.01765	4.897	4.915	290.29	895.0	1185.3
325	96.16		0.01771	4.583	4.601	295.49	891.1	1186.6
330	103.03		0.01776	4.292	4.310	300.69	887.1	1187.8
335	110.31		0.01782	4.021	4.039	305.91	883.2	1189.1
340	117.99		0.01788	3.771	3.789	311.14	879.2	1190.3
345	126.10		0.01793	3.539	3.557	316.38	875.1	1191.5
350	134.62		0.01799	3.324	3.342	321.64	871.0	1192.6
355	143.58		0.01805	3.126	3.144	326.91	866.8	1193.7
360	153.01		0.01811	2.940	2.958	332.19	862.5	1194.7
365	162.93		0.01817	2.768	2.786	337.48	858.2	1195.7
370	173.33		0.01823	2.607	2.625	342.79	853.8	1196.6
375	184.23		0.01830	2.458	2.476	348.11	849.4	1197.5
380	195.70		0.01836	2.318	2.336	353.45	844.9	1198.4
385	207.71		0.01843	2.189	2.207	358.80	840.4	1199.2
390	220.29		0.01850	2.064	2.083	364.17	835.7	1199.9
395	233.47		0.01857	1.9512	1.9698	369.56	831.0	1200.6
400	247.25		0.01864	1.8446	1.8632	374.97	826.2	1201.2
405	261.67		0.01871	1.7445	1.7632	380.40	821.4	1201.8
410	276.72		0.01878	1.6508	1.6696	385.83	816.6	1202.4

v = specific volume, ft$_3$/lb. h = enthalpy, Btu/lb.

TABLE C.1 (*cont.*)

Temp. Fahr. t	Absolute pressure Lb/in.² p	In. Hg 32°F	Specific volume Sat. Liquid v_f	Evap. v_{fg}	Sat. Vapor v_g	Enthalpy Sat. Liquid h_f	Evap. h_{fg}	Sat. Vapor h_g
415	292.44		0.01886	1.5630	1.5819	391.30	811.7	1203.0
420	308.82		0.01894	1.4806	1.4995	396.78	806.7	1203.5
425	325.91		0.01902	1.4031	1.4221	402.28	801.6	1203.9
430	343.71		0.01910	1.3303	1.3494	407.80	796.5	1204.3
435	362.27		0.01918	1.2617	1.2809	413.35	791.2	1204.6
440	381.59		0.01926	1.1973	1.2166	418.91	785.9	1204.8
445	401.70		0.01934	1.1367	1.1560	424.49	780.4	1204.9
450	422.61		0.01943	1.0796	1.0990	430.11	774.9	1205.0
455	444.35		0.0195	1.0256	1.0451	435.74	769.3	1205.0
460	466.97		0.0196	0.9745	0.9941	441.42	763.6	1205.0
465	490.43		0.0197	0.9262	0.9459	447.10	757.8	1204.9
470	514.70		0.0198	0.8808	0.9006	452.84	751.9	1204.7
475	539.90		0.0199	0.8379	0.8578	458.59	745.9	1204.5
480	566.12		0.0200	0.7972	0.8172	464.37	739.8	1204.2
485	593.28		0.0201	0.7585	0.7786	470.18	733.6	1203.8
490	621.44		0.0202	0.7219	0.7421	476.01	727.3	1203.3
495	650.59		0.0203	0.6872	0.7075	481.90	720.8	1202.7
500	680.80		0.0204	0.6544	0.6748	487.80	714.2	1202.0
505	712.19		0.0206	0.6230	0.6436	493.8	707.5	1201.3
510	744.55		0.0207	0.5932	0.6139	499.8	700.6	1200.4
515	777.96		0.0208	0.5651	0.5859	505.8	693.6	1199.4
520	812.68		0.0209	0.5382	0.5591	511.9	686.5	1198.4
525	848.37		0.0210	0.5128	0.5338	518.0	679.2	1197.2
530	885.20		0.0212	0.4885	0.5097	524.2	671.9	1196.1
535	923.45		0.0213	0.4654	0.4867	530.4	664.4	1194.8
540	962.80		0.0214	0.4433	0.4647	536.6	656.7	1193.3
545	1003.6		0.0216	0.4222	0.4438	542.9	648.9	1191.8
550	1045.6		0.0218	0.4021	0.4239	549.3	640.9	1190.2
555	1088.8		0.0219	0.3830	0.4049	555.7	632.6	1188.3
560	1133.4		0.0221	0.3648	0.3869	562.2	624.1	1186.3
565	1179.3		0.0222	0.3472	0.3694	568.8	615.4	1184.2
570	1226.7		0.0224	0.3304	0.3528	575.4	606.5	1181.9
575	1275.7		0.0226	0.3143	0.3369	582.1	597.4	1179.5
580	1326.1		0.0228	0.2989	0.3217	588.9	588.1	1177.0
585	1378.1		0.0230	0.2840	0.3070	595.7	578.6	1174.3
590	1431.5		0.0232	0.2699	0.2931	602.6	568.8	1171.4
595	1486.5		0.0234	0.2563	0.2797	609.7	558.7	1168.4
600	1543.2		0.0236	0.2432	0.2668	616.8	548.4	1165.2

TABLE C.1 (*cont.*)

Temp. Fahr. t	Absolute pressure Lb/in.^{2} p	Absolute pressure In. Hg 32°F	Specific volume Sat. Liquid v_f	Specific volume Evap. v_{fg}	Specific volume Sat. Vapor v_g	Enthalpy Sat. Liquid h_f	Enthalpy Evap. h_{fg}	Enthalpy Sat. Vapor h_g
605	1601.5		0.0239	0.2306	0.2545	624.1	537.7	1161.8
610	1661.6		0.0241	0.2185	0.2426	631.5	526.6	1158.1
615	1723.4		0.0244	0.2068	0.2312	638.9	515.3	1154.2
620	1787.0		0.0247	0.1955	0.2202	646.5	503.7	1150.2
625	1852.4		0.0250	0.1845	0.2095	654.3	491.5	1145.8
630	1919.8		0.0253	0.1740	0.1993	662.2	478.8	1141.0
635	1989.0		0.0256	0.1638	0.1894	670.4	465.5	1135.9
640	2060.3		0.0260	0.1539	0.1799	678.7	452.0	1130.7
645	2133.5		0.0264	0.1441	0.1705	687.3	437.6	1124.9
650	2208.8		0.0268	0.1348	0.1616	696.0	422.7	1118.7
655	2286.4		0.0273	0.1256	0.1529	705.2	407.0	1112.2
660	2366.2		0.0278	0.1167	0.1445	714.4	390.5	1104.9
665	2448.0		0.0283	0.1079	0.1362	724.5	372.1	1096.6
670	2532.4		0.0290	0.0991	0.1281	734.6	353.3	1087.9
675	2619.2		0.0297	0.0904	0.1201	745.5	332.8	1078.3
680	2708.4		0.0305	0.0810	0.1115	757.2	310.0	1067.2
685	2800.4		0.0316	0.0716	0.1032	770.1	284.5	1054.6
690	2895.0		0.0328	0.0617	0.0945	784.2	254.9	1039.1
695	2992.7		0.0345	0.0511	0.0856	801.3	219.1	1020.4
700	3094.1		0.0369	0.0389	0.0758	823.9	171.7	996.5
705	3199.1		0.0440	0.0157	0.0597	870.2	77.6	947.8
705.34*	3206.2		0.0541	0	0.0541	910.3	0	910.3

*Critical temperature. v = specific volume, ft3/lb. h = enthalpy, Btu/lb.
Source: Combustion Engineering, Inc.

Appendix D

PHYSICAL PROPERTIES OF VARIOUS ORGANIC AND INORGANIC SUBSTANCES

General Sources of Data for Tables on the Physical Properties, Heat Capacities, and Thermodynamic Properties in Appendices D, E, and F

1. Kobe, Kenneth A., and Associates. "Thermochemistry of Petrochemicals." Reprint from *Petroleum Refiner*, Houston: Gulf Publishing Company, Jan. 1949–July 1958. (Enthalpy tables D.2–D. and heat capacities of several gases in Table E.1, Appendix E.)

2. Lange, N. A. *Handbook of Chemistry*, 12th ed. New York: McGraw-Hill, 1979.

3. Maxwell, J. B. *Data Book on Hydrocarbons*. New York: Van Nostrand Reinhold, 1950.

4. Perry, J. H., and C. H. Chilton, eds. *Chemical Engineers' Handbook*, 5th ed. New York: McGraw-Hill, 1973.

5. Rossini, Frederick D., et al. "Selected Values of Chemical Thermodynamic Properties." From *National Bureau of Standards Circular 500*. Washington, DC: U.S. Government Printing Office, 1952.

6. Rossini, Frederick D., et al. "Selected Values of Physical and Thermodynamic Properties of Hydrocarbons and Related Compounds." American Petroleum Institute Research Project 44, 1953 and subsequent years.

7. Weast, Robert C. *Handbook of Chemistry and Physics*, 59th ed. Boca Raton, FL: CRC Press, 1979.

TABLE D.1 Physical Properties of Various Organic and Inorganic Substances* (Additional compounds are in the disk in the back of this book provided by Professor Carl L. Yaws)

To convert to kcal/g mol multiply by 0.2390; to Btu/lb mol multiply by 430.2.

Compound	Formula	Formula Wt	Sp Gr	Melting Temp. (K)	$\Delta\hat{H}$ Fusion (kJ/g mol)	Normal b.p. (K)	$\Delta\hat{H}$ Vap. at b.p. (kJ/g mol)	T_c (K)	p_c (atm)	$\hat{V}_c$ (cm³/g mol)	z_c
Acetaldehyde	C_2H_4O	44.05	0.7831[18/4°]	149.5		293.2		461.0			
Acetic acid	CH_3CHO_2	60.05	1.049	289.9	12.09	390.4	24.4	594.8	57.1	171	0.200
Acetone	C_3H_6O	58.08	0.791	178.2	5.69	329.2	30.2	508.0	47.0	213	0.238
Acetylene	C_2H_2	26.04	0.9061(A)	191.7	3.7	191.7	17.5	309.5	61.6	113	0.274
Air			1.00					132.5	37.2		
Ammonia	NH_3	17.03	0.817[−79°] 0.597(A)	195.40	5.653	239.73	23.35	405.5	111.3	72.5	0.243
Ammonium carbonate	$(NH_4)_2CO_3 \cdot H_2O$	114.11		(decomposes at 331 K)							
Ammonium chloride	NH_4CL	53.50	2.53[17°]			(decomposes at 623 K)					
Ammonium nitrate	NH_4NO_3	80.05	1.725[25°]	442.8	5.4	(decomposes at 483.2 K)					
Ammonium sulfate	$(NH_4)_2SO_4$	132.14	1.769	786		(decomposes at 786 K after melting)					
Aniline	C_6H_7N	93.12	1.022	266.9		457.4		699	52.4		
Benzaldehyde	C_6H_5CHO	106.12	1.046	247.16		452.16	38.40				
Benzene	C_6H_6	78.11	0.879	278.693	9.837	353.26	30.76	562.6	48.6	260	0.274
Benzoic acid	$C_7H_6O_2$	122.12	1.316[28/4°]	395.4		523.0					
Benzyl alcohol	C_7H_8O	108.13	1.045	257.8		478.4					
Boron oxide	B_2O_3	69.64	1.85	723	22.0						
Bromine	Br_2	159.83	3.119[20°] 5.87(A)	265.8	10.8	331.78	31.0	584	102	144	0.306
1, 2-Butadiene	C_4H_6	54.09	0.652[20°]	136.7		283.3		446			
1, 3-Butadiene	C_4H_6	54.09	0.621	164.1		268.6		425	42.7	221	0.271
Butane	$n\text{-}C_4H_{10}$	58.12	0.579	134.83	4.661	272.66	22.31	425.17	37.47	255	0.374
iso-Butane	$iso\text{-}C_4H_{10}$	58.12	0.557	113.56	4.540	261.43	21.29	408.1	36.0	263	0.283

*Sources of data are listed at the beginning of Appendix D.

Sp gr = 20°C/4°C unless specified. Sp gr for gas referred to air (A).

TABLE D.1 *(cont.)*

Compound	Formula	Formula Wt	Sp Gr	Melting Temp. (K)	$\Delta\hat{H}$ Fusion (kJ/g mol)	Normal b.p. (K)	$\Delta\hat{H}$ Vap. at b.p. (kJ/g mol)	T_c (K)	p_c (atm)	$\hat{V}_c$ (cm^3/g mol)	z_c
1-Butene	C_4H_8	56.10	0.60	87.81	3.848	266.91	21.92	419.6	39.7	240	0.277
Butyl phthalate	*see* Dibutyl phthalate										
n-Butyric acid	n-$C_4H_8O_2$	88.10	0.958	267		437.1		628	52.0	290	0.293
iso-Butyric acid	iso-$C_4H_8O_2$	88.10	0.949	226		427.7		609			
Calcium arsenate	$Ca_3(AsO_4)_2$	398.06		1723							
Calcium carbide	Ca_2C_2	64.10	2.22$^{18°}$	2573							
Calcium carbonate	$CaCO_3$	100.09	2.93	(decomposes at 1098 K)							
Calcium chloride	$CaCl_2$	110.99	2.152$^{15°}$	1055	28.4						
	$CaCl_2 \cdot H_2O$	129.01									
	$CaCl_2 \cdot 2H_2O$	147.03									
	$CaCl_2 \cdot 6H_2O$	219.09	1.78$^{17°}$	303.4	37.3	(−6H$_2$O at 473 K)					
Calcium cyanamide	$CaCN_2$	80.11	2.29								
Calcium cyanide	$Ca(CN)_2$	92.12									
Calcium hydroxide	$Ca(OH)_2$	74.10	2.24	(−H$_2$O at 853 K)							
Calcium oxide	CaO	56.08	2.62	2873	50	3123					
Calcium phosphate	$Ca_3(PO_4)_2$	310.19	3.14	1943							
Calcium silicate	$CaSiO_3$	117.17	2.915	1803	48.62						
Calcium sulfate (gypsum)	$CaSO_4 \cdot 2H_2O$	172.18	2.32	(−1½H$_2$O at 301° K)							
Carbon	C	12.010	2.26	3873	46.0	4473					
Carbon dioxide	CO_2	44.01	1.53(A) 1.229 satd. liq. at 5162 kPa	217.0$^{5.2atm}$	8.32	(sublimes at 195 K)		304.2	72.9	94	0.275

Name	Formula		Sp gr								
Carbon disulfide	CS_2	76.14	$1.261^{22/20°}$, 2.63(A)	161.1	4.39	319.41	26.8	552.0	78.0	170	0.293
Carbon monoxide	CO	28.01	0.968(A)	68.10	0.837	81.66	6.042	133.0	34.5	93	0.294
Carbon tetrachloride	CCl_4	153.84	1.595	250.3	2.5	349.9	30.0	556.4	45.0	276	0.272
Chlorine	Cl_2	70.91	2.49(A)	172.16	6.406	239.10	20.41	417.0	76.1	124	0.276
Chlorobenzene	C_6H_5Cl	112.56	1.107	228		405.26	36.5	632.4	44.6	308	0.265
Chloroform	$CHCl_3$	119.39	$1.489^{20°}$	209.5		334.2		536.0	54.0	240	0.294
Chromium	Cr	52.01	7.1								
Copper	Cu	63.54	8.92	1356.2	13.0	2855	305				
Cumene	C_9H_{12}	120.19	0.862	177.125	7.1	425.56	37.5	636	31.0	440	0.260
Cupric sulfate	$CuSO_4$	159.61	$3.605^{15°}$	(decomposes at 873 K)							
Cyclohexane	C_6H_{12}	84.16	0.779	279.83	2.677	353.90	30.1	553.7	40.4	308	0.274
Cyclopentane	C_5H_{10}	70.13	0.745	179.71	0.6088	322.42	27.30	511.8	44.55	260	0.27
Decane	$C_{10}H_{22}$	142.28	$0.730^{20°}$	243.3		447.0		619.0	20.8	602	0.2476
Dibutyl phthalate	$C_8H_{22}O_4$	278.34	$1.045^{21°}$			613					
Diethyl ether	$(C_2H_5)_2O$	74.12	$0.708^{25°}$	156.86	7.301	307.76	26.05	467	35.6	281	0.261
Ethane	C_2H_6	30.07	1.049(A)	89.89	2.860	184.53	14.72	305.4	48.2	148	0.285
Ethanol	C_2H_6O	46.07	0.789	158.6	5.021	351.7	38.6	516.3	63.0	167	0.248
Ethyl acetate	$C_4H_8O_2$	88.10	0.901	189.4		350.2		523.1	37.8	286	0.252
Ethyl benzene	C_8H_{10}	106.16	0.867	178.185	9.163	409.35	36.0	619.7	37.0	360	0.260
Ethyl bromide	C_2H_5Br	108.98	1.460	154.1		311.4		504	61.5	215	0.320
Ethyl chloride	CH_3CH_2Cl	64.52	$0.903^{10°}$	134.83	4.452	285.43	25	460.4	52.0	199	0.274
3-Ethyl hexane	C_8H_{18}	114.22	0.7169			391.69	34.3	567.0	26.4	466	0.264
Ethylene	C_2H_4	28.05	0.975(A)	103.97	3.351	169.45	13.54	283.1	50.5	124	0.270
Ethylene glycol	$C_2H_6O_2$	62.07	$1.113^{19°}$	260	11.23	470.4	56.9				
Ferric oxide	Fe_2O_3	159.70	5.12	1833		(decomposes at 1833 K)					
Ferric sulfide	Fe_2S_3	207.90	4.3	(decomposes)							

*Sources of data are listed at the beginning of Appendix D.

Sp gr = 20°C/4°C unless specified. Sp gr for gas referred to air (A).

TABLE D.1 *(cont.)*

Compound	Formula	Formula Wt	Sp Gr	Melting Temp. (K)	$\Delta\hat{H}$ Fusion (kJ/g mol)	Normal b.p. (K)	$\Delta\hat{H}$ Vap. at b.p. (kJ/g mol)	T_c (K)	p_c (atm)	$\hat{V}_c$ (cm³/g mol)	z_c
Ferrous sulfide	FeS	87.92	4.84	1466		(decomposes)					
Formaldehyde	H_2CO	30.03	0.815⁻²⁰°	154.9		253.9	24.5				
Formic acid	CH_2O_2	46.03	1.220	281.46	12.7	373.7	22.3				
Glycerol	$C_3H_8O_3$	92.09	1.260⁵⁰°	291.36	18.30	563.2					
Helium	He	4.00	0.1368(A)	3.5	0.02	4.216	0.084	5.26	2.26	58	0.304
Heptane	C_7H_{16}	100.20	0.684	182.57	14.03	371.59	31.69	540.2	27.0	426	0.260
Hexane	C_6H_{14}	86.17	0.659	177.84	13.03	341.90	28.85	507.9	29.9	368	0.264
Hydrogen	H_2	2.016	0.06948(A)	13.96	0.12	20.39	0.904	33.3	12.8	65	0.304
Hydrogen chloride	HCl	36.47	1.268(A)	158.94	1.99	188.11	16.15	324.6	81.5	87	0.266
Hydrogen fluoride	HF	20.01	1.15	238		293		503.2			
Hydrogen sulfide	H_2S	34.08	1.1895(A)	187.63	2.38	212.82	18.67	373.6	88.9	98	0.284
Iodine	I_2	253.8	4.93²⁰°	386.5	15	457.4		826.0			
Iron	Fe	55.85	7.7	1808		3073	353				
Iron oxide	Fe_3O_4	231.55	5.2	1867	138	(decomposes at 1867 K after melting)					
Lead	Pb	207.21	11.337²⁰°	600.6	5.10	2023	180				
Lead oxide	PbO	223.21	9.5	1159	11.7	1745	213				
Magnesium	Mg	24.32	1.74	923	9.2	1393	132				
Magnesium chloride	$MgCl_2$	95.23	2.325²⁵°	987	43.1	1691	137				
Magnesium hydroxide	$Mg(OH)_2$	58.34	2.4	(decomposes at 623 K)	77.4						
Magnesium oxide	MgO	40.32	3.65	3173		3873					
Mercury	Hg	200.61	13.546²⁰°								
Methane	CH_4	16.04	0.554(A)	90.68	0.941	111.67	8.180	190.7	45.8	99	0.290
Methanol	CH_3OH	32.04	0.792	175.26	3.17	337.9	35.3	513.2	78.5	118	0.222
Methyl acetate	$C_3H_6O_2$	74.08	0.933	174.3		330.3		506.7	46.3	228	0.254

Name	Formula										
Methyl amine	CH_5N	31.06	$0.699^{-11°}$	180.5		$266.3^{758\text{mm}}$		429.9	73.6		
Methyl chloride	CH_3Cl	50.49	1.785(A)	175.3		249		416.1	65.8	143	0.276
Methyl ethyl ketone	C_4H_8O	72.10	0.805	186.1		352.6					
Methyl cyclohexane	C_7H_{14}	98.18	0.769	146.58	6.751	374.10	31.7	572.2	34.32	344	0.251
Molybdenum	Mo	95.95	10.2								
Napthalene	$C_{10}H_8$	128.16	1.145	353.2		491.0					
Nickel	Ni	58.69	$8.90^{20°}$	1725		3173	30.30				
Nitric acid	HNO_3	63.02	1.502	231.56	10.47	359					
Nitrobenzene	$C_6H_5O_2N$	123.11	1.203	278.7		483.9					
Nitrogen	N_2	28.02	12.5(D)	63.15	0.720	77.34	5.577	126.2	33.5	90	0.291
Nitrogen dioxide	NO_2	46.01	1.448	263.86	7.334	294.46	14.73	431.0	100.0	82	0.232
Nitrogen (nitric) oxide	NO	30.01	1.0367(A)	109.51	2.301	121.39	13.78	180	64.0	58	0.251
Nitrogen pentoxide	N_2O_5	108.02	$1.63^{18°}$	303		320					
Nitrogen tetraoxide	N_2O_4	92	$1.448^{20°}$	263.7		294.3		431.0	99.0		
Nitrogen trioxide	N_2O_3	76.02	$1.4472°$	171		276.5					
Nitrous oxide	N_2O	44.02	$1.226^{-89°}$, 1.530(A)	182.1		184.4		309.5	71.7	96.3	0.272
n-Nonane	C_9H_{20}	128.25	0.718	219.4		423.8		595	23		
n-Octane	C_8H_{18}	114.22	0.703	216.2		398.7		595.0	22.5	543	0.250
Oxalic acid	$C_2H_2O_4$	90.04	1.90	(decomposes at 459 K)							
Oxygen	O_2	32.00	1.1053(A)	54.40	0.443	90.19	6.820	154.4	49.7	74	0.290
n-Pentane	C_5H_{12}	72.15	$0.630^{18°}$	143.49	8.393	309.23	25.77	469.8	33.3	311	0.269
iso-Pentane	iso-C_5H_{12}	72.15	$0.621^{19°}$	113.1		300.9		461.0	32.9	308	0.268
1-Pentane	C_5H_{10}	70.13	0.641	107.96	4.937	303.13		474	39.9		
Phenol	C_6H_5OH	94.11	$1.071^{25°}$	315.66	11.43	454.56		692.1	60.5		

*Sources of data are listed at the beginning of Appendix D.

Sp gr = 20°C/4°C unless specified. Sp gr for gas referred to air (A).

TABLE D.1 *(cont.)*

Compound	Formula	Formula Wt	Sp Gr	Melting Temp. (K)	$\Delta\hat{H}$ Fusion (kJ/g mol)	Normal b.p. (K)	$\Delta\hat{H}$ Vap. at b.p. (kJ/g mol)	T_c (K)	p_c (atm)	$\hat{V}_c$ (cm³/g mol)	z_c
Phenyl hydrazine	$C_6H_8N_2$	108.14	1.097²³°	292.76	16.43	51.66					
Phosphoric acid	H_3PO_4	98.00	1.834¹⁸°	315.51	10.5	$(-\tfrac{1}{2}H_2O$ at 486 K)					
Phosphorus (red)	P_4	123.90	2.20	863	81.17	863	41.84				
Phosphorus (white)	P_4	123.90	1.82	317.4	2.5	553	49.71				
Phosphorus pentoxide	P_2O_5	141.95	2.387	(sublimes at 523 K)							
Propane	C_3H_8	44.09	1.562(A)	85.47	3.524	231.09	18.77	369.9	42.0	200	0.277
Propene	C_3H_6	42.08	1.498(A)	87.91	3.002	255.46	18.42	365.1	45.4	181	0.274
Propionic acid	$C_3H_6O_2$	74.08	0.993	252.2		414.4		612.5	53.0		
n-Propyl alcohol	C_3H_8O	60.09	0.804	146		370.2		536.7	49.95	220	0.251
iso-Propyl alcohol	C_3H_8O	60.09	0.785	183.5		355.4		508.8	53.0	219	0.278
n-Propyl benzene	C_9H_{12}	120.19	0.862	173.660	8.54	432.38	38.2	638.7	31.3	429	0.257
Silicon dioxide	SiO_2	60.09	2.25	1883	8.54	2503					
Sodium bisulfate	$NaHSO_4$	120.07	2.742	455							
Sodium carbonate (sal soda)	$Na_2CO_3 \cdot 10H_2O$	286.15	1.46	306.5		$(-H_2O$ at 306.5 K)					
Sodium carbonate (soda ash)	Na_2CO_3	105.99	2.533	1127	33.4	(decomposes)					
Sodium chloride	$NaCl$	58.45	2.163	1081	28.5	1738	171				
Sodium cyanide	$NaCN$	49.01		835	16.7	1770	155				
Sodium hydroxide	$NaOH$	40.00	2.130	592	8.4	1663					
Sodium nitrate	$NaNO_3$	85.00	2.257	583	15.9	(decomposes at 653 K)					
Sodium nitrite	$NaNO_2$	69.00	2.168⁰°	544		(decomposes at 593 K)					
Sodium sulfate	Na_2SO_4	142.05	2.698	1163	24.3						
Sodium sulfide	Na_2S	78.05	1.856	1223	6.7						
Sodium sulfite	Na_2SO_3	126.05	2.633¹⁵°	(decomposes)							

Name	Formula	Sp gr	Mol wt								
Sodium thiosulfate	$Na_2S_2O_3$	1.667	158.11								
Sulfur (rhombic)	S_8	2.07	256.53	386	10.0	717.76	84				0.269
Sulfur (monoclinic)	S_8	1.96	256.53	392	14.17	717.76	84				0.262
Sulfur chloride (mono)	S_2CL_2	1.687	135.05	193.0		411.2	36.0				
Sulfur dioxide	SO_2	2.264(A)	64.07	197.68	7.402	263.14	24.92	430.7	77.8	122	0.263
Sulfur trioxide	SO_3	2.75(A)	80.07	290.0	24.5	316.5	41.8	491.4	83.8	126	0.230
Sulfuric acid	H_2SO_4	$1.834^{18°}$	98.08	283.51	9.87			(decomposes at 613 K)			
Toluene	$C_6H_5CH_3$	0.866	92.13	178.169	6.619	383.78	33.5	593.9	40.3	318	
Water	H_2O	$1.00^{4°}$	18.016	273.16	6.009	373.16	40.65	647.4	218.3	56	
m-Xylene	C_8H_{10}	0.864	106.16	225.288	11.57	412.26	34.4	619	34.6	390	0.27
o-Xylene	C_8H_{10}	0.880	106.16	247.978	13.60	417.58	36.8	631.5	35.7	380	0.26
p-Xylene	C_8H_{10}	0.861	106.16	286.423	17.11	411.51	36.1	618	33.9	370	0.25
Zinc	Zn	7.140	65.38	692.7	6.673	1180	114.8				
Zinc sulfate	$ZnSO_4$	$3.74^{15°}$	161.44	(decomposes at 1013 K)							

*Sources of data are listed at the beginning of Appendix D.
Sp gr = 20°C/4°C unless specified. Sp gr for gas referred to air (A).

TABLE D.2   Enthalpies of Paraffinic Hydrocarbons, C_1–C_6 (J/g mol) (at 1 atm)
To convert to Btu/lb mol, multiply by 0.4306.

K	C_1	C_2	C_3	n-C_4	i-C_4	n-C_5	n-C_6
273	0						
291	630	912	1,264	1,709	1,658	2,125	2,545
298	879	1,277	1,771	2,394	2,328	2,976	3,563
300	950	1,383	1,919	2,592	2,522	3,222	3,858
400	4,740	7,305	10,292	13,773	13,623	17,108	20,463
500	9,100	14,476	20,685	27,442	27,325	34,020	40,622
600	14,054	22,869	32,777	43,362	43,312	53,638	64,011
700	19,585	32,342	46,417	61,186	61,220	75,604	90,123
800	25,652	42,718	61,337	80,600	80,767	99,495	118,532
900	32,204	53,931	77,404	101,378	101,754	125,101	148,866
1,000	39,204	65,814	94,432	123,428	123,971	152,213	181,041
1,100	46,567	78,408	112,340	146,607	147,234	180,665	214,764
1,200	54,308	91,504	131,042	170,707	171,418	210,246	249,868
1,300	62,383	105,143	150,331	195,727	196,480	240,872	286,143
1,400	70,709	119,202	170,205	221,375	222,212	272,378	323,465
1,500	79,244	133,678	190,581	247,650	248,571	304,595	361,539
1,600	88,031						
1,800	106,064						
2,000	124,725						
2,200	143,804						
2,500	173,050						

TABLE D.3   Enthalpies of Other Hydrocarbons (J/g mol) (at 1 atm)
To convert to Btu/lb mol, multiply by 0.4306.

K	Ethylene	Propylene	1-Butene	iso-Butene	Acety-lene	Benzene
273	0	0	0	0	0	0
291	753	1,104	1,536	1,538	769	1,381
298	1,054	1,548	2,154	2,154	1,075	1,945
300	1,125	1,665	2,313	2,322	1,163	2,109
400	6,008	8,882	12,455	12,367	5,899	11,878
500	11,890	17,572	24,765	24,468	11,125	24,372
600	18,648	27,719	38,911	38,425	16,711	39,150
700	26,158	39,049	54,643	53,889	22,556	55,840
800	34,329	51,379	71,755	70,793	28,686	74,015
900	43,053	64,642	89,997	88,826	35,074	93,471
1,000	52,258	78,742	109,286	107,947	41,635	113,972
1,100	61,923	93,470	129,494	123,846	48,388	135,394
1,200	71,964	108,825	150,456	148,866	55,271	157,611
1,300	82,341	124,683	172,087	170,414	62,342	180,456
1,400	92,968	141,000	194,304	188,363	69,622	203,844
1,500	103,888	157,736	217,065	215,183	76,944	227,777

TABLE D.4   Enthalpies of Nitrogen and Some of its Oxides (J/g mol) (at 1 atm)
To convert to Btu/lb mol, multiply by 0.4306.

K	N_2	NO	N_2O	NO_2	N_2O_4
273	0	0	0	0	0
291	524	537	681	658	1,384
298	728	746	951	917	1,937
300	786	801	9,660	985	2,083
400	3,695	3,785	13,740	4,865	10,543
500	6,644	6,811	18,179	9,070	19,915
600	9,627	9,895	22,919	13,564	30,124
700	12,652	13,054	27,924	18,305	
800	15,756	16,292	33,154	23,242	
900	18,961	19,597	38,601	28,334	
1,000	22,171	22,970	44,258	33,551	
1,100	25,472	26,392	50,115	38,869	
1,200	28,819	29,861	56,170	44,266	
1,300	32,216	33,371	62,425	49,731	
1,400	35,639	36,915	68,868	55,258	
1,500	39,145	40,488	75,504	60,826	
1,750	47,940	49,505			
2,000	56,902	58,634			
2,250	65,981	67,856			
2,500	75,060	77,127			

TABLE D.5   Enthalpies of Sulfur Compounds (J/g mol) (at 1 atm)
To convert to Btu/lb mol, multiply by 0.4306.

K	$S_2(g)$	SO_2	SO_3	H_2S	CS_2
273	0	0	0	0	0
291	579	706	899	607	807
298	805	984	1,255	845	1,125
300	869	1,064	1,338	909	1,217
400	4,196	5,234	6,861	4,372	5,995
500	7,652	9,744	13,033	7,978	11,108
600	11,192	14,514	19,832	11,752	16,455
700	14,790	19,501	27,154	15,706	21,974
800	18,426	24,647	34,748	19,840	27,631
900	22,087	29,915	42,676	24,145	33,388
1,000	25,769	35,275	50,835	28,610	39,220
1,100	29,463	40,706	59,203	33,216	45,103
1,200	33,174	46,191	67,738	37,953	51,044
1,300	36,898	51,714	76,399	42,802	57,027
1,400	40,630	57,320		47,739	63,052
1,500	44,371	62,927		52,802	69,119
1,600	48,116	68,533		57,906	75,186
1,700	51,881	74,182		63,094	81,295
1,800	55,605	79,872		68,324	87,361
1,900	59,370	85,520			
2,000	63,136	91,253			
2,500	82,006	119,871			
3,000	100,959	148,657			

TABLE D.6   Enthalpies of Combustion Gases* (J/g mol)†

K	N_2	O_2	Air	H_2	CO	CO_2	H_2O
273	0	0	0	0	0	0	0
291	524	527	523	516	525	655	603
298	728	732	726	718	728	912	837
300	786	790	784	763	786	986	905
400	3,695	3,752	3,696	3,655	3,699	4,903	4,284
500	6,644	6,811	6,660	6,589	6,652	9,204	7,752
600	9,627	9,970	9,673	9,518	9,665	13,807	11,326
700	12,652	13,225	12,736	12,459	12,748	18,656	15,016
800	15,756	16,564	15,878	15,413	15,899	23,710	18,823
900	18,961	19,970	19,116	18,384	19,125	28,936	22,760
1,000	22,171	23,434	22,367	21,388	22,413	34,308	26,823
1,100	25,472	26,940	25,698	24,426	25,760	39,802	31,011
1,200	28,819	30,492	29,078	27,509	29,154	45,404	35,312
1,300	32,216	34,078	32,501	30,626	32,593	51,090	39,722
1,400	35,639	37,693	35,953	33,789	36,070	56,860	44,237
1,500	39,145	41,337	39,463	36,994	39,576	62,676	48,848
1,750	47,940	50,555	48,325	45,275	48,459	77,445	60,751
2,000	56,902	59,914	57,320	53,680	57,488	92,466	73,136
2,250	65,981	69,454	66,441	62,341	66,567	107,738	85,855
2,500	75,060	79,119	75,646	71,211	75,772	123,176	98,867
2,750	84,265	88,910	84,935	80,290	85,018	138,699	112,089
3,000	93,512	98,826	94,265	89,453	94,265	154,347	125,520
3,500	112,131	119,034	113,135	108,030	112,968	185,895	152,799
4,000	130,875	141,410	132,172	127,528	131,796	217,777	180,414

†To convert to cal/g mol multiply by 0.2390.

*Pressure = 1 atm.

SOURCE: Page 30 of Reference in Table D.7.

TABLE D.7 Enthalpies of Combustion Gases* (Btu/lb mol)

°R	N_2	O_2	Air	H_2	CO	CO_2	H_2O
492	0.0	0.0	0.0	0.0	0.0	0.0	0.0
500	55.67	55.93	55.57	57.74	55.68	68.95	64.02
520	194.9	195.9	194.6	191.9	194.9	243.1	224.2
537	313.2	315.1	312.7	308.9	313.3	392.2	360.5
600	751.9	758.8	751.2	744.4	752.4	963	867.5
700	1,450	1,471	1,450	1,433	1,451	1,914	1,679
800	2,150	2,194	2,153	2,122	2,154	2,915	2,501
900	2,852	2,931	2,861	2,825	2,863	3,961	3,336
1,000	3,565	3,680	3,579	3,511	3,580	5,046	4,184
1,100	4,285	4,443	4,306	4,210	4,304	6,167	5,047
1,200	5,005	5,219	5,035	4,917	5,038	7,320	5,925
1,300	5,741	6,007	5,780	5,630	5,783	8,502	6,819
1,400	6,495	6,804	6,540	6,369	6,536	9,710	7,730
1,500	7,231	7,612	7,289	7,069	7,299	10,942	8,657
1,600	8,004	8,427	8,068	7,789	8,072	12,200	9,602
1,700	8,774	9,251	8,847	8,499	8,853	13,470	10,562
1,800	9,539	10,081	9,623	9,219	9,643	14,760	11,540
1,900	10,335	10,918	10,425	9,942	10,440	16,070	12,530
2,000	11,127	11,760	11,224	10,689	11,243	17,390	13,550
2,100	11,927	12,610	12,030	11,615	12,050	18,730	14,570
2,200	12,730	13,460	12,840	12,160	12,870	20,070	15,610
2,300	13,540	14,320	13,660	12,890	13,690	21,430	16,660
2,400	14,350	15,180	14,480	13,650	14,520	22,800	17,730
2,500	15,170	16,040	15,300	14,400	15,350	24,180	18,810

*Pressure = 1 atm.

SOURCE: Page 30 Kobe, K.A., et al., *Thermochemistry of Petrochemicals*, Reprint No. 44 from the *Petroleum Refiner*, Gulf Publ. Co., Houston, TX (1958).

Appendix E

HEAT CAPACITY EQUATIONS

TABLE E.1   Heat Capacity Equations for Organic and Inorganic Compounds (at Low Pressures)*

Forms: (1) $C_p^\circ = a + b(T) + c(T)^2 + d(T)^3$;

         (2) $C_p^\circ = a + b(T) + c(T)^{-2}$.

         Units of C_p° are J/(g mol)(K or °C).

         To convert to cal/(g mol)(K or °C) = Btu/(lb mol)(°R or °F), multiply by 0.2390.

Note: $b \cdot 10^2$ means the vaue of b is to be multiplied by 10^{-2}, e.g., 20.10×10^{-2} for acetone.

Compound	Formula	Mol. Wt.	State	Form	T	a	$b \cdot 10^2$	$c \cdot 10^5$	$d \cdot 10^9$	Temp. Range (in T)
Acetone	CH_3COCH_3	58.08	g	1	°C	71.96	20.10	-12.78	34.76	0–1200
Acetylene	C_2H_2	26.04	g	1	°C	42.43	6.053	-5.033	18.20	0–1200
Air		29.0	g	1	°C	28.94	0.4147	0.3191	-1.965	0–1500
			g	1	K	28.09	0.1965	0.4799	-1.965	273–1800
Ammonia	NH_3	17.03	g	1	°C	35.15	2.954	0.4421	-6.686	0–1200
Ammonium sulfate	$(NH_4)_2SO_4$	132.15	c		K	215.9				275–328
Benzene	C_6H_6	78.11	l	1	K	-7.27329	77.054	-164.82	1,897.9	279–350
			g	1	°C	74.06	32.95	-25.20	77.57	0–1200
Isobutane	C_4H_{10}	58.12	g	1	°C	89.46	30.13	-18.91	49.87	0–1200
n-Butane	C_4H_{10}	58.12	g	1	°C	92.30	27.88	-15.47	34.98	0–1200
Isobutene	C_4H_8	56.10	g	1	°C	82.88	25.64	-17.27	50.50	0–1200
Calcium carbide	CaC_2	64.10	c	2	K	68.62	1.19	-8.66×10^{10}	—	298–720
Calcium carbonate	$CaCO_3$	100.09	c	2	K	82.34	4.975	-12.87×10^{10}	—	273–1033
Calcium hydroxide	$Ca(OH)_2$	74.10	c	1	K	89.5				276–373
Calcium oxide	CaO	56.08	c	2	K	41.84	2.03	-4.52×10^{10}		10273–1173
Carbon	C	12.01	c†	2	K	11.18	1.095	-4.891×10^{10}		273–1373
Carbon dioxide	CO_2	44.01	g	1	°C	36.11	4.233	-2.887	7.464	0–1500
Carbon monoxide	CO	28.01	g	1	°C	28.95	0.4110	0.3548	-2.220	0–1500
Carbon tetrachloride	CCl_4	153.84	l	1	K	12.285	0.01095	-318.26	3,425.2	273–343
Chlorine	Cl_2	70.91	g	1	°C	33.60	1.367	-1.607	6.473	0–1200
Copper	Cu	63.54	c	1	K	22.76	0.06117			273–1357

†Graphite.   ‡Rhombic.   §Moinoclinic.   (at 1 atm)

TABLE E.1 (*cont.*)

Compound	Formula	Mol. Wt.	State	Form	T	a	b · 10²	c · 10⁵	d · 10⁹	Temp. Range (in T)
Cumene (Isopropyl benzene)	C_9H_{12}	120.19	g	1	°C	139.2	53.76	−39.79	120.5	0–1200
Cyclohexane	C_6H_{12}	84.16	g	1	°C	94.140	49.62	−31.90	80.63	0–1200
Cyclopentane	C_5H_{10}	70.13	g	1	°C	73.39	39.28	−25.54	68.66	0–1200
Ethane	C_2H_6	30.07	g	1	°C	49.37	13.92	−5.816	7.280	0–1200
Ethyl alcohol	C_2H_6O	46.07	l	—	K	−325.137	0.041379	−1,403.1	1.7035×10^4	250–400
			g	1	°C	61.34	15.72	−8.749	19.83	0–1200
Ethylene	C_2H_4	28.05	g	1	°C	40.75	11.47	−6.891	17.66	0–1200
Ferric oxide	Fe_2O_3	159.70	c	2	K	103.4	6.711	-17.72×10^{10}	—	273–1097
Formaldehyde	CH_2O	30.03	g	1	°C	34.28	4.268	0.0000	−8.694	0–1200
Helium	He	4.00	g	1	°C	20.8				All
n-Hexane	C_6H_{14}	86.17	l	—	K	31.421	0.97606	−235.37	3,092.7	273–400
			g	1	°C	137.44	40.85	−23.92	57.66	0–1200
Hydrogen	H_2	2.016	g	1	°C	28.84	0.00765	0.3288	−0.8698	0–1500
Hydrogen bromide	HBr	80.92	g	1	°C	29.10	−0.0227	0.9887	−4.858	0–1200
Hydrogen chloride	HCl	36.47	g	1	°C	29.13	−0.1341	0.9715	−4.335	0–1200
Hydrogen cyanide	HCN	27.03	g	1	°C	35.3	2.908	1.092		0–1200
Hydrogen sulfide	H_2S	34.08	g	1	°C	33.51	1.547	0.3012	−3.292	0–1500
Magnesium chloride	$MgCl_2$	95.23	c	1	K	72.4	1.58			273–991
Magnesium oxide	MgO	40.32	c	2	K	45.44	0.5008	-8.732×10^{10}		273–2073
Methane	CH_4	16.04	g	1	°C	34.31	5.469	0.3661	−11.00	0–1200
			g	—	K	19.87	5.021	1.268	−11.00	273–1500
Methyl alcohol	CH_3OH	32.04	l	—	K	−259.25	0.03358	−1.1639	1.4052×10^4	273–400
			g	1	°C	42.93	8.301	−1.87	−8.03	0–700
Methyl cyclohexane	C_7H_{14}	98.18	g	1	°C	121.3	56.53	−37.72	100.8	0–1200
Methyl cyclopentane	C_6H_{12}	84.16	g	1	°C	98.83	45.857	−30.44	83.81	0–1200
Nitric acid	HNO_3	63.02	l	1	—	110.0				25

Nitric oxide	NO	30.01	g	1	29.50	°C	0.8188	−0.2925	0.3652	0–3500
Nitrogen	N_2	28.02	g	1	29.00	°C	0.2199	0.5723	−2.871	0–1500
Nitrogen dioxide	NO_2	46.01	g	1	36.07	°C	3.97	−2.88	7.87	0–1200
Nitrogen tetraoxide	N_2O_4	92.02	g	1	75.7	°C	12.5	−11.3		0–300
Nitrous oxide	N_2O	44.02	g	1	37.66	°C	4.151	−2.694	10.57	0–1200
Oxygen	O_2	32.00	g	1	29.10	°C	1.158	−0.6076	1.311	0–1500
n-Pentane	C_5H_{12}	72.15	l	1	33.24	K	192.41	−236.87	17,944	270–350
			g	1	114.8	°C	34.09	−18.99	42.26	0–1200
Propane	C_3H_8	44.09	g	1	68.032	°C	22.59	−13.11	31.71	0–1200
Propylene	C_3H_6	42.08	g	1	59.580	°C	17.71	−10.17	24.60	0–1200
Sodium carbonate	Na_2CO_3	105.99	c	1	121	K				288–371
Sodium carbonate ·$10H_2O$	Na_2CO_3·$10H_2O$	286.15	c	1	535.6	K				298
Sulfur	S	32.07	c‡	1	15.2	K	2.68			273–368
			c§	1	18.5	K	1.84			368–392
Sulfuric acid	H_2SO_4	98.08	l	1	139.1	°C	15.59			10–45
Sulfur dioxide	SO_2	64.07	g	1	38.91	°C	3.904	−3.104	8.606	0–1500
Sulfur trioxide	SO_3	80.07	g	1	48.50	°C	9.188	−8.540	32.40	0–1000
Toluene	C_7H_8	92.13	l	1	1.8083	K	81.222	−151.27	1,630	270–370
			g	1	94.18	°C	38.00	−27.86	80.33	0–1200
Water	H_2O	18.016	l	1	18.2964	K	47.212	−133.88	1,314.2	273–373
			g	1	33.46	°C	0.6880	0.7604	−3.593	0–1500

†Graphite. ‡Rhombic. §Monoclinic.

HEATS OF FORMATION AND COMBUSTION

TABLE F.1   Heats of Formation and Heats of Combustion of Compounds at 25°C*†

Standard states of products for $\Delta\hat{H}_c^{\circ}$ are $CO_2(g)$, $H_2O(l)$, $N_2(g)$, $SO_2(g)$, and $HCl(aq)$. To convert to Btu/lb mol, multiply by 430.6.

Compound	Formula	Mol. wt.	State	$\Delta\hat{H}_f^{\circ}$ (kJ/g mol)	$\Delta\hat{H}_c^{\circ}$ (kJ/g mol)
Acetic acid	CH_3COOH	60.05	l	−486.2	−871.69
			g		−919.73
Acetaldehyde	CH_3CHO	40.052	g	−166.4	−1192.36
Acetone	C_3H_6O	58.08	aq, 200	−410.03	
			g	−216.69	−1821.38
Acetylene	C_2H_2	26.04	g	226.75	−1299.61
Ammonia	NH_3	17.032	l	−67.20	
			g	−46.191	−382.58
Ammonium carbonate	$(NH_4)_2CO_3$	96.09	c		
			aq	−941.86	
Ammonium chloride	NH_4Cl	53.50	c	−315.4	
Ammonium hydroxide	NH_4OH	35.05	aq	−366.5	
Ammonium nitrate	NH_4NO_3	80.05	c	−366.1	
			aq	−339.4	
Ammonium sulfate	$(NH_4)SO_4$	132.15	c	−1179.3	
			aq	−1173.1	
Benzaldehyde	C_6H_5CHO	106.12	l	−88.83	
			g	−40.0	

TABLE F.1   (*cont.*)

Compound	Formula	Mol. wt.	State	$\Delta \hat{H}^{\circ}_{f}$ (kJ/g mol)	$\Delta \hat{H}^{\circ}_{c}$ (kJ/g mol)
Benzene	C_6H_6	78.11	l	48.66	−3267.6
			g	82.927	−3301.5
Boron oxide	B_2O_3	69.64	c	−1263	
			l	−1245.2	
Bromine	Br_2	159.832	l	0	
			g	30.7	
n-Butane	C_4H_{10}	58.12	l	−147.6	−2855.6
			g	−124.73	−2878.52
Isobutane	C_4H_{10}	58.12	l	−158.5	−2849.0
			g	−134.5	−2868.8
1-Butene	C_4H_8	56.104	g	1.172	−2718.58
Calcium arsenate	$Ca_3(AsO_4)_2$	398.06	c	−3330.5	
Calcium carbide	CaC_2	64.10	c	−62.7	
Calcium carbonate	$CaCO_3$	100.09	c	−1206.9	
Calcium chloride	$CaCl_2$	110.99	c	−794.9	
Calcium cyanamide	$CaCN_2$	80.11	c	−352	
Calcium hydroxide	$Ca(OH)_2$	74.10	c	−986.56	
Calcium oxide	CaO	56.08	c	−635.6	
Calcium phosphate	$Ca_3(PO_4)_2$	310.19	c	−4137.6	
Calcium silicate	$CaSiO_3$	116.17	c	−1584	
Calcium sulfate	$CaSO_4$	136.15	c	−1432.7	
			aq	−1450.5	
Calcium sulfate (gypsum)	$CaSO_4 \cdot 2H_2O$	172.18	c	−2021.1	
Carbon	C	12.01	c	0	−393.51
			Graphite (β)		
Carbon dioxide	CO_2	44.01	g	−393.51	
			l	−412.92	
Carbon disulfide	CS_2	76.14	l	87.86	−1075.2
			g	115.3	−1102.6
Carbon monoxide	CO	28.01	g	−110.52	−282.99
Carbon tetrachloride	CCl_4	153.838	l	−139.5	−352.2
			g	−106.69	−384.9
Chloroethane	C_2H_5Cl	64.52	g	−105.0	−1421.1
			l	−41.20	−5215.44
Cumene (isopropylbenzene)	$C_6H_5CH(CH_3)_2$	120.19	g	3.93	−5260.59
			c	−769.86	
Cupric sulfate	$CuSO_4$	159.61	aq	−843.12	
			c	−751.4	
Cyclohexane	C_6H_{12}	84.16	g	−123.1	−3953.0
Cyclopentane	C_5H_{10}	70.130	l	−105.8	−3290.9
			g	−77.23	−3319.5

TABLE F.1   (*cont.*)

Compound	Formula	Mol. wt.	State	$\Delta \hat{H}_f^\circ$ (kJ/g mol)	$\Delta \hat{H}_c^\circ$ (kJ/g mol)
Ethane	C_2H_6	30.07	g	−84.667	−1559.9
Ethyl acetate	$CH_3CO_2C_2H_5$	88.10	l	−442.92	−2274.48
Ethyl alcohol	C_2H_5OH	46.068	l	−277.63	−1366.91
			g	−235.31	−1409.25
Ethyl benzene	$C_6H_5 \cdot C_2H_5$	106.16	l	−12.46	−4564.87
			g	29.79	−4607.13
Ethyl chloride	C_2H_5Cl	64.52	g	−105	
Ethylene	C_2H_4	28.052	g	52.283	−1410.99
Ethylene chloride	C_2H_3Cl	62.50	g	31.38	−1271.5
3-Ethyl hexane	C_8H_{18}	114.22	l	−250.5	−5470.12
			g	−210.9	−5509.78
Ferric chloride	$FeCl_3$		c	−403.34	
Ferric oxide	Fe_2O_3	159.70	c	−822.156	
Ferric sulfide	FeS_2	*see* Iron sulfide	*see* Iron sulfide		
Ferrosoferric oxide	Fe_3O_4	231.55	c	−1116.7	
Ferrous chloride	$FeCl_2$		c	−342.67	−303.76
Ferrous oxide	FeO	71.85	c	−267	
Ferrous sulfide	FeS	87.92	c	−95.06	
Formaldehyde	H_2CO	30.026	g	−115.89	−563.46
n-Heptane	C_7H_{16}	100.20	l	−224.4	−4816.91
			g	−187.8	−4853.48
n-Hexane	C_6H_{14}	86.17	l	−198.8	−4163.1
			g	−167.2	−4194.753
Hydrogen	H_2	2.016	g	0	−285.84
Hydrogen bromide	HBr	80.924	g	−36.23	
Hydrogen chloride	HCl	36.465	g	−92.311	
Hydrogen cyanide	HCN	27.026	g	130.54	
Hydrogen sulfide	H_2S	34.082	g	−20.15	−562.589
Iron sulfide	FeS_2	119.98	c	−177.9	
Lead oxide	PbO	223.21	c	−219.2	
Magnesium chloride	$MgCl_2$	95.23	c	−641.83	
Magnesium hydroxide	$Mg(OH)_2$	58.34	c	−924.66	
Magnesium oxide	MgO	40.32	c	−601.83	
Methane	CH_4	16.041	g	−74.84	−890.4
Methyl alcohol	CH_3OH	32.042	l	−238.64	−726.55
			g	−201.25	−763.96
Methyl chloride	CH_3Cl	50.49	g	−81.923	−766.63[†]
Methyl cyclohexane	C_7H_{14}	98.182	l	−190.2	−4565.29
			g	−154.8	−4600.68
Methyl cyclopentane	C_6H_{12}	84.156	l	−138.4	−3937.7
			g	−106.7	−3969.4

TABLE F.1   (*cont.*)

Compound	Formula	Mol. wt.	State	$\Delta \hat{H}_f^\circ$ (kJ/g mol)	$\Delta \hat{H}_c^\circ$ (kJ/g mol)
Nitric acid	HNO_3	63.02	l	−173.23	
			aq	−206.57	
Nitric oxide	NO	30.01	g	90.374	
Nitrogen dioxide	NO_2	46.01	g	33.85	
Nitrous oxide	N_2O	44.02	g	81.55	
n-Pentane	C_5H_{12}	72.15	l	−173.1	−3509.5
			g	−146.4	−3536.15
Phosphoric acid	H_3PO_4	98.00	c	−1281	
			aq ($1H_2O$)	−1278	
Phosphorus	P_4	123.90	c	0	
Phosphorus pentoxide	P_2O_5	141.95	c	−1506	
Propane	C_3H_8	44.09	l	−119.84	−2204.0
			g	−103.85	−2220.0
Propene	C_3H_6	42.078	g	20.41	−2058.47
n-Propyl alcohol	C_3H_8O	60.09	g	−255	−2068.6
n-Propylbenzene	$C_6H_5 \cdot CH_2 \cdot C_2H_5$	120.19	l	−38.40	−5218.2
			g	7.824	−5264.5
Silicon dioxide	SiO_2	60.09	c	−851.0	
Sodium bicarbonate	$NaHCO_3$	84.01	c	−945.6	
Sodium bisulfate	$NaHSO_4$	120.07	c	−1126	
Sodium carbonate	Na_2CO_3	105.99	c	−1130	
Sodium chloride	NaCl	58.45	c	−411.00	
Sodium cyanide	NaCN	49.01	c	−89.79	
Sodium nitrate	$NaNO_3$	85.00	c	−466.68	
Sodium nitrite	$NaNO_2$	69.00	c	−359	
Sodium sulfate	Na_2SO_4	142.05	c	−1384.5	
Sodium sulfide	Na_2S	78.05	c	−373	
Sodium sulfite	Na_2SO_3	126.05	c	−1090	
Sodium thiosulfate	$Na_2S_2O_3$	158.11	c	−1117	
Sulfur	S	32.07	c (rhombic)	0	
			c (monoclinic)	0.297	
Sulfur chloride	S_2Cl_2	135.05	l	−60.3	
Sulfur dioxide	SO_2	64.066	g	−296.90	
Sulfur trioxide	SO_3	80.066	g	−395.18	
Sulfuric acid	H_2SO_4	98.08	l	−811.32	
			aq	−907.51	
Toluene	$C_6H_5CH_3$	92.13	l	11.99	−3909.9
			g	50.000	−3947.9

TABLE F.1   (*cont.*)

Compound	Formula	Mol. wt.	State	$\Delta \hat{H}_f^\circ$ (kJ/g mol)	$\Delta \hat{H}_c^\circ$ (kJ/g mol)
Water	H_2O	18.016	l	−285.840	
			g	−241.826	
m-Xylene	$C_6H_4(CH_3)_2$	106.16	l	−25.42	−4551.86
			g	17.24	−4594.53
o-Xylene	$C_6H_4(CH_3)_2$	106.16	l	−24.44	−4552.86
			g	19.00	−4596.29
p-Xylene	$C_6H_4(CH_3)_2$	106.16	l	−24.43	−4552.86
			g	17.95	−4595.25
Zinc sulfate	$ZnSO_4$	161.45	c	−978.55	
			aq	−1059.93	

*Sources of data are given at the beginning of Appendix D, References 1, 4, and 5.
†Standard state HCl(g).

Appendix G

VAPOR PRESSURES

TABLE G.1 Vapor Pressures of Various Substances

Antoine equation:

$$\ln (p^*) = A - \frac{B}{C + T}$$

where p^* = vapor pressure, mm Hg

T = temperature, K

A, B, C = constants

Name	Formula	Range (K)	A	B	C
Acetic acid	$C_2H_4O_2$	290–430	16.8080	3405.57	−56.34
Acetone	C_3H_6O	241–350	16.6513	2940.46	−35.93
Ammonia	NH_3	179–261	16.9481	2132.50	−32.98
Benzene	C_6H_6	280–377	15.9008	2788.51	−52.36
Carbon disulfide	CS_2	288–342	15.9844	2690.85	−31.62
Carbon tetrachloride	CCl_4	253–374	15.8742	2808.19	−45.99
Chloroform	$CHCl_3$	260–370	15.9732	2696.79	−46.16
Cyclohexane	C_6H_{12}	280–380	15.7527	2766.63	−50.50
Ethyl acetate	$C_4H_8O_2$	260–385	16.1516	2790.50	−57.15
Ethyl alcohol	C_2H_6O	270–369	18.5242	3578.91	−50.50
Ethyl bromide	C_2H_5Br	226–333	15.9338	2511.68	−41.44
n-Heptane	C_7H_{16}	270–400	15.8737	2911.32	−56.51
n-Hexane	C_6H_{14}	245–370	15.8366	2697.55	−48.78
Methyl alcohol	CH_4O	257–364	18.5875	3626.55	−34.29
n-Pentane	C_5H_{12}	220–330	15.8333	2477.07	−39.94
Sulfur dioxide	SO_2	195–280	16.7680	2302.35	−35.97
Toluene	$C_6H_5CH_3$	280–410	16.0137	3096.52	−53.67
Water	H_2O	284–441	18.3036	3816.44	−46.13

SOURCE: R. C. Reid, J. M. Prausnitz, and T. K. Sherwood. *The Properties of Gases and Liquids*, 3rd ed., Appendix A. New York: McGraw-Hill, 1977.

Appendix H

HEATS OF SOLUTION AND DILUTION

TABLE H.1   Integral Heats of Solution and Dilution at 25°C

Formula	Description	State	$-\Delta \hat{H}_f^\circ$ (kJ/g mol)	$-\Delta \hat{H}_{soln}^\circ$ (kJ/g mol)	$-\Delta \hat{H}_{dil}^\circ$ (kJ/g mol)
NaOH	crystalline II		426.726		
	in 3 H$_2$O	aq	455.612	28.869	28.869
	4 H$_2$O	aq	461.156	34.434	5.564
	5 H$_2$O	aq	464.486	37.739	3.305
	10 H$_2$O	aq	469.227	42.509	4.769
	20 H$_2$O	aq	469.591	42.844	.334
	30 H$_2$O	aq	469.457	42.718	.125
	40 H$_2$O	aq	469.340	42.593	.125
	50 H$_2$O	aq	469.252	42.509	.083
	100 H$_2$O	aq	469.059	42.342	.167
	200 H$_2$O	aq	469.026	42.258	.083
	300 H$_2$O	aq	469.047	42.300	.041
	500 H$_2$O	aq	469.097	42.383	.083
	1,000 H$_2$O	aq	469.189	42.467	.083
	2,000 H$_2$O	aq	469.285	42.551	.083
	10,000 H$_2$O	aq	469.448	42.718	.041
	50,000 H$_2$O	aq	469.528	42.802	.083
	∞ H$_2$O	aq	469.595	42.886	.083
H$_2$SO$_4$		liq	811.319		
	in 0.5 H$_2$O	aq	827.051	15.731	15.731
	1.0 H$_2$O	aq	839.394	28.074	12.343
	1.5 H$_2$O	aq	848.222	36.902	8.823
	2 H$_2$O	aq	853.243	41.923	5.021
	3 H$_2$O	aq	860.314	48.994	7.071
	4 H$_2$O	aq	865.376	54.057	5.063
	5 H$_2$O	aq	869.351	58.032	3.975
	10 H$_2$O	aq	878.347	67.027	8.995
	25 H$_2$O	aq	883.618	72.299	5.272
	50 H$_2$O	aq	884.664	73.345	1.046
	100 H$_2$O	aq	885.292	73.973	0.628
	500 H$_2$O	aq	888.054	76.734	2.761
	1,000 H$_2$O	aq	889.894	78.575	1.841
	10,000 H$_2$O	aq	898.388	87.069	2.636
	100,000 H$_2$O	aq	904.957	93.637	6.568
	500,000 H$_2$O	aq	906.630	95.311	1.674
	∞ H$_2$O	aq	907.509	96.190	0.879

SOURCE: F. D. Rossini et al. "Selected Values of Chem. Thermo. Properties." *National Bureau of Standards Circular 500.* Washington, DC: U.S. Government Printing Office, 1952.

670

ENTHALPY-CONCENTRATION DATA

Appendix I

TABLE I.1 Enthalpy-concentration Data for the Single-phase Liquid Region and Also the Saturated Vapor of the Acetic Acid-water System at 1 Atmosphere

Reference state: Liquid water at 32°F and 1 atm; solid acid at 32°F and 1 atm.

Liquid or Vapor		Enthalpy—Btu/lb Liquid Solution							Enthalpy
Mole Fraction Water	Weight Fraction Water	20°C 68°F	40°C 104°F	60°C 140°F	80°C 176°F	100°C 212°F	Satu- rated Liquid		Saturated Vapor Btu/lb
0.00	0.0	93.54	111.4	129.9	149.1	169.0	187.5		361.8
0.05	0.01555	93.96	112.2	130.9	150.4	170.6	186.9		
0.10	0.03225	93.82	112.3	131.5	151.3	172.0	186.5		374.6
0.20	0.0698	92.61	111.9	131.7	152.2	173.8	185.2		395.3
0.30	0.1140	90.60	110.7	131.3	152.6	175.0	183.8		423.7
0.40	0.1667	87.84	108.9	130.6	152.6	176.1	182.9		461.4
0.50	0.231	83.96	106.3	129.1	152.7	177.1	182.4		510.5
0.55	0.268	81.48	104.5	128.1	152.5	177.5	182.3		
0.60	0.3105	78.53	102.5	126.9	152.1	178.0	182.0		573.4
0.65	0.358	75.36	100.2	125.5	151.6	178.3	181.9		
0.70	0.412	71.72	97.71	123.9	151.1	178.8	181.7		656.0
0.75	0.474	67.59	94.73	122.2	149.9	179.3	182.1		
0.80	0.545	62.88	91.43	120.2	149.7	179.9	181.6		767.3
0.85	0.630	57.44	87.56	117.8	148.9	180.5	181.6		
0.90	0.730	51.03	83.02	115.1	147.8	180.8	181.7		921.6
0.95	0.851	43.74	77.65	111.7	146.4	181.2	181.5		
1.00	1.00	36.06	71.91	107.7	143.9	180.1	180.1		1150.4

SOURCE: Data calculated from miscellaneous literature sources and smoothed.

TABLE I.2 Vapor-liquid Equilibrium Data for the Acetic Acid-water System; Pressure = 1 Atmosphere

x Mole Fraction Water in the Liquid	y Mole Fraction Water in the Vapor
0.020	0.035
0.040	0.069
0.060	0.103
0.080	0.135
0.100	0.165
0.200	0.303
0.300	0.425
0.400	0.531
0.500	0.627
0.600	0.715
0.700	0.796
0.800	0.865
0.900	0.929
0.940	0.957
0.980	0.985

SOURCE: Data from L. W. Cornell and R. E. Montonna. *Ind. Eng. Chem.*, 25 1933: 1331–1335.

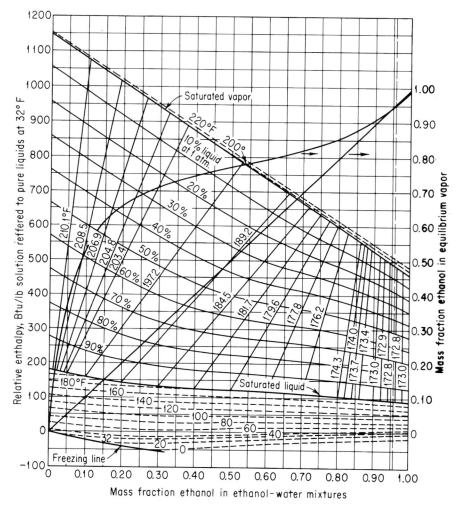

Figure I.1 Enthalpy-composition diagram for the ethanol-water system, showing liquid and vapor phases in equilibrium at 1 atm.

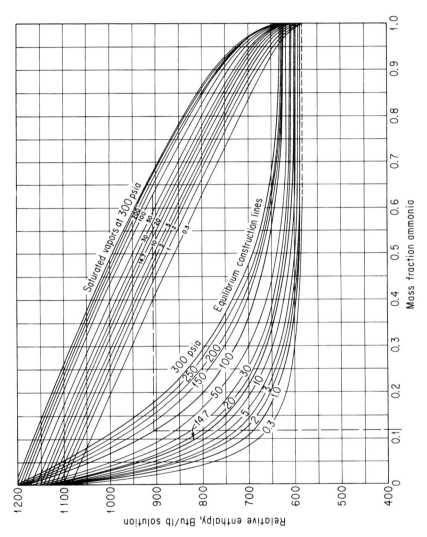

Figure I.2 Enthalpy-concentration chart for NH_3–H_2O

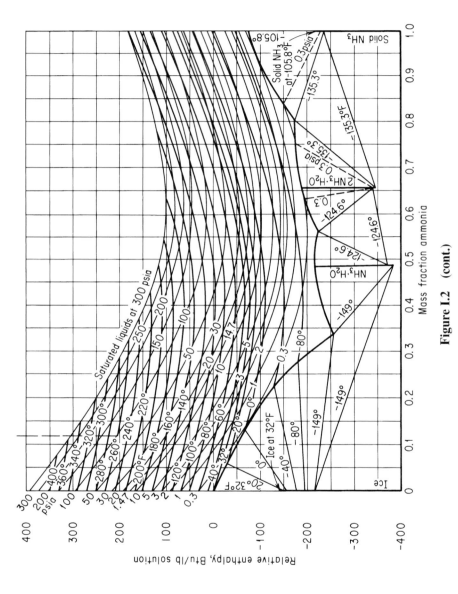

Figure I.2 (cont.)

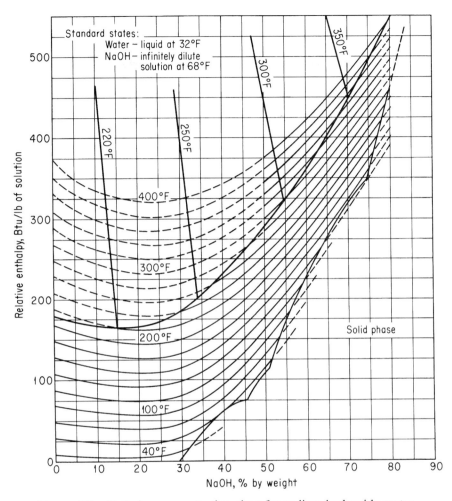

Figure I.3 Enthalpy-concentration chart for sodium hydroxide-water.

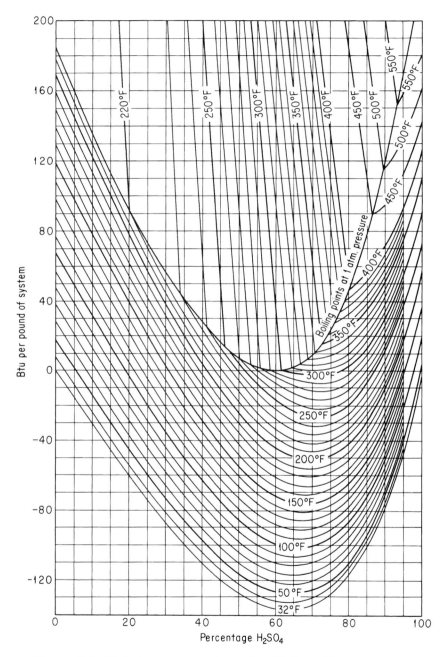

Figure I.4 Enthalpy-concentration of sulfuric acid-water system relative to pure components. (Water and H_2SO_4 at 32°F and own vapor pressure.) (Data from International Critical Tables, © 1943 O. A. Hougen and K. M. Watson.)

Appendix J

THERMODYNAMIC CHARTS

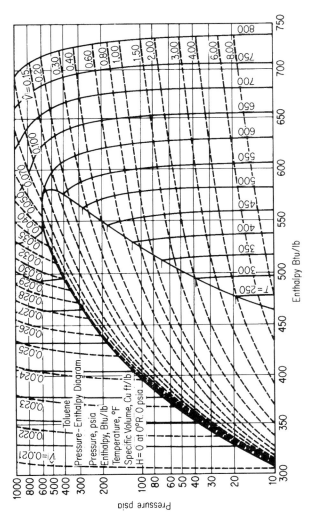

Figure J.1 Pressure-enthalpy chart for toluene.

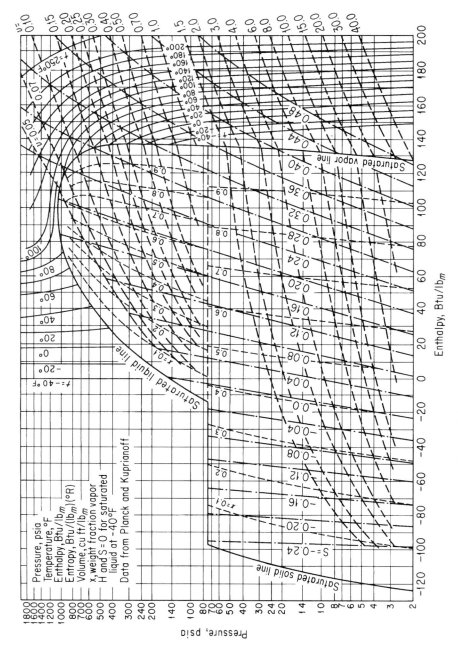

Figure J.2 Pressure-enthalpy chart for carbon dioxide.

Enthalpy, Btu/lb$_m$

Pressure, psia

678

Appendix K

PHYSICAL PROPERTIES OF PETROLEUM FRACTIONS

In the early 1930s, tests were developed which characterized petroleum oils and petroleum fractions, so that various physical characteristics of petroleum products could be related to these tests. Details of the tests can be found in *Petroleum Products and Lubricants,* an annual publication of the Committee D-2 of the American Society for Testing Materials.[1] These tests are not scientifically exact, and hence the procedure used in the tests must be followed faithfully if reliable results are to be obtained. However, the tests have been adopted because they are quite easy to perform in the ordinary laboratory and because the properties of petroleum fractions can be predicted from the results. The specifications for fuels, oils, and so on, are set out in terms of these tests plus many other properties, such as the flashpoint, the percent sulfur, and the viscosity.

Over the years various phases of the initial work have been extended and development of a new characterization scheme using the psuedocompound approach is evolving. Daubert[2] summarizes the traditional and new methods insofar as predicting molecular weights, pseudocritical temperature and pressure, acentric factor, and characterization factors.

In this appendix we present the results of the work of Smith and Watson and associates,[3] who related petroleum properties to a factor known as the *characterization factor* (sometimes called the *UOP characterization factor*). It is defined as

$$K = \frac{(T_B)^{1/3}}{S}$$

where K = *UOP* characterization factor
      T = cubic average boiling point, °R
      S = specific gravity at 60°F/60°F

[1]Report of Committee D-2, ASTM, Philadelphia, annually.
[2]T. E. Daubert. "Property Predictions." *Hydrocarbon Proc.* (March 1980): 108–110.
[3]R. L. Smith and K. M. Watson. *Ind. Eng. Chem., 29* (1937): 1408; K. M. Watson and E. F. Nelson. *Ind. Eng. Chem., 25* (1933): 880; K. M. Watson, E. F. Nelson, and G. B. Murphy. *Ind. Eng. Chem., 27* (1935): 1460.

Other averages for boiling points are used in evaluating K and the other physical properties in this Appendix. (Refer to Daubert[4] or Miquel[5] for details.) This factor has been related to many of the other simple tests and properties of petroleum fractions, such as viscosity, molecular weight, critical temperature, and percentage of hydrogen, so that it is quite easy to estimate the factor for any particular sample. Furthermore, tables of the UOP characterization factor are available for a wide variety of common types of petroleum fractions as shown in Table K.1 for typical liquids.

In Table K.2 are shown the source, boiling-point basis, and any special limitations of the various charts in this appendix. For use in computer-aided calculations, refer to the Appendix of the Supplement on the disk in the back of the book.

Table K.1   Typical UOP Characterization Factors

Type of Stock	K	Type of stock	K
Pennsylvania crude	12.2–12.5	Propane	14.7
Mid-Continent crude	11.8–12.0	Hexane	12.8
Gulf Coast crude	11.0–11.8	Octane	12.7
East Texas crude	11.9	Natural gasoline	12.7–12.8
California crude	10.98–11.9	Light gas oil	10.5
Benzene	9.5	Kerosene	10.5–11.5

[4]M. R. Riazi and T. E. Daubert. *Ind. Eng. Chem. Res., 26* (1987): 755–759.
[5]J. Miquel and F. Castells. *Hydrocarbon Processing* (Dec., 1993): 101–105.

Table K.2   Information Concerning Charts in Appendix K

1. *Specific heats of hydrocarbon liquids*
   Source: J. B. Maxwell. *Data Book on Hydrocarbons* (p. 93). New York: Van Nostrand Reinhold, 1950 (original from M. W. Kellogg Co.).
   Description: A chart of C_p (0.4 to 0.8) vs. t (0 to 1000°F) for petroleum fractions from 0 to 120° API.
   Boiling-point basis: Volumetric average boiling point, which is equal to graphical integration of the differential ASTM distillation curve (Van Winkle's "exact method").
   Limitations: This chart is not valid at temperatures within 50°F of the pseudocritical temperatures.

2. *Vapor pressure of hydrocarbons*
   Source: Maxwell, *Data Book on Hydrocarbons,* p. 42.
   Description: Vapor pressure (0.002 to 100 atm) vs. temperature (50 to 1200°F) for hydrocarbons with normal boiling points of 100 to 1200°F (C_4H_{10} and C_5H_{12} lines shown).
   Boiling-point basis: Normal boiling point (pure hydrocarbons).
   Limitations: These charts apply well to all hydrocarbon series except the lowest-boiling members of each series.

3. *Heat of combustion of fuel oils and petroleum fractions*
   Source: Maxwell, *Data Book on Hydrocarbons,* p. 180.
   Description: Heats of combustion above 60°F (17,000 to 25,000 Btu/lb) vs. gravity (0 to 60°API) with correction for sulfur and inerts included (as shown on chart).

4. *Properties of petroleum fractions*
   Source: O. A. Hougen and K. M. Watson. *Chemical Process Principles Charts* (Chart 3). New York: Wiley, 1946.
   Description: °API(-10 to 90°API) vs. boiling point (100 to 1000°F) with molecular weight, critical temperature, and K factors as parameters.
   Boiling-point basis: Use cubic average boiling point when using the K values; use mean average boiling point when using the molecular weights.

5. *Heat of vaporization of hydrocarbons and petroleum fractions at* 1.0 *atm pressure*
   Source: Hougen and Watson, *Chemical Process Principles Chart,* Chart 68.
   Description: Heats of vaporization (60 to 180 Btu/lb) vs. mean average boiling point (100 to 1000°F) with molecular weight and API gravity as parameters.
   Boiling-point basis: Mean average boiling point.

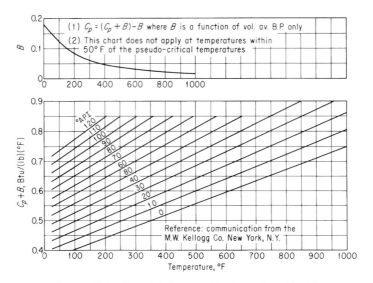

Figure K.1   Specific heats of hydrocarbon liquids.

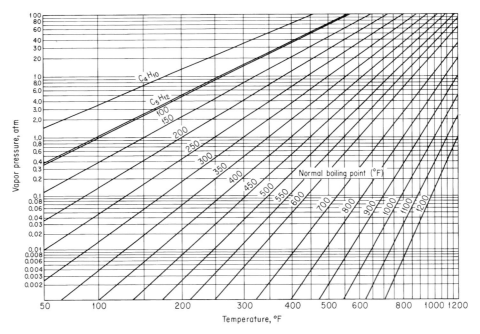

Figure K.2 Vapor pressure of hydrocarbons.

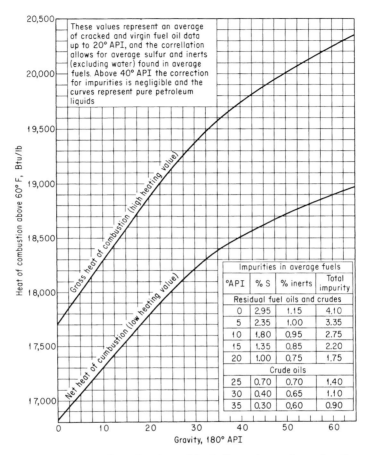

Figure K.3   Heat of combustion of fuel oils and petroleum fractions.

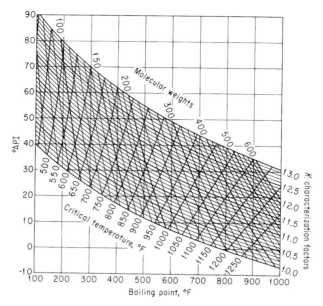

Figure K.4 Properties of petroleum fractions.

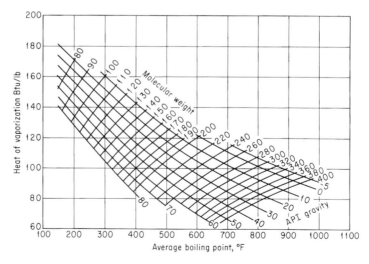

Figure K.5 Heats of vaporization of hydrocarbon and petroleum fractions at 1.0 atmosphere pressure.

Appendix L

SOLUTION OF SETS OF EQUATIONS

L.1 INDEPENDENT LINEAR EQUATIONS

This appendix contains a brief summary of methods of solving linear and nonlinear equations. It is only a summary; for details consult one of the numerous texts on numerical analysis that can be found in any library.

If you write several *linear* material balances, say m in number, they will take the form

$$a_{11}x_1 + a_{12}x_2 + \ldots + a_{1n}x_n = b_1$$

$$a_{21}x_1 + a_{22}x_2 + \ldots + a_{2n}x_n = b_2 \qquad \text{(L1.)}$$

$$\ldots$$

$$a_{m1}x_1 + a_{m2}x_2 + \ldots + a_{mn}x_n = b_m$$

or in compact matrix notation

$$\mathbf{Ax = b} \qquad \text{(L1.a)}$$

where $x_1, x_2, \ldots, x_n$ represent the unknown variables, and the a_{ij} and b_i represent the constants and known variables. As an example of Eq. (L.1), we can write the three component mass balances corresponding to Figure L.1:

$$0.50(100) = 0.80(P) + 0.05(W)$$

$$0.40(100) = 0.05(P) + 0.925(W) \qquad \text{(L.2)}$$

$$0.10(100) = 0.15(P) + 0.025(W)$$

With m equations in n unknown variables, three cases can be distinguished:

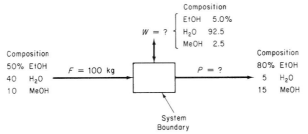

Figure L.1.   Data for a material balance.

1. There is no set of x's that satisfies Eq. (L.1).
2. There is a unique set of x's that satisfies Eq. (L.1).
3. There is an infinite number of sets of x's that satisfy Eq. (L.1).

Figure L.2 represents each of the three cases geometrically in two dimensions. Case 1 is usually termed inconsistent, whereas cases 2 and 3 are consistent; but to an engineer who is interested in the solution of practical problems, case 3 is as unsatisfying as case 1. Hence case 2 will be termed *determinate*, and case 3 will be termed *indeterminate*.

To ensure that a system of equations represented by (L.1) has a unique solution, it is necessary to first show that (L.1) is consistent, that is, that the coefficient matrix **A** and the augmented matrix [**A, b**] must have the same rank r. Then, if $n = r$, the system (L.1) is determinate, whereas if $r < n$, as may be the case, then the number $(n - r)$ variables must be specified in some manner or determined by optimization procedures. If the equations are independent, $m = r$. By calculating the order of the largest nonzero determinant in a given matrix you can determine

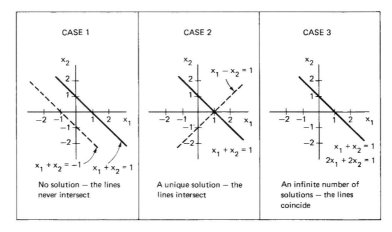

Figure L.2.   Types of solutions of linear equations.

the rank of a matrix. If the matrix is not square, the nonzero determinant is sought for the largest of any square matrices that can be formed from $\mathbf{A}$. You need to check all of the possible submatrices of $\mathbf{A}$. We use integers as elements in the matrices below. Noninteger elements in the matrices may lead to less clear results in determining whether the value of a determinant is zero or not.

Let r = number of ***independent*** equations
   n = number of variables whose values are unknown.

Homogenous Equations (all the right hand coefficients are zero) Ax = 0

$\boxed{r = n}$ The only solution is $x_1 = x_2 = \ldots = 0$ (a unique solution), the trivial solution.

Example:

$$\left.\begin{array}{l} x_1 + x_2 - x_3 = 0 \\ 2x_1 + 4x_2 - x_3 = 0 \\ 3x_1 + 2x_2 + 2x_3 = 0 \end{array}\right\} \text{the only solution is } x_1 = x_2 = x_3 = 0$$

$$\det\ \begin{bmatrix} 1 & 1 & -1 \\ 2 & 4 & -1 \\ 3 & 2 & 2 \end{bmatrix} = 1(8+2) - 2(2+2) + 3(-1+4) = 11 \neq 0 \Big\} \begin{array}{l}\text{a unique solution exists} \\ \text{because } r = 3,\ n = 3\end{array}$$

$\det\quad [\mathbf{A},\mathbf{b}] \neq 0$ (because $\mathbf{b} = 0$)   hence $r = 3$.

$\boxed{r < n}$ multiple non-unique solutions exist.

Example 1:

$$\left.\begin{array}{l} x_1 + x_2 - x_3 = 0 \\ x_1 - 5x_2 - 3x_3 = 0 \end{array}\right\} \begin{array}{l}\text{you can combine to get } 3x_2 - 2x_3 = 0 \\ \text{which has multiple solutions.}\end{array}$$

Examine the largest square matrices of $\mathbf{A}$. For example

$$\det\ \begin{bmatrix} 1 & 1 & -1 \\ 1 & -5 & 3 \end{bmatrix} = -5 - 1 = -6 \neq 0 \text{ so the rank is 2, but } n = 3\Big\} \text{multiple solutions.}$$

The determinant of the augmented matrix has the same rank as $\mathbf{A}$.

Example 2:

$$\begin{array}{l} x_1 + x_2 - x_3 = 0 \\ 2x_1 - 3x_2 + x_3 = 0 \\ x_1 - 4x_2 + 2x_3 = 0 \end{array}$$

$$\det \begin{bmatrix} 1 & 1 & -1 \\ 2 & -3 & 1 \\ 1 & -4 & 2 \end{bmatrix} = 1(-6+4) - 2(2-4) + 1(1-3) = -2+4-2 = 0$$

Hence the rank is 2 as is rank of the augmented matrix but n still equals 3. Thus, multiple solutions exist.

$\boxed{r > n}$ Only the trivial solution exists.

$$\left. \begin{array}{l} x + y = 0 \\ x + 4y = 0 \\ 3x + 2y = 0 \end{array} \right\} \text{ the only solution is the trivial solution } x = y = 0$$

The rank of $\mathbf{A} = \begin{bmatrix} 1 & 1 \\ 1 & 4 \\ 3 & 2 \end{bmatrix}$ is 2 and $n = 2$. The rank of the augmented matrix is also 2. A unique solution exists, but it is the trivial solution.

Nonhomogenous Equations Ax = b

$\boxed{r = n}$ A unique solution exists.

Example 1:

$$2x_1 + x_2 - 2x_3 = 0$$
$$3x_1 + 2x_2 + 2x_3 = 1 \qquad \text{3 equations, 3 variables, hence a solution}$$
$$5x_1 + 4x_2 + 3x_3 = 4 \qquad \text{exists: } x_1 = 1, x_2 = 3, x_3 = -3$$

$$\det \begin{bmatrix} 2 & 1 & -2 \\ 3 & 2 & 2 \\ 5 & 4 & 3 \end{bmatrix} \neq 0, \text{ so the rank } = 3; \text{ the augmented matrix has the same rank.}$$

Example 2

$$x_1 + 2x_2 - 3x_3 = 4$$
$$x_1 + 3x_2 + x_3 = 11 \qquad \text{4 equation, 3 variables}$$
$$2x_1 + 5x_2 - 4x_3 = 13$$
$$2x_1 + 6x_2 + 2x_3 = 22$$

The rank of the coefficient matrix is 3 (maximum would be 3) and the rank of the augmented matrix is 3 (the maximum would be 4) so that a unique solution exists ($x_1 = 1$, $x_2 = 3$, $x_3 = 1$). Note that only 3 of the equations are independent.

Example 3:

$$x_1 + 2x_2 - 3x_3 = 1$$
$$3x_1 - x_2 + 2x_3 = 7 \quad \Big\} \text{ 3 equations, 3 variables}$$
$$5x_1 + 3x_2 - 4x_3 = 2$$

Although it superficially appears that $n = r$, such is not the case because

$$\det \begin{bmatrix} 1 & 2 & -3 \\ 3 & -1 & 2 \\ 5 & 3 & -4 \end{bmatrix} = 0, \text{ so the rank is 2.}$$

In the augmented matrix $\begin{bmatrix} 2 & -3 & 1 \\ -1 & 2 & 7 \\ 3 & -4 & 2 \end{bmatrix}$ Does the matrix have rank of 3?
a 3 by 3 matrix exists:

$$\det \begin{bmatrix} 2 & -3 & 1 \\ -1 & 2 & 7 \\ 3 & -4 & 2 \end{bmatrix} = -3 \neq 0, \text{ hence the rank is 3, and thus there is no unique solution. This is really a case of } r < n; \text{ one equation is not independent.}$$

$\boxed{r < n}$ Multiple solutions exist.

Example 1:

$$x_1 + 2x_2 + x_3 = 3$$
$$2x_1 + x_2 - x_3 = 1$$

The rank of the coefficient matrix and the augmented matrix are both equal to 2, and $n = 3$. Multiple solutions exist.

Example 2

$$-2x_1 + 5x_2 + 7x_3 = 6$$
$$-x_1 + x_2 - 2x_3 = 1 \quad \Big\} \text{ 3 equations, 3 variables}$$
$$x_1 + 2x_2 + x_3 = 3$$

$$\det \begin{bmatrix} -2 & 5 & 7 \\ -1 & 1 & 2 \\ 1 & 2 & 1 \end{bmatrix} = 0, \text{ so the rank of the coefficient matrix is 2.}$$

$$\det \begin{bmatrix} 5 & 7 & 6 \\ 1 & 2 & 1 \\ 2 & 1 & 3 \end{bmatrix} = 0,$$ so the rank for this matrix is 2. The rank of other combinations of the columns of the augmented matrix give same result. Thus, $r = 2$ and $n = 3$.

Multiple solutions exist.

$\boxed{r > n}$ No solution exists.

Example 1:

$$x_1 + 2x_2 = 4$$

$$x_1 + 3x_2 = 1 \quad \text{3 (inconsistent) equations, 2 variables}$$

$$2x_1 + 5x_2 = 3$$

The rank of the coefficient matrix is 2.
The rank of the augmented matrix = ?

$$\det \begin{bmatrix} 1 & 2 & 4 \\ 1 & 3 & 1 \\ 2 & 5 & 3 \end{bmatrix} \neq 0, \text{ hence the rank is 3 and no solution exists.}$$

Example 2:

$$\left. \begin{array}{l} x_1 + 2x_2 = 4 \\ x_1 + 3x_2 = 1 \\ 2x_1 + 5x_2 = 5 \end{array} \right\} \text{3 equations, 2 variables}$$

The rank of the coefficient matrix = 2; the rank of the augmented matrix = ?

$$\det \begin{bmatrix} 1 & 2 & 4 \\ 1 & 3 & 1 \\ 2 & 5 & 5 \end{bmatrix} = 0, \text{ hence the rank is 2. This is really a case of the } r = n \text{ as one equation is not independent. A unique solution exists.}$$

Next, suppose that you are interested in solving n linear independent equations in n unknown variables:

$$\left. \begin{array}{l} a_{11}x_1 + a_{12}x_2 + \cdots + a_{1n}x_n = b_1 \\ a_{21}x_1 + a_{22}x_2 + \cdots + a_{2n}x_n = b_2 \\ \cdots \\ a_{n1}x_1 + a_{n2}x_2 + \cdots + a_{nn}x_n = b_n \end{array} \right\} \mathbf{ax} = \mathbf{b} \qquad (\text{L.3})$$

In general there are two ways to solve Eq. (L.3) for $x_1, \ldots, x_n$: elimination techniques and iterative techniques. Both are easily executed by computer programs. In the pocket in the back cover of this book you will find a disk containing computer programs that can be used in solving sets of linear equations. We shall illustrate the Gauss-Jordan elimination method. Other techniques can be found in texts on matrices, linear algebra, and numerical analysis.

The essence of the Gauss-Jordan method is to transform Eq. (L.3) into Eq. (L.4) by sequential nonunique elementary operations on Eq. (L.3):

$$x_1 + 0 + \cdots + 0 = b_1'$$
$$0 + x_2 + \cdots + 0 = b_2'$$
$$\cdots$$
$$0 + 0 + \cdots + x_n = b_n'$$

(L.4)

Equation (L.4) has a solution for $x_1, \ldots, x_n$ that can be obtained by inspection.

To illustrate the elementary operations that are required to execute the Gauss-Jordan method, consider the following independent set of three equations in three unknowns:

$$4x_1 + 2x_2 + x_3 = 15 \qquad (1)$$
$$20x_1 + 5x_2 - 7x_3 = 0 \qquad (2)$$
$$8x_1 - 3x_2 + 5x_3 = 24 \qquad (3)$$

The augmented matrix is

$$\begin{bmatrix} 4 & 2 & 1 & | & 15 \\ 20 & 5 & -7 & | & 0 \\ 8 & -3 & 5 & | & 24 \end{bmatrix}$$

Take the a_{11} element as a pivot. To make it 1 and the other elements in the first column zero, carry out the following elementary operations shown in order for each row:

(a) Subtract $(\frac{20}{4})$ Eq. (1), from Eq. (2).
(b) Subtract $(\frac{8}{4})$ Eq. (1), from Eq. (3)
(c) Multiply Eq. (1) by $\frac{1}{4}$.

to get

$$\begin{array}{ll} & \textit{new eq. no.} \\ \begin{bmatrix} 1 & \frac{1}{2} & \frac{1}{4} & | & \frac{15}{4} \\ 0 & -5 & -12 & | & -75 \\ 0 & -7 & 3 & | & -6 \end{bmatrix} & \begin{array}{l} (1a) \\ (2a) \\ (3a) \end{array} \end{array}$$

Carry out the following elementary operations to make the pivot element a_{22} equal to 1 and the other elements in the second column equal to zero:

(d) subtract $[(\frac{1}{2})/-5]$ of Eq. (2a), from Eq. (1a).

(e) Subtract $(-7/-5)$ of Eq. (2a), from Eq. (3a).

(f) Multiply Eq. (2a) by $(1/-5)$.

to obtain

new eq. no.

$$\begin{bmatrix} 1 & 0 & -\frac{19}{20} & -\frac{15}{4} \\ 0 & 1 & \frac{12}{5} & 15 \\ 0 & 0 & \frac{99}{5} & 99 \end{bmatrix} \quad \begin{matrix} \text{(1b)} \\ \text{(2b)} \\ \text{(3b)} \end{matrix}$$

Another series of elementary operations (left for you to propose) leads to a 1 for the element a_{33} and zeros for the other two elements in the third column:

$$\begin{bmatrix} 1 & 0 & 0 & 1 \\ 0 & 1 & 0 & 3 \\ 0 & 0 & 1 & 5 \end{bmatrix}$$

The solution to the original set of equations is

$$x_1 = 1$$

$$x_2 = 3$$

$$x_3 = 5$$

as can be observed from the augmentation column.

To obtain good accuracy and avoid numerical errors, the choice of the pivot should be made by scanning all the eligible coefficients and choosing the one with the greatest magnitude for the next pivot. For example, you might choose a_{21} for the first pivot, and then find that a_{31} would be the next pivot and finally a_{22} the last pivot to give

$$\begin{bmatrix} 0 & 1 & 0 & 3 \\ 1 & 0 & 0 & 1 \\ 0 & 0 & 1 & 5 \end{bmatrix}$$

By exchanging rows 1 and 2, you get exactly the form as (L.4).

Determine the number of independent components (which is the same as the number of independent material balances) for a process involving the following two competing reactions:

$$CO + 2H_2 \rightarrow CH_3OH$$

$$CO + 3H_2 \rightarrow CH_4 + H_2O$$

Solution

Prepare a matrix in which the rows are the atomic species for which the balances are to be made and the columns are chemical compounds entering and leaving the process. Each element in the matrix is the number of atoms in the chemical compound.

$$
\begin{array}{ccccc}
 & H_2 & H_2O & CO & CH_4 & CH_3OH \\
H & \begin{bmatrix} 2 & 2 & 0 & 4 & 4 \\
O & 0 & 1 & 1 & 0 & 1 \\
C & 0 & 0 & 1 & 1 & 1 \end{bmatrix}
\end{array}
$$

Transform the matrix by elementary operations so that there are 1's on the main diagonal starting at the a_{11} position, and only zeros below the main diagonal. The sum of the diagonal elements [starting at the a_{11} position] is the *rank* of the matrix that is equivalent to the number of components. Do you get three for the matrix above?

The number of independent components is not always equal to the number of atomic species, as shown in the next example.

Determine the number of independent components for a process involving the following reaction:

$$ SO_3 + H_2O \rightarrow H_2SO_4 $$

Solution

Form the species matrix and determine its rank.

$$
\begin{array}{cccc}
 & H_2O & SO_3 & H_2SO_4 \\
H & \begin{bmatrix} 2 & 0 & 2 \\
S & 0 & 1 & 1 \\
O & 1 & 3 & 4 \end{bmatrix}
\end{array}
$$

Are you able to make the transformation to

$$
\begin{array}{cccc}
 & H_2O & SO_3 & H_2SO_4 \\
H & \begin{bmatrix} 1 & 0 & 1 \\
S & 0 & 1 & 1 \\
O & 0 & 0 & 0 \end{bmatrix}
\end{array}
$$

Note that the rank of the matrix is 2, not 3; hence only two independent components exist for independent material balances.

L.2 NONLINEAR INDEPENDENT EQUATIONS

The precise criteria used to ascertain if a linear system of equations is determinate cannot be neatly extended to nonlinear systems of equations. Furthermore, the solution of sets of nonlinear equations requires the use of computer codes that may fail to solve your problem for one or more of a variety of reasons, a few of which are mentioned below. The problem to be solved can be written as

$$\left.\begin{array}{l} f_1(x_1, \ldots, x_n) = 0 \\ f_2(x_1, \ldots, x_n) = 0 \\ \quad \cdots \\ f_n(x_1, \ldots, x_n) = 0 \end{array}\right\} \quad \mathbf{f(x)} = \mathbf{0} \qquad (L.5)$$

where each function $f_i(x_1, \ldots, x_n)$ corresponds to a nonlinear function containing one or more of the variable whose values are unknown.

Figure L.3 classifies the major general methods of solving systems of nonlinear equations. Within each category and as combinations of categories you can find innumerable variations and submethods in the literature and available as computer codes. In the disk in the pocket in the back of this book is a simple Fortran code that makes use of the basic Newton method and a code for the secant method.

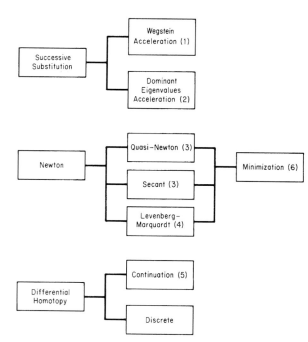

Figure L.3   General categories of techniques to solve nonlinear equations. From: 1. Wegstein, J. H. *Commu. ACM.*, *1*, (1958): 9; 2. Orbach, O., and C. M. Crowe. *Can. J. Chem. Eng.*, *49* (1971): 509; 3. Dennis, J. E. and R. B. Schnabel. *Numerical Methods for Unconstrained Optimization and Nonlinear Equations.* Englewood Cliffs, NJ, Prentice-Hall, 1983; 4. Marquardt, D. *SIAM J Appld. Math.*, *11*, (1963): 431; 5. Davidenko, D., *Ukrain Math.*, *5* (1953): 196; 6. Edgar, T. F. and D. M. Himmelblau. *Optimization of Chemical Processes.* New York: McGraw-Hill, 1988.

Newton's Method

Refer to equations (L.5). For a single equation (and variable), $f(x) = 0$, Newton's method uses the expansion of $f(x)$ in a first-order Taylor series about a reference point (a starting guess for the solution) x_0.

$$f(x) \simeq f(x_0) + \frac{df(x_0)}{dx}(x - x_0) \tag{L.6}$$

Note that Eq. (L.6) is a *linear equation* that is tangent to $f(x)$ at x_0. Examine Figure L.4. The right-hand side of Eq. (L.6) is equated to zero and the resulting equation solved for $(x - x_0)$.

$$x - x_0 = -\frac{f(x_0)}{df(x_0)/dx} \tag{L.7}$$

For example, suppose that $f(x) = 4x^3 - 1 = 0$, hence $df(x)/dx = 12x^2$. The sequence of steps to apply Newton's method using Eq. (L.7) starting at $x_0 = 3$ would be

$$x_1 = x_0 - \frac{4x^3 - 1}{12x^2}$$

$$= 3 - \frac{107}{108} = 2.009259$$

$$x_2 = 2.00926 - \frac{31.4465}{48.4454} = 1.36015$$

Additional iterations yield the following values for x_k:

k	x_k
0	3.00000
1	2.009259
2	1.3601480
3	0.9518103
4	0.7265254
5	0.6422266
6	0.6301933
7	0.6299606
8	0.6299605
9	0.6299605

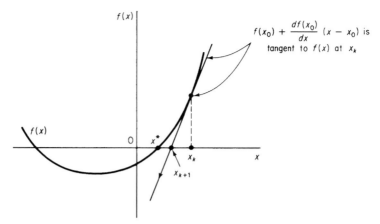

Figure L.4   Newton's method applied to the solution of $f(x) = 0$ starting at x_k. x_{k+1} is the next reference point for linearization. x^* is the solution.

so that the solution is given with increasing precision as k increases.

   Suppose that two independent equations in two variables whose values are to be determined are

$$f_1(x_1, x_2) = 0$$
$$f_2(x_1, x_2) = 0 \tag{L.8}$$

To apply Newton's method, expand each equation as a first-order Taylor series to get a set of linear equations at the point (x_{10}, x_{20}).

$$f_1(x_1, x_2) \simeq f_1(x_{10}, x_{20}) + \frac{\partial f_1(x_{10}, x_{20})}{\partial x_1}(x_1 - x_{10}) + \frac{\partial f_1(x_{10}, x_{20})}{\partial x_2}(x_2 - x_{20}) \tag{L.9}$$

$$f_2(x_1, x_2) \simeq f_2(x_{10}, x_{20}) + \frac{\partial f_2(x_{10}, x_{20})}{\partial x_1}(x_1 - x_{10}) + \frac{\partial f_2(x_{10}, x_{20})}{\partial x_2}(x_2 - x_{20})$$

Let the partial derivatives be designed by the constants $a_{ij} = \partial f_i/\partial x_j$ to simplify the notation, and let $(x_i - x_{i0}) = \Delta x_i$. Then, after equating the right-hand side of Eqs. (L.9) to zero, they become

$$a_{11}\,\Delta x_1 + a_{12}\,\Delta x_2 = -f_1(x_{10}, x_{20})$$
$$a_{21}\,\Delta x_1 + a_{22}\,\Delta x_2 = -f_2(x_{10}, x_{20}) \tag{L.10}$$

These equations are linear and can be solved by a linear equation solver to get the next reference point (x_{11}, x_{21}). Iteration is continued until a solution of satisfactory precision is reached. Of course, a solution may not be reached, as illustrated in Figure L.5c, or may not be reached because of round-off or truncation errors. If

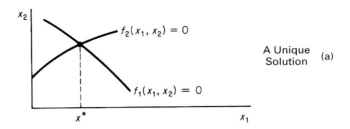

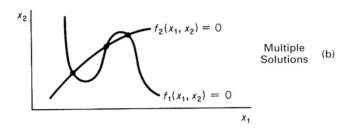

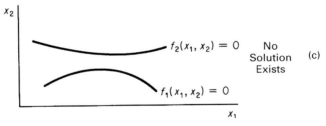

Figure L.5   Possible cases for the solution of two independent nonlinear equations $f_1(x_1, x_2) = 0$ and $f_2(x_1, x_2) = 0$ in two unknown variables.

the Jacobian matrix [see Eq. (L.11) below] is singular, the linearized equations, may have no solution or a whole family of solutions, and Newton's method probably will fail to obtain a solution. It is quite common for the Jacobian matrix to become ill-conditioned because if x_0 is far from the solution or the nonlinear equations are badly scaled, the correct solution will not be obtained.

The analog of (L.10) in matrix notation is

$$\mathbf{J}_k(\mathbf{x} - \mathbf{x}_k) = -\mathbf{f}(\mathbf{x}_k) \tag{L.11}$$

where $\mathbf{J}$ is the Jacobian matrix (the matrix whose elements are composed of the first partial derivatives of the equations with respect to the variables). For two equations

$$\mathbf{J} = \begin{bmatrix} \partial f_1(\mathbf{x})/\partial x_1 & \partial f_1(\mathbf{x})/\partial x_2 \\ \partial f_2(\mathbf{x})/\partial x_1 & \partial f_2(\mathbf{x})/\partial x_2 \end{bmatrix} \qquad \begin{aligned} \mathbf{x} &= \begin{bmatrix} \Delta x_1 \\ \Delta x_2 \end{bmatrix} \\ \mathbf{f} &= \begin{bmatrix} f_1 \\ f_2 \end{bmatrix} \end{aligned}$$

Quasi-Newton Methods

A quasi-Newton method in general is one that imitates Newton's method. If $f(x)$ is not given by a formula, or the formula is so complicated that analytical derivatives cannot be formulated, you can replace df/dx in Eq. (L.7) with a finite difference approximation

$$x_{k+1} = x_k - \frac{f(x)}{[f(x+h) - f(x-h)]/2h} \tag{L.12}$$

A central difference has been used in Eq. (L.12) but forward differences or any other difference scheme would suffice *as long as the step size* h *is selected to match the difference formula and the computer (machine) precision* for the computer on which the calculations are to be executed.

Other than the problem of the selection of the value of h, the only additional disadvantage of a quasi-Newton method is that additional function evaluations are needed on each iteration k. Eq. (L.12) can be applied to sets of equations if the partial derivatives are replaced by finite difference approximations.

Secant Methods

In the secant method the approximate model analogous to the right hand side of Eq. (L.6) (equated to zero) is

$$f(x_k) + m(x - x_k) = 0 \tag{L.13}$$

where m is the slope of the line connecting the point x_k and a second point x_q, given by

$$m = \frac{f(x_q) - f(x_k)}{x_q - x_k}$$

Thus the secant method imitates Newton's method (in this sense the secant method is also a quasi-Newton method (see Figure L.6)

Secant methods start out by using two points x_k and x_q spanning the interval of x, points at which the values of $f(x)$ are of opposite sign. The zero of $f(x)$ is predicted by

$$\tilde{x} = x_q - \frac{f(x_q)}{\dfrac{f(x_q) - f(x_k)}{x_q - x_k}} \tag{L.14}$$

The two points retained for the next step are $\tilde{x}$ and either x_q or x_k, the choice being made so that the pair of values $f(\tilde{x})$, and either $f(x_k)$ or $f(x_q)$, have opposite signs to maintain the bracket on x^*. (This variation is called "regula falsi" or the method of false position.) In Figure L.6, for the $(k + 1)$st stage, $\tilde{x}$ and x_q would be selected as the end points of the secant line. Secant methods may seem crude, but they work well in practice. The details of the computational aspects of a sound algorithm to solve multiple equations by the secant method are too lengthy to outline here (particularly the calculation of a new Jacobian matrix from the former one; instead refer to Dennis and Schnabel[1]).

The application of Eq. (L.14) yields the following results for $f(x) = 4x^3 - 1 = 0$ starting at $x_k = -3$ and $x_q = 3$. Some of the values of $f(x)$ and x during the search are shown below; note that x_q remains unchanged in order to maintain the bracket with $f(x) > 0$.

k	x_q	x_k	$f(x_k)$
0	3	−3	−109.0000
1	3	0.0277778	− 0.9991
2	3	0.055296	− 0.9992
3	3	0.0825434	− 0.9977
4	3	0.1094966	− 0.9899
5	3	0.1361213	− 0.9899
20	3	0.4593212	− 0.6124
50	3	0.6223007	− 0.0360
100	3	0.6299311	− 1.399×10^{-4}
132	3	0.6299597	− 3.952×10^{-6}

Brent's and Brown's Methods

Brent's[2] and Brown's[3] methods are variations of Newton's method that improve convergence. The calculation of the elements in $\mathbf{J}_k$ in Eq. (L.11) and the solving of the linear equations are intermingled. Each row of $\mathbf{J}_k$ is obtained as needed

[1]Dennis, J. E. and R. B. Schnabel. *Numerical Methods for Unconstrained Optimization and Nonlinear Equations* (Appendix A) Englewood Cliffs, NJ: Prentice-Hall, 1983.

[2]Brent, R. P. *SIAM J. Num. Anal.*, *10* (1973): 327.

[3]Brown, K. M., PhD Dissertation, Purdue University, 1966.

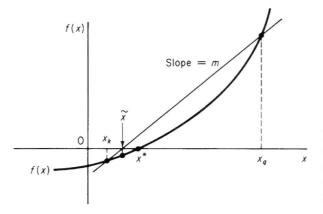

Figure L.6   Secant Method for the solution of $f(x) = 0$. x^* is the solution, $\tilde{x}$ the approximate to x^*, and x_q and x_k the starting points for iteration k of the secant method.

using the latest information available. Then one more step in the solution of the linear equations is executed. Brown's method is an extension of Gaussian elimination; Brent's method is an extension of QR factorization. Computer codes are generally implemented by using numerical approximations for the partial derivatives in $\mathbf{J}_k$.

Powell Hybrid and Levenberg-Marquardt Methods

Powell[4] and Levenberg[5]-Marquardt[6] calculated a new point $\mathbf{x}^{(k+1)}$ from the old one $\mathbf{x}^{(k)}$ by (note in the next two equations the superscript (k) is used instead of a subscript k to denote the stage of iteration so that the notation is less confusing)

$$\mathbf{x}^{(k+1)} = \mathbf{x}^{(k)} + \Delta x^{(k)} \tag{L.15}$$

where $\Delta x^{(k)}$ was obtained by solving the set of linear equations

$$\sum_{j=1}^{n} [\mu^{(k)} I_{ij} + \sum_{t=1}^{n} J_{ti}^{(k)} J_{tj}^{(k)}] \Delta x_j^{(k)} = -\sum_{t=1}^{n} J_{ti}^{(k)} f_t(x^{(k)}) \qquad i = 1, \ldots, n \tag{L.16}$$

where I_{ij} is an element of the unit matrix $\mathbf{I}$ and $\mu^{(k)}$ is a non-negative parameter whose value is chosen to reduce the sum of squares of the deviations (f_i-0) on each stage of the calculations. Powell used numerical approximations for the elements of $\mathbf{J}$. In matrix notation Eq. (L.16) can be derived by premultiplying Eq. (L.11) by $\mathbf{J}_k^T$.

[4]Powell, M.J.D. In *Numerical Methods for Nonlinear Algebraic Equations*, ed. P. Rabinowitz, Chapt. 6. New York: Gordon Breach, 1970.

[5]Levenberg, K. *Quart. Appld. Math.*, 2 (1944): 164.

[6]Marquardt, D. W. *J. SIAM*, 11 (1963): 431.

$$\mathbf{J}_k^T \mathbf{J}_k \Delta \mathbf{x}_k = -\mathbf{J}_k^T f(\mathbf{x}_k)$$

and adding a weighting factor $\mu_k \mathbf{I}$ to the left hand side to insure $\mathbf{J}_k^T \mathbf{J}$ is positive definite.

Minimization Methods

The solution of a set of nonlinear equations can be accomplished by minimizing the sum of squares of the deviations between the function values and zero.

$$\text{Minimize } F = \sum_{i=1}^{n} f_i^2(\mathbf{x}) \qquad i = 1, \ldots, m$$

Edgar and Himmelblau[7] list a number of codes to minimize F, including codes that enable you to place constraints on the variables.

Method of Successive Substitutions

Successive substitution (or resubstitution) starts by solving each equation $f_i(x)$ for a single (different) output variable. For example, for three equations, you solve for an output variable (f_2 for x_1, f_3 for x_3, and f_1 for x_2)

$$f_1(x) = 3x_1 + x_2 + 2x_3 - 3 = 0$$

$$f_2(x) = -3x_1 + 5x_2^2 + 2x_1 x_3 - 1 = 0$$

$$f_3(x) = 25x_1 x_2 + 20x_3 + 12 = 0$$

and rearrange them as follows

$$\left. \begin{aligned} x_1 &= F_1(x) = \frac{5x_2^2}{3} + \frac{2x_3 x_1}{3} - \frac{1}{3} \\ x_2 &= F_2(x) = -3x_1 - 2x_3 + 3 \\ x_3 &= F_3(x) = -\frac{25x_1 x_2}{20} - \frac{12}{20} \end{aligned} \right\} \quad \mathbf{x} = \mathbf{F}(\mathbf{x}) \qquad (\text{L.17})$$

An initial vector (x_{10}, x_{20}, x_{30}) is guessed and then introduced into the right-hand side of (L.17) to get the next vector (x_{11}, x_{21}, x_{31}), which is in turn introduced into the right-hand side and so on. In matrix notation the iteration from k to $k+1$ is

$$\mathbf{x}_{k+1} = \mathbf{F}(\mathbf{x}_k) \qquad (\text{L.18})$$

[7]Edgar, T. F., and D. M. Himmelblau. *Optimization of Chemical Processes*. (Chapters 6 and 8). New York: McGraw-Hill, 1988.

For the procedure of successive substitution to be guaranteed to converge, the value of the largest absolute eigenvalue of the Jacobian matrix of $\mathbf{F(x)}$ evaluated at each iteration point must be less than (or equal to) one. If more than one solution exists for Eqs. (L.17), the starting vector and the selection of the variable to solve for in an equation controls the solution located. Also, different arrangements of the equations and different selection of the variable to solve for may yield different convergence results.

The Wegstein and Dominant Eigenvalue methods listed in Figure L.3 are useful techniques to speed up convergence (or avoid nonconvergence) of the method of successive substitutions. Consult the references cited in Figure L.3 for the specific details.

Wegstein's method, which is used in many flowsheeting codes, accelerates the convergence of the method of successive substitutions on each iteration. In the secant method, the approximate slope is

$$m = \frac{f(x_k) - f(x_{k-1})}{x_k - x_{k-1}},$$

where x_k is the value of x on the kth (current) iteration and x_{k-1} is the value of x on the $(k - 1)$st (previous) iteration. The equation of a line through x_k and $f(x_k)$ with slope m is

$$f(x) - f(x_k) = m(x - x_k).$$

In successive substitutions we solve $x = f(x)$. Introduce $x = f(x)$ into the equation for the line

$$x - f(x_k) = m(x - x_k)$$

and solve for x:

$$x = \frac{1}{1-m} f(x_k) - \frac{m}{1-m} x_k.$$

Let $t = \dfrac{1}{1-m}$. Then for the $(k+1)$st (next) iteration,

$$x_{k+1} = (1 - t)x_k + tf(x_k).$$

For the solution of several equations simultaneously, each x is treated independently, a procedure that may possibly cause some instability if the x's interact. In such cases, an upper limit should be placed on t, say $0 \le t \le 1$. The Wegstein algorithm is for stage k.

1. Calculate x_k from the previous stage.
2. Evaluate $f(x_k)$.

3. Calculate m and t.

4. Calculate x_{k+1}.

5. Set $x_k \rightarrow x_{k+1}$ and repeat the above, starting with 2.

6. Terminate when $x_{k+1} - x_k <$ tolerance assigned.

Homotopy (Continuation) Methods

Homotopy methods[8] can be viewed as methods that widen the domain of convergence of any particular method of solving nonlinear equations, or, alternatively, as a method of obtaining starting guesses satisfactorily close to the desired solution. A set of functions $\mathbf{F}(\mathbf{x})$ is modified as follows to be a linear combination of a parameter t:

$$\mathbf{F}(\mathbf{x},t) = \mathbf{F}(\mathbf{x}) + (1 - t)\mathbf{F}(\mathbf{x}_0) = \mathbf{0} \qquad 0 \le t \le 1 \qquad \text{(L.19)}$$

where t is a scalar parameter such that when t is a fixed number the trajectory $\mathbf{F}(\mathbf{x},t) = 0$ occurs [a mapping of $\mathbf{F}(\mathbf{x})$] and when $t = 1$ the trajectory of the set of equations reaches the desired solution $\mathbf{F}(\mathbf{x}^*) = 0$. With this definition $\mathbf{x}$ describes a curve in space (a continuous mapping called a homotopy) with one end point at a known value (the starting guess) for $\mathbf{x}$, namely $\mathbf{x}_0$, and the other end point at a solution of $\mathbf{F}(\mathbf{x}) = \mathbf{0}$, $\mathbf{x}^*$.

Lin et al. outline the procedure, which is first to determine $\mathbf{x}$ and t as functions of the arc length of the homotopy trajectory. Then Eq. (L.19) is differentiated with respect to the arc length to yield an initial value problem in ordinary differential equations. Starting at $\mathbf{x}_0$ and t_0, the initial value problem is transformed by using Euler's method to a set of linear algebraic equations that yield the next step in the trajectory. The trajectory may reach some or all of the solutions of $\mathbf{F}(\mathbf{x}) = \mathbf{0}$; hence several starting points may have to be selected to create paths to all the solutions, and many undesired solutions (from a physical viewpoint) will be obtained. A number of practical matters to make the technique work can be found in the review by Seydel and Hlavacek.[9]

SUPPLEMENTARY REFERENCE

RHEINBOLT, W. C. *Numerical Analysis of Parameterized Nonlinear Equations.* New York: Wiley-Interscience, 1986.

[8]M. Kubicek. "Algorithm 502." *ACM Trans. Math. Software,* 2 (1976):98; W. J. Lin, J. D. Seader, and T. L. Wayburn. *AIChE J.,* 33 (1987):33.

[9]R. Seydel, and V. Hlavacek. *Chem. Eng. Sci.,* 42 (1987): 1281.

Appendix M

FITTING FUNCTIONS TO DATA

Frequently, you would like to estimate the best values of the coefficients in an equation from experimental data. The equation might be a theoretical law or just a polynomial, but the procedure is the same. Let y be the dependent variable in the equation, b_i be the coefficients in the equation, and x_i be the independent variables in the equation so that the model is of the form

$$y = f(b_0, b_1, \ldots; x_0, x_1, \ldots) \tag{M.1}$$

Let ϵ represent the error between the observation of y, Y, and the predicted value of y using the values of x_i and the estimated values of the coefficients b_i:

$$Y = y + \epsilon \tag{M.2}$$

The classical way to get the best estimates of the coefficients is by least squares, that is, by minimizing the sum of the squares of the errors (of the deviations) between Y and y for all the j sets of experimental data:

$$\text{Minimize } F = \sum_{j=1}^{p} (\epsilon_j)^2 = \sum_{j=1}^{p} (Y_j - f_j)^2 \tag{M.3}$$

Let us use a model linear in the coefficients with one independent variable x

$$y = b_0 + b_1 x \tag{M.4}$$

(in which x_0 associated with b_0 always equals 1 in order to have an intercept) to illustrate the principal features of the least-squares method to estimate the model coefficients. The objective function is

$$F = \sum_{j=1}^{p} (Y_j - y_j)^2 = \sum_{j=1}^{p} (Y_j - b_0 - b_1 x_j)^2 \tag{M.5}$$

704

There are two unknown coefficients, b_0 and b_1, and p known pairs of experimental values of Y_j and x_j. We want to minimize F with respect to b_0 and b_1. Recall from calculus that you take the first partial derivatives of F and equate them to zero to get the necessary conditions for a minimum.

$$\frac{\partial F}{\partial b_0} = 0 = 2 \sum_{j=1}^{p} (Y_j - b_0 - b_1 x_j)\,(-1) \tag{M.6a}$$

$$\frac{\partial F_0}{\partial b_1} = 0 = 2 \sum_{j=1}^{p} (Y_j - b_0 - b_1 x_j)\,(-x_j) \tag{M.6b}$$

Rearrangement yields a set of linear equations in two unknowns, b_0 and b_1:

$$\sum_{j=1}^{p} b_0 + \sum_{j=1}^{p} b_1 x_j = \sum_{j=1}^{p} Y_j$$

$$\sum_{j=1}^{p} b_0 x_j + \sum_{j=1}^{p} b_1 x_j^2 = \sum_{j=1}^{p} x_j Y_j$$

The summation $\sum_{j=1}^{p} b_0$ is $(p)(b_0)$ and in the other summations the constants b_0 and b_1 can be removed from within the summation signs so that

$$b_0(p) + b_1 \sum_{j=1}^{p} x_j = \sum_{j=1}^{p} Y_j \tag{M.7a}$$

$$b_0 \sum_{j=1}^{p} x_j + b_1 \sum_{j=1}^{p} x_j^2 = \sum_{j=1}^{p} x_j Y_j \tag{M.7b}$$

The two linear equations above in two unknowns, b_0 and b_1, can be solved quite easily for b_0 the intercept and b_1 the slope.

EXAMPLE M.1 Application of Least Squares

Fit the model $y = \beta_0 + \beta_1 x$ to the following data (Y is the measured response and x the independent variable).

x	Y
0	0
1	2
2	4
3	6
4	8
5	10

Solution

The computations needed to solve Eqs. (M.7) are

$$\sum x_j = 15 \qquad \sum x_j Y_j = 110$$

$$\sum Y_j = 30 \qquad \sum x_j^2 = 55$$

Then

$$6b_0 + 15b_1 = 30$$
$$15b_0 + 55b_1 = 110$$

Solution of these two equations yields

$$b_0 = 0 \qquad b_1 = 2$$

and the model becomes $\hat{Y} = 2x$, where $\hat{Y}$ is the predicted value for a given x.

The least-squares procedure outlined above can be extended to any number of variables as long as the model is linear in the coefficients. For example, a polynomial

$$y = a + bx + cx^2$$

is linear in the coefficients (but not in x), and would be represented as

$$y = b_0 + b_1 x_1 + b_2 x_2$$

where $a = b_0$, $b = b_1$, $c = b_2$, $x_1 = x$, and $x_2 = x^2$. Linear equations equivalent to Eqs. (M.7) with several independent variables can be solved via a computer. If the equation you want to fit is nonlinear in the coefficients such as

$$y = b_0 e^{b_1 x} + b_2 x^2$$

you can minimize F in Eq. (M.3) directly by the computer program on the disk in the back of this book, or by using a nonlinear least-squares computer code taken from a library of computer codes.

Additional useful information can be extracted from a least-squares analysis if four basic assumptions are made in addition to the presumed linearity of y in x:

1. The x's are deterministic variables (not random).

2. The variance of ϵ_j is constant or varies only with the x's.

3. The observations Y_j are mutually statistically independent.

4. The distribution of Y_j about y_j given x_j is a normal distribution.

For further details, see Box, Hunter, and Hunter[1] or Box and Draper.[2]

[1]G. E. P. Box, W. G. Hunter, and J. S. Hunter. *Statistics for Experimenters*. New York: Wiley-Interscience, 1978.

[2]G. E. P. Box and N. R. Draper. *Empirical Model Building and Response Surfaces*. New York: Wiley, 1987.

Appendix N

ANSWERS TO SELECTED PROBLEMS

Chapter 1

1.1 (a) 4.17×10^9 m^3

1.4 (b) B: ft^3/lb$_m$; C: (ft^3/lb$_m$)^2; (c) B* = (.016/MW)B

1.8 (a) 0.14 (ft)(lb$_f$)/lb (b) 29.37 gal/min

1.10 (f) 20°C or 28°C; (g) J/s; (h) 250 N

1.16 (a) 2.40×10^5 (lb$_f$)(ft)/min; (b) 6000 watts

1.20 Only if the coefficient 24/1000 also includes the units ft$^{0.26}$/yr

1.25 Yes it is consistent

1.30 (a) 100.09 g/g mol; (b) 0.0999 g/g mol; (c) 20.0 lb mol CaCO$_3$

1.33 (a) 380.9 g; (b) 4×10^4 g; (c) 1.26×10^{-3} lb mol

1.37 (a) 1.26 lb and 0.0352 lb; (b) 129.8 kg and 5.448×10^3 g

1.40 129.5 gal

1.44 0.879

1.47 Discharge is just below the limit

1.50 Yes

1.54 Yes

1.57 (a) 133 µg/L; (b) 63.4%

1.60 (a) 18.6 lb/lb mol; (b) CH$_4$: 0.688, C$_2$H$_4$: 0.151, C$_2$H$_6$: 0.161

1.63 C: 0.51, S: 0.03, N: 0.06, H: 0.21, O: 0.19

1.67 263K, 14°F, 474°R

1.70 (a) 50°F, (b) 510°R

1.74 (a) 140°F, 600°R, 333K, 60°C; (b) 77°F, 537°R, 298K, 25°C

1.77 1.014×10^4 kPa

1.81 (a) 115 ft (difference); (b) No

1.84 655 mm Hg

1.87 0.996 kPa gage

1.91 Procedure is ok

1.94 $p = 15.84 + 0.4751h$ in psia

1.97 24 in. Hg

1.101 No

1.104 244 kPa

1.107 (a) 912 lb CO_2

1.110 (a) 609 ton S

1.114 12.4%

1.117 (a) 43.2% $CaCO_3$; (b) 0.305 kg CO_2

1.120 108.4 kg

1.123 (a) 1.9%; (b) 40%

1.126 (a) CO is limiting; (b) H_2O is excess; (c) 0.514; (d) 0.60; (e) 0.0576; (f) 0.60

Chapter 2

2.1 The hydrometer is broken.

2.5 Possible options are largest area pen, longest pen, closest to the house (that will form one side), etc.

2.7 No. Morpholine has an autoignition temperature of 590°F.

2.11 Some possible methods: Acoustic emissions, flame ionization detectors (portable), examine fluid turbidity, etc.

2.15a $X_1 = 1, X_2 = 3, X_3 = -3$

2.18a Yes

Chapter 3

3.2 (a) closed; (b) open

3.5 Open system, steady state (after start up) or unsteady state as contaminant decreases.

3.8 18.6 kg

3.11 (a) Yes (but the rank of the coefficient matrix and the augmented matrix differ so that no solution exists).

3.14 No. There are only 3 independent equations.

3.17 (a) 3 (F,P,W); (b) 0 (if use summation of mole fractions equals one); (c) No (unless one stream is selected as the basis).

3.20 No (except a least squares solution).

3.24 26.5 lb

3.28 47.8 kg

3.31 A = 0.60; B = 0.35; C = 0.05

3.34 $F \cong P = 227$ kg/min

3.37 160 kg

3.42 $R = 1.97$

3.44 (a) N_2: 79%, CO_2: 21%; (b) CO_2: 14%, N_2: 79%, O_2: 7%

3.47 Analysis is not correct.

3.50 50%

3.53 C_2H_6: 0.21%, O_2: 14.41%, N_2: 76.05%, CO_2: 3.32%, CO: 0.41%, H_2O: 5.60%

3.57 No; O_2 in the fuel has been ignored.

3.61 Alcohol: 3.06%; propenoic acid: 2.89%

3.64 NH_4OH: 232 g; $Cu(NH_3)_4Cl_2$: 337 g

3.67 9%

3.69 (a) 8; (b) 8 (assuming summation of fractions is not counted).

3.73 For an air leak, have 207 kg mol/100 kg mol of feed.

3.76 (a) 4,166 lb/hr (b) S: 0.0143, W: 0.0357

3.79 (a) 53%, (b) 11.1%

3.81 7670 kg/hr

3.84 (a) R = 12.45 lb gal/day, (b) oil plus dirt $\omega = 4.43 \times 10^{-4}$

3.88 0.2 mol R/mol $SiCl_4$

3.92 8.33 kg

3.95 1.035 mol R/mol product

3.98 (a) 14.3 kg mol, (b) 29.7 kg mol

Chapter 4

4.2 (a) 2.88×10^5 ft^3/hr, (b) 5.49×10^5 lb

4.5 0.83 atm

4.8 51.3 psia

4.12 695 m^3/min

4.15 10.5%

4.18 (a) 1.987 cal/(g mol)(k), (b) 1.987 Btu/(lb mol)(°R), (c) 10.73 (psia)(ft^3)/(lb mol)(°R)

4.21 1.49

4.24 17.7 mm Hg

4.27 (b) 2.03×10^3 m^3

4.30 (a) 4.38 m^3 air/m^3 ammonia

4.33 0.314 H to C

4.37 15.6%

4.39 1.95×10^6 Pa

4.41 For n, T, and V fixed, the predicted pressure will be higher by the ideal gas law than the true pressure for $Z < 1$ and lower for $Z > 1$.

4.45 0.708 (a decrease)

4.48 About 1500 ft^3/hr

4.51 177 atm

4.54 2.26×10^3 kPa

4.56 a = 209.3 atm (ft^3/lb mol)^2, b = 0.4808 ft^3/lb mol

4.61 2.66 g mol

4.63 (a) VDV: 10.60, (b) RW: 10.61

4.67 Via Van der Waals: n_2/n_1 = 0.19, V = 0.075 m^3

4.69 (a) $p*$ = 805 mm Hg

4.73 (a) 126 kPa via Cox chart

4.76 **(a)** 104.8 kPa, **(b)** 0.0349
4.80 **(a)** $-11.5°C$ lower **(b)** 15.4°C higher concentration
4.83 0.0193 m^3/min
4.86 40°C (104°F)
4.90 C2: 1.2%, C3: 10.2%, iso-C4: 28.4%, n-C4: 31.9%, iso-C5: 28.6%
4.94 About 120°F
4.97 122°F
4.100 62.2 to 81.7°C
4.104 **(a)** 1, **(b)** 1
4.109 **(a)** 12.3%, **(b)** 14.9%, **(c)** about 75°F
4.112 **(a)** Yes, **(b)** 53.0 ft^3
4.115 32.4 kg mol (940 kg) air
4.118 298 kPa
4.121 0.766 mol recovered/mol feed
4.125 **(a)** 0.946 lb H_2O, **(b)** 958 ft^3
4.129 **(a)** 349 kg mol H_2O/hr, **(b)** 4640 kg H_2O condenses
4.133 **(a)** 0.21, **(b)** 96.5 lb H_2O

Chapter 5

5.2 **(a)** 1.00 Btu/(lb)(°F), **(b)** -0.0179 Btu/lb_m
5.6 **(a)** 7.17×10^8 m^2, **(b)** No
5.10 **(a)** $Q > 0$, $W = 0$, $\Delta E > 0$, $\Delta U > 0$
5.14 **(a)** system is can plus water; $Q = 0$, $W \neq 0$; **(b)** system is H_2 plus bomb; $Q \neq 0$, $W = 0$
5.17 **(a)** intensive, **(b)** intensive, **(c)** intensive, **(d)** extensive
5.21 $\Delta H = \Delta U = 0$
5.24 **(a)** $V = 36.5$ ft^3 @ 100 atm, 40°F; $p = 180$ atm; $\Delta H = 32,000$ Btu
5.28 Loss = 8.75×10^8 Btu/hr; % loss = 61.8%
5.32 $C_p = 8.3810 + 7.9891 \times 10^{-3}T - 1.5055 \times 10^{-6}T^2$ where T is in °C
5.35 **(a)** $C_p = 6.852 + 1.62 \times 10^{-3}T - 0.26 \times 10^{-6}T^2$ where T is in °C
5.38 3789 cal/g mol (14,600 J/g mol)
5.41 9.7×10^7 J
5.44 0.162
5.47 1.54×10^{-3} kWh
5.52 2.91×10^7 J/kg mol
5.55 $-354, 880$ kJ
5.58 **(a)** 31,180 J; **(b)** 504 min; **(c)** 31°C
5.63 Q lost = 3.19×10^4 Btu
5.66 $W = 0$, $Q = 0$, $\Delta U = 0$
5.70 6.6 Btu
5.74 -41.7 kW
5.78 **(a)** -17Btu/lb, work done; **(b)** 38°F; **(c)** 40 Btu/lb removed
5.81 22.1 lb

5.84 Not adiabatic

5.87 332 K

5.91 (a) −2.85 kJ, (b) −102.2 kJ, (c) 183.6 kJ, (d) 3.75 kJ

5.96 (a) −23.096 kJ/g mol NH_3

5.99 (a) HHV = 11,800 Btu/lb, LHV = 11,770 Btu/lb

5.103 −5218 kJ/g mol

5.108 Reaction at 25°C and 1 atm with $Q = \Delta H$: 0.76L/g at 35°C (body temp.) at 35°C with O_2 used in reaction

5.111 $Q = −147.23$ kJ/g mol CH_3OH

5.114 Reduction of 2000 to 700 lb/hr

5.119 143,400 J/g mol CO_2 removed

5.124 625 kJ/hr

5.128 $T \cong 2200$ K

5.133 −1313 Btu/lb mol = work done on surroundings

5.137 72.8 hp

5.140 80%

5.142 (a) −33,000 Btu/lb mol; (b) −289.65 k cal/g mol $CaCl_2$; (c) −1440 Btu/lb mol $CaCl_2$

5.147 28%

5.148 (a) 465 lb of 73% NaOH added

5.151 (a) 0.079 kg mol H_2O/kg mol air; (b) 87.1 kPa; (c) 37°C

5.154 (a) 0.01813 lb H_2O/lb air; (b) 0.031 lb H_2O/lb air; (c) 1.14%

Chapter 6

6.1 Degrees of freedom = C + 4

6.5 Degrees of freedom = 2C + 2N + 5; specify p(N), Q(N), F(C + 2), S(C + 2), N(1) for each stage

6.12 Degrees of freedom = 16

6.13 1,129.5 kg/feed

6.15 B = 240,270, C = 95,930, D = 10,270, E = 44,750, F = 4,480, A = 336,200 all lb/hr

6.18 (a) 90°F; (b) 2,800 lb of 40% NaCl produced/hr.

6.23 (a) 3000 lb (NH_3), 11,000 lb ($NH_4)_2SO_4$, 7,500 lb Benzol, 2,500 lb Toluol, 1,500 lb Pyridine; (c) 2660 lb 40% NaOH; (d) 3720 lb 50% H_2SO_4

6.26 Heat duty for exchangers (in MM kJ/hr): No. 1 = 82.04, No. 2 = 120.33, No. 3 = 165.35 (all temperatures 60°C)

Chapter 7

7.2 230.5 min

7.8 37,700 lb

7.11 121.5 s

7.17 moles $C_6H_{12} = e^{-kt}$

7.21 28 min

7.24 (b) 129°F for half-full

NOMENCLATURE *

A = area

A = constant

a = acceleration

$\mathbf{A}$ = coefficient matrix

a = constant in general

a = constant in van der Waals' equation (see Table 4.2)

a, b, c = constants in heat capacity equation

$\mathcal{A}S$ = absolute saturation

$°API$ = specific gravity of oil defined in Eq. (1.3)

B = constant

b = constant in van der Waals' equation (see Table 4.2)

$\dot{B}$ = rate of energy transfer accompanying $\dot{w}$

(c) = crystalline

C = constant in Eq. (1.1)

C = number of chemical components in the phase rule

C_p = heat capacity at constant pressure

C_s = humid heat

C_v = heat capacity at constant volume

D = diameter

D = distillate product

E = total energy in system $= U + K + P$

E_v = irreversible conversion of mechanical energy to internal energy

F = force

F = number of degrees of freedom in the phase rule

F = feed stream

(g) = gas

g = acceleration due to gravity

g_c = conversion factor of $\dfrac{32.174(\text{ft})(\text{lb}_\text{m})}{(\sec^2)(\text{lb}_\text{f})}$

$\hat{H}$ = enthalpy per unit mass or mole

h = distance above reference plane

H = enthalpy, with appropriate subscripts, relative to a reference enthalpy

$\mathcal{H}$ = humidity, lb water vapor/lb dry air

(l) = liquid

h_c = heat transfer coefficient

$\Delta\hat{H}$ = enthalpy change per unit mass or mole

ΔH = enthalpy change, with appropriate subscripts

ΔH_{rxn} = heat of reaction

*Units are discussed in text.

ΔH_{soln} = heat of solution
ΔH_v = heat of vaporization
ΔH_c^o = standard heat of combustion
ΔH_f^o = standard heat of formation
K = characterization factor
K = degree Kelvin
K = kinetic energy
k_g' = mass transfer coefficient
K_i = vapor/liquid equilibrium, Eq. (4.14)
L = moles of liquid, Eq. (4.20)
l = distance
lb = pound, as a mass (without subscript)
lb_f = pound, as a force
lb_m = pound, as a mass
m = mass of material
m = number of equations
$\dot{m}$ = rate of mass transport through defined surfaces
mol. wt. = molecular weight
n = number of moles
n = number of unknown variables
N_d = degrees of freedom
N_{Re} = Reynolds number
N_r = number of independent constraints
N_{sp} = number of chemical species
N_v = total number of variables
$\mathcal{P}$ = number of phases in the phase rule
p = partial pressure (with a suitable subscript for p)
p = pressure
p_c' = pseudocritical pressure
p^* = vapor pressure
p_c = critical pressure
P = potential energy
P = product
p_r' = pseudoreduced pressure
p_r = reduced pressure = p/p_c
p_t = total pressure in a system
Q = heat transferred
$\dot{Q}$ = rate of heat transferred (per unit time)
q = volumetric flow rate
(s) = solid
R = recycle stream
R = universal gas constant
r = rank of a matrix
$\dot{r}_A$ = rate of generation of consumption of component A (by chemical reaction)

$\mathcal{R}\mathcal{H}$ = relative humidity

$\mathcal{R}\mathcal{S}$ = relative saturation

S = cross-sectional area perpendicular to material flow

s = second

sp gr = specific gravity

T = absolute temperature or temperature in general

t = time

T_c' = pseudocritical temperature

T_r' = pseudoreduced temperature

T_b = normal boiling point (in K or °R)

T_c = critical temperature (absolute)

T_{DB} = dry-bulb temperature

T_f = melting point (in K)

T_r = reduced temperature = T/T_c

T_{WB} = wet-bulb temperature

U = internal energy

$\hat{V}$ = humid volume per unit mass or mole

$\hat{V}$ = specific volume (volume per unit mass or mole)

V = system volume of fluid volume in general

v = velocity

V_c = critical volume

$\hat{V}_{ci}$ = ideal critical volume = RT_c/p_c

$\hat{V}_{ci}$ = pseudocritical ideal volume

V_g = volume of gas

V_l = volume of liquid

V_r = reduced volume = V/V_c

V_{ri} = ideal reduced volume = V/V_{ci}

$\dot{W}$ = rate of work done by system (per unit time)

W = waste stream

W = work done by or on the system

$\dot{w}_A, \dot{w}$ = rate of mass flow of component A and total mass flow, respectively, through system boundary other than a defined surface

x = mass or mole fraction in general

x = mass or mole fraction in the liquid phase for two-phase systems

x = unknown variable

y = mass or mole fraction in the vapor phase for two-phase systems

z = compressibility factor

z_c = critical compressibility factor

z_j = mole fraction of j

Greek letters:

α, β, γ = constants

Δ = difference between exit and entering stream; also used for final minus initial times or small time increments

ρ = density
ρ_A, ρ = mass of component A, or total mass, respectively, per unit volume
ρ_L = liquid density
μ = viscosity
ϵ = error
ρ_v = vapor density
ω = mass fraction
θ = time, or dimensionless time
λ = constant in equations of state

Subscripts:

A, B = components in a mixture
c = critical
i = any component
i = ideal state
r = reduced state
t = total
$1, 2$ = system boundaries

Overlays

$\hat{}$ = per unit mass or per mole
$\cdot$ = per unit time (a rate)

INDEX

USING THE CD-ROM

Viewing the Contents of the CD-ROM

To view the contents of the CD-ROM, do the following:

1. Place the CD-ROM in your CD-ROM drive.
2. In Windows, open the File Manager.
3. Click on the drive icon for your CD-ROM drive to open a file list window.
4. The following directories will appear:

acrobat	(contains Adobe Acrobat Reader 2.1 for Windows)
fortran	(contains Fortran programs)
physprop	(contains the Physical Properties database)
polymath	(contains PolyMath program and documentation)
workbook	(contains the Supplementary Problems Workbook)

Installing Acrobat Reader for Use With the Supplementary Problems Workbook and PolyMath Documentation

The Workbook files and the documentation for PolyMath use the Adobe Acrobat ".PDF" format. To view these files, you must install the Acrobat Reader onto your hard disk drive from the CD. To view the Acrobat installation instructions, do the following:

1. In File Manager, open the "acrobat" directory by double clicking on it.
2. Double click on the file README_R.TXT. This will open the document with the Windows NotePad program.
3. Follow the installation instructions on the screen.

The Supplementary Problems Workbook files are arranged in the "workbook" directory as follows:

front.pdf contains all front matter including a table of contents

Chap1.pdf through *Chap5.pdf* contain all of the chapters of the workbook

append.pdf contains both Appendices A and B

index.pdf contains the index

The files can all be opened with the Acrobat Reader, viewed on screen, or printed. When viewing graphics in the workbook, we recommend increasing the viewing percentage to greater than 100% for clarity.

The PolyMath documentation can be found in the "polymath" directory in the file: *polyma30.pdf*

Installing PolyMath

Instructions for installing the PolyMath software are also on the CD-ROM. To view these instructions, do the following:

1. In File Manager, open the "polymath" directory by double clicking on it.
2. Double click on the file README.TXT. This will open the document with the Windows Note-Pad program.
3. When you have read the README.TXT file, you should follow the procedure described in step 2 above to read the file INSTALL.TXT for installation instructions.
4. Follow the installation instructions in each of these files.

   (NOTE: You must install PolyMath from the DOS prompt. Double click on the DOS prompt icon on the Windows desktop (usually found in the Main window) to go to DOS.)

Using the Fortran Files

The Fortran files are located in the "fortran" directory and can be accessed through DOS.

1. Change the drive to the CD-ROM drive.
2. Type **cd \fortran** and press ENTER
3. To begin, at the d:\fortran>, type **start** and press ENTER. (Note: d: is your CD-ROM drive)
4. Follow the instructions as they appear on screen.

Using the Yaws Physical Properties Database

The Yaws Physical Properties database is located in the "physprop" directory. It can be accessed at the DOS prompt.

1. In the File Manager, go to the CD-ROM drive.
2. Open the "physprop" directory and locate the file "hfgas.exe".
3. Double click on the file "hfgas.exe" to start the database.
4. Follow the instructions as they appear on screen.
5. To stop printing on screen, type an *upper case* "X" (a lower case x will continue printing).
6. To quit the database, type "5".

Technical Support

Technical support for this CD-ROM is available at 1-800-842-2958.

LICENSE AGREEMENT AND LIMITED WARRANTY

READ THE FOLLOWING TERMS AND CONDITIONS CAREFULLY BEFORE OPENING THIS DISK PACKAGE. THIS LEGAL DOCUMENT IS AN AGREEMENT BETWEEN YOU AND PRENTICE-HALL, INC. (THE "COMPANY"). BY OPENING THIS SEALED DISK PACKAGE, YOU ARE AGREEING TO BE BOUND BY THESE TERMS AND CONDITIONS. IF YOU DO NOT AGREE WITH THESE TERMS AND CONDITIONS, DO NOT OPEN THE DISK PACKAGE. PROMPTLY RETURN THE UNOPENED DISK PACKAGE AND ALL ACCOMPANYING ITEMS TO THE PLACE YOU OBTAINED THEM FOR A FULL REFUND OF ANY SUMS YOU HAVE PAID.

1. **GRANT OF LICENSE:** In consideration of your payment of the license fee, which is part of the price you paid for this product, and your agreement to abide by the terms and conditions of this Agreement, the Company grants to you a nonexclusive right to use and display the copy of the enclosed software program (hereinafter the "SOFTWARE") on a single computer (i.e., with a single CPU) at a single location so long as you comply with the terms of this Agreement. The Company reserves all rights not expressly granted to you under this Agreement.

2. **OWNERSHIP OF SOFTWARE:** You own only the magnetic or physical media (the enclosed disks) on which the SOFTWARE is recorded or fixed, but the Company retains all the rights, title, and ownership to the SOFTWARE recorded on the original disk copy(ies) and all subsequent copies of the SOFTWARE, regardless of the form or media on which the original or other copies may exist. This license is not a sale of the original SOFTWARE or any copy to you.

3. **COPY RESTRICTIONS:** This SOFTWARE and the accompanying printed materials and user manual (the "Documentation") are the subject of copyright. You may not copy the Documentation or the SOFTWARE, except that you may make a single copy of the SOFTWARE for backup or archival purposes only. You may be held legally responsible for any copying or copyright infringement which is caused or encouraged by your failure to abide by the terms of this restriction.

4. **USE RESTRICTIONS:** You may not network the SOFTWARE or otherwise use it on more than one computer or computer terminal at the same time. You may physically transfer the SOFTWARE from one computer to another provided that the SOFTWARE is used on only one computer at a time. You may not distribute copies of the SOFTWARE or Documentation to others. You may not reverse engineer, disassemble, decompile, modify, adapt, translate, or create derivative works based on the SOFTWARE or the Documentation without the prior written consent of the Company.

5. **TRANSFER RESTRICTIONS:** The enclosed SOFTWARE is licensed only to you and may not be transferred to any one else without the prior written consent of the Company. Any unauthorized transfer of the SOFTWARE shall result in the immediate termination of this Agreement.

6. **TERMINATION:** This license is effective until terminated. This license will terminate automatically without notice from the Company and become null and void if you fail to comply with any provisions or limitations of this license. Upon termination, you shall destroy the Documentation and all copies of the SOFTWARE. All provisions of this Agreement as to warranties, limitation of liability, remedies or damages, and our ownership rights shall survive termination.

7. **MISCELLANEOUS:** This Agreement shall be construed in accordance with the laws of the United States of America and the State of New York and shall benefit the Company, its affiliates, and assignees.

8. **LIMITED WARRANTY AND DISCLAIMER OF WARRANTY:** The Company warrants that the SOFTWARE, when properly used in accordance with the Documentation, will operate in substantial conformity with the description of the SOFTWARE set forth in the Documentation. The Company does not warrant that the SOFTWARE will meet your requirements or that the operation of the SOFTWARE will be uninterrupted or error-free. The Company warrants that the media on which the SOFTWARE is delivered shall be free from defects in materials and workmanship under normal use for a period of thirty (30) days from the date of

your purchase. Your only remedy and the Company's only obligation under these limited warranties is, at the Company's option, return of the warranted item for a refund of any amounts paid by you or replacement of the item. Any replacement of SOFTWARE or media under the warranties shall not extend the original warranty period. The limited warranty set forth above shall not apply to any SOFTWARE which the Company determines in good faith has been subject to misuse, neglect, improper installation, repair, alteration, or damage by you. EXCEPT FOR THE EXPRESSED WARRANTIES SET FORTH ABOVE, THE COMPANY DISCLAIMS ALL WARRANTIES, EXPRESS OR IMPLIED, INCLUDING WITHOUT LIMITATION, THE IMPLIED WARRANTIES OF MERCHANTABILITY AND FITNESS FOR A PARTICULAR PURPOSE. EXCEPT FOR THE EXPRESS WARRANTY SET FORTH ABOVE, THE COMPANY DOES NOT WARRANT, GUARANTEE, OR MAKE ANY REPRESENTATION REGARDING THE USE OR THE RESULTS OF THE USE OF THE SOFTWARE IN TERMS OF ITS CORRECTNESS, ACCURACY, RELIABILITY, CURRENTNESS, OR OTHERWISE.

IN NO EVENT, SHALL THE COMPANY OR ITS EMPLOYEES, AGENTS, SUPPLIERS, OR CONTRACTORS BE LIABLE FOR ANY INCIDENTAL, INDIRECT, SPECIAL, OR CONSEQUENTIAL DAMAGES ARISING OUT OF OR IN CONNECTION WITH THE LICENSE GRANTED UNDER THIS AGREEMENT, OR FOR LOSS OF USE, LOSS OF DATA, LOSS OF INCOME OR PROFIT, OR OTHER LOSSES, SUSTAINED AS A RESULT OF INJURY TO ANY PERSON, OR LOSS OF OR DAMAGE TO PROPERTY, OR CLAIMS OF THIRD PARTIES, EVEN IF THE COMPANY OR AN AUTHORIZED REPRESENTATIVE OF THE COMPANY HAS BEEN ADVISED OF THE POSSIBILITY OF SUCH DAMAGES. IN NO EVENT SHALL LIABILITY OF THE COMPANY FOR DAMAGES WITH RESPECT TO THE SOFTWARE EXCEED THE AMOUNTS ACTUALLY PAID BY YOU, IF ANY, FOR THE SOFTWARE.
SOME JURISDICTIONS DO NOT ALLOW THE LIMITATION OF IMPLIED WARRANTIES OR LIABILITY FOR INCIDENTAL, INDIRECT, SPECIAL, OR CONSEQUENTIAL DAMAGES, SO THE ABOVE LIMITATIONS MAY NOT ALWAYS APPLY. THE WARRANTIES IN THIS AGREEMENT GIVE YOU SPECIFIC LEGAL RIGHTS AND YOU MAY ALSO HAVE OTHER RIGHTS WHICH VARY IN ACCORDANCE WITH LOCAL LAW.

ACKNOWLEDGMENT

YOU ACKNOWLEDGE THAT YOU HAVE READ THIS AGREEMENT, UNDERSTAND IT ,AND AGREE TO BE BOUND BY ITS TERMS AND CONDITIONS. YOU ALSO AGREE THAT THIS AGREEMENT IS THE COMPLETE AND EXCLUSIVE STATEMENT OF THE AGREEMENT BETWEEN YOU AND THE COMPANY AND SUPERSEDES ALL PROPOSALS OR PRIOR AGREEMENTS, ORAL, OR WRITTEN, AND ANY OTHER COMMUNICATIONS BETWEEN YOU AND THE COMPANY OR ANY REPRESENTATIVE OF THE COMPANY RELATING TO THE SUBJECT MATTER OF THIS AGREEMENT.

Should you have any questions concerning this Agreement or if you wish to contact the Company for any reason, please contact in writing at the address below or call the at the telephone number provided.

Robin Short
PTR Customer Service
Prentice Hall PTR
One Lake Street
Upper Saddle River, New Jersey 07458